International Conference on Physics of Materials and Nanotechnology (ICPN 2023)

IOP Conference Series: Materials Science and Engineering
Volume 1300

Mangalore, India
21-23 September 2023

ISBN: 978-1-7138-8876-5
ISSN: 1757-8981

Printed from e-media with permission by:

Curran Associates, Inc.
57 Morehouse Lane
Red Hook, NY 12571

Some format issues inherent in the e-media version may also appear in this print version.

Institute of Physics
Dirac House, Temple Back
Bristol BS1 6BE UK

Phone: 44 1 17 929 7481
Fax: 44 1 17 920 0979

techtracking@iop.org

Additional copies of this publication are available from:

Curran Associates, Inc.
57 Morehouse Lane
Red Hook, NY 12571 USA
Phone: 845-758-0400
Fax: 845-758-2633
Email: curran@proceedings.com
Web: www.proceedings.com

TABLE OF CONTENTS

Preface

Peer Review Statement

Computational Insights into the Electronic Structure of Functionalized Monolayer hBN Supercapacitor Electrodes 1
T Sruthi, Vincent Mathew

Optical Properties of Lawsone Dye Doped with Zinc Oxide for the Fabrication of Dye Sensitized Solar Cell 8
D S Suresh, Abdullah Ba Shbil, S P Vijaykumar, Sapna Sharanappa, H Ganesha

Simultaneous Synthesis of Carbon Quantum Dots and Porous Carbon Using One Step Hydrothermal Method for Supercapacitor Applications 14
Sapna Sharanappa, S P Vijaykumar, D S Suresh, H Ganesha, Abdullah Ba Shbil, M Basappa

Natural Light Harvesting Materials: A Study on Optoelectronic Properties of Potential Natural Dyes 19
D Harshitha, H M Mahesh, C G Renuka

Performance Analysis of A3B2X9 (Cs3Bi2I9) Based Perovskite Photovoltaic Tandem Structure with Crystalline Silicon (c-Si) 25
Shreyus Goutham Kumar, Tadi Surya Teja Reddy, N Suraj, C R Prashanth, Gajanan V Honnavar

Deep Learning Neural Network (DNN) Coupled Laser Induced Breakdown Spectroscopy for Quantitative Analysis of Copper Alloys 31
Leya Pauly, Lekha Mary John, M S Suryan Sivadas, K K Anoop

Crystal Structure, Molecular Mechanics and in Silico Analyses of Piperizine Derivative Against Human Mammary Carcinoma Cells Inhibition. 37
H Keshav Kumar, S Ananda, Dukanya, V Keerthikumara, Basappa, M Mahendra

Electrical Nature and Surface Enhanced Raman Spectroscopy of Ag Nanoparticles Decorated Graphene Sheet 43
P S Sachidanand, Manu Mohan, S Reshmi, Gauthami Viswan, Kuntala Bhattacharjee

Tailored Synthesis of CN-PPy Composite for Enhanced Electrochemical Functionality 49
S P Vijaykumar, S Sapna, D S Suresh, H Ganesha, Abdullah Ba Shbil

Nilgiri Tree Peels-Derived Activated Porous Carbon for High Performance Supercapacitor Applications 55
Abdullah Ba Shbil, Y S Nagaraju, H Ganesha, S Veeresh, D S Suresh, S P Vijaykumar, Sapna Sharanappa

Dielectric Analysis of Amine Modified Nano-Graphite: Effects of Thickness, Temperature and Frequency 61
Bhanu Pratap Singh, Pushpendra Kumar, S. P. Mahapatra

Synthesis and Characterization of Barium Hexagonal Ferrite (BaFe12O19) 69
C. S Naveen, Anand Rohidas, S Raghu, B. N Havareddi, R Ramanna, S Mattyappannavar

Determination of Effective Atomic Number, Electron Density and Kerma of Some Ferroelectric Materials Using Mass Attenuation Coefficients in the Energy Range 1 keV -100 GeV 74
A. Pushpa, I. B Bhagyalaxmi, B. Shivaleela, G. G Shivraj

Synthesis and Electrical Properties of Reused-Graphite/Polyaniline Composites 87
Basheerabegum Faniband, S Sarvesha Chandra, R Sreekanth, T Aravinda, Jayadev Pattar, Sanna Kotrappanavar Nataraj, M Nagaraja

Effect of Temperature on Electrical Resistivity of EVA-LDPE Nanocomposites for Photovoltaic Encapsulating Application: An Activation Energy Method 93
Moumita Naskar, Prabhat Kumar Maiti

PH Responsive Properties of Methyl Red/PVA Polymers for Smart Food Packaging Applications 99
K Swaroop, S G Ashwini, M S Murari, C S Naveen, A K Vinayaka, B M Santhosh, B N Veerabhadraswamy, I Yashodhara

Study the Effect of Cobalt Chloride (CoCl2) on the Structural, Optical, and Electrical Properties of Polymer Electrolyte (PVA-CoCl2) Films 105
K V Roopa, S Udayaraj, M Annappa, Gavisiddayya Mathad, K Subramanya, H Devendrappa

Alternating Current Response Study on PANI/V2O5 Composite at Room Temperature 111
M Annappa, B Chethan, S Udayaraj, M Gavisiddayya, K Vroopa, H Devendrappa, K Subramanya

Ftir, Uv-Vis and Conductivity Studies in a Copolymer of Indole and Carbazole 117
B Raghavendra, T Sankarappa, Amarkumar Malge, Mohansingh Heerasingh, Ashwini Devidas

Structural and Optical Properties of PMMA-MgO Nanocomposite Film 125
K N Shruthi, V Ramaraja Varma, Mohan Kumar, Sushma, Ganesh Sanjeev

Structural, Physical and Optical Studies of Sc3+ Doped Lithium Aluminum Borate Glasses 131
J Nagaraju, B Eraiah

Situ Synthesis and Characterization of polypyrrole/ZnO Nanocomposites for Optical and Photocatalytic Activity 137
V Mahadevi, S Veeresh, B T Vijaykumr, Reshma, Priyanka Kolhar, Bharati Basavaraj, Basavaraja Sannakki

Investigations on Impact of YSZ Nanoparticles on Morphology & Electrical Conductivity of PEDOT: PSS 143
C J Vijaykumar, S S Bulla, C K Chavan, R F Bajantri

Optical and Structural Properties of TiO2 Doped Calcium Borate Glasses........ 151
B Pruthviraja, B Eraiah

Synthesis, Characterization, Dielectric, and Electrical Conductivity Studies of Zinc Oxide Nanoparticles........ 158
A Sharma, P Kumar, S P Mahapatra

Wide Band Gap Tuning of Mg Doped ZnO Thin Films for Optoelectronic Applications. 167
K Mohammed Salman, Mohamed Zikriya, C. G. Renuka

Effect of Aspect Ratio on the Nonlinear Optical Properties of ZnO Nanorods 173
M Ramya, Swathy Satheesh, Sreelakshmi Sreedhar, Jitha P Marydasan, M Kailasnath

The Mechanical and Tribological Properties of Composites of Nanographene and an Aluminium-Metal Matrix Were Studied, as Was Their Corrosion Behaviour, in Order to Develop Materials Suitable for Use as Shielding Materials 178
Romala Rajesh, V. V. S. Prasad, Maduthuri Venkatesh, P Surya Nagendra, Gopalapalli Madhav Pavan Surya

Switching Behavior of the Composite Low Dimensional Structural Hybrids of Carbon After UV Exposure 185
Sonia Saini, Kuntala Bhattacharjee, Girish M Gouda

UV-Assisted Green Synthesis of Silver Nanoparticles and Their Characterizations 191
K Kavyasree, H Basappa, J M Lavita, S B Suma, Y Sangappa

Performance of Hybrid Nanofluids of Mineral Oil and Vegetable Oil Based Ester 197
Prabhat Kumar Maiti, Manish Kumar, Moumita Naskar

Influence of Zirconium Substitution on Electrical Properties of Nickel-Zinc Ferrites 203
Soumya Sajjan, Ananda Kulkarni, Shivraj G. Gounhalli

TiO2 / Graphitic Carbon Nitride Z-Scheme Catalyst for Binary Dye Degradation by Persulphate Activation 209
Shalini Viswanathan, Anooj Anil Pallikkara, Fermil Muhammed, Majji Kavya Sri, K Aparna

Predicting Shock Hugoniot and Equation of State of Single-Crystal Molybdenum by Molecular Dynamics Simulation 215
Anuradha Singla, Aditi Ray

Structural and Optical Properties of AgO Nanoparticles Synthesized Using Solid State Combustion 221
Thejas Ramakrishnaiah, D Vinay, Ganesh S Hegde, T S Siddivinayaka, K T Vasudevan, Abhiram Jagannathan

Green Synthesis and Characterization of Zirconium Oxide with Antimicrobial Activities 227
B N Veerabhadraswamy, H K Pradeep, K Swaroop, K M Manoj, M V Dhanush Nadigar, M P Akash Patel, N P Bhagya

Virtual Substrate Method: Synthesis and Growth Kinetics of 2D Metal Oxide Nanosheets 233
S M Aiswarya, Somesh Chandra, Ashish Jain, Shamima Hussain, S Ganesamoorthy, G M Bhalerao, R Govindaraj

Physicochemical Investigation of Synthesized Bismuth and Silver-Doped Bismuth Nanoferrites, and Their Dielectric Properties 239
Amrutesh Kannolli, P Avinash

Synthesis and Characterization of NiFe2 O4 Nanoparticles by Tartaric Acid Assisted Sol-Gel Auto-Combustion Method 247
Prabhakar Ningayya Patil, Sarvesh Kumar, V M Jali, B Sahoo

Investigation on the Crystal Structure, Magnetic, and Electrical Transport Properties of Magnetron Sputtered Co2FeGe Heusler Alloy Films 253
P. V. Midhunlal, Joshy Joseph, J. Arout Chelvane, A. Rajani Kanth, N. Harish Kumar

Effect of Sn on Physical and Optical Properties of Lithium Zinc Borate Glasses 259
K Mohamed Haneen, Akshay K Thammaiah, R Abhishek, C B Shivamurthy, Vivek Thirumalaimurugan, Jagadeesha B Gangadaraiah, Abhiram Jagannathan

Unveiling the Structural and Optical Properties of Samarium Titanates Pyrochlore, Sm2Ti2O7 265
Rachana Sain, Chandan Upadhyay

Effect of Immersion Liquids on Physical Properties of Pr3+-Ions Doped Lithium Zinc Borate Glasses 272
T Uthayakumar, Namrata Yaduvanshi, A Madhu, N Srinatha

Dielectric Relaxation Studies in Sm2O3 Doped Boro-Zinc-Vanadate Glasses 278
D Ashwini, T Sankarappa, M Amarkumar, H Mohansingh, J Pallavi, D Arvind, B Raghavedra

Electronic and Optical Properties of Na2O Doped P2O5-TeO2CoO-ZnO Glasses 286
J G Pallavi, T. Sankarappa, M Amarkumar, H Mohansingh, D Ashwini

Synthesis and Characterization of Zinc Doped Nickel Nanoferrites Via Sol-Gel Auto-Combustion Technique 295
Ashwini Rayar, G D Prasanna

Author Index

IOP Conf. Series: Materials Science and Engineering 1300 (2024) 011001 doi:10.1088/1757-899X/1300/1/011001

IOP Material Science & Engineering Conference Proceedings

Third International Conference on Physics of Materials & Nanotechnology (ICPN-2023)

held during

September 21-23, 2023

Chief Editor

Dr. Devendrappa H

ICPN-2023

Organized by,

Department of Physics

Mangalore University, Mangalagangothri – 574199

Karnataka, India.

IOP Conf. Series: Materials Science and Engineering 1300 (2024) 011001 doi:10.1088/1757-899X/1300/1/011001

PREFACE

Third International Conference on Physics of Materials & Nanotechnology (ICPN-2023)

On behalf of Organizing committee, it's my great pleasure to bring this beautiful collection of research articles which were presented in the Third International Conference on Physics of Materials and Nanotechnology (ICPN-2023) which was organized by Department of Physics, Mangalore University, Mangalagangothri, Mangalore (Karnataka), India during September 21st to 23rd, 2023. This conference will serve as an ideal platform for young researchers, academician and students from all over the country as well as from the globe to share their ideas and experiences in Physics, Nanotechnology and Nano science research. This conference provided meaningful platform for young researchers to update and upgrade themselves. The main themes of the conferences Advanced Energy and Solar Materials, 1-D, 2-D and Quantum dot Materials, Advanced Electronic Materials, Polymers, Conducting Polymer and Thin films, Solid State Materials and their properties, Sensors and actuators, Carbon related materials, Metal Oxide & Sulphide Materials, Ferroelectrics and dielectric materials, Soft, Magnetic and smart materials, Nano Material, Nanotechnology, Chemistry of Materials, Alloys, Semiconductor and optical Materials and Crystal growth & applications etc. I am glad that experts in the various field of Nanotechnology are here to present and discussed their ideas and latest development in the field. ICPN 2023 has received a great response throughout the country and globe. It includes 14 invited talks, 11 oral presentations and more than 120 poster presentations. We have received more than 130 papers for the publications in this issue, after reviewing and editing processes, 47 papers were accepted for the publication out of 74 submitted papers. We thank the authors for their contribution and the reviewers for their comments to improve the quality of article. I would like to express my gratitude to all members of organizing committee, staff, research scholars and postgraduate students of the institute for their willingness, commitment, dedication and relentless efforts in ensuring that conference was successful.

IOP Conf. Series: Materials Science and Engineering 1300 (2024) 011001 doi:10.1088/1757-899X/1300/1/011001

Editorial Board;

IOP Conf. Series: Materials Science and Engineering 1300 (2024) 011002 doi:10.1088/1757-899X/1300/1/011002

Peer review statement

All conference organisers/editors are required to declare details about their peer review. Therefore, please provide the following information:

- **Type of peer review: Single-blind / Double-blind / Triple-blind / Open / Other (please describe)**
  Single blind
- **Conference submission management system: Morresises**
- **Number of submissions received: 74**
- **Number of submissions sent for review: 74**
- **Number of submissions accepted: 47**
- **Acceptance Rate (Number of Submissions Accepted / Number of Submissions Received X 100): 63.51 %**
- **Average number of reviews per paper: 2**
- **Total number of reviewers involved: 5**
- **Contact person for queries:**

Dr. Vijeth

Email: vijethhebri@gmail.com

Affiliation: Assistant Professor Department of Physics

Nagaland University (A Central University), Lumami, Zunheboto Dist. Nagaland

Dr. Ganesha H

Email: ganeshkumar4006@gmail.com

Affiliation: Department of Chemistry

Kasetsart University, 50 Ngamwongwan Rd, Lat Yao, Chatuchak, Bangkok 10900

IOP Conf. Series: Materials Science and Engineering 1300 (2024) 012001 doi:10.1088/1757-899X/1300/1/012001

Computational insights into the electronic structure of functionalized monolayer hBN supercapacitor electrodes

Sruthi T and Vincent Mathew

Computational Nanoscience Lab, Department of Physics, Central University of Kerala, Kasaragod, Kerala-671316, India

E-mail: drsruthi2023@gmail.com

Abstract. In this study, we examined a collection of functionalized hBN monolayers' electronic structure and quantum capacitance. Different ad-atom adsorption techniques have been used to functionalize the hBN monolayer. Density functional theory calculations are performed to precisely determine the electrical structure of an ad-atom doped hBN monolayer with a range of doping concentrations. The quantum capacitance of every functionalized system was then calculated. Above 500 $\mu F/cm^2$, a noticeable quantum capacitance has been seen. Our calculations demonstrate that ad-atom doping at hollow position of hBN honeycomb lattice greatly increases the quantum capacitance of hBN monolayer. The microscopic cause of this system's increased quantum capacitance has been investigated. Our DFT-based calculation shows that the system has a very large quantum capacitance due to the creation of new electronic states in the vicinity of the band edge and the Fermi level shift caused by the ad-atom adsorption.

Introduction

One of the biggest obstacles to halting climate change is the process of producing enough energy without using fossil fuels by using renewable sources instead. Thus, it is imperative in the current context to generate green energy on a large scale using sustainable energy sources. However, technology must be created that can effectively transform and preserve the energy produced. Supercapacitors, fuel cells, and batteries are the three main types of modern technology used to convert and store energy. Supercapacitors have many advantages, including a high power density, long lifespan, greater stability, and the ability to be used in a wide range of temperatures[1, 2]. Large amounts of electrical energy can be temporarily stored by supercapacitors and released as needed. Their high production costs and low energy density, however, impede their development[3]. To increase the use of supercapacitors in more complex applications, these issues must be resolved. Research is being done extensively to alter the electrode material of supercapacitors in an effort to solve the problem of poor energy density[4-6]. A noticeably high ion density in the electrode material is one of the fundamental conditions for a highly effective supercapacitor. The ideal alternative for supercapacitor electrode applications may consequently be two-dimensional materials because they often have a large specific area. Graphene is a 2D substance that is one atom thick and composed of carbon atoms organised in a honeycomb lattice. It exhibits exceptional electrical and mechanical capabilities. Studies have been done using graphene, graphene that has undergone chemical modification, and composites based on

IOP Conf. Series: Materials Science and Engineering 1300 (2024) 012001 doi:10.1088/1757-899X/1300/1/012001

graphene as electrode materials for supercapacitors[7–10]. Its limited scalability [11, 12] hinders its broad application, necessitating the search for substitute, suitable two-dimensional materials for supercapacitor electrodes.

Due to its insufficiently large bandgap (5.95 eV), hexagonal boron nitride (h-BN), a 2D material, has not been thought to be an appropriate material for supercapacitor electrodes (Cassabois et al. 2016).Chemical modification changes h-BN from a broad bandgap semiconductor to a metallic system with DOS at the Fermi level. So, in an effort to make these materials viable for supercapacitor applications, I functionalized them with the appropriate adatoms.

Methodology

The structural optimisation and electronic structure information of different functionalized systems are obtained using first-principle DFT calculations, which are included into the Vienna ab initio simulation(VASP) software[13,14]. The exchange-correlation energy was described by the generalized-gradient approximation with the Perdew-Burke-Ernzerhof parameterization [15]. For accurate results, the energy cut-off value (>400 eV) was considered to be extremely high. Calculations were performed using 3×3×1 supercells of hBN unit cells, which are composed of nine nitrogen and nine boron atoms, to examine the effects of different ad-atom doping on the quantum capacitance[16-18]. Comparison of ground-state energies reveals the stable adsorption location of ad-atoms on hBN monolayer. The stable configurations chosen for doping are hollow position in hBN monolayer. To avoid interaction with the layers of hBN unit cells' periodic pictures, a huge empty space (height>10Å) was taken into consideration along the out of the plane direction. The maximum energy tolerance for convergence was 10^{-6} H for geometrical optimisation. We employed a 6×6×1 Monkhorst-Pack grid to sample the Brillouin zone. To obtain the accurate D(E), and atom projected density of states (PDOS), a denser 20×20×1 Monkhorst-Pack grid was utilised. An electrical double-layer capacitor's total capacitance, C_T, is represented by the following formula:

$$\frac{1}{C_T} = \frac{1}{C_Q} + \frac{1}{C_D} \tag{1}$$

where C_D and C_Q denote, respectively, double-layer and quantum capacitance. Materials' quantum capacitance is determined by measuring the pace at which excessive charges (ions) shift when the applied voltage changes [31]. It is therefore directly related to the electronic energy configuration of the electrode materials and can be defined as the derivative of the net excess charge on the electrode or substrate with respect to the electrostatic potential. i.e.

$$C_Q = \frac{dQ}{d\phi} \tag{2}$$

where ϕ is the electrode's chemical potential and Q is the electrode's excessive charge. An electronic DOS weighted average up to the Fermi level, or E_F, is used to determine the overall charge. Therefore, the excessive charge on the electrode (Q) can be explained by an integral term related to the Fermi-Dirac distribution function f(E) and the electronic density of state D(E). So we can express Q as

$$Q = e\int_{-\infty}^{+\infty} D(E)[f(E) - f(E-\phi)]dE \tag{3}$$

Therefore, if the DOS is known, the channel's C_Q at a given temperature T may be

$$C_Q = \frac{dQ}{d\phi} = \frac{e^2}{4kT}\int_{-\infty}^{+\infty} D(E)sech^2\frac{E-e\phi}{2kT}dE \tag{4}$$

IOP Conf. Series: Materials Science and Engineering 1300 (2024) 012001 doi:10.1088/1757-899X/1300/1/012001

Chemical potential, electron charge, and Boltzmann constant are represented by the symbols ϕ, e and k, respectively.

Results and discussions

It is proven that equation (4) includes the DOS which is found close to the Fermi energy. As the single-layer hBN has no states near the Fermi level and has a large band gap semiconductor [32], it displays zero quantum capacitance (Figure 1). Ad-atom doping can be used to modify a material's electrical structure. The production of quantum capacitance by monolayer hBN could occur if the alteration in electronic structure leads to the collection of states at the Fermi level.

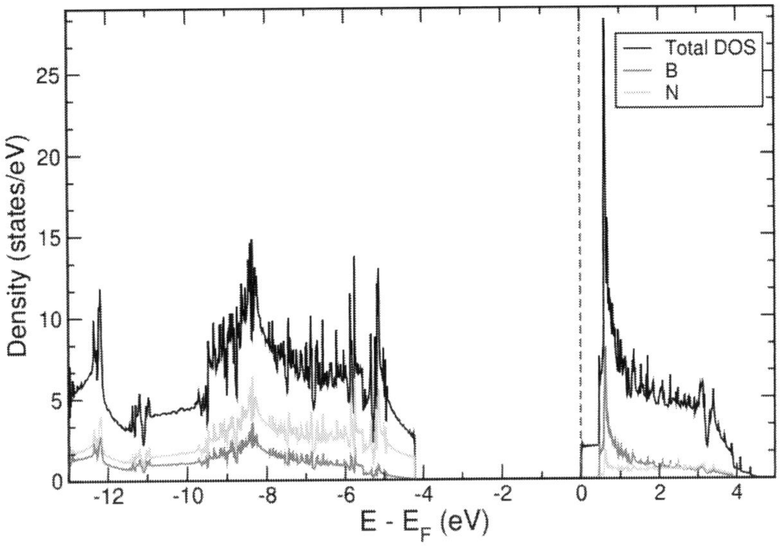

Figure 1. (color online) Atom projected density of states for monolayer hBN. The doped atom's DOS is represented by the coloured curve. E=0 is the Fermi energy represented by the vertical blue dashed line.

Chemical modification of monolayer hBN at Hollow position

On the hBN lattice, there may be three possible adsorption sites: the top (on the vertical top of the B and N atoms), hollow (in the sheet's hexagonal middle), and bridge site (between the B and N atoms). By contrasting ground-state energies, ad-atoms' stable adsorption position on monolayer hBN is found. According to our calculations, the hollow sites are the best places for ad-atom adsorption.

The average adsorption energy E_{ad} has been determined using the equation to investigate the stability of doped hBN structures.The stability of doped hBN structures has been examined by calculating the average adsorption energy E_{ad} utilising the equation.

$$E_{ad} = \frac{1}{n}[E_{tot} - E_{hBN} - nE_{at}] \qquad (5)$$

where n is the unit cell's ad-atom count, E_{at} is the ad-atom's energy per atom. E_{tot} is the functionalized hBN unit cell's total energy, and the entire energy of pure hBN within a single unit cell is E_{hBN}. The adsorption energies for the different functionalized hBN are listed in Table 1. These atoms can readily bind to the pure hBN surface, according to the relatively high adsorption energies.

Table 1. Energy of adsorption per atom that adsorbs on monolayer hBN.

ad-atom	E_{ad}(eV)	ad-atom	E_{ad}(eV)
Br	-2.019	Li	-2.042
Si	-2.120	P	-0.202

We have studied how doping different atoms at the hollow position affects the quantum capacitance of monolayer hBN. The optimized geometric structure of configurations is shown in figure 2.

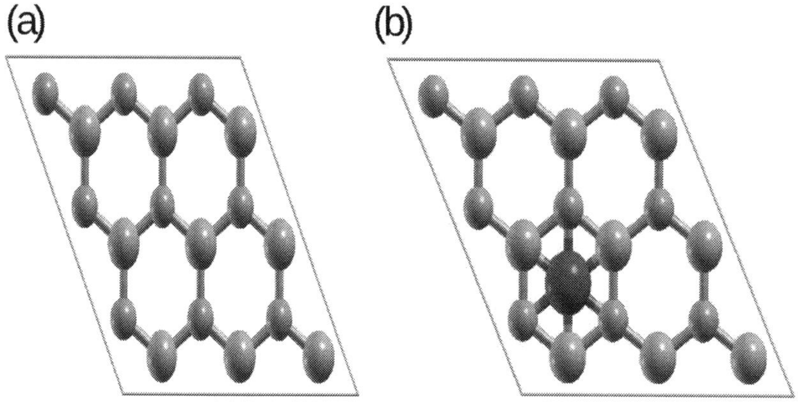

Figure 2. (color online) Structure of (a) Pristine hBN monolayer and (b)Br functionalized hBN at hollow site. Light brown balls and blue balls represent B and N atoms respectively. Dark brown ball represent the bromin atom which doped at hollow site.

The DOS and electronic band structure of the charge carrier in the functionalized-hBN configurations can be used to predict its characteristics. The electrical structure's tuning could result in a significant improvement in C_Q.

Figure 3 plots the computed C_Q for various ad-atom doped hBN monolayers, and table 2 lists the values. As illustrated in figure 3 for Br-functionalized hBN, significant amounts of DOS build up close to the Fermi level. As a result, high-quantum capacitance values of 1341.71 was found in this system. These localised DOS at the Fermi level cause a rise in C_Q, which is greatest for Br, Si, Li and P when compared to other ad-atoms.

Table 2. C_Q value computed at Fermi energy for different ad-atom doped hBN monolayer.

Configuration	C_Q (μF/cm^2)	Configuration	C_Q (μF/cm^2)
Br−hBN	1341.71	Li−hBN	643.90
Si−hBN	852.12	P−hBN	601.10

A considerable charge reorganisation caused by chemical modification is one way to achieve a high C_Q value. As a result, we used Bader charge analysis to examine the charge redistribution

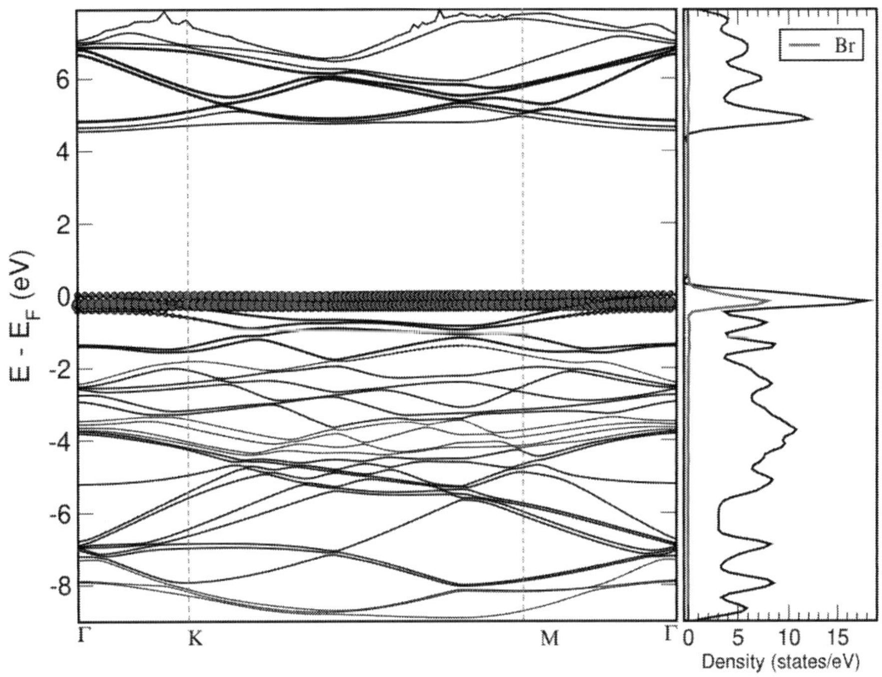

Figure 3. (color online) The coloured curve in the DOS and the circle in the band structure represent the electronic band and DOS for Br doped hBN at the hollow site, respectively. Doped atom contributions are displayed.

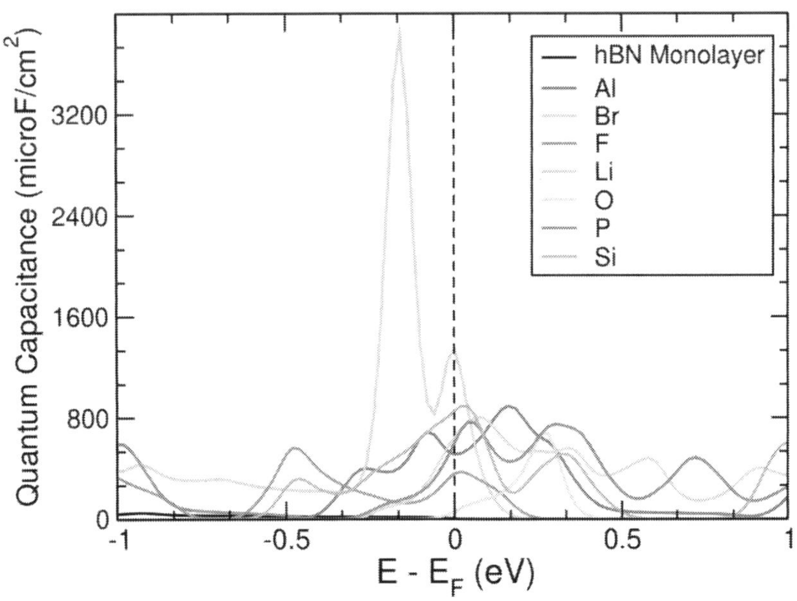

Figure 4. (color online) The energy variations in quantum capacitance for hBN functionalized with various ad-atoms at the Hollow site are shown by the varied colour trends.

in the chemically altered hBN. As shown in table 3, we discovered that there was a considerable quantity of charge transfer between the ad-atoms and the hBN monolayer.

Table 3. Charge transfer data for different ad-atom functionalized monolayer hBNs.

System	Transferred charge (e)
Br−hBN	-0.3075
Si−hBN	+2.3639
Li−hBN	-1.2576
P−hBN	+0.5672

Conclusion

We have investigated the functionalized hBN monolayer monolayer's quantum capacitance. Our theoretical analysis demonstrates that adding ad-atoms to the monolayer hBN sheet can greatly increase the quantum capacitance (C_Q) of hBN electrodes. The results demonstrate the improvement in quantum capacitance caused by altering the hBN monolayer's electrically insulating behaviour to a metallic one. These adatoms form a strong connection with the hBN monolayer, and the optimised shape exhibits metallic behaviour.

Above 500 μF/cm^2, a noticeable quantum capacitance has been observed. These calculations demonstrate that doping the honeycomb lattice's hollow site with various adatoms increases the quantum capacitance of the hBN monolayer. The functionalized hBN monolayer showed a large increase in quantum capacitance, which is mostly due to the doping effect's creation of new electronic states close to the Fermi level. In ad-atom doped monolayer hBN, a significant charge transfer and redistribution occurs. Ad-atoms' extra charge carriers alter the monolayer hBN's carrier concentration, shifting the Fermi level as a result.

Acknowledgments
Authors acknowledge the computing facility of Central University of Kerala. I am thankful for the post doctoral fellowship granted by the Kerala State Council for Science, Technology and Environment (KSCSTE).

References

[1] Kresse G and Furthmüller J 1996 *Phys. Rev. B* 54 11169
[2] Sruthi T and Kartick T 2019 *Bulletin of Materials Science* 42 257
[3] Choi B G, Hong J, Hong W H, Hammond P T and Park H 2011 *ACS Nano* 5 7205–13
[4] Sruthi T and Kartick T 2021 *Physica B: Condensed Matter* 604 412676
[5] Kresse G and Furthmüller J 1996 *Comput. Mater. Sci.* 6 15–50
[6] Perdew J P, Burke K and Ernzerhof M 1996 *Phys. Rev. Lett.* 77 3865
[7] Huang X, Qi X, Boey F and Zhang H 2012 *Chem. Soc. Rev.* 41 666–86
[8] Sruthi T and Kartick T 2019 *J. Phys.: Condens. Matter* 31 475502
[9] Hirunsit P, Liangruksa M and Khanchaitit P 2016 *Carbon* 108 7–20
[10] Yang X, Cheng C, Wang Y, Qiu L and Li D 2013 *Science* 341 534–7
[11] Kim T Y, Lee H W, Stoller M, Dreyer D R, Bielawski C W, Ruoff R S and Suh K S 2011 *ACS Nano* 5 436–42
[12] Zhu Y et al 2011 Science 332 1537–41
[13] Wang G, Zhang L and Zhang J 2012 *Chem. Soc. Rev.* 41 797–828
[14] Reina A, Jia X, Ho J, Nezich D, Son H, Bulovic V, Dresselhaus M S and Kong J 2009 *Nano Lett.* 9 30–5
[15] Xu Q, Yang G M, Fan X and Zheng W T 2019 *Phys. Chem. Chem. Phys.* 21 4276–85
[16] Stoller M D, Park S, Zhu Y, An J and Ruoff R S 2008 *Nano Lett.* 8 3498–502

[17] Sruthi T, Nayana Devaraj and Kartick T 2021 *Electronic Structure* 3
[18] Simon P and Gogotsi Y 2008 *Nature Materials* 7 845–54

IOP Conf. Series: Materials Science and Engineering 1300 (2024) 012002 doi:10.1088/1757-899X/1300/1/012002

Optical Properties of Lawsone dye doped with Zinc Oxide for the Fabrication of Dye Sensitized Solar Cell

D S Suresh*, Abdullah Ba Shbil, S P Vijaykumar, Sapna Sharanappa, H Ganesha.

Department of Physics, Mangalore University, Mangalagangothri-574199, Karnataka, India
*Corresponding author E-mail: sureshds.cta@gmail.com

Abstract: Due to their availability of basic ingredients and affordable cost, natural dyes are important in many applications. One of such natural dyes, Lawsone which is naturally extracted from henna leaves. The Zinc Oxide (ZnO) was intermixed with lawsone dye to their matching of energy levels with Titanium dioxide (TiO_2) enables efficient charge transfer phenomenon. The optical study illustrates Lawsone dye - Zinc Oxide (Ld-ZnO) composite to utilize in energy conversion application. The FTIR spectra demonstrates the presence of Zn-O, C-O, C=O and hydroxyl groups are evident from near 500, 1633 and around 3250 cm^{-1} respectively. The UV-Visible spectra reveal maximum absorbance around narrow and broad peak at 368 nm for Lawsone dye (Ld) and Ld-ZnO composite respectively. The lawsone dye has better emission property that causes recombination of charge carriers reduces the performance of Dye Sensitized Solar Cells (DSSCs). The DSSCs based on lawsone dye (Ld) and Ld-ZnO based material exhibits PCE of 0.12% and 0.17% respectively. This optical study suggests natural extraction of lawsone dye from henna leaves with suitable semi-conducting material was promising candidate to improve performance of DSSCs.

Keywords: Henna leaves; Dye sensitized solar cell; Natural dyes; Zinc oxide; Lawsone dye;

1. Introduction

The need for clean and renewable energy sources has intensified recently as the globe faces against the terrible effects of climate change and the depletion of finite fossil fuels. Among the many promising renewable energy technologies, dye-sensitized solar cells (DSSCs) have emerged as a viable choice for solar energy capture due to its affordability, environmental friendliness and efficiency [1]. Plants carry out photosynthesis, which is replicated by the distinct photo electrochemical mechanism employed by DSSCs [2]. This creative technique converts solar energy into electricity by using pigments that absorb light. The *Lawsonia inermis* plant's leaves are the source of the Lawsone dye, which has been utilized for generations in a variety of innovative and cultural endeavors. Given its historical relevance and artistic appeal, it is a compelling candidate for the field of renewable energy. However, because of its inherent spectrum limits and instability when exposed to UV light, the direct use of henna in DSSCs has demonstrated low efficiency [3].

Charge recombination is the process by which photo generated electrons and holes recombine before contributing to the electrical output is one of the primary challenges in DSSCs. The interfacial recombination between TiO_2 and lawsone dye based DSSCs can result in considerable electron-hole recombination losses, resulting in decreased total cell efficiency. This problem can be lessened through establishing appropriate interfacial layers. It is well known that lawsone dye is sensitive to the environment, particularly to extended UV light exposure. Because of its sensitivity, the dye may eventually deteriorate, reducing light absorption and creating instability in the performance of DSSCs [4, 5]. Researchers have turned to nano-engineering techniques and coupled lawsone with semiconducting metal oxides such as zinc oxide (ZnO) to get around these issues and improve DSSC performance. A promising area of renewable energy technology is the integration of lawsone dye with zinc oxide in dye-sensitized solar cells [6].

This novel approach aims to study the unique optical properties of both lawsone dye and ZnO to improve the photo-electro-chemical efficiency of DSSCs. The exploration of charge transfer phenomenon in this hybrid material since the energy level of TiO_2 NPs matching with Zinc oxide. The charges enable this to hold tremendous potential in mechanism of DSSCs. The potential synergy of lawsone and ZnO offers an approach to address the challenges faced by traditional silicon-based solar cells. This study will evolve into the fundamental principles, properties and potential applications of Lawsone-ZnO dye as photo-sensitizer in DSSCs, laying the groundwork for future advancements in solar energy conversion and paving the way for a more sustainable and environmentally conscious energy landscape.

2. Experimental methods

2.1 Materials

The Henna leaves are collected from university campus garden. The ITO glass substrates, Zinc Oxide (ZnO), Titanium di-oxide (TiO_2), Graphite was purchased from Sigma Aldrich. The ethylene glycol, acetone, isopropyl alcohol and distilled water were used as solvents.

2.2 Synthesis of Ld-ZnO (Lawsone dye- Zinc Oxide) Composite

The synthesis of Lawsone dye – Zinc Oxide (Ld-ZnO) prepared through natural dye extraction method. The small pieces of Henna leaves which is washed and dried. They were boiled at 120°C and dye solution was extracted through filtration process. The heat treatment of dye permits much more light absorption to use as dye sensitizer in DSSCs. The 20ml of Lawsone dye solution and 20 ml of aqueous Zinc oxide solution was intermixed through probe sonication for 30 minutes. The Lawsone dye - Zinc oxide (Ld-ZnO) composite synthesis as shown in scheme 1.

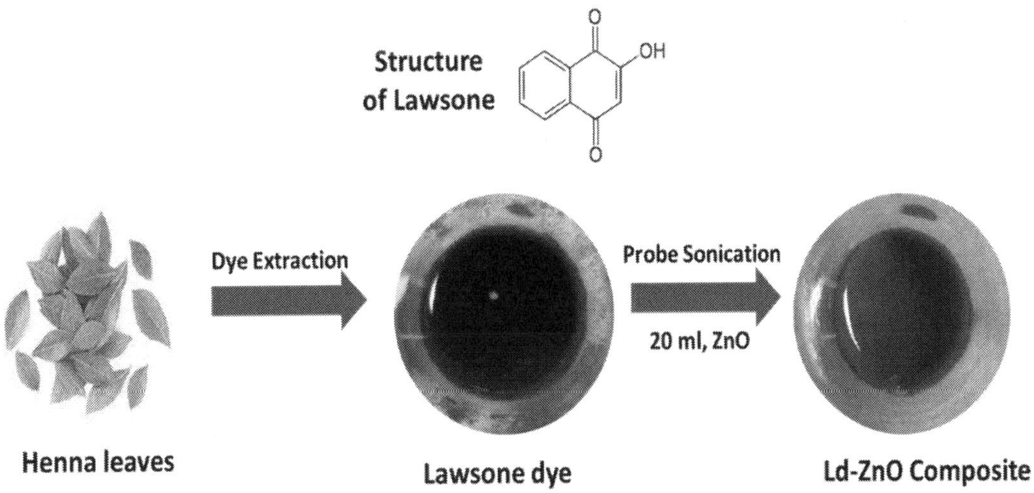

Scheme 1: The schematic representation of Ld-ZnO (Lawsone dye- Zinc Oxide) Composite synthesis.

2.3 Fabrication of Dye sensitized solar cell

The ITO glass substrates were washed through ultra-sonication with distilled water, acetone and iso-propyl alcohol each. The synthesized Titanium di-oxide (TiO_2) NPs was deposited on ITO substrates through doctor-blading method that can dried at 350° C [7]. The TiO_2 coated glass substrates are dipped in solutions of lawsone dye with zinc oxide and Lawsone dye are separately for 30 hours in dark environment. The graphite coated on ITO that can acts as counter electrode. Both ITO substrates arrange one over other to fabricate DSSC for Lawsone dye with ZnO and lawsone dye by placing few drops of iodide electrolyte between them respectively [8].

3. Results and Discussion

3.1 FTIR Analysis

The intermolecular interactions are studied by Fourier Transform Infrared Spectrometer (ATR Alpha Bruker) model. It is an important tool to identify chemical bonding and stretching vibrational group measurements carried out in the range 450-4000cm^{-1}.

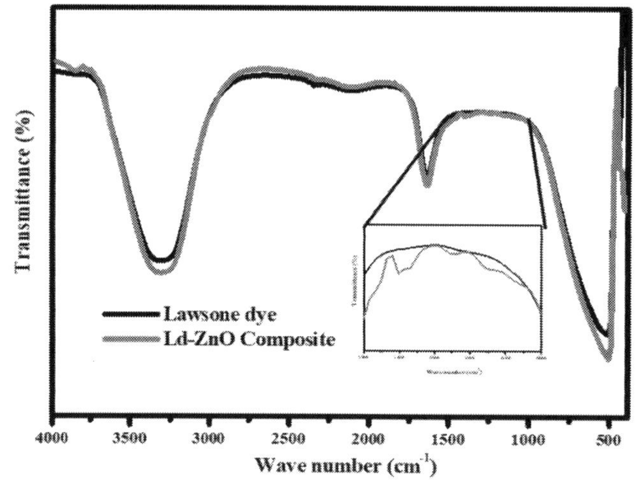

Figure 1: FTIR Spectra of (a) Lawsone dye (Ld) and (b) Ld-ZnO composite.

The broad absorption peak appeared at around 3250 cm^{-1} was assigned to stretching vibration of hydroxyl (-OH) group and it is due to also the presence of water content in dye [9]. An auxochrome is a functional group of molecules that connects to a chromophore that alters the wavelength of that the chromophore's absorption of light. The skeleton C=C stretching vibrations of an aromatic ring revealed from the peak at 1636 cm^{-1} [10]. The magnified FTIR spectrum shows the presence of ZnO with lawsone dye alters the energy levels and also reveals together well blended in the composite.

3.2 Optical analysis

The light absorption properties of synthesized lawsone dye and Ld-ZnO composite are studied by using Perkin Elmer lambda 35 UV-Visible Spectrometer.

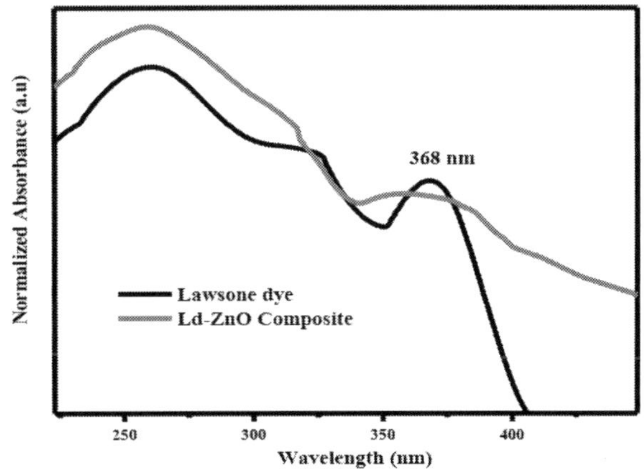

Figure 2: UV-Visible spectra Lawsone dye and Ld-ZnO Composite.

As per the structure and FTIR analysis, the lawsone dye and its composite with zinc oxide consists of auxochrome functional group such as hydroxyl (-OH) present in it. These groups can cause green color of dye that enables broad absorption of light [11]. The broad peak near 250 nm shows solvent content present in the dye [12]. The more absorption of light from Ld-ZnO composite can promise to utilize this material in energy conversion applications.

3.3 Photo-voltaic study of Lawsone dye based DSSC

The figure 3 (a) reveals J-V characteristics was measured using class AAA Solar simulator. The above optical analysis evident that Ld and Ld-ZnO composite has efficient optical properties to use as sensitizer in fabricated DSSC. The light absorption from Lawsone dye and its composite can enable charge transport through iodide electrolyte, Titanium dioxide and graphite coated ITO substrate electrodes. Due to the device structure aligned with well blended energy levels of TiO_2 and ZnO, the photon and electron interaction causes increase in photo-current in Ld-ZnO composite due to efficient charge transport takes place between them [13].

Material	Open Circuit Voltage (V_{OC})	Short Circuit Current density (J_{SC})	Fill Factor (FF)	Power Conversion Efficiency (%)
Lawsone dye	0.5	0.3061	0.67	0.12
Ld-ZnO Composite	0.556	0.4626	0.72	0.17

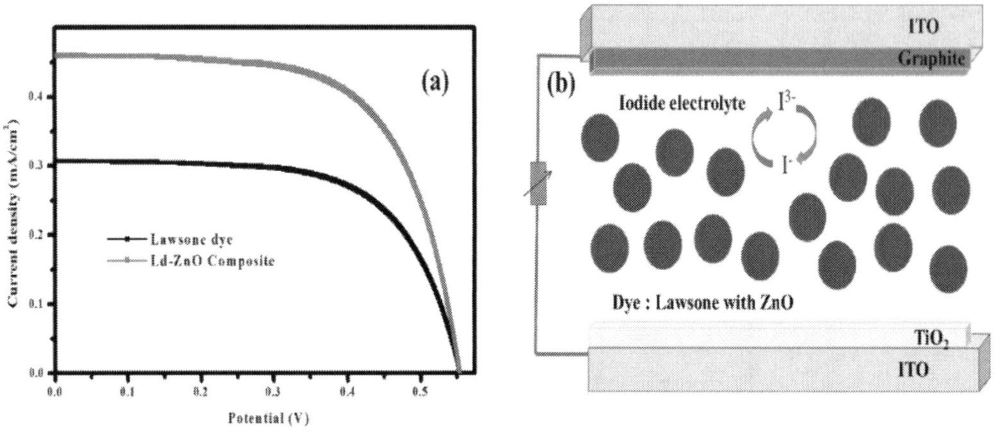

Figure 3: a) J-V characteristics and b) Mechanism of fabricated DSSCs.

The solar parameters of DSSC based on lawsone and its composite as reported in Table 1. The Ld-ZnO composite based DSSC has best performance due to its charge transport phenomenon of ZnO and TiO_2 upon irradiation causes the high photo-current density. The graphite as counter have lower efficiency compared to other cathode can enhance efficiency to nearly 1% [14]. The Ld-ZnO material is promising candidate to utilize as transport layer in energy conversion applications.

4. Conclusion

Lawsone is one of the pigments that is naturally derived from henna through the process of natural extraction. Lawsone and Zinc oxide were combined because of their similar energy levels of ZnO with TiO2, which provide effective optical properties. C=C, C-C, and -OH interactions appear in the FTIR spectrum and are assigned at around 500 cm^{-1}, 1633 cm^{-1}, and approximately 3300 cm^{-1} respectively. The Lawsone dye (Ld) and Ld-ZnO composite exhibit maximal absorbance at a narrow and broad peak at 368 nm respectively according to the UV-visible spectra. Dye-sensitized solar cells (DSSCs) perform less efficiency due to weak binding between lawsone dye with TiO_2 electrode. Lawsone dye (Ld) and Ld-ZnO-based materials used in the DSSCs show PCE values of 0.12% and 0.17% respectively. According to this optical investigation, a viable option to enhance the performance of DSSCs is the natural extraction of lawsone dye from henna plants using an appropriate semi-conducting material.

Acknowledgement:

This work was financially supported by Science and Engineering Research Board (SERB), New Delhi (SERB/F/11204/2018-2019 dated 15 February, 2019).

References

1. Bera, S., Sengupta, D., Roy, S., & Mukherjee, K. (2021). Research into dye-sensitized solar cells: a review highlighting progress in India. *Journal of Physics: Energy*, *3*(3), 032013.
2. Amogne, N. Y., Ayele, D. W., & Tsigie, Y. A. (2020). Recent advances in anthocyanin dyes extracted from plants for dye sensitized solar cell. *Materials for Renewable and Sustainable Energy*, *9*, 1-16.

3. Hatta, f. A. M. The use of malay cultural ethnobotany as natural colourants in eco-friendly dyeing. *Ehmap*, 183.
4. Tripathi, B., Yadav, P., & Kumar, M. (2014). Charge transfer and recombination kinetics in dye-sensitized solar cell using static and dynamic electrical characterization techniques. *Solar energy*, *108*, 107-116.
5. Magiswaran, K., Norizan, M. N., Mohamad, I. S., Mahmed, N., Idris, S. N., & Sobri, S. A. (2021). Charge Recombination in Zinc Oxide-Based Dye-Sensitized Solar Cell: A Mini Review. *International Journal of Nanoelectronics & Materials*, *14*.
6. Boro, B., Gogoi, B., Rajbongshi, B. M., & Ramchiary, A. (2018). Nano-structured TiO2/ZnO nanocomposite for dye-sensitized solar cells application: A review. *Renewable and Sustainable Energy Reviews*, *81*, 2264-2270.
7. Sinha, D., De, D., Goswami, D., & Ayaz, A. (2018). Fabrication of DSSC with nanostructured ZnO photo anode and natural dye sensitizer. *Materials Today: Proceedings*, *5*(1), 2056-2063.
8. Saeidi, M., Abrari, M., & Ahmadi, M. (2019). Fabrication of dye-sensitized solar cell based on mixed tin and zinc oxide nanoparticles. *Applied Physics A*, *125*, 1-9.
9. Safie, N. E., Ludin, N. A., Su'ait, M. S., Hamid, N. H., Sepeai, S., Ibrahim, M. A., & Teridi, M. A. M. (2015). Preliminary study of natural pigments photochemical properties of curcuma longa l. and lawsonia inermis l. as tio 2 photoelectrode sensitizer. *Malaysian Journal of Analytical Sciences*, *19*(6), 1243-1249.
10. Bekele, E. T., & Sintayehu, Y. D. (2022). Recent Progress, Advancements, and Efficiency Improvement Techniques of Natural Plant Pigment-Based Photosensitizers for Dye-Sensitized Solar Cells. *Journal of Nanomaterials*, *2022*.
11. Chanana, G., & Batra, K. (2022). Modelling natural dye molecules lawsone and purpurin in different solvents for DSSC applications: a DFT and TD-DFT study. *Molecular Simulation*, *48*(9), 760-775.
12. Khadtare, S. S., Jadkar, S. R., Salunke-Gawali, S., & Pathan, H. M. (2013). Lawsone sensitized ZnO photoelectrodes for dye sensitized solar cells. *Journal of Nano Research*, *24*, 140-145.
13. Khadtare, S. S., Ware, A. P., Salunke-Gawali, S., Jadkar, S. R., Pingale, S. S., & Pathan, H. M. (2015). Dye sensitized solar cell with lawsone dye using a ZnO photoanode: experimental and TD-DFT study. *RSC Advances*, *5*(23), 17647-17652.
14. Sathyajothi, S., Jayavel, R., & Dhanemozhi, A. C. (2017). The fabrication of natural dye sensitized solar cell (Dssc) based on TiO2 using henna and beetroot dye extracts. *Materials Today: Proceedings*, *4*(2), 668-676.

Simultaneous synthesis of carbon quantum dots and porous carbon using one step hydrothermal method for supercapacitor applications

Sapna Sharanappa*, Vijaykumar S P, Suresh D S, Ganesha H, Abdullah Ba Shbil, Basappa M.

Department of Physics, Mangalore University, Mangalagangothri-574199, India
*Corresponding author E-mail: sapnaawanti97@gmali.com

Abstract: The present work reports a one-step hydrothermal carbonization process used to simultaneously produce carbon quantum dots and porous carbon. The synthesized materials were characterized by using a Fourier transform infrared spectrometer (FT-IR) to study the chemical interactions, the optical properties were studied with the help of UV-visible absorption spectroscopy, and the morphology analysis of CQDs and porous carbon was examined using FESEM, in which spherical morphology confirms the formation of CQDs, while XRD confirms the amorphous nature of the porous carbon. The application of CQDs and porous materials is evaluated as a supercapacitor material by using the cyclic voltammetry technique. The specific capacitance of porous carbon was 161.33 Fg^{-1} at 0.01mVs^{-1} scan rate. This makes CQDs and porous carbon a promising candidate for applications in energy storage devices such as supercapacitors.

Keywords: Carbon quantum dots; Porous carbon; Specific capacitance; Energy storage;

1. Introduction

Carbon Quantum Dots are the zero-dimensional nanomaterials that are most attracted to extensively researched materials having a tremendous application in various fields such as energy storage, bio-sensors, catalysts, drug delivery, and bio-imaging due to their outstanding properties like the plain synthetic route, unique structures of graphitic cores, amorphous shells, and various organic groups, and special merits including large surface area surface states, good dispersion in solvents, superior conductivity, and low cost of production over carbon nanomaterials such as graphene and carbon nanotubes [1]. Porous carbon, in contrast to CQDs, has been studied for a long time. As an excellent adsorbent, it possesses a very large specific surface area, a high pore volume, and multiple active sites for doping, which have broad applications in pollutant adsorption, supercapacitors, hydrogen storage, and catalysis [2]. Researchers have been able to tailor the properties of porous carbon to meet specific application requirements. For example, by optimising the pore size and surface chemistry, porous carbon can be designed to selectively adsorb certain pollutants or enhance catalytic activity. Additionally, advancements in carbon doping techniques have further expanded the range of applications for porous carbon materials [3]. The current work focuses on the one-step hydrothermal carbonization process used for the simultaneous production of carbon quantum dots and porous carbon. The prepared materials were carefully investigated with the help of different characterization techniques. These materials exhibited excellent electrochemical performance, with high specific

IOP Conf. Series: Materials Science and Engineering 1300 (2024) 012003 doi:10.1088/1757-899X/1300/1/012003

capacitance and good cycling stability. Furthermore, the porous structure of the carbon material provided a large surface area, enhancing the accessibility of electrolyte ions and improving overall energy storage capacity. Overall, our findings emphasise the possibility of employing carbon quantum dots and porous carbon as improved electrode materials in different energy storage devices [4].

2. Experimental methods

2.1 Materials used

Citric acid (CA) (Sigma-Aldrich, 99%), Urea (CH_4N_2O) from National Chemicals India, Deion-ized (DI) water was used throughout the experiment.

2.2 Synthesis of CQDs and Porous carbon

CQDs and porous carbon were synthesised using citric acid and urea through a one-step hydrothermal carbonization method. In a typical procedure, 1 g of citric acid and 0.5 g of urea were added to 20 ml of DI water and stirred for about 30 minutes to get a clear suction. Then, the resulting solution was transferred to a 50 ml Teflon-lined autoclave, and the autoclave was heated at 150 °C for 18 hours using a hot air oven. After being cooled down to room temperature, the resultant brown aqueous solution is a carbon quantum dot, and the final filtrate product is dried at 60 °C for 12 hours to obtain porous carbon.

3. Characterization

Chemical compositions were studied using the Bruker Alpha ATR FT-IR spectrometer, optical behaviour was studied using the double beam computer interface Perkin Elmer Lambda 350 UV-Visible Spectrometer, structural analysis was carried out using the Rigaku X-ray diffractometer (XRD), surface morphology was examined using Sigma Zeiss Field Emission Scanning Electron Microscopy (FESEM), and electrochemical performance was carried out using a PC-based CHI 660E electrochemical workstation.

4. Results and Discussion

4.1 FT-IR spectra and UV-Visible

The FTIR spectra of CQDs in Figure 1(a) show that the peaks at 1043, 1211, and 1387 cm^{-1} are assigned to the C-H, C-N, and C=N vibration bonds [5]. The two principle characteristic peaks at 1532 and 1671 cm^{-1} are observed for both CQDs and porous carbon belonging to the C=C and C=O stretching vibrations, which confirm the formation of porous carbon [6]. The peak at 2343 cm^{-1} is due to atmospheric CO_2, and the additional peaks at 3588, 3790, and 3860 cm^{-1} confirm the presence of O-H groups on the surface of CQDs and porous carbon. These functional groups enhance the hydrophilicity and stability of blue luminescent CQDs in an aqueous system [7].

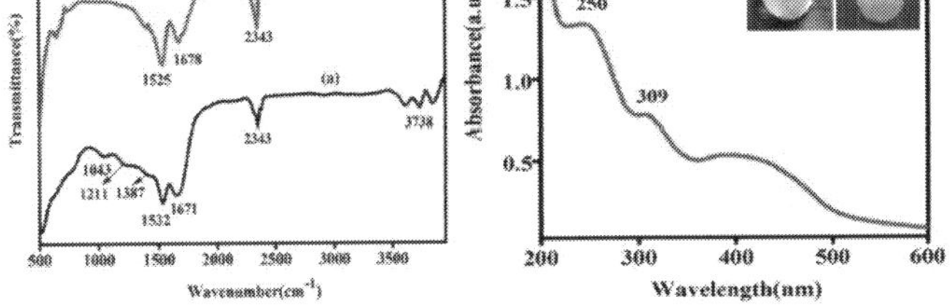

Figure 1: (a) FTIR spectra of CQDs, (b)Porous carbon, and (c) UV-visible spectra of CQDs

UV-visible absorption spectra of CQDs in aqueous media are presented in Figure 1(c). It is clear that CQDs show optical absorption peaks at 250 and 309 nm in the UV region, which correspond to the π-π^* electronic transition of the conjugated C=C and C=O bonds in CQDs [8]. The prepared CQDs in aqueous solution show blue luminescence on exposure to UV light at 365 nm, as shown in the inset of Figure 1c, which confirms the formation of CQDs [9].

4.2 XRD analysis

Figure 2(a) shows the XRD spectra of CQDs. CQDs show a broad characteristic intense peak at 19.6^0 with interlayer spacing of 0.45 nm attributed to the (002) hkl crystal plane, which is higher than that of graphitic lattice d-spacing. This confirms the poor crystalline nature of CQDs [10].

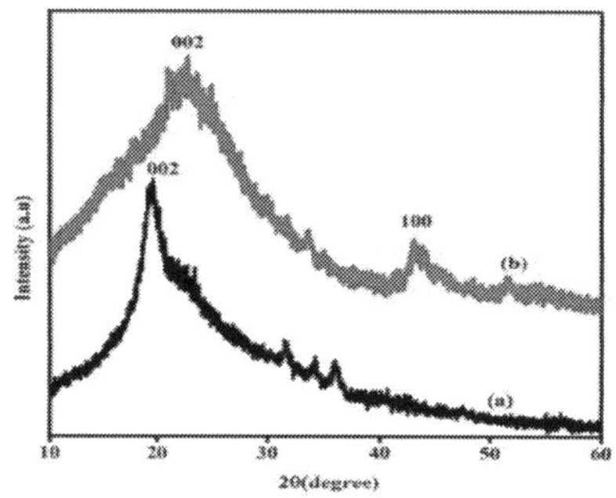

Figure 2: (a) XRD spectra of CQDs, (b)Porous carbon,

Figure 2(b) shows the XRD spectra of porous carbon. The two broad diffraction peaks at 22.7^0 and 43.3^0 indicate the (002) and (001) crystal planes of amorphous carbon [11].

4.3 FESEM Image analysis

Figure 3 (a) depicts the FESEM morphology of CQDs. Without the addition of any surface passivating agent, the CQDs nucleate, develop, and self-assemble into tiny spheres under hydrothermal carbonization conditions. This spherical morphology verifies the successful synthesis of carbon quantum dots [12].

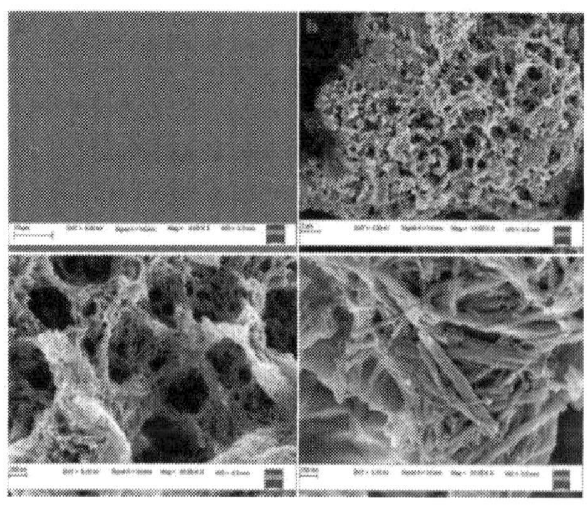

Figure 3: FESEM images of a) TCQDs and Porous carbon (b, c and d)

The morphology of porous carbon is shown in Figure 3 (b, c, and d). The porous carbon possessed continuous sponge- and rod-like structural networks, which improved the structural integrity and connectivity of electron transport [13]. The interconnected pore structure observed from all SEM images of porous carbon provides a large specific surface area for energy storage applications [14].

4.4 Electrochemical performances

Cyclic voltammetry is a powerful technique used to study the electrochemical behaviour of materials. By measuring the current response as a function of applied voltage, it can reveal valuable insights into the oxidation and reduction processes occurring at the material's surface [15]. The electrochemical behaviour of CQDs and porous carbon is measured using three electrode configurations with potential windows ranging from -2 to 2 V and -0.8 to 0.8 V, respectively, as shown in Figure 4 (a, & b). It can be observed that CV curves retain an identical shape without any distortion, and the current under the CV curve mechanism increases with increasing scan rate from 10 to 50 mVs^{-1} [16-17].

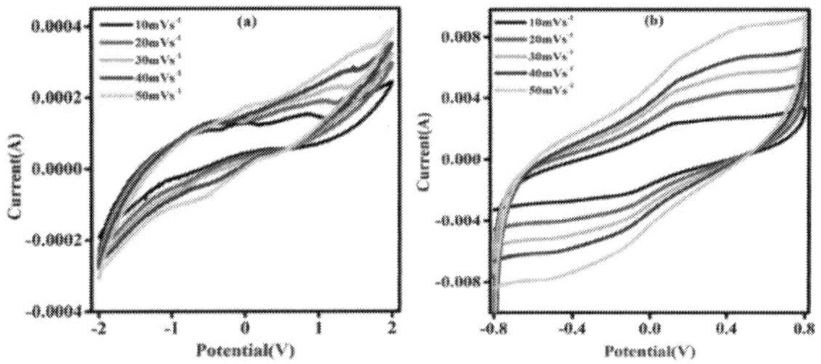

Figure 4: Cyclic voltammetry curves of (a) TCQDs and (b) TCQDs/PPy composite.

This implies that the electrode surface electrochemical response is independent of the scan rate. Additionally, the increasing current indicates that the reaction is becoming more efficient at higher scan rates. Furthermore, the FESEM images provide visual evidence of the porous carbon's large surface area, confirming its potential for enhanced performance in electrochemical processes. This increase in surface area allows for more active sites for electrochemical reactions to occur, resulting in a higher electrochemical activity of porous carbon than that of CQDs [18].

The specific capacitance of porous carbon is calculated by using the below equation [19].

$$Csp = \frac{\int IdV}{mX\Delta VX\vartheta}$$

Where, Csp is specific capacitance, I is current response, ΔV is the voltage window, ϑ is scan rate and m is mass of the deposited electrode material.

The calculated specific capacitance of porous carbon is 161.33 Fg^{-1} at 10 mVs^{-1} scan rate. This high specific capacitance indicates that porous carbon has a high ability to store electrical charge. This makes it a promising material for applications in energy storage devices such as supercapacitors [4].

5. Conclusion

In this work, CQDs and porous carbon are successfully prepared simultaneously by the hydrothermal carbonization method. The prepared materials are confirmed by FTIR, UV-visible, XRD, and FESEM characterization techniques. The cyclic voltammetry study suggests that CQDs and porous carbon could be promising candidates for applications in energy storage. Additionally, the low scan rate suggests that the material's capacitance remains stable even at slower charging and discharging rates, further highlighting its potential for long-term energy storage solutions.

Acknowledgement

Authors acknowledge to the PURSE Lab., Mangalore University for providing FESEM characterization facility. Author also acknowledged to the Commission of OBC board, Government of Karnataka for providing financial assistance.

References

1. Yadav, P. K., Chandra, S., Kumar, V., Kumar, D., & Hasan, S. H. (2023). Catalysts, **13(2)**, 422.
2. Balakumar, V., Ramalingam, M., Sekar, K., Chuaicham, C., & Sasaki, K. (2021). Chemical Engineering Journal, **426**, 131739.
3. Borchardt, L., Zhu, Q.-L., Casco, M. E., Berger, R., Zhuang, X., Kaskel, S., Feng, X., & Xu, Q. (2017). Materials Today, **20(10)**, 592–610.
4. Zhai, Z., Zhang, L., Du, T., Ren, B., Xu, Y., Wang, S., Miao, J., & Liu, Z. (2022). Materials & Design, 221, 111017.
5. Arkan, E., Barati, A., Rahmanpanah, M., Hosseinzadeh, L., Moradi, S., & Hajialyani, M. (2018). Advanced Pharmaceutical Bulletin, **8(1)**, 149–155.
6. Fedoseeva, Y. V., Lobiak, E. V., Shlyakhova, E. V., Kovalenko, K. A., Kuznetsova, V. R., Vorfolomeeva, A. A., Grebenkina, M. A., Nishchakova, A. D., Makarova, A. A., Bulusheva, L. G., & Okotrub, A. V. (2020). Nanomaterials, **10(11)**, 2163.
7. Pillar-Little, T. J., Wanninayake, N., Nease, L., Heidary, D. K., Glazer, E. C., & Kim, D. Y. (2018). Carbon, **140**, 616–623.
8. Chen, Y., Qin, X., Yuan, C., & Wang, Y. (2020). RSC Advances, **10(4)**, 1989–1994.
9. Zhou, C., Wu, S., Qi, S., Song, W., & Sun, C. (2021). Journal of Analytical Methods in Chemistry, 1–11.
10. Beck, M. L., Freihaut, B., Henry, R., Pierce, S., & Bayer, W. L. (1975). British Journal of Haematology, **29(1)**, 149–156.
11. Gao, Y., Qiu, X., Wang, X., Chen, X., Gu, A., & Yu, Z. (2020). Nanotechnology, **31(15)**, 155702.
12. Mai, X.-D., Phan, Y. T. H., & Nguyen, V.-Q. (2020). Advances in Materials Science and Engineering, 1–5.
13. Cetinbas, F. C., Ahluwalia, R. K., Kariuki, N. N., De Andrade, V., & Myers, D. J. (2020). Journal of The Electrochemical Society, **167(1)**, 013508.
14. Ban, C.-L., Xu, Z., Wang, D., Liu, Z., & Zhang, H. (2019). ACS Sustainable Chemistry & Engineering, **7(12)**, 10742–10750.
15. Navashree, N., & Parthasarathy, P. (2023). Materials Today: Proceedings, S2214785323027992.
16. Lengle, E. E., Frerman, F. E., & Grossberg, S. E. (1976). Archives of Biochemistry and Biophysics, **177(1)**, 157–169.
17. Vijaykumar, S.P., Sapna, S., Suresh, D.S., Ganesha, H., Veeresh, S., Nagaraju, Y.S. and Devendrappa, H. (2023). Diamond and Related Materials, **(139)**, 110257.
18. Ab. Rahim, A. H., Ramli, N., Nordin, A. N., & Abd. Wahab, Mohd. F. (2021). Carbon Trends, **5**, 100101.
19. Sharanappa S, Vijaykumar SP, Suresh DS, Shbil AB, Ganesha H, Veeresh S, NagarajuYS, Devendrappa H. (2023). Journal of Energy Storage, **74**,109371.

Natural Light Harvesting Materials: A Study on Optoelectronic Properties of Potential Natural Dyes

D Harshitha[1] H M Mahesh[2] and C G Renuka[1*]

[1] Department of Physics, Jnanabharathi campus, Bangalore University, Bengaluru-560056, India.
[2] Department of Electronics Science, Jnanabharathi campus, Bangalore University, Bengaluru-560056, India.

* Corresponding author: renubub..@gmail.com

Abstract: Natural dyes are being highlighted by research and industry trends; the majority of plant species offer abundant sources of colouring compounds. They can be harvested using affordable technology and used in cutting-edge applications. Hence, an examination of the optoelectronic characteristics of Carissa caranda fruit dye has been looked into in this paper. Anthocyanins present in Carissa caranda were extracted by solvent extraction of the fruits with ethanol and methanol solvents, maintaining pH at 7 under room temperature. UV-visible spectroscopy analyzed the dye extracts, and the optoelectronic properties such as optical energy band gap, extinction coefficient, and refractive index, are studied. UV-Vis investigations revealed large absorption peaks in the visible area and obvious band gaps. Carissa caranda natural dye had the lowest direct bandgap of 2.98eV and an indirect bandgap of 1.93eV for ethanolic extraction. It was found that the optical absorption in the dyes obeyed both direct and indirect transitions between the molecular energy levels. FTIR spectroscopy has been used to confirm the composition of the natural dyes. The findings here may be particularly significant for organic electronics applications, including O-LEDs and sensors.

Keywords: Carissa caranda, Ethanol, Methanol, Optoelectronic parameters.

1. Introduction

Natural colours produced from flora are safe since they are non-toxic, biodegradable, and non-carcinogenic [1]. They act as an alternative dye that is better for the environment and does not cause pollution or wastewater problems [2]. Colours can be extracted from a wide variety of plant, animal, and insect sources. Over 500 different plant species have been used to create natural dyes in India [3]. Due to the special characteristics these materials offer, such as flexibility, optoelectronic tuneability, and cost-effectiveness, the application of organic dyes has recently attracted attention from both industry and academia [4-6]. Since the dawn of time, plant sources have provided a plethora of coloured dyes, many of which are highly fluorescent [7]. Therefore, materials with light-sensitive optical properties, such as nonlinear optical materials, have garnered increased interest. Carissa carandas is a species of flowering shrub in the family Apocyanaceae, commonly known as Karonda. [11]. The carissa fruit has been a highly potent cosmeceutical agent [8,9]. karonda leaves shows Antioxidant activity and DNA damage inhibition, anti-inflammatory properties [10,11], organoleptic qualities [12]. Karonda fruits are employed in the Catalytic Application of Silver Nanoparticles [13], textile dyeing [14], the food industry, confectionaries [15], and nutraceutical supplements [16]. However, the optoelectronic properties of the flower extracts have not yet been studied, though. Due to these factors, two different solvents were used in the current study to extract the dye from Carissa caranda fruits. The UV-visible and FTIR spectroscopy have been used to analyzed. Exploring the feasibility of these natural dyes for prospective use in organic electronics, in particular, is made relevant by the results that have been described.

2. Experimental

2.1 Materials and methods

The Carissa caranda fruits of consistent dark red colour having no physical damage were collected from the lush campus of Bangalore University. The fruits were thoroughly cleaned with distilled water, then the seeds were removed and the fruits were taken for the extraction of the colour. The extraction solvents ethanol, and methanol, were of AR grade and purchased from S.D. Fine chemicals. An FTIR (Fourier transform infrared) spectrophotometer (Bruker alpha I ATR FTIR), and ocean optics UV-Vis-NIR spectrometer (USB4J00022) are used to evaluate the dye extract.

2.2 Extraction of natural dye

It was established from the literature that ethanol, because of its increased solubility, was the optimum solvent for color extraction [15]. We have extracted dye in two different solvents Ethanol and Methanol, by employing the same extraction procedure (Figure 1). Investigations into the activation and inactivation of dye components with absorption in both the solvent media were conducted- For the karonda dye extraction, 10g of deseeded fruits were pounded into a paste in a mortar. The pulp was then extracted and transferred to a beaker containing 100 mL of methanol. The mixture was then allowed to rest for 24 hours in a dark room to improve the dye extraction. After filtration, further analysis was carried out.

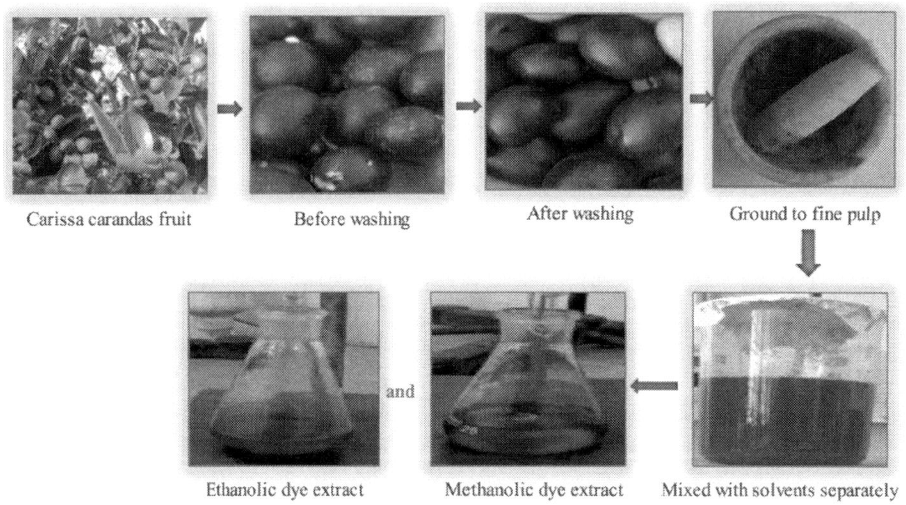

Figure 1. Pictorial representation of Karonda dye extraction with methanol and ethanol solvents

3. Results and discussion

3.1 The UV-Vis absorption spectroscopy

The most prominent characterization technique to measure the absorbance of a material is UV-visible spectrophotometry. Furthermore, bandgap and other optical properties of the material can be investigated. From the absorbance spectra, the electronic transitions and band structure pertaining to that material can be calculated [17]. Figure 2(a) shows the absorption spectra of the Karonda dyes extracted in ethanol and methanol solvents. The primary anthocyanin pigment in karonda fruit dye was identified by HPLC analysis as cyanidin 3-O-glucoside [18]. Figure 2(b) displays the dye's absorption coefficient over the 200–1200 nm wavelength range.

The dye's absorption coefficient in the 200–1200 nm wavelength range is shown in Figure 2(b). It is determined using the absorbance values by

$$\alpha = \frac{2.303A}{t} \qquad (1)$$

It is calculated using the values of the absorbance where A is the absorbance and t is the cuvette thickness. Although there is a slight shift in the peak positions, the dye extracts in both solvents exhibit a comparable absorbance nature, demonstrating broad absorption in the UV area. The UV absorbance peaks at 301 and 347 nm in methanolic extracts and at 315 and 348 nm in ethanolic extracts are related to the Quercetin pigments of the anthocyanin group, which correlate to the π to π * transition. Similar visible peaks may also be seen in cyanidin-3-O-glucoside [18] pigments of the anthocyanin, which are found at 526 and 539 nm and correlate to the n to π* transition. According to earlier research, the quercetin pigment exhibits a UV absorbance peak at 301 (nm) and a cyanidin-3-O-glucoside peak at 535 (nm), respectively. However, we saw somewhat different peaks than those previously reported due to the variable climatic circumstances in which plants are cultivated.

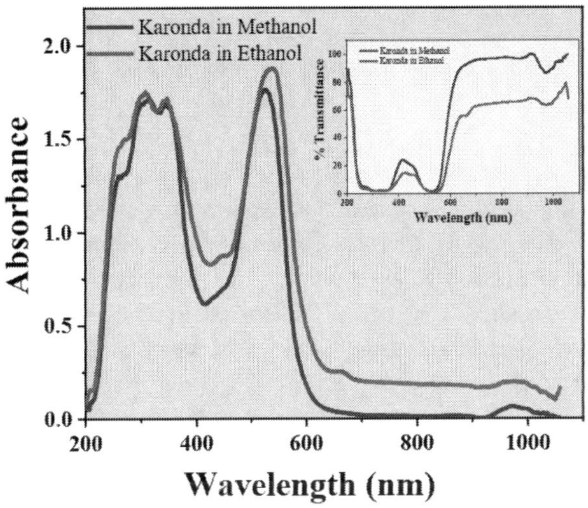

Figure 2(a) Optical absorption spectra of extracted dye.

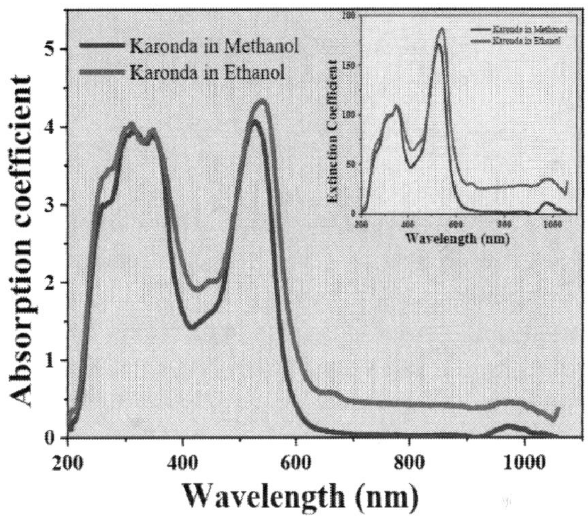

Figure 2(b) Absorption coefficient spectra of dye.

The inset graph in Figure 2(a) shows the transmittance spectra of the karonda dye extracts in both solvents. Transmittance is very low in the UV-visible regions. Anisotropic behavior is seen in both the dye samples. Similarly, the inset graph in Figure 2(b) presents the Extinction coefficient spectra which tell us about how the energy is lost by the incident electromagnetic wave as a result of scattering and absorption per unit thickness of the medium. The band gap energy E_g, is dependent on the photon energy hv, and the absorption coefficient, is determined by Tauc's relation [19].

$$\alpha hv = \alpha_o\left(hv - E_g\right)^n \qquad (2)$$

Where α_o is an energy-independent constant, E_g is the energy gap, and n shows the nature of the optical transition. Using the linear region as an extrapolation to the X-axis, as illustrated in the below figures 3(a) & 3(b), the optical bandgap E_g is computed. The methanolic and ethanolic extracts' direct energy gap of 3.18eV and 2.98eV and indirect band gap, respectively, were estimated to be 2.02eV and 1.93eV respectively. Using the linear region as an extrapolation to the X-axis, as illustrated in the below figures 3(a) & 3(b), the optical bandgap E_g is computed. The methanolic and ethanolic extracts' direct energy gap of 3.18eV and 2.98eV and indirect band gap, respectively, were estimated to be 2.02eV and 1.93eV respectively. The mixture of pigments in the karonda dye extracts is what causes both bandgaps to exist. The dye extracted from the ethanol solvents shows less direct and indirect bandgaps compared to the methanolic dye extract.

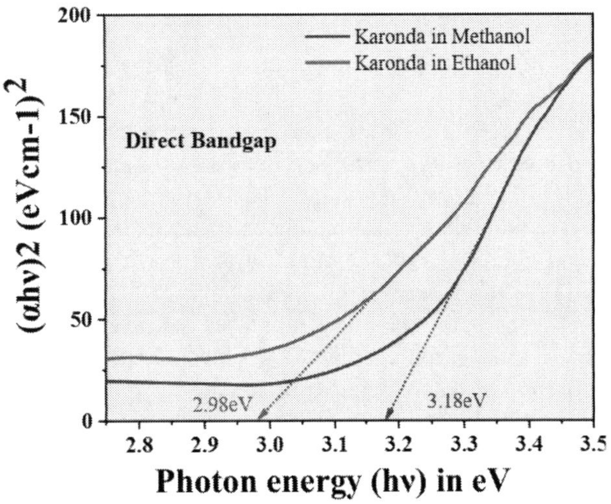

Figure 3(a). Direct bandgap using Tauc's relation.

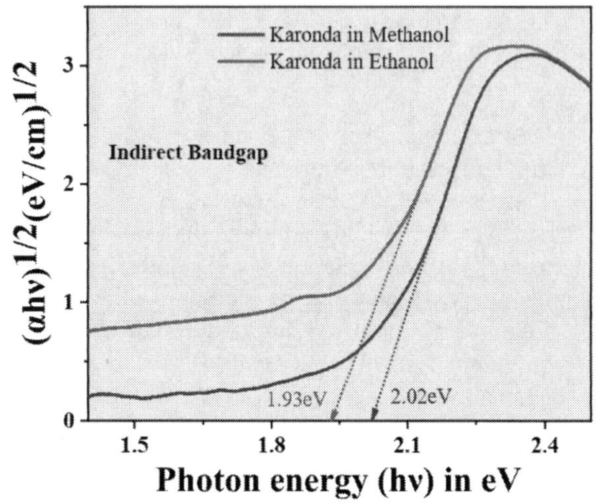

Figure 3(b). Indirect bandgap using Tauc's relation.

Figure 4(a) presents the Urbach energy measuring the density's broadness as well as how steeply the onset absorption from the top and bottom of HOMO and LOMO approaches the band edge. Plotting ln verses allows us to visualize the Urbach energy from the straight-line slope. Urbach energy for the methanolic-based extraction is determined to be 107.99MeV, lower than that for the ethanolic-based extraction at 139.93MeV. This difference is related to variations in the internal fields, which are further related to the interaction of the chemical group or a reduced capacity to transmit weak bonds into defects. One of the key variables to analyze when figuring out the optical characteristics sample is its refractive index. The variations in the refractive index for natural dyes for methanol and ethanol were shown in Figure 4(b) as a function of incident photon wavelength (nm).

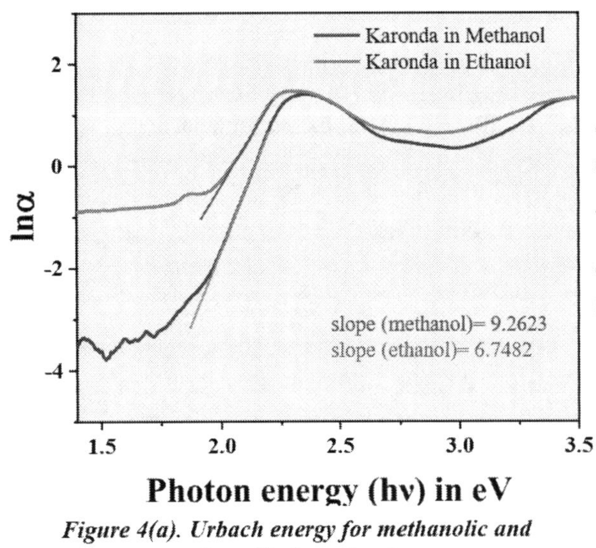

Figure 4(a). Urbach energy for methanolic and ethanolic dye extracts.

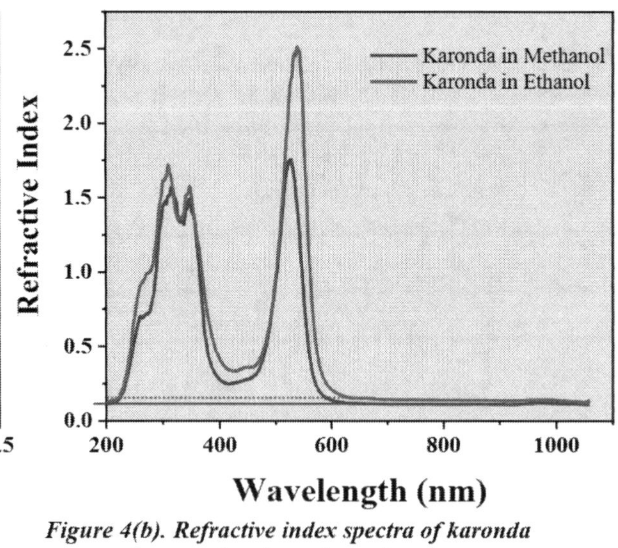

Figure 4(b). Refractive index spectra of karonda dye as a function of wavelength.

IOP Conf. Series: Materials Science and Engineering 1300 (2024) 012004 doi:10.1088/1757-899X/1300/1/012004

The refractive index (n) at a specific wavelength can be determined according to Swanepoel's formula [20].

$$n = \sqrt{N + \sqrt{N^2 + s^2}} \quad \textbf{(3)}$$

$$\mathrm{N} = (2s\,(\frac{T_M + T_m}{T_M T_m})) + \frac{s^2+1}{2} \quad \textbf{(4)}$$

Here, T_M and T_m stand for the upper and lower envelop transmittances, respectively, at a given wavelength. S= 1.51 is wavelength-dependent and can be computed using the transmittance spectra. The dye's refractive indices, 1.5138 for methanol and 1.5160 for ethanol are hardly different from one another. The components of each dye's refractive index showed higher values for photon energy at the visible wavelength ranges. Significant linear behaviour was seen between 600 and 1100 nm, whereas both the dyes displayed divergences between 200 and 600 nm [21].

3.2 FTIR (Fourier Transform Infrared) Spectroscopy

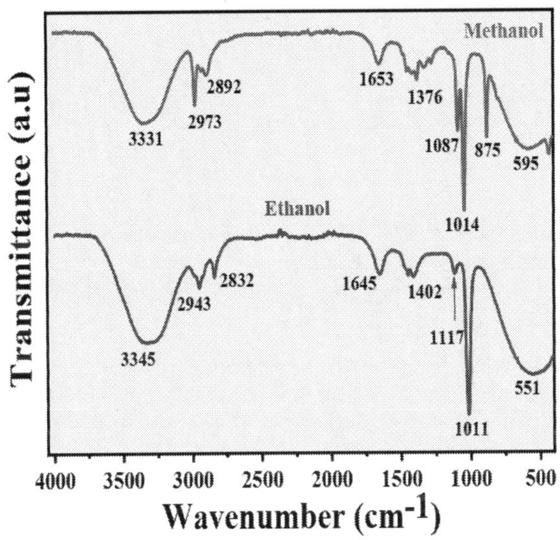

Figure 5. FTIR spectrum of methanolic and ethanolic extracts of Carissa carandas natural dye

Absorption range (cm^{-1})	Appearance	Group	Class of Compound
3331-3250	Strong	C-H stretching	Alkyne
2840-3000	Medium	C-H stretching	Alkane
1653-1648	Medium	C=C stretching	Alkene
1648-1638	Strong	C=C stretching	Alkene
1420-1330	Medium	O-H bending	Alcohol
1087-1124	Strong	C-O stretching	Secondary alcohols
860-880	Strong	C=C bending	Alkene
551-600	Weak	C-Cl	Halogen

Table 1. Absorption range, groups, and class of compounds present in the methanolic and ethanolic dye extracts of karonda fruits.

Figure 5 presents the FTIR spectra of the methanolic and ethanolic extract of karonda ye in the wavenumber region of 4000 to 400 cm^{-1}. This spectrum consists of various functional groups with intense absorption bonds along the wavenumber range. The absorption range and corresponding class of compounds for both dye extracts are shown in Table 1. The absorbance peaks in the spectra are the results of anthocyanins and their derivatives and of phenolic groups. The phenolic compounds of both methanolic and ethanolic extracts are the reason for bands at 2840 to 3331cm ^{1} responsible for the C-H stretching of alkanes and alkynes [22]. The small shoulder bumps around 1640 to 1087cm^{-1} are due to the C=C stretching of alkanes of anthocyanin derivatives. 1420 to 1330cm^{-1} bands were because of O-H bending or vibrations of alcohols. 1087 to 1124cm^{-1} bands were corresponding to strong C-O stretching of secondary alcohols. 600 to 860cm^{-1} band range of FTIR spectra is due to the presence of C=C bending of alkenes and C-Cl bonds of halogens. Hence, FTIR spectroscopy yields partial confirmation of the chemical components present in the karonda dye extracts.

4. Conclusions

This work has shown that dye extraction can be accomplished by utilizing low-cost techniques. Natural dye has been extracted employing methanol and ethanol solvents from the fruits of Carissa carandas also known as karonda. UV-Vis and FTIR spectroscopy investigations revealed that Anthocyanin derivatives and phenols are the main dye components of Karonda fruits. Optical properties such as direct and indirect bandgaps, extinction coefficient(k), and refractive index (n) are calculated from the absorbance data. Ethanolic dye extract revealed lower bandgaps compared to the dye extracted in methanol. The identification of various functional groups using FTIR spectroscopy serves as a partial confirmation of the chemical constituents that are in charge of the extracted dye having good optical characteristics. The findings offered here are essential for establishing the role of dye in the field of organic electronics such as organic photovoltaics, sensors, and UV absorbers.

5. Acknowledgments

The author C G Renuka gratefully acknowledges the Bangalore University, Bengaluru of the financial support provided with project number UNI.ORDER.NO.DEV:D2a:BU_RP.2020-21

6. References

[1] Wang, C, et al. 2013 *Coloration technology*, **129(1)**: 32-38.
[2] Narayana Swamy, et al. 2014 *Journal of The Institution of Engineers (India): Series E*, **95**: 11-17.
[3] Vettumperumal, R, et al. 2018 *Optik*, **159**: 108-114.
[4] Hemmatzadeh, R., & Mohammadi, A. 2013 *Journal of theoretical and applied physics*, **7**:1-7.
[5] Hemmatzadeh, R., & Jamali, A. 2015 *Journal of Renewable and Sustainable Energy*, *7*(1).
[6] Delgado-Vargas, F, et al. 2000 *Critical reviews in food science and nutrition*, **40(3)**: 173-289.
[7] Singh, V., & Mishra, A. K. 2015 *Scientific reports*, **5(1)**: 11118.
[8] Khuanekkaphan, M, et al. 2021 *Journal of Advanced Pharmaceutical Technology & Research*, **12(2)**: 162.
[9] Anunthawan, T., & Rimlumduan, T. 2023 *Journal of Applied Pharmaceutical Science*, **13(10)**: 217-224.
[10] Verma, K, et al. 2015 *Journal of Taibah University for Science*, **9(1)**: 34-40.
[11] Neimkhum, W, et al. 2021 *Antioxidants*, **10(9)**: 1345.
[12] Navya, D. V, et al. 2020 *Journal of Pharmacognosy and Phytochemistry*, **9(6)**: 994-997.
[13] Anupama, N., & Madhumitha, G. 2017 *Inorganic and Nano-Metal Chemistry*, **47(1)**: 116-120.
[14] Manicketh, T. J, et al. 2021 *Natural Product Research*, **35(21)**: 4159-4163.
[15] Rafique, N, 2023 *Food Science & Nutrition*.
[16] Krishna, H, et al. 2007 *IJMFM AP*, **3**: 28-33.
[17] Imran Khan, et al. Bhati 2018 *Optik* **170**: 43–47
[18] Sarkar, R., et al. 2018 *LWT*, ***93***: 673-678.
[19] Tauc, J. 1974. *In Amorphous and liquid semiconductors* (pp. 159-220). Boston, MA: Springer US.
[20] Shaaban, et al, 2009 *Physica B: Condensed Matter*, ***404*(20)**: 3571-3576.
[21] Rauf, M. A, et al, 2009. *Spectrochimica Acta Part A: Molecular and Biomolecular Spectroscopy*, *72*(**1**): 133-137.
[22] Yang, Rong, et al. 2021 Scientific Reports 11.1: 14331

IOP Conf. Series: Materials Science and Engineering 1300 (2024) 012005 doi:10.1088/1757-899X/1300/1/012005

Performance analysis of $A_3B_2X_9$ ($Cs_3Bi_2I_9$) based perovskite photovoltaic tandem structure with crystalline Silicon (c-Si)

Shreyus Goutham Kumar[1], Tadi Surya Teja Reddy[1], Suraj N[1], Prashanth C R[2] and Gajanan V Honnavar[3,*]

[1]Dept. of Electronics and Communication, PES University, EC Campus, Hosur Road, Near Electronic City, Bangalore 560100, India.
[2]Dept. of Electronics and Telecommunication, Dr Ambedkar Institute of Technology, Affiliated to VTU, Bengaluru, Karnataka 560056, India.
[3]Dept. of Science & Humanities, PES University, EC Campus, Hosur Road, Near Electronic City, Bangalore 560100, India.

*Corresponding author's email: gajanan.honnavar@pes.edu

Abstract. The efficiency of solar cells with single-junction utilizing organic-inorganic hybrid perovskites have attained a value more than 25.5%. The device power conversion efficiency (PCE) can be improved further either by optimizing the absorber layer (Perovskite film) or by investigating the novel device structures such as tandem based solar cells with perovskite and silicon. This combination of top cell (Perovskite solar cell) and bottom cell (Silicon solar cell) can improve the PCE which surpasses the Shockley-Queisser limit of single-junction solar cells by utilizing a wider range of solar spectrum. This paper presents an optimization and simulation of standalone $Cs_3Bi_2I_9$ perovskite solar cell which was later integrated with a c-Si solar cell to simulate a tandem structure using SCAPS-1D software. The aim is to investigate the performance enhancement of the perovskite solar cell by optimizing it and stacking it on top of a high-efficiency c-Si solar cell using a four-terminal (4T) structure. At a short-circuit current density of 16.165 mA/cm^2 and an open-circuit voltage of 1.41 V, the simulation findings demonstrate that the $Cs_3Bi_2I_9$ perovskite solar cell exhibits a high-power conversion efficiency of 20.37%. The tandem structure demonstrates an enhanced power conversion efficiency of 31.59% which is significantly higher than the efficiency of the individual cells. The $Cs_3Bi_2I_9$ perovskite solar cell is a great choice for application in tandem systems with c-Si solar cells for high-efficiency, according to the simulation results. This study provides valuable insights for the development of efficient perovskite/c-Si tandem solar cells.

1. Introduction

Numerous attempts to create multi-junction cells (tandem cells), which are made up of solar cells with various band gaps, have been made to outperform solar cells with a single junction. The Shockley-Queisser power conversion efficiency (PCE) limit of 31% for single-junction devices is still being thoroughly researched for a variety of tandem solar device topologies (multijunction cells made up of various bandgaps). This is a workable strategy for addressing the projected rise in the world's energy demand from 15 TW in 2011 to 30 TW in the future (2050) [1]–[3] Wide bandgap (WBG) material for the top absorber and narrow bandgap (NBG) material for the bottom absorber are appropriate for tandem systems. When Si served as the bottom absorber and III-V compound semiconductor-based materials as the top absorber in the current generation of MJ-TSCs, the conversion efficiency was found to be 39%

under a 1 sun condition [4]. Scaling up in the commercial sector is challenging due to the high cost and precision manufacturing required for these III-V compounds [5]. Over the past ten years, research teams studying III-V materials have been impacted by the emergence of perovskite materials.

Due to their many beneficial features, including bandgap tunability, longer carrier diffusion length, smaller carrier effective mass, superior absorption coefficient, affordability, and ease of fabrication, perovskite materials have established themselves as a potential substitute. In order to improve conversion efficiency and lower the levelized cost of electricity (LCOE) of the solar cell (SC) technology, two-junction TSCs are being studied [5], [6]. Mailoa et al. created the first two-terminal (2-T) perovskite/Si TSC in 2015, and it had a PCE of 13.7% [7]. Zheng et al. improved it by using SnO_2 as the top cell's electron transport layer (ETL) and interconnecting layer (ICL), and the resulting PCEs on 16cm^2 TSC were 17.10% [8]. Due to its distinct characteristics, the $A_3B_2X_9$ perovskite structure is a preferred option in perovskite solar cell technology more than the ABX_3 structure [9]. First off, the $A_3B_2X_9$ perovskite structure has improved energy conversion efficiency due to its higher light absorption efficiency. Second, compared to the ABX_3 structure, the $A_3B_2X_9$ perovskite structure has a longer carrier lifetime, allowing it to maintain its energy conversion efficiency for a longer period of time. In comparison to the ABX_3 structure, the $A_3B_2X_9$ perovskite structure is also more stable and less likely to deteriorate. Additionally, compared to the ABX_3 structure, the $A_3B_2X_9$ perovskite structure has a higher defect tolerance, which leads to improved performance and increased efficiency.

Using SCAPS-1D software, we have simulated Pb-free PSCs with a device architecture of ($FTO/Cd_{0.5}Zn_{0.5}S/Cs_3Bi_2I_9/CuSbS_2/Au$). SCAPS (a Solar Cell Capacitance Simulator) is a 1-D solar cell simulation program developed at the Department of Electronics and Information Systems (ELIS) of the University of Gent, Belgium [10]. Additionally, we used c-Si as the bottom sub-cell and the simulated perovskite structure as the top sub-cell to simulate a TPSC. The device architecture of Pb-free PSCs ($FTO/Cd_{0.5}Zn_{0.5}S/Cs_3Bi_2I_9/CuSbS_2/Au$) with experimental validation has not yet been reported in any publications, to our knowledge.

2. Device structure

While analyzing the performance of tandem solar cells, the top and bottom sub-cells are simulated separately. The tunnel junction is assumed to be flawless and both the optical and electrical losses at the interfaces are ignored in the standard multi-junction model [11]–[14]. Perovskite solar cells are frequently used as the top cell and silicon solar cells as the bottom cell in tandem solar cell systems as shown in Figure 1. The device architecture shown in Figure 1. is the optimized architecture which provides a maximum efficiency. In this study, the top cell consists of $Cs_3Bi_2I_9$ (Band gap: 1.9 eV, Thickness: 1.00 μm) as the absorber layer [15][16]. A hole-transport layer (HTL) and an electron-transport layer sandwich the perovskite layer (ETL). $CuSbS_2$ (Band gap: 1.580 eV, Thickness: 0.100 μm), the material that makes up the HTL, has great stability and outstanding hole-transport characteristics [17]. $Cd_{0.5}Zn_{0.5}S$ (Band gap: 2.8 eV, Thickness: 0.150 μm), which has a narrow bandgap and high electron mobility, is the material used to make the ETL [18]. Crystalline silicon (c-Si) cells, which have a high open-circuit voltage, low resistance, and long-term stability, make up the bottom cell [19]. To guarantee the greatest amount of energy is captured, the bandgap and voltage of the c-Si cell are tuned to match those of the perovskite cell. It is a highly efficient photovoltaic device due to the combination of the perovskite cell's high absorption coefficient, the c-Si cell's high open-circuit voltage and low resistance, the low bandgap and high electron mobility of the ETL, the HTL's good hole-transport properties, and the low bandgap and high electron mobility of the ETL. The back contact (Au) is only for the stand-alone cell. For the tandem structure ETL ($CuSbS_2$) is in contact with n+Si.

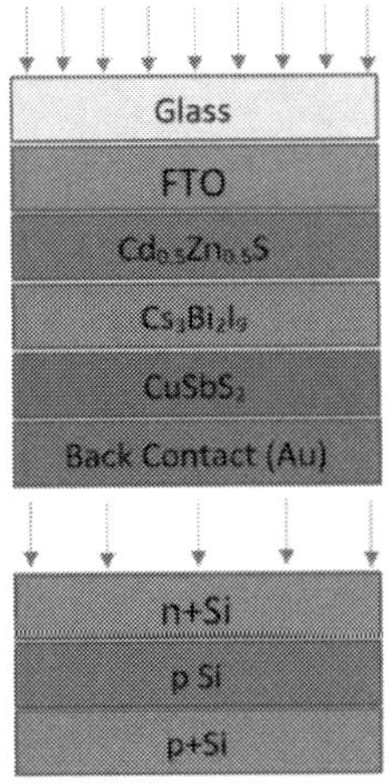

Figure 1. Structure of Proposed Standalone and Tandem Solar Cell

3. Results and Discussion

This study utilized the SCAPS-1D software to simulate and analyse the efficiency of tandem solar cells, with a top cell made of $Cs_3Bi_2I_9$ perovskite and a bottom cell made of crystalline silicon. A tandem device is a series of two-cell combinations that form a tandem structure with two terminals. Thus, the tandem cell's open-circuit voltage, V_{oc}, is the summation of its sub-cell voltages. The lowest of the junction currents, however, will set a limit on the entire device's short-circuit current J_{sc} [20]. The following subsections will include simulations of both the entire tandem device and individual cells. Table 1 gives the physical parameters used for simulation of the top cell and bottom cell.

Table 1. Input parameters for different layers used for SCAPS – 1D simulation

Parameters/Layers	FTO [18]	$Cd_{0.5}Zn_{0.5}S$ [18]	$Cs_3Bi_2I_9$ [16]	$CuSbS_2$ [17]	n+Si [19]	p Si [19]	p+Si [19]
Thickness(μm)	0.500	0.150	1	0.100	0.500	300	10
E_g (eV)	3.5	2.8	1.9	1.58	1.12	1.12	1.12
χ (eV)	4	3.9	3.55	4.2	4.05	4.05	4.05
ϵ_r	9	10	9.68	14.6	11.9	11.9	11.9
N_c (cm^{-3})	$2.2x10^{18}$	$1x10^{18}$	$4.98x10^{19}$	$2x10^{18}$	$2.8x10^{19}$	$2.8x10^{19}$	$2.8x10^{19}$
N_v (cm^{-3})	$1.8x10^{19}$	$1x10^{18}$	$2.11x10^{19}$	$1x10^{19}$	$1.04x10^{19}$	$1.04x10^{19}$	$1.04x10^{19}$
μ_n (cm^2/Vs)	20	100	4.3	49	1400	1400	1400
μ_p (cm^2/Vs)	10	25	1.7	49	450	450	450
N_d (cm^{-3})	$2x10^{19}$	$1x10^{17}$	$1x10^{9}$	0	$1x10^{20}$	0	0
N_a (cm^{-3})	-	0	$1x10^{9}$	$1.38x10^{18}$	0	$1x10^{16}$	$1x10^{20}$
N_t (cm^{-3})	$1x10^{15}$	$1x10^{15}$	$1x10^{13}$	$1x10^{15}$	-	-	-

3.1. Effect of the thickness and band-gap of the Absorber Layer

The overall cell's performance is considerably dependent on the perovskite layer's thickness. Using a numerical simulation, its impact on the overall performance of the cell is also examined. The consequences of varying the thickness in the range of 100 nm to 2000 nm are depicted in Figure 2. The Figure 2. shows that the J_{sc} and PCE values rise with the increase in the perovskite layer's thickness, while the V_{oc} and FF fall. More photons are absorbed by the layer with the increase in the perovskite layer's thickness, raising the J_{sc} value. So, an increase in excess carrier concentration leads to an increase in J_{sc}. On the other hand, with the increase in the perovskite layer's thickness, the solar cell's internal power depletion and series resistance value also increases, causing FF and V_{oc} to continuously decrease. As the thickness approaches and surpasses 1000 nm, the PCE hits its maximum and then starts to decline, as shown in Figure 2.

The band gap of $Cs_3Bi_2I_9$ falls between 1.9 eV and 2.2 eV [15]. We can see from the simulated results presented in Figure 3. that 1.9 eV is the ideal band gap for high efficiency as in the graph we can observe that with the increase in the band gap, the PCE decreases. Therefore, we set the absorber layer's band gap to 1.9 eV.

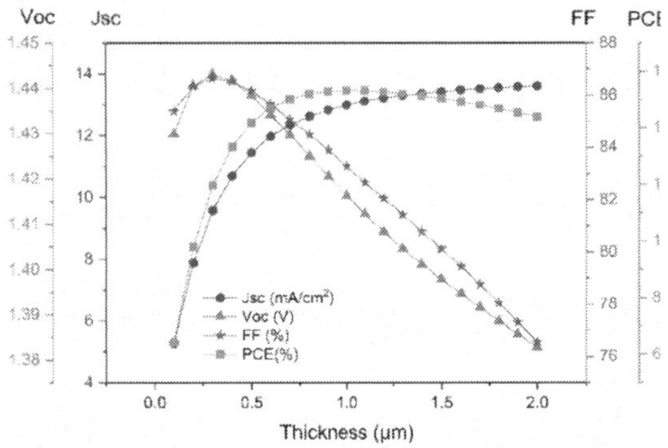

Figure 2. Effect of Thickness of Absorber Layer

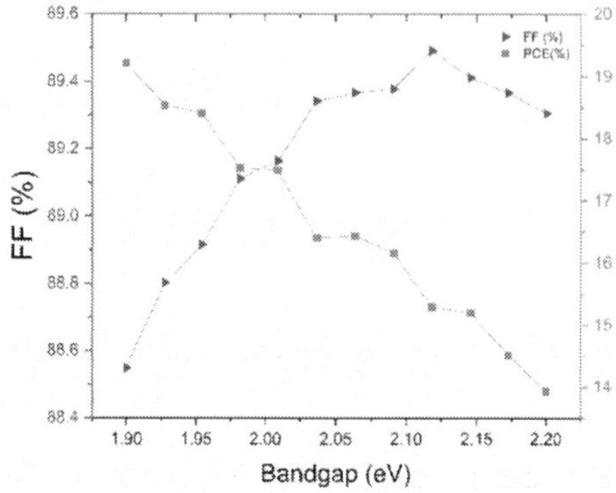

Figure 3. Effect of Band Gap

3.2. Effect of Defect Density of Absorber Layer

Defect density measurements from this investigation vary from 10^9 to 10^{13} cm^{-3} [21]. It is clear from Figure 4. that the PCE and FF decrease as the absorber layer defect density increases. A nonradiative Shockley-Read-Hall recombination center is a deep energy level defect. As a result, as the number of defects in the absorber layer increases, charge recombination increases, leading to an increase in Voc and Jsc while the short minority carrier lifespan decreases. Furthermore, the essential p-n junction does not form as the PSC transitions to semi-insulation, which results in poor cell performance when the absorber's defect density is lower than or equal to the doping density [20], [21]. The density of defects in the absorber layer must be decreased, if not eliminated, to achieve the greatest PSC performance. A substantial reduction in defect density using current product innovations is also still extremely difficult. According to our study, for maximum cell effectiveness, the perovskite layer's defect density shouldn't be higher than 10^{15} cm^{-3}.

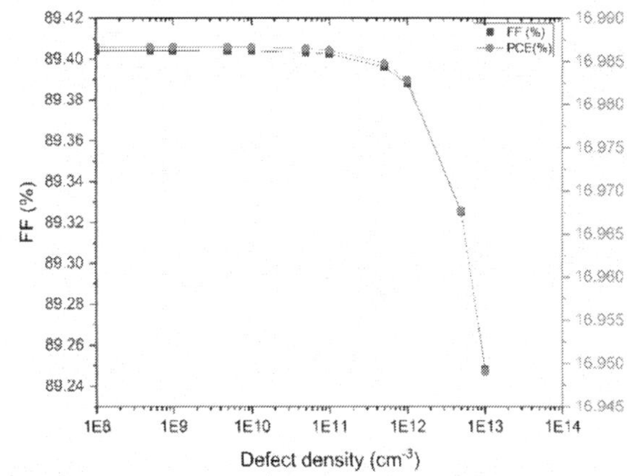

Figure 4. Effect of Defect Density

3.3. Simulation of Top cell (FTO/$Cd_{0.5}Zn_{0.5}S$/$Cs_3Bi_2I_9$/$CuSbS_2$/Au) and Bottom Cell (c-Si)

Using SCAPS - 1D, standalone devices with active layers of $Cs_3Bi_2I_9$ have been simulated. The top device's layer thicknesses are as follows: The thicknesses of the active layer (also known as the perovskite layer) is 1000 nm, the ETL ($Cd_{0.5}Zn_{0.5}S$) is 150 nm, the HTL ($CuSbS_2$) is 100 nm, and the ITO layer (FTO) is 500 nm. Table 1. provides a list of other parameters that were used in this simulation. SCAPS - 1D software simulates the stand-alone top perovskite solar cell using these parameters. An optimum efficiency of 20.37% is obtained for the top cell in our simulation. The results of this simulation and the optimum values of various parameters are listed in Table 2.

The bottom cell for the tandem device in this work is a c-Si solar cell. Here, n+ Si of 0.5 nm, p Si of 300 nm, and p+ Si of 10 nm have been used [19]. The electrical properties of the separate cells must match in order to determine how efficient tandem solar cells will be. This synchronization makes sure that each layer's absorption of light is effectively turned into electrical energy. The electrical resistance and open circuit voltage between cells can be matched by modifying the layer thickness, doping levels, and material composition of each layer to achieve the current matching condition in tandem solar cells. Tandem solar cells are a potential technique for the production of renewable energy because they can reach high conversion efficiencies by preserving this current matching state [22]. An optimum efficiency of 22.78% for the bottom cell is observed as indicated in Table 2.

Table 2. Optimized parameters

Optimized Parameters	Optimized Layer / Values	PCE (%)
ETL	$Cd_{0.5}Zn_{0.5}S$	13.21
HTL	$CuSbS_2$	14.89
Absorber layer Thickness	1.00 µm	15.33
Defect Density of Absorber Layer	1.00E+13	16.95
Band gap of Absorber Layer	1.9 eV	20.37
Tandem	c-Si(Bottom cell)	22.78

3.4 Simulation of Tandem Solar cell

The tandem cell was simulated using the SCAPS-1-D simulation program using a fundamental approach. In our research, we have used mechanically stacked two-terminal tandem cells, which may be understood as a series connection of two diodes. Additionally, we have implemented the current matching condition between the top and bottom cells, as described in reference [22].The cell with the lower J_{sc} dominates the current-limiting criterion of the entire tandem arrangement, even if the voltage is derived by summing the voltages of individual cells. Either maximum power current density (JMP) variation or J_{sc} variation is used to match the current. The $Cs_3Bi_2I_9$/c-Si tandem structure's current matching profile is depicted in Figure 5. We adjust the bottom cell's thickness to determine the current matching point. A maximum efficiency of 31.59% is observed for the tandem structure. The results tabulated in Table 3. in comparison with the reference PSC [21] shows an improved efficiency of 31.59% from a reported efficiency of 11.54%.

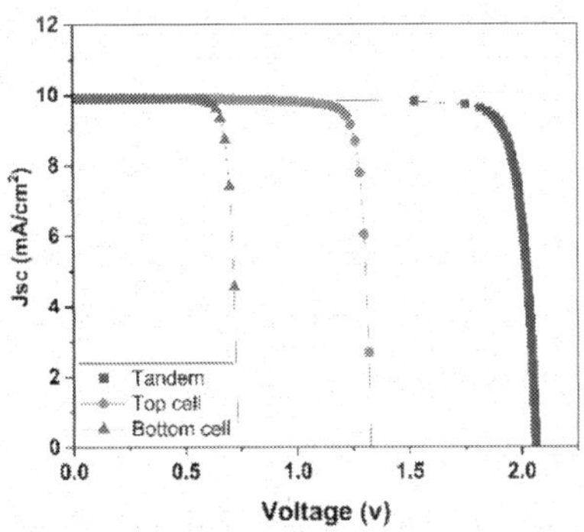

Figure 5. JV characteristic curve

Table 3. Results of Simulations

Cell/Parameters	Voc	Jsc	FF (%)	PCE (%)
Reference [16]	1.22	12.19	77.73	11.54
$Cs_3Bi_2I_9$ (Top Cell)	1.34	16.14	88.55	20.37
c-Si (Bottom Cell)	0.74	21.28	85.25	22.78
$Cs_3Bi_2I_9$ on c-Si (Tandem)	2.06	9.92	86.07	31.59

4. Conclusion

In this work, non-lead PSCs with absorber layer made of $Cs_3Bi_2I_9$ perovskite are investigated using SCAPS modelling. First, it is decided which materials are best for the HTL and ETL, which are $CuSbS_2$ and $Cd_{0.5}Zn_{0.5}S$, respectively. The effects of thickness of absorber layer, its density of defect, and band-gap are then investigated with regard to PSC performance. PSC setup is glass substrate/FTO/$Cd_{0.5}Zn_{0.5}S$/$Cs_3Bi_2I_9$/$CuSbS_2$/Au. According to modelling studies, the ideal thickness of absorber layer is 1000 nm. The simulation also concludes that the absorber layer's ideal defect density is $1x10^{13}$ cm^{-3}, and that any higher defect density will result in solar cell performance reduction because additional recombination sites will occur. The research's findings will contribute to the creation of effective, non-lead PSCs and the expansion of solar wind wave and other forms of renewable energy.

References

[1] Sharma, Surbhi, Soumen Basu, Nagaraj P. Shetti, and Tejraj M. Aminabhavi. *Science of the Total Environment* **713** (2020):136633.

[2] Glatt, Moritz F., Li Yi, Gülsüm Mert, Barbara S. Linke, and Jan C. Aurich. *Journal of Cleaner Production* **206** (2019): 727-740.

[3] Seker, Sukran, and Nezir Aydin. *International Journal of Hydrogen Energy* **45**, no. 32 (2020): 15855-15868.

[4] Green, Martin, Ewan Dunlop, Jochen Hohl-Ebinger, Masahiro Yoshita, Nikos Kopidakis, and Xiaojing Hao. *Progress in photovoltaics: research and applications* **29**, no. 1 (2021): 3-15.

[5] Leijtens, Tomas, Kevin A. Bush, Rohit Prasanna, and Michael D. McGehee. *Nature Energy* **3**, no. 10 (2018): 828-838.

[6] Yu, Zhengshan J., Joe V. Carpenter III, and Zachary C. Holman. *Nature Energy* **3**, no. 9 (2018): 747-753.

[7] Mailoa, Jonathan P., Colin D. Bailie, Eric C. Johlin, Eric T. Hoke, Austin J. Akey, William H. Nguyen, Michael D. McGehee, and Tonio Buonassisi. *Applied Physics Letters* **106**, no. 12 (2015).

[8] Zheng, Jianghui, Cho Fai Jonathan Lau, Hamid Mehrvarz, Fa-Jun Ma, Yajie Jiang, Xiaofan Deng, Anastasia Soeriyadi et al. *Energy & Environmental Science* **11**, no. 9 (2018): 2432-2443.

[9] Ahmad, Khursheed, Shagufi Naz Ansari, Kaushik Natarajan, and Shaikh M. Mobin. *ACS Applied Energy Materials* **1**, no. 6 (2018): 2405-2409.

[10] Burgelman, Marc, Peter Nollet, and Stefaan Degrave. *Thin solid films* **361** (2000): 527-532.

[11] Kim, Kihwan, Jihye Gwak, Seung Kyu Ahn, Young-Joo Eo, Joo Hyung Park, Jun-Sik Cho, Min Gu Kang, Hee-Eun Song, and Jae Ho Yun. *Solar energy* **145** (2017): 52-58.

[12] Kim, Kihwan, Jin Su Yoo, Seoung Kyu Ahn, Young-Joo Eo, Jun-Sik Cho, Jihye Gwak, and Jae Ho Yun. *Solar Energy* **155** (2017): 167-177.

[13] Gupta, Goutam Kumar, and Ambesh Dixit. *Optical Materials* **82** (2018): 11-20.

[14] Madan, Jaya, Rahul Pandey, and Rajnish Sharma. *Solar energy* **197** (2020): 212-221.

[15] Hong, Ki-Ha, Jongseob Kim, Lamjed Debbichi, Hyungjun Kim, and Sang Hyuk Im. *The Journal Of Physical Chemistry C* **121**, no. 1 (2017): 969-974.

[16] Ahmad, Khursheed, Mohd Quasim Khan, Rais Ahmad Khan, and Haekyoung Kim. *Optical Materials* **128** (2022): 112458.

[17] Teimouri, R., and R. Mohammadpour. *Superlattices and Microstructures* **118** (2018):116-122.

[18] Baig, Faisal, Yousaf Hameed Khattak, Bernabe Mari, Saira Beg, Abrar Ahmed, and Khurram Khan. *Journal of Electronic Materials* **47** (2018): 5275-5282.

[19] Sherman, Devonte, Jesus Marquez, Yves Ramirez, Mariano Urbina, Andres Meza, and Deidra Hodges.In *2021 IEEE 48th Photovoltaic Specialists Conference (PVSC)*, pp. 2640-2642. IEEE, 2021.

[20] Larsen-Olsen, Thue T., Thomas R. Andersen, Birgitta Andreasen, Arvid PL Böttiger, Eva Bundgaard, Kion Norrman, Jens W. Andreasen, Mikkel Jørgensen, and Frederik C. Krebs. *Solar Energy Materials and Solar Cells* **97** (2012): 43-49.

[21] Le Corre, Vincent M., Elisabeth A. Duijnstee, Omar El Tambouli, James M. Ball, Henry J. Snaith, Jongchul Lim, and L. Jan Anton Koster. *ACS energy letters* **6**, no. 3 (2021): 1087-1094.

[22] Ulbrich, C., C. Zahren, A. Gerber, B. Blank, Tz Merdzhanova, A. Gordijn, and U. Rau. *International journal of photoenergy* **2013** (2013).

IOP Conf. Series: Materials Science and Engineering 1300 (2024) 012006 doi:10.1088/1757-899X/1300/1/012006

Deep Learning Neural Network (DNN) Coupled Laser Induced Breakdown Spectroscopy for Quantitative Analysis of Copper Alloys

Leya Pauly[1], Lekha Mary John[1], M S Suryan Sivadas[1] and K K Anoop[1, *]

[1]Applied Optics Division, Department of Physics, Cochin University of Science and Technology, Kochi-682022, India.
*E-mail: anoopkk@cusat.ac.in

Abstract. Conventional Laser Induced Breakdown Spectroscopy (LIBS) quantitative analysis, employing both calibration and calibration-free techniques, is challenged by spectral overlapping, self-absorption, and spectral broadening effects, leading to decreased accuracy. Recently, integration of machine learning (ML) algorithms with LIBS has been increasingly employed to tackle these challenges. This article explores the augmentation of LIBS with deep learning neural networks (DNN) to enhance accuracy of quantitative analysis of multi-elemental copper alloys. Sufficient training data was acquired by simulating optical emission spectra for bronze (Cu-Sn) and admiralty brass (Cu-Zn-Sn) alloys under standard laser produced plasmas conditions, encompassing different alloy concentrations, electron temperatures, and densities. We designed a regularized DNN structure, fine-tuned using a validation dataset to optimize quantitative results. The model's accuracy was assessed with test dataset. The quantitative results demonstrated reduced loss as training spectra increased from 500 to 5000 for both alloys. The decline in mean squared error, from 2.793×10^{-3} to 4.283×10^{-5} for bronze and from 3.245×10^{-2} to 5.598×10^{-4} for admiralty brass alloys, as training data increased from 500 to 5000, underscores the proposed DNN model's potential for metallurgical alloy quantification.

1. Introduction

Laser-Induced Breakdown Spectroscopy (LIBS) was swiftly introduced and suggested as a powerful analytical tool shortly after the advent of laser technology in 1960. LIBS functions by directing an intense pulsed laser beam onto a sample's surface, initiating material breakdown and generating microplasma [1]. The resulting emission lines signify specific atomic, ionic, or molecular transitions, uniquely identifying and characterizing the sample's composition.

LIBS quantitative analysis can be approached through two primary methods: the standard calibration approach and the calibration-free methods [1–3]. The most commonly used method involves constructing a calibration curve, which establishes a link between integrated line intensity and known concentrations, enabling the determination of unknown concentrations [1]. The calibration-free Laser Induced Breakdown Spectroscopy (CF-LIBS) as an alternative method, is based on a physical model assuming Local Thermodynamic Equilibrium (LTE) and utilizes intensities of lines to initially estimate plasma parameters, which in turn facilitates the determination of chemical compositions [2,3]. CF-LIBS depends upon the intensity of emitted spectral lines, and any deviation in this intensity leads to errors in the estimations of analyte concentrations. In most practical situations, the spectral line intensity exhibits fluctuations departing from the actual value, predominantly due to self-absorption, detector inefficiencies, and variations in laser pulse energy. Consequently, the accuracy of CF-LIBS is often not up to the expectation.

The above-mentioned drawbacks can be minimized by using machine learning (ML) methods that consider the complete measured spectrum as input. Recently, there has been widespread use of these ML techniques to speed up both the qualitative and quantitative analysis of LIBS spectra [4–8]. The utilization of these models, coupled with feature engineering techniques has demonstrated remarkable proficiency in precisely identifying materials of comparable chemical compositions. Sun et al. [4] developed a multivariate calibration model using machine learning algorithms and used SelectKBest algorithm for feature selection, aiming to determine trace elements in soil samples. Tian et al. [5] proposed a non-linear support vector machine (SVM) model for the phosphorus determination in seafood. Zhang et al. [6] determined the minor metal elements like Mn, Cr, and Ni in steel by employing multivariate models rooted in back-propagation neural networks (BPNN). Additional regression algorithms, such as random forest regression (RFR), have seen extensive utilization as well [7,8].

In all these examples, machine learning techniques require manual feature selection as part of the regression process. Deep learning, as a subfield within machine learning, has the inherent ability to automatically extract vital feature representations, streamlining machine learning workflows considerably, and it holds the potential to provide a more accurate and efficient approach. Eynde et al. [9] introduced deep learning regression technique to estimate the concentrations of alloying elements in aluminium reference samples and post-consumer scrap pieces. It is noteworthy that deep learning is an emerging procedure for spectral analysis, offering the advantage of avoiding self-absorption and matrix effects, and although there is relatively less work reported in this area, it holds promise for advancing spectral analysis methodologies [9].

LIBS is a key tool for alloy analysis in quality control. Copper-tin-zinc alloys are versatile materials used across industries, demanding precise element quantification for material quality assessment. In this study, a deep learning neural network (DNN) model was designed and employed in combination with LIBS for the efficient quantitative analysis of multielemental copper-tin-zinc alloys for metallurgical purposes. The methodology described in this paper marks a vital starting point for forthcoming endeavors aimed at examining LIBS experimental data.

2. Deep Learning Neural Network Approach

As an initial step, we simulated optical emission spectra (OES) of bronze and admiralty brass alloys at typical laser produced plasma (LPP) conditions to acquire a sufficient volume of data for training the models. The process of simulating the optical emission spectrum consists of four distinct stages. Initially, we compute the concentration of each ionization state within the plasma, then apply the Boltzmann distribution law to determine the intensity of individual emission lines. Following this, we gave gaussian broadening to each line, and finally, graphed the spectrum based on this broadening [10]. A comprehensive explanation of these various stages in the simulation can be found in our prior research work [11].

2.1. Dataset Preparation

Using simulation approach, we generated datasets that exhibited characteristics aligned with the underlying assumptions:

i. 500 and 5000 normalised spectral datasets were generated separately for both bronze and admiralty brass alloys.

ii. The spectral range was 200-1100 nm and resolution (resolution refers to the ability to distinguish and separate closely spaced spectral lines in a spectrum) used was 0.7 nm.

iii. To determine the elemental composition for each bronze spectrum, we initiated the process by selecting a random uniform value for the copper (Cu) concentration ranging from 0% to 100%. Subsequently, the tin (Sn) concentration was adjusted to complete the total concentrations to 100%. The same method was applied to generate datasets for admiralty brass alloy.

iv. The electron temperature (T_e) was then selected randomly within the range of 0.5 to 1.5 eV, and the electron number density (N_e) was chosen between 1.0×10^{16} and 1.0×10^{18} cm^{-3}. These T_e and N_e values are typical for laser induced plasmas.

2.2. Data Preprocessing

As part of the preprocessing step, the input feature (intensities) is scaled to fall within the [0,1] range, ensuring that all emission lines are placed on a uniform intensity scale. With the goal in mind, we adjust each input value using the max-min formula as described in equation (1),

$$I_{norm} = \frac{I - I_{min}}{I_{max} - I_{min}} \tag{1}$$

where I represent the original intensity, I_{norm} is the normalized intensity, and I_{max} and I_{min} denote the highest and lowest possible values of emission intensity, respectively.

2.3. Deep Learning Neural Network Model Development

Here, we introduced a deep learning neural network (DNN) approach for dealing with the problem of finding out the composition of the elements present in the simulated spectra of Bronze (Cu-Sn) and Admiralty brass (Cu-Zn-Sn) alloys used in metallurgical field without using the time-consuming calibration curve approach. Scikit-learn, a python library for machine learning available as open-source and Keras, a deep learning framework, make the process of data analysis more accessible and efficient [12]. A deep neural network involves two key stages: firstly, the feed-forward. Secondly, there is the backpropagation step, where the model parameters are updated by assessing the disparity between the target and predicted values. The proposed deep feed-forward neural network model consists of two hidden layers, with 128 and 180 hidden neurons in each hidden layer for bronze and admiralty brass alloys respectively, to implement the fitting model. The deep neural network has the spectral lines as input and two and three outputs (elements and their corresponding concentrations as output) for bronze and admiralty brass alloys respectively. As outlined in table 1, the activation functions for hidden neurons and output neurons are the ReLU function (Rectified Linear Unit) and the Softmax function, respectively. The proposed DNN model (figure 1) was used for training 500 and 5000 simulated spectra of both bronze and admiralty brass alloys separately for quantitative analysis. The network model that was created is trained using 60% of the available data, validated with 20%, and tested with the remaining 20%.

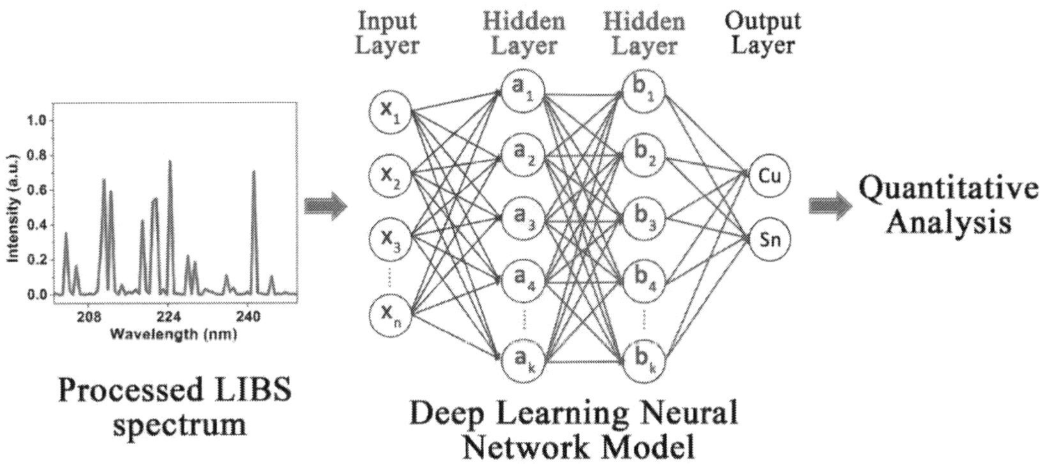

Figure 1. Schematic representation of the deep learning neural network model (DNN) coupled LIBS for quantitative analysis of bronze alloy.

Table 1. Parameters used for deep neural network model.

Parameters	**Values**	
	Bronze	Admiralty brass
Number of hidden layers	2	2
Number of hidden neurons in each hidden layer	128	180
Number of output neurons	2	3
Transfer Functions	ReLU (Hidden Layers), Softmax (Output Layer)	

3. Results and Discussion

The proposed DNN model was used for predicting the composition of simulated bronze and admiralty brass alloy samples. The DNN data are represented as a matrix during the input process, as depicted in figure 1. The input tensor undergoes fully connected dense layers, and activation functions to generate prediction results. The model parameters are subsequently fine-tuned iteratively using stochastic gradient descent in the backpropagation step. While constructing the network model, we ascertain the most effective regression by adjusting DNN parameters and structure and evaluating model performance on the verification dataset. In each iteration, we evaluated the network's performance by computing the Mean Square Error (MSE) on the test dataset following the formula provided in equation (2).

$$MSE = \frac{1}{N}\sum_{i=1}^{N}|t_i - t_0|^2 \tag{2}$$

where t_i and t_0 (vectors) represent the desired and estimated concentration values for spectrum i and N is the number of test spectra included in the evaluation. As the training dataset size grew from 500 to 5000 spectra, the mean square error value for bronze alloy decreased from 2.793×10^{-3} to 4.283×10^{-5}. Likewise, the MSE for admiralty brass exhibited a similar trend, decreasing from 3.245×10^{-2} to 5.598×10^{-4} as the training dataset expanded from 500 to 5000 spectra. As the spectral data used for training increased, the proposed deep neural network model, tend to exhibit improved accuracy and robustness in the alloy quantification. This observation strongly indicates an enhanced predictive capacity of the proposed model for determining the elemental composition within bronze and admiralty brass alloys as the training dataset size increases.

The accuracy and loss rate of the proposed DNN model tend to converge after 1000 training epochs of 500 and 5000 number of spectra which clearly indicates the increase in the accuracy of the proposed model for efficient quantitative analysis of bronze and admiralty brass alloy samples. The prediction (predicted wt% values) versus ground truth (true wt% values) and loss function graphs corresponding to 500 and 5000 number of spectra for bronze and admiralty brass alloys respectively, are depicted in figure 2.

The findings demonstrate the robustness and benefits of utilizing DNN models. Clearly, the proposed DNN model demonstrates a strong ability in accurately determining the elemental composition of simulated Cu-Zn-Sn alloys. Although the models did well with simulated data, future research should aim to create DNN models using real experimental data.

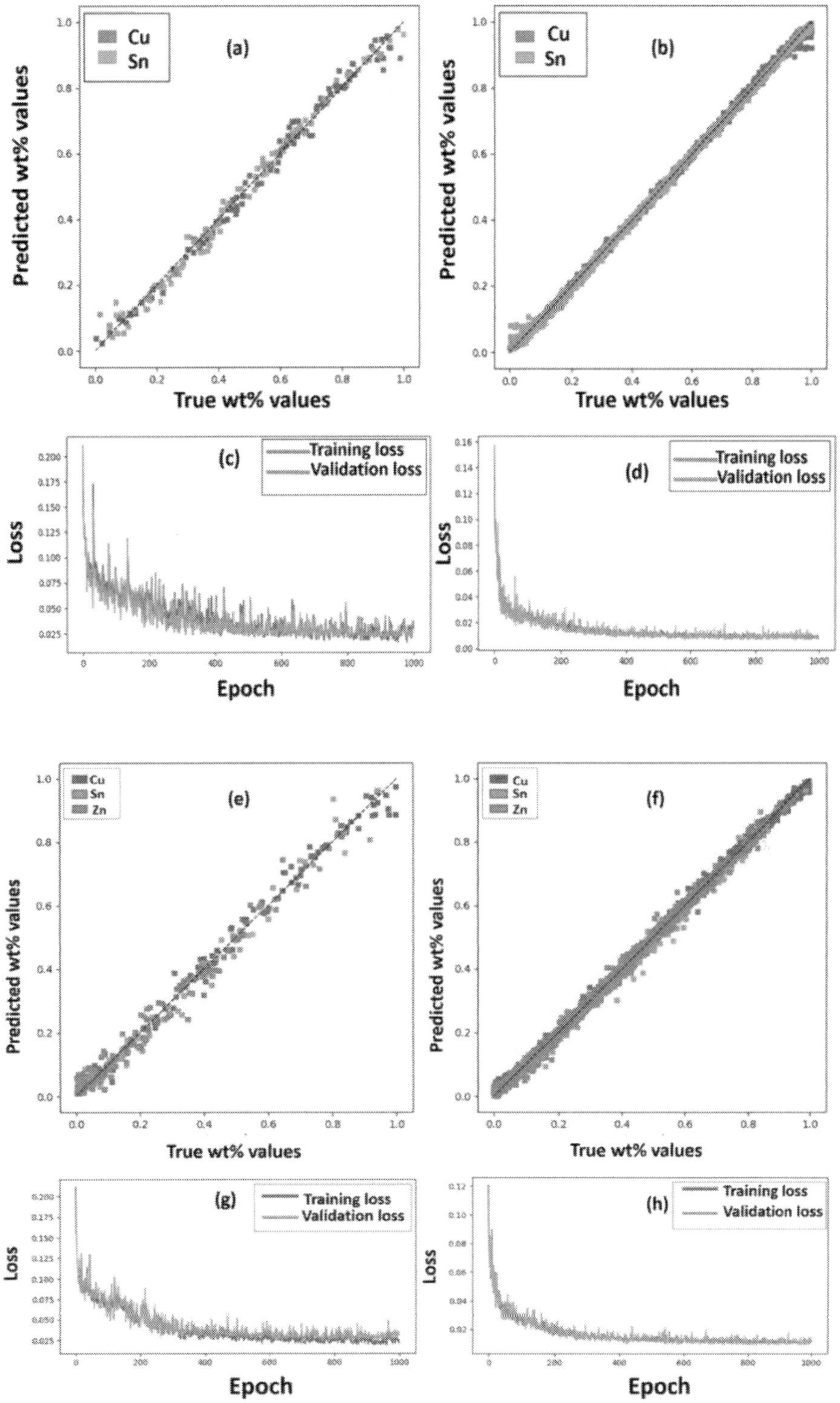

Figure 2. True wt% values vs. Predicted wt% values for 500 and 5000 spectra of bronze alloy (a,b) and admiralty brass alloy (e,f) respectively. The loss function curve of learning process for 500 and 5000 spectra of bronze alloy (c,d) and admiralty brass alloy (g,h) respectively.

4. Conclusion

We illustrated an effective approach that leverages a combination of laser-induced breakdown spectroscopy and deep neural networks for efficient multi-elemental quantitative analysis of alloys used in metallurgical purposes. This approach is utilized to ascertain the composition of copper alloys by analysing simulated datasets generated through computational methods. The proposed deep neural network model is a feed-forward deep neural network with spectral line data as input neurons, two hidden layers, and two and three output neurons for bronze and admiralty brass alloys respectively. To utilize network for alloy composition analysis, we used 60% of the simulated spectra for training the neural network, 20% for validation and the remaining 20% for testing. The decrease in MSE values for both the alloys, as the number of alloy spectra used for training increased from 500 to 5000 clearly demonstrates that regression accuracy improves with a larger dataset. This indicates that the proposed DNN model is highly effective in determining the composition of alloys employed in the metallurgical industry.

5. References

[1] Cremers D A and Radziemski L J 2013 *Handbook of Laser-Induced Breakdown Spectroscopy* (Wiley)

[2] Tognoni E, Cristoforetti G, Legnaioli S and Palleschi V 2010 Calibration-Free Laser-Induced Breakdown Spectroscopy: State of the art *Spectrochim. Acta Part B At. Spectrosc.* **65** 1–14

[3] Mary John L, Issac R, Sankararaman S and K K A 2022 Multi-element Saha Boltzmann Plot (MESBP) coupled calibration-free laser-induced breakdown spectroscopy (CF-LIBS): An efficient approach for quantitative elemental analysis *J. Anal. At. Spectrom.* **37**

[4] Sun C, Tian Y, Gao L, Niu Y, Zhang T, Li H, Zhang Y, Yue Z, Delepine-Gilon N and Yu J 2019 Machine Learning Allows Calibration Models to Predict Trace Element Concentration in Soils with Generalized LIBS Spectra *Sci. Rep.* **9** 1–18

[5] Tian Y, Chen Q, Lin Y, Lu Y, Li Y and Lin H 2021 Quantitative determination of phosphorus in seafood using laser-induced breakdown spectroscopy combined with machine learning *Spectrochim. Acta Part B At. Spectrosc.* **175** 106027

[6] Zhang Y, Sun C, Gao L, Yue Z, Shabbir S, Xu W, Wu M and Yu J 2020 Determination of minor metal elements in steel using laser-induced breakdown spectroscopy combined with machine learning algorithms *Spectrochim. Acta Part B At. Spectrosc.* **166** 105802

[7] Ruan F, Qi J, Yan C, Tang H, Zhang T and Li H 2017 Quantitative detection of harmful elements in alloy steel by LIBS technique and sequential backward selection-random forest (SBS-RF) *J Anal Spectrom* **32** 2194–9

[8] Liang J, Li M, Du Y, Yan C, Zhang Y, Zhang T, Zheng X and Li H 2020 Data fusion of laser induced breakdown spectroscopy (LIBS) and infrared spectroscopy (IR) coupled with random forest (RF) for the classification and discrimination of compound salvia miltiorrhiza *Chemom. Intell. Lab. Syst.* **207** 104179

[9] Van Den Eynde S, Diaz-Romero D J, Zaplana I and Peeters J 2022 Deep Learning Regression for Quantitative LIBS Analysis of Aluminium Scrap *SSRN Electron. J.*

[10] Weisstein, Eric W. Gaussian Function -- from Wolfram MathWorld

[11] Sivadas M S S, John L M and Anoop K K 2022 Simulation of Optical Emission Spectra of Cu-Sn-Zn Alloy Plasmas for Laser-Induced Breakdown Spectroscopy Applications *IOP Conf. Ser. Mater. Sci. Eng.* **1221** 012027

[12] Pedregosa et al. 2011 Scikit-learn: Machine Learning in Python *Journal of Machine Learning Research* **12** 2825–30

Acknowledgment

Leya Pauly and K K Anoop acknowledges the Rashtriya Uchchatar Shiksha Abhiyan (RUSA) 2.0 Major Project T1C [CUSAT/PL(UGC).A1/2314/2023].

Crystal structure, molecular mechanics and *In silico* analyses of piperizine derivative against human mammary carcinoma cells inhibition.

H Keshav Kumar[1], S Ananda[1], Dukanya[2], V Keerthikumara[1], Basappa[3], M Mahendra[1*]

[1]Department of Studies in Physics, University of Mysore, Manasagangotri, Mysuru 570 006, Karnataka.

[2] PG Department of Chemistry, JSS College for Women, Saraswathipuram, Mysuru 570 009, Karnataka.

[3]Department of Studies in Organic Chemistry, University of Mysore, Manasagangotri, Mysuru 570 006, Karnataka.

*Corresponding author: mahendra@physics.uni-mysore.ac.in

Abstract. In the present study, the single crystal of novel piperazine derivative4-(2, 3-chlorophenyl) piperazine-1-yl) (2-hydroxyphenyl) methanone (KDM) is grown by using the solvent evaporation method. The 3D structure of the molecule is confirmed by the single-crystal X-ray diffraction method. The study revealed that the molecular system is crystallized in the orthorhombic system with space group P_{bca}. The supramolecular crystal architecture establishes the stability of a compound *via* short contacts and halogen-hydrogen interactions. The Hirshfeld surface analysis were performed to evaluate the numerous intermolecular interactions based on the anisotropy of the topology. The Frontier molecular orbital (FMO) analysis and Molecular electrostatic potential (MEP) plots are investigated to understand the electronic structure properties of compounds using Density Functional Theory (DFT). *In silico* molecular docking, analysis is carried out to predict the best binding pose of the compound in the active site pocket of the BCl-XL/BAK protein-protein interface. Further, *in vitro* cytotoxicity studies against human breast cancer (MCF-7) cell lines of similarly designed piperazine-based derivatives showed prominent results. The results of the current study revealed that the compounds under investigation possess potential anti-cancer properties.

1.Introduction

Cancer treatment has remained dishearteningly low despite adopting multimodality therapy in the medical field. The 5-year survival rate typically ranges from 6% to 14% in males and 7% to 18% in females [1]. Targeted medications are aimed at oncogenic driving mutations in adenocarcinoma treatments and treatment with radiation, surgery, and chemotherapy [2].
Among the B-cell lymphoma (BCL) proteins' family, BCL-XL belongs to the family with anti-apoptotic function. It inhibits cell death by preventing the release of cytochrome c and apoptotic factors from the mitochondria into the cytoplasm. The release of apoptotic factors is a crucial step in initiating the cell death process. On the other hand, BCL-2-associated death promoter (BAK) is a pro-apoptotic that promotes cell death. When BAK is activated, it forms pores (channels) in the mitochondrial membrane, releasing cytochrome c with other apoptotic factors into the cytoplasm. The interaction between the BCL-XL and BAK is a critical aspect of the regulation of apoptosis, a process of programmed cell death that plays an significant role in tissue homeostasis maintainence and preventing the survival of damaged or abnormal cells [12]. When BCL-XL and BAK interact, BAK's pro-apoptotic activity can be neutralized by preventing apoptosis and blocking the cytochrome's release from the mitochondria and other apoptotic factors. These released factors activate caspases, which are proteases that initiate cellular

destruction during apoptosis. Hence, the balancing between pro-apoptotic and anti-apoptotic BCL family proteins is crucial for determining whether a cell will undergo apoptosis or survive. Disruption of this balance, such as increase of expression in anti-apoptotic proteins or reduce of expression of pro-apoptotic proteins contributing to cancer development and resistance to cancer treatments.
A small molecule interaction with the interface of BCL-XL and BAK modulates the sensitive balance between cell survival and apoptosis. These interactions leads to implications for cellular function, tissue homeostasis, and disease processes, particularly in cancer therapy and the regulation of programmed cell death [13]. Scientists and biologists are continually intrigued by molecules with small molecular masses because of their propensity to significantly influence the operation of macromolecules that make up living systems. Small-molecule modifies the protein function and provides the framework for chemically changing the state of disease and 'probes' in chemical genetics to analyze biological systems [3, 4]. Among heterocyclics, piperazine is a significant bioactive natural product. It is precious in drug discovery across therapeutic applications, including antitumor, antibacterial, antifungal, and anti-Alzheimer medications [5]. Given this, we herein suggest a novel class of piperazine derivative for the inhibition of cancer cells and show its potential as a suitable candidate for drug development.

2.Experimental and Computational Methods

The block-shaped single crystal KDM having dimensions 0.21×0.24×0.32 cm was subjected to X-ray diffraction (XRD) using a Bruker D8 Diffractometer using Apex II CCD Detector equipped with monochromatic MoKα radiation (Wavelength $\lambda = 0.7103$ Å). The structure was further solved by direct methods and refined using SHELXS and SHELXL, respectively [6, 7]. The geometrical figures were generated and visualized by PLATON and MERCURY software [8]. The computational calculation was performed using the Gaussian 09 program [9] package at the B3LYP method level with a standard 6-311++ G(d, p) basis set. AutoDock Tools (ADT) (v1.5.7) [10] software package was employed to perform *in silico* molecular docking analysis.

3.Results and Discussions

3.1 Single Crystal Structure X-ray Analysis

The single crystal X-ray analysis unveiled the confirmation of 3D structure of the KDM in orthorhombic system with P_{bca} space group, respectively. The geometrical lattice parameters were found to be a=10.0827 Å, b= 13.0922 Å, c= 24.6781 Å, $\alpha = \beta = \gamma = 90°$, Volume (V) = 3257.6 Å^3 and Z= 8. The ORTEP view with a 50% probability of the KDM molecule is shown in Figure 1(a). The packing diagram of molecule along b-axis exhibited halogen mediated hydrogen bond interactions as shown in Figure 1(b). The halogen-hydrogen interactions are obsereved among atoms CL8...H3 and Cl7...H10A with 2.7 and 2.9Å, respectively. The KDM is also stabilized by C-H...O type inermolecular interactions.

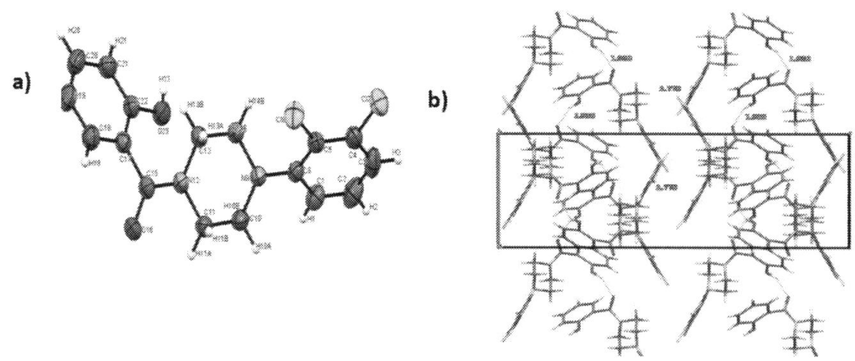

Figure 1: a) ORTEP drawn at 50% probability b) Packing diagram with halogen and hydrogen bond interactions along b-axis

3.2 *Hirshfeld Surface Analysis*

The packing modes investigations, intermolecular interaction, and molecular shape corresponding to the crystalline environment are explored using Hirshfeld surface analysis (Figure 2a). The information is obtained based on electron density distribution by mapping Hirshfeld surfaces with d_{norm}. The 2D fingerprint plots are generated by combining d_e and d_i distances with the help of Crystal Explorer 17.5 [11]. The fingerprint plots resulted in different regions of the area with H...H (38.2%), Cl-H...H-Cl (22.2%), C-H...H-C (20.2%), and O-H...H-O (12.7%) interactions (Figure 2c). The most noticed interaction in comparison is the H...H interaction, and the least is N-H...H-N (1.1%) interaction. The shape index diagram depicts the π-π interaction in the molecule encircled in black (Figure 2b).

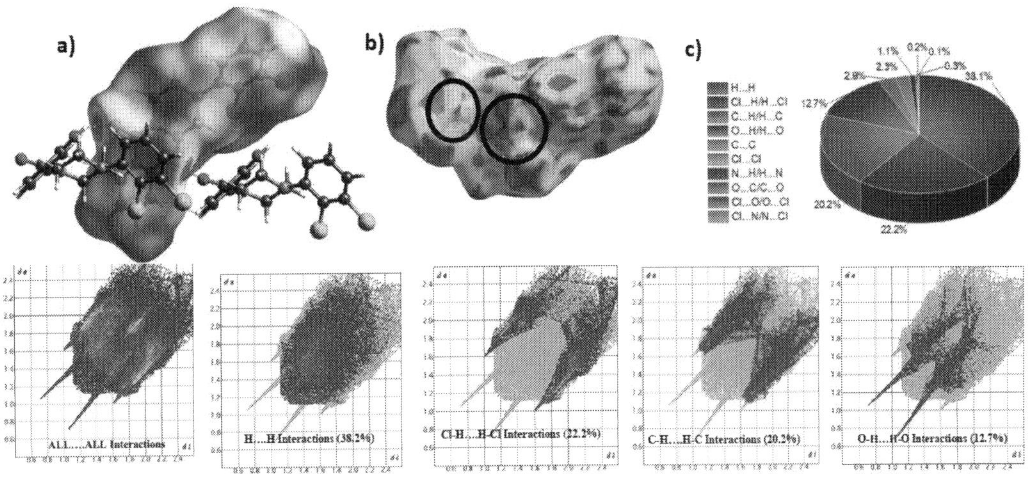

Figure 2 : a) Hirshfeld surface mapped with d_{norm} showing prominent interactions b) Shape index c) 2D Fingerprint plots showing percentage of intermolecular interactions

3.3. *Density Functional Theory Calculations*

3.3.1 *Frontier Molecular Orbital (FMO) and Molecular Electrostatic Potential (MEP) Analysis*

The quantum chemical characteristics of the highest occupied molecular orbital (HOMO) and the lowest unoccupied molecular orbital (LUMO) show the donation and acceptance of electrons, respectively. The computations are carried out using basis level theory B3LYP/6-311++G (d, p). The physicochemical properties, such as kinetic stability, chemical hardness/softness, and reactivity of the molecule, are characterized by the HOMO-LUMO energy gap (Table 1). The HOMO and LUMO orbitals of the KDM molecule are shown in Figure 3. The energy gap is found to be 4.744eV. The electronegativity and electrophilicity of KDM are observed to be 2.831 and 3.672 eV, respectively. The MEP plot was employed to understand the molecule's predicted reactive sites, such as electrophilic or nucleophilic sites. An electron isosurface map with an electrostatic potential surface indicates the molecule's size, shape, charge density, and reactive sites. Figure 4 depicts the resulting MEP map of the KDM compound.

IOP Conf. Series: Materials Science and Engineering 1300 (2024) 012007 doi:10.1088/1757-899X/1300/1/012007

Table 1 : Global Chemical Reactivity Descriptor Parameters

Molecule Properties	KDM
Energy in a. u.	-1838.07
E_{HOMO} in eV	-6.044
E_{LOMO} in eV	-1.300
$\Delta E_{LUMO-HOMO}$ in eV	4.744
Ionization potential (I)	6.044
Electron affinity (A)	1.300
Global Hardness (η)	2.372
Softness (S)	0.210
Chemical potential (μ)	-3.672
Electronegativity (χ)	3.672
Electrophilicity (ψ)	2.831

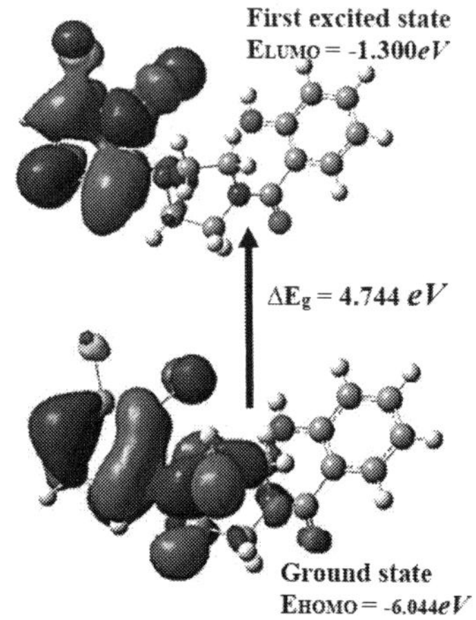

Figure 3: Frontier molecular orbital (FMO) plot

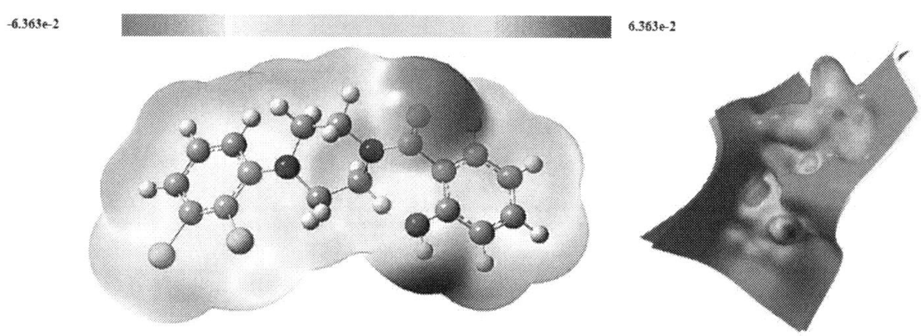

Figure 4: Molecular electrostatic potential (MEP) plot

3.4.In Silico Molecular Docking Analysis

In silico molecular docking, the analysis determined the optimal binding pose of the KDM molecule in the active site B-cell lymphoma-extra-large (BCL-XL) and BCL-2-associated death promoter (BAK) protein-protein interface. The crystal structure of BCL-XL bounded to BAK was downloaded from the RCSB Protein data bank having PDB ID: 5fmk (https://www.rcsb.org/). Figure 5 illustrates the binding pose of the KDM at the binding site BCL-XL/BAK interface, along with various interactionsbetween neighbouring aminoacids. The binding energy resulted in -8.5 kcal/mol having an inhibition constant value of 1.56 μM. The interaction formed a strong hydrogen bonding between the O-H group and GLU129 and ASP133 amino acids, respectively, having bond lengths of 1.98Å and 1.92Å. The KDM exhibited hydrophobic interactions among ARG132, PHE131, LEU 130, VAL126 and GLU129 amino acids. Further π-π stacking interactions were obsereved between piperizne and dichloro benzene ring moieties with LEU130 and VAL126, respectively. As a result,the molecule binds effectively with BCL-XL. Therefore, the synthetic KDM effectively inhibits BAK phosphorylation in BCL-XL, preventing cancer cell proliferation. Based on the core structural geometry of KDM, Girimanchanaika et al., reported the design and synthesis of eighteen molecules and were further sent for cytotoxic studies against MCF-7 cell lines, which showed prominent anti-cancer activities [12].

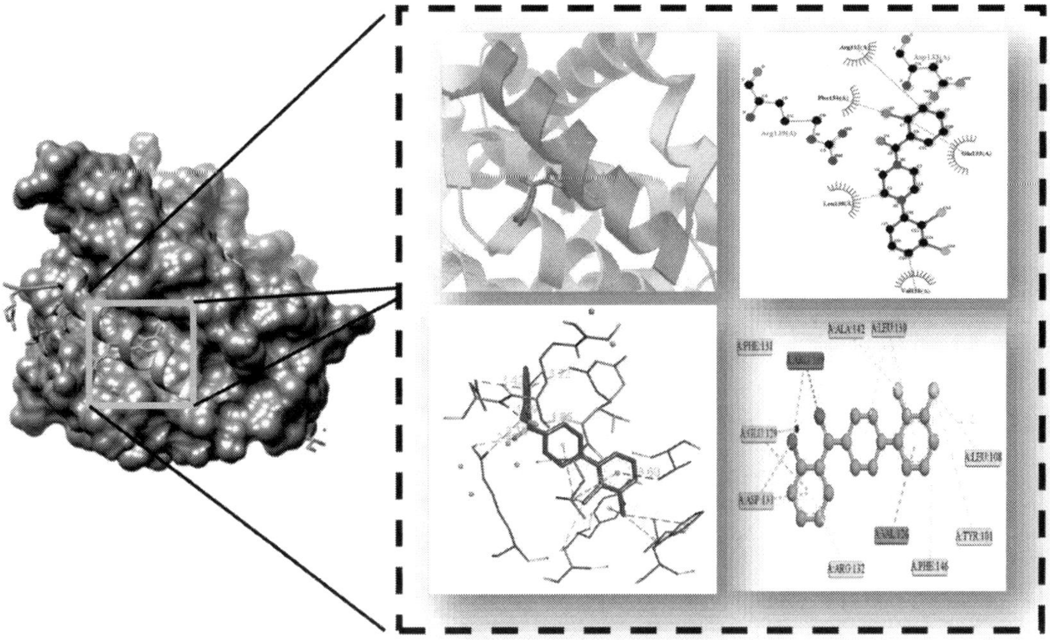

Figure 5: Binding interactions of KDM with BCL-XL protein at the binding site associated with BAK protein

Conclusion

The three-dimensional crystal structure of 4-(2, 3-chlorophenyl) piperazine-1-yl) (2-hydroxyphenyl) methanone (KDM) is determined using a single X-ray diffraction technique. The compound resulted in prominent hydrogen bond interactions in the experiment and DFT. Frontier molecular orbital (FMO) and molecular electrostatic potential (MEP) analyses of KDM were performed by DFT at the B3LYP/6311++G (d, p) basis set to understand the chemical reactive sites and global descriptor parameters of the compound. The KDM was firmly bound into the BCL-XL/BAK interface pocket, which deserves a potential candidate for cancer cell growth inhibition.

Acknowledgments

The author, Keshav Kumar H., would like to thank the Sophisticated Analytical Instrumentation Facility (SAIF), IIT Madras, for the SCXRD facility.

References

[1] Youlden, D. R., Cramb, S. M.,and Baade, P. D. (2008). The International Epidemiology of Lung Cancer: geographical distribution and secular trends. *Journal of thoracic oncology*, *3*(8), 819-831.
[2] Zhang, M., Wang, B., and Lobie, P. E. (2019). A novel small-molecule inhibitor of trefoil factor 3 (TFF3) potentiates MEK1/2 inhibition in lung adenocarcinoma. *Oncogenesis*, *8*(11), 65.
[3] Spring, D. R. (2005). Chemical genetics to chemical genomics: small molecules offer big insights. *Chemical Society Reviews*, *34*(6), 472-482.

[4] Kawasumi, M., and Nghiem, P. (2007). Chemical genetics: elucidating biological systems with small-molecule compounds. *Journal of investigative dermatology*, *127*(7), 1577-1584.
[5] Zhang, R. H., Guo, H. Y., and Quan, Z. S. (2021). Piperazine skeleton in the structural modification of natural products: a review. *Journal of enzyme inhibition and medicinal chemistry*, *36*(1), 1165-1197.
[6] Bruker, A. X. S. Inc. (2004) APEX 2. Bruker Advanced X-ray Solutions, Madison, Wisconsin, USA.
[7] Sheldrick, G. M. (2015). Crystal structure refinement with SHELXL. *Acta Crystallographica Section C: Structural Chemistry*, *71*(1), 3-8.
[8] Spek, A. L. (1990). PLATON, is an integrated tool for the analysis of the results of a single crystal structure determination. *Acta Crystallographica Section A: Foundations of Crystallography*, 46(s1), c34-c34
[9] Gaussian 09, Revision A.02, M. J. Frisch Gaussian, Inc., Wallingford CT, 2016.
[10] Morris, G. M., Huey, R., Lindstrom, W., Sanner, M. F., Belew, R. K., Goodsell, D. S. and Olson, A. J. (2009) Autodock4 and AutoDockTools4: automated docking with selective receptor flexibility. *J. Computational Chemistry* 2009, **16**: 2785-91.
[11] Turner, M. J., McKinnon, J. J., Wolff, S. K., Grimwood, D. J., Spackman, P. R., Jayatilaka, D., and Spackman, M. A. (2017). CrystalExplorer17.
[12] Girimanchanaika, S. S., Dukanya, D., Swamynayaka, A., Govindachar, D. M., Madegowda, M., Periyasamy, G and Basappa, B. (2021). Investigation of npb analogs that target phosphorylation of bad-ser99 in human mammary carcinoma cells. *International Journal of Molecular Sciences*, *22*(20), 11002.
[13] Pandey, V., Wang, B., Mohan, C. D., Raquib, A. R., Rangappa, S., Srinivasa, V., and Lobie, P. E. (2018). Discovery of a small-molecule inhibitor of specific serine residue BAD phosphorylation. *Proceedings of the National Academy of Sciences*, 115(44), E10505-E10514.

Electrical nature and surface enhanced Raman spectroscopy of Ag nanoparticles decorated graphene Sheet

Sachidanand P S[1,2], **Manu Mohan**[1,2], **Reshmi S**[2], **Gauthami Viswan**[3] **and Kuntala Bhattacharjee**[1,2*]

[1]Department of Physics, Indian Institute of Space Science and Technology Valiyamala, Trivandrum 695547, Kerala, India
[2]Institute of Physics, Sachivalaya Marg, Sainik School (PO), Bhubaneswar 751005, Odisha, India
[3]University of Nebraska-Lincoln, 1400 R St, Lincoln, NE 68588, USA
*Corresponding email: kbhattacharjee@iopb.res.in; kuntala.iopb@gmail.com

Abstract. The interaction between graphene and silver nanoparticles (AgNPs) is investigated by studying the surface-enhanced Raman spectroscopy (SERS) that shows a large enhancement of Raman signal from the AgNPs coated graphene compared to the bare graphene. In addition, the electrical nature of the sample is examined using a two-probe probe station to acquire current-voltage (I-V) characteristics exhibiting a semi-conducting current behaviour of the AgNPs-graphene heterojunctions, in contrast to a semi-metallic electronic behaviour observed for the pristine graphene. We propose band gap engineering of graphene from semi-metallic to semi-conducting via breaking of A-B sub-lattice symmetry due to deposition of AgNPs which is believed to be associated by causing deformations at the regions where graphene is in close proximity with the AgNPs. Furthermore, there is also a possibility of doping induced by AgNPs to graphene and thereby, emergence of a bandgap due to tuning of the electronic structure. In this work, we report these investigations which possess profound implications for application of graphene as a semi-conducting material.

1. Introduction

Graphene, the veteran of the 2D materials family, has attracted a major attention due to its promising nanoelectronic applications. The sp^2 hybridized hexagonal carbon lattice of graphene and its atomic thickness contribute to exciting properties such as anomalous quantum hall effect, high Fermi velocity as the charge carriers mimic massless Dirac-fermions [1], Klein tunnelling [2] etc. However, a major drawback of graphene is its zero bandgap that prevents any semiconductor applications and also, light-matter interaction of graphene is very weak owing to its atomic thickness. One of the subtle ways to investigate the light-matter interaction is to coat graphene with metal nanoparticles (NPs) of silver (Ag), gold (Au) etc. that are well known for their excellent plasmonic properties. These studies are based on excitations of the conduction electrons of metal NPs [3, 4] with the incident electromagnetic field to excite the localized surface plasmons (LSPs) and induce a resonance called localized surface plasmon resonance (LSPR) [5] that significantly enhances the Raman signal. This method emerged as a powerful analytical technique known as surface enhanced Raman spectroscopy (SERS). Raman spectroscopy has played a crucial role in studying and understanding the fundamental properties of graphitic materials including graphene over years [6, 7]. Moreover, it has grown as a dynamic tool to characterize the electrons and phonons in graphene [8]. Herein, we study the interaction of light with

AgNPs coated graphene by investigating the SERS spectra. Unlike the case of graphene-Ag nanocomposites [9], AgNPs spin coated over graphene in this study shows better Raman signal with higher intensity compared to bare graphene. The changes in the Raman spectra of graphene invoke the proof for the charge transfer between graphene and AgNPs. This must be reflected in the electronic behaviour of the sample, and indeed we observe different trends in the current-voltage (I-V) characteristics acquired using a two-probe probe station. In this letter, we address these effects with a qualitative understanding of the underlying physical phenomena.

2. Experimental techniques

Single layer of graphene on the copper (Cu) foil was procured from Sigma Aldrich India which was grown by chemical vapour deposition (CVD) method. We adapted a polymer supported chemical process to transfer the graphene from the Cu foil [10]. During the growth process of graphene by CVD, carbon was also deposited at the bottom side of the foil and as a result, removal of the residual carbon was necessary. This was carried out by an etching process using nitric acid-distilled water solution in the ratio of 1:3, followed by separating the graphene sheet from the upper side of the Cu foil through a PMMA assisted transfer method. The single layer graphene was then placed on a 2 micron thick SiO_2/Si substrate. This transfer method is described elsewhere [11]. AgNPs of diameter 40 nm and concentration 0.02 mg/ml were procured from Sigma Aldrich which were deposited by spin coating with 20 μL of AgNP solution at 2500 rpm speed and 500 rpm/s acceleration on one half of the graphene sheet. The other half of the graphene was covered using a glass mask. Optical microscopy, Raman spectroscopy and I-V measurements were performed on both bare graphene and AgNPs decorated graphene layer. Optical microscopy measurements were performed using a compound microscope: model MX61-F of OLYMPUS Corporation. Raman spectra were acquired using a Renishaw Invia Raman microscope with laser excitation of 532 nm for a frequency range 100-3200 cm^{-1}. Raman signals of pristine graphene and AgNPs decorated graphene were measured separately. For the I-V measurements, Au electrodes were made using a thermal evaporation system via shadow mask technique. I-V curves were acquired using a two-probe probe station across bare graphene and AgNPs-graphene using an Agilent B2912A source/measure unit.

3. Results and discussion

3.1. Optical Microscopy

The optical microscopy image of the sample is shown in figure 1. The transparent graphene layer and the spin coated AgNP film on the graphene sheet are marked in the figure.

Figure 1. Optical image of the sample: pristine graphene (left) and AgNPs coated graphene (right).

3.2. Raman Scattering

The measured Raman spectra of pristine graphene (red) and AgNPs-graphene (black) are shown in figure 2. Laser excitation produces two prominent Raman peaks for monolayer graphene at around 1577 cm^{-1} and 2674 cm^{-1}. The former is known as the G band and the latter is called the 2D or G' band. There is also a defect induced D band which is expected around half the frequency of the 2D band. This band is prominent at the edges of the sample or for the specimen with disorders. We observe a very feeble signature of D peak at around 1341 cm^{-1} for pristine graphene, signifying a very small amount of defects or disorders present in the sample (figure 2). The small peak like feature evident at around 2450 cm^{-1} is the G* band representing double resonant Raman scattering in graphene involving one transverse optical (TO) and one longitudinal acoustical (LA) phonon modes [12]. The Raman shifts of the AgNPs-graphene reveal a higher intensity of the bands compared to the pristine graphene which is clearly visible in figure 2. This increase in the intensity after the deposition of the AgNPs on graphene is due to surface enhanced Raman scattering (SERS). In SERS, the enhancement of the Raman peaks is mainly due to the surface plasmon resonance (SPR) of the plasmonic materials by laser excitation of the sample in the far-field and the inelastic scattering of the Raman dipole in the near field [13-16]. SPR is usually associated with metal films whose dimensions are much greater than the wavelength of the incident light, allowing the plasmon resonance to propagate on the surface of the film. Excitations of conduction electron in metallic nanostructures with dimensions smaller than that of incident light (AgNP- 40 nm and Laser- 532 nm) is called localized surface plasmon resonance (LSPR), which can get coupled with electromagnetic fields leading to the confinement of light to very small volumes [17]. Thus, the electromagnetic field gets enhanced due to the resonance between the confined light and the sharp corner of the metal nanoparticle or the very small gap between the nanoparticles called "hot spot" [18]. This enhancement of the electric field, in turn, amplifies the weak Raman signal due to the small cross-sectional area in the sample. In our case, the SERS hotspots may be identified as both the sharp edges of AgNPs and the narrow gap between the AgNPs and the graphene.

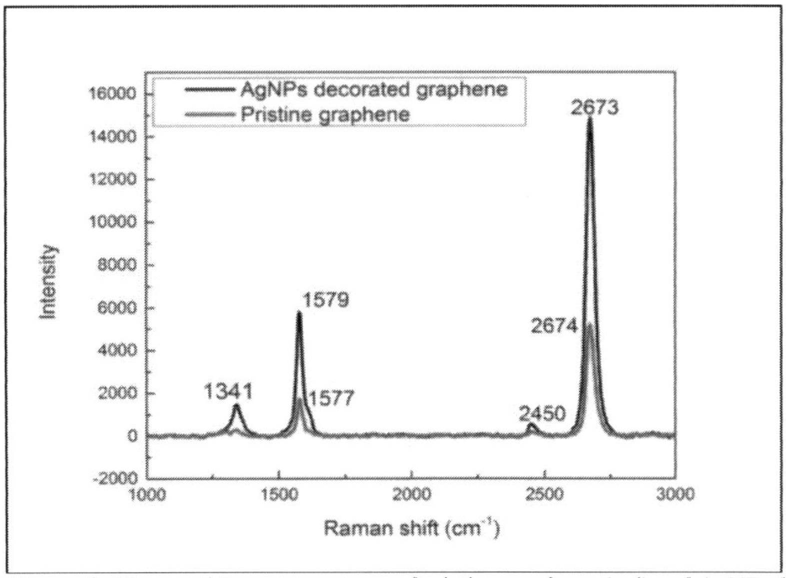

Figure 2. Measured Raman spectrum of pristine graphene (red) and AgNPs decorated graphene (black).

Comparing the Raman peaks of graphene at the areas with and without AgNPs, in addition to the enhanced intensity, we observe a spectral shift from 2674 cm^{-1} to 2673 cm^{-1} for the 2D peak. For the G line, the observed spectral shift is from 1577 cm^{-1} to 1579 cm^{-1}. A similar trend was reported earlier in a study by Alexander Urich et.al., about Ag nanoisland enhanced Raman scattering in graphene [19].

The Raman scattering process of graphene is mainly interpreted based on phonon dispersion of carbon atoms. Since the unit cell of the graphene consists of two carbon atoms, there are six identified phonon dispersion bands in graphene. The G band arises due to the first order Raman scattering processes which is associated with doubly degenerate in-plane transverse optic (iTO) phonon mode or longitudinal optic (LO) phonon mode. In fact, this is the only band that arises from the normal first order Raman scattering process in graphene. The 2D and D bands originate from the second order process. The 2D band is associated with two iTO phonons near the K-point and the D band is always defect induced. In addition there are other weak bands such as G* band which is associated with one iTO and one longitudinal acoustic (LA) phonons. These are the most prominent bands in the Raman spectrum of graphene.

3.3. I-V measurements

Figure 3(a) depicts the schematic of the two probe measurement system. In figure 3(b), we show the I-V characteristics of the sample measured separately for pure graphene and AgNPs decorated graphene.

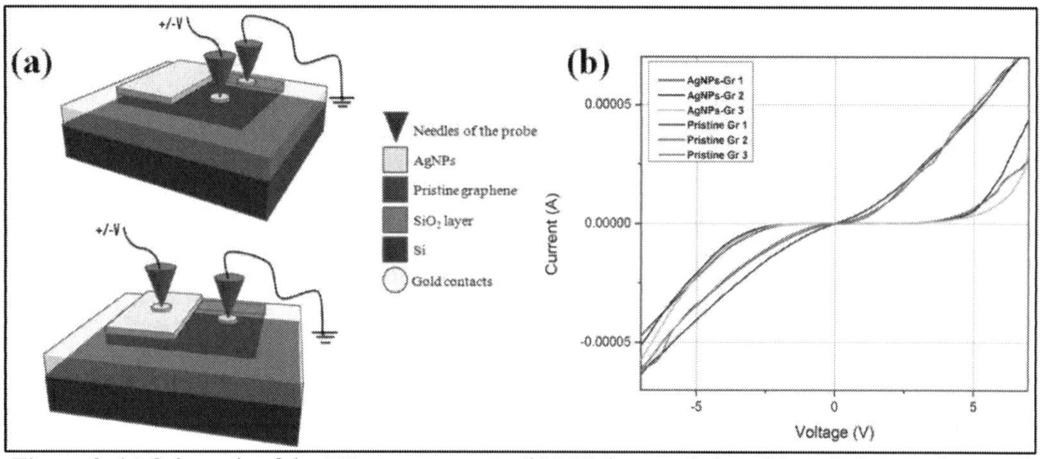

Figure 3. (a) Schematic of the I-V measurements, (b) I-V characteristics of the pristine graphene and AgNPs decorated graphene

I-V curve acquired from bare graphene reveals usual semi-metallic characteristic, whereas AgNPs coated graphene exhibits semiconducting behaviour (figure 3b). Using first principles density functional theory (DFT) calculations, K.S. Subrahmanyam et.al [9] have investigated, the changes in the structural and electronic properties of graphene through deposition of Ag, Au, Pd and Pt nanoclusters. They have reported a local breakage of A-B sublattice symmetry of graphene due to deformation at the regions where graphene is in close proximity to the metal nanocluster. This causes an opening of a band gap near the Fermi level of graphene. They have proposed the possibility of placing the Fermi level of graphene in this energy gap by appropriate tuning of the carrier concentration so that the system could behave as a semiconductor. In the I-V curve, one can also see that the current has been reduced for AgNPs-graphene possibly due to the quantum tunnelling effects induced by defects in the sample. The intensity of the D peak in the measured Raman spectrum of AgNPs-graphene is enhanced (figure 2), indicating presence of defects. This could induce an energy barrier at the AgNPs-graphene interface and a current transfer through the interface will be due to quantum tunnelling with a low yield of current.

4. Conclusion

In this work, we investigated the electronic and electrical nature of AgNPs coated graphene by performing SERS and two probe I-V measurements. An enhanced Raman signal from AgNPs-graphene is due to LSPs as a result of the light-matter interaction causing excitations of the conduction electrons of AgNPs (3, 4), and generating a resonance called LSPR. I-V curves obtained from the AgNPs-graphene interface exhibit a semi-conducting current behaviour of the heterojunction, whereas graphene reveals usual semi-metallic behaviour. This can be understood due to the deposition of AgNPs and breaking of A-B sub-lattice symmetry of graphene, causing defects and deformation at the interface. This will facilitate charge transfer via direct and tunnelling process and will modulate the electronic structure of graphene. This work is of significant importance to use graphene for semi-conductor applications by proposing an easy and cost-effective method for engineering graphene's band gap.

5. References

[1] K. S. Novoselov , A. K. Geim , S. V. Morozov , D. Jiang , M. I. Katsnelson , I. V. Grigorieva , S. V. Dubonos , A. A. Firsov , Nature 2005, 438, 197 .
[2] M. I. Katsnelson , K. S. Novoselov , A. K. Geim , Nat. Phys. 2006, 2, 620 .
[3] T. J. Echtermeyer, L. Britnell, P. K. Jasnos, A. Lombardo, R. V. Gorbachev, A. N. Grigorenko, A. K. Geim, A. C. Ferrari, and K. S. Novoselov, Nat. Commun. 2, 458 (2011).
[4] Y. Liu, R. Cheng, L. Liao, H. Zhou, J. Bai, G. Liu, L. Liu, Y. Huang, and X. Duan, Nat. Commun. 2, 579 (2011).
[5] Kelly KL, Coronado E, Zhao LL, Schatz GC. The optical properties of metal nanoparticles: the influence of size, shape, and dielectric environment. J PhysChem B. 2003;107(3):668–77.
[6] M.S. Dresselhaus, G. Dresselhaus, K. Sugihara, I.L. Spain, H.A. Goldberg, Graphite Fibers and Filaments, in: Springer Series in Materials Science, vol. 5, Springer-Verlag, Berlin, 1988.
[7] M.A. Pimenta, G. Dresselhaus, M.S. Dresselhaus, L.G. Cançado, A. Jorio, R. Saito, Pys. Chem. Chem. Phys. 9 (2007) 1276–1291.
[8] J.-C. Charlier, P.C. Eklund, J. Zhu, A.C. Ferrari, in: Ado Jorio, M.S. Dresselhaus, G. Dresselhaus (Eds.), Electron and Phonon Properties of Graphene: Their Relationship with Carbon Nanotubes, Topics in the Synthesis, Structure, Properties and Applications, in: Springer Series on Topics in Applied Physics, vol.111, Springer-Verlag, B.
[9] K.S. Subrahmanyam, Arun K. Manna, Swapan K. Pati, C.N.R. Rao, A study of graphene decorated with metal nanoparticles, Chemical Physics Letters, Volume 497, Issues 1–3, 2010.
[10] Mohamed Tarek Ghoneim. Efficient transfer of graphene A physical and electrical performance perspective. Thesis Presentation at King Abdullah University of Science and Technology Thuwal, Kingdom of Saudi Arabia, 2012.
[11] Sachidanand P S, MoS_2 nanostructures as highly transparent material: Optical transmittance measurements and surface plasmonic studies of Graphene. Thesis presentation at Amrita Vishwa Vidyapeetham, Kerala, India, 2018.
[12] L.M. Malard, M.A. Pimenta, G. Dresselhaus, M.S. Dresselhaus, Raman spectroscopy in graphene, Physics Reports, Volume 473, Issues 5–6, 2009.
[13] Tang, L.; Liu, Y.; Liu, G.; Chen, Q.; Li, Y.; Shi, L.; Liu, Z.; Liu, X. A Novel SERS Substrate Platform: Spatially Stacking Plasmonic Hotspots Films. Nanoscale Res. Lett. 2019, 14, 94.
[14] Liu, G. Q.; Yu, M. D.; Liu, Z. Q.; Liu, X. S.; Huang, S.; Pan, P. P.; Wang, Y.; Liu, M. L.; Gu, G. One-Process Fabrication of Metal Hierarchical Nanostructures with Rich Nanogaps for Highly-Sensitive Surface-Enhanced Raman Scattering. Nanotechnology 2015, 26, 185702.
[15] Schlücker, S. Surface-Enhanced Raman Spectroscopy: Concepts and Chemical Applications. Angew. Chem., Int. Ed. 2014, 53, 4756 4795.
[16] Zeman, E. J.; Schatz, G. C. An Accurate Electromagnetic Theory Study of Surface Enhancement Factors for Silver, Gold, Copper, Lithium, Sodium, Aluminum, Gallium, Indium, Zinc, and Cadmium. J. Phys. Chem. 1987, 91, 634 643.
[17] S. A. Maier, Plasmonics: Fundamentals and Applications (Springer, NewYork, 2007).
[18] Lee, J.; Hua, B.; Park, S.; Ha, M.; Lee, Y.; Fan, Z.; Ko, H. Tailoring Surface Plasmons of High-Density Gold Nanostar Assemblies on Metal Films for Surface-Enhanced Raman Spectroscopy. Nanoscale 2014, 6, 616–623.

[19] Alexander Urich, Andreas Pospischil, Marco M. Furch, Danie Dietze, Karl Unterrainer, Thomas Mueller,. (2012). Silver nanoisland enhanced Raman interaction in graphene. Applied Physics Letters. 101. 153113. 10.1063/1.4758696.

Acknowledgments

K.B. sincerely thanks DST Mobility fellowship for overall scientific funding and support and for providing professional and personal stability. We specially thank Mr. Satish Kumar, senior technical assistant at IIST for helping us with the Raman measurements.

Tailored Synthesis of CN-PPy Composite for Enhanced Electrochemical Functionality

Vijaykumar S P*, Sapna S, Suresh D S, Ganesha H, Abdullah Ba Shbil.

Department of Physics, Mangalore University, Mangalagangothri-574199, India
*Corresponding author E-mail: vijaykumar34139@gmail.com

Abstract: This study details the synthesis of pure carbon nitride (CN) and composite polypyrrole through an in-situ polymerization method. The prepared samples were comprehensively characterized using Fourier Transform Infrared Spectroscopy (FT-IR), Field Emission Scanning Electron Microscopy (FESEM), and X-ray Diffraction (XRD). These analyses confirmed the chemical bonding, structural, and morphological characteristics of the products. Electrochemical measurements, conducted through cyclic voltammetry (CV) and Linear Sweep Voltammetry (LSV), revealed significant current rates in both anodic and cathodic peaks, indicating superior electrochemical properties. This work not only details the synthesis process but also provides a thorough understanding of the structural, morphological, and electrochemical aspects, exhibiting the materials' potential for various applications in electrochemical devices.

Keywords: Urea, cyclic voltammetry, Polypyrrole, Carbonization method.

1. Introduction

Significant progress has been made in science and nanotechnology worldwide in a number of fields in recent years. Among these are supercapacitors, solid state batteries, fuel cells, electrochemical sensors, bio-sensors, electrochemical corrosion, hydrogen evolution, bio-imaging, photovoltaic, biological electrochemistry, and electro-catalysis. Since the 20th century, most of the world's grid electricity has been produced using these technologies.[1] Supercapacitors, a type of energy storage device, fill a vital role in the energy storage industry by filling the void left by batteries and conventional capacitors. Their low cost, long cycle life, fast recharge capacity, high PD, and environmental friendliness have all drawn a lot of attention [2-3]. and electrochemical sensors are implanted or external devices that are frequently used as diagnostic tools to identify human and animal diseases, as well as to monitor environmental pollution, test the presence of drugs in food and drink, as a result, growing concerns about environmental damage, global warming, and the energy crisis have spurred the development of sustainable alternative energy sources like wind and solar energy. To bridge these areas, the initial focus is on certain conducting polymers, such as derivatives of polypyrrole (PPY), polyaniline (PANI), polythiophene (PTH), and poly 3,4 ethylene dioxy thiophene (PEDOT), which are widely used in various industrial applications and also used in our daily lives due to their promising advantages, such as high machinability, lightweight, and low cost compared to metal and ceramic materials. These materials are our primary focus for the

development of supercapacitor applications. Specifically, by adjusting the moieties and structures of side chains and backbones, polymer materials' properties can be precisely controlled to meet a range of application needs [4]. Conducting polymers, like polypyrrole, polyaniline, and poly(3,4-ethylene-dioxythiophene), have demonstrated great potential as active materials and electrodes for a variety of energy and electronic devices. And among various carbon materials, such as graphite and carbon nanotubes, the 2D material based on graphitic carbon nitride (CN) has generated the most discussion recently because of its remarkable performance as well as its unique architecture and properties [5–6]. Because of its high sensitivity to analysts, quick response to external stimulations, excellent fluorescence quenching abilities, light and electricity conversion properties, biocompatibility, and high stability, the CN nano-sheet is a promising candidate as a modified electrode for sensors to detect various substances. New types of sensors to detect various materials have also been revealed by the CN loaded with metal oxides [7]. This is because of their high levels of sensitivity, affordability, easy testing procedures, mobility, and fast reaction times. The precursors that are most commonly used to create CN are melamine, dicyandiamide, cyanamide, urea, thiourea, and ammonium thiocyanate [8]. The synthesis and physical properties of CN were first covered. After that, we talked about how CN can be combined with various 2D metal oxides, metal sulfides, and conducting polymers to increase its efficiency and make the composites useful for a range of electrochemical applications.
In this study, carbon nitride (CN) were synthesized via urea-assisted carbonization, while a polypyrrole composite was prepared through in-situ polymerization. Characterization using FTIR, XRD, and SEM confirmed the successful fabrication. Electrochemical properties were investigated through cyclic voltammetry (CV) and linear sweep voltammetry (LSV), providing insights into their performance. The combination of urea as a precursor and in-situ polymerization enhances the material's structural and morphological properties. This dual synthesis approach, along with detailed characterization and electrochemical analysis, offers a comprehensive understanding of the materials for potential applications in diverse electrochemical devices.

2. Experimental methods

2.1. Materials

Urea (NH_2CONH_2), Pyrrole monomer (C_4H_5N) (99%), Ferric chloride ($FeCl_3$) (97%), Methyl Orange ($C_{14}H_{14}N_3NaO_3S$) (85%) were bought from Sigma Aldrich and All the reagents were used as received without any purification.

2.2. Synthesis of CN

CN is prepared via simple pyrolysis method, ten gram of Urea was placed into alumina crucible with cover and heated at 550°C for 2 hours in a semi closed system, by ramping time of 2°C/min. when the sample was cooled to room temperature sample was removed from muffle furnace the resultant yellowish powder is the desired CN, which is collected and ground into powder form for further use.

2.3. Synthesis of CN/PPy

The micelles self-degraded template method was used to prepare the CN/PPy composite. Typically, 50 milliliters of distilled water were used to dissolve 0.15 grams of CN powder. The CN solution was then vigorously stirred continuously while 0.48 grams of FeCl3 and 0.098 grams of methyl orange were added. Following a half hour, 0.15 milliliters of pyrrole monomer were gradually added to the aforementioned solution via injection while being constantly stirred. To finish the polymerization reaction, which was indicated by a color change brown to black, the mixture was stirred constantly for 24 hours. The resulting CN/PPy residue was filtered through Whatman paper after 24 hours, repeatedly cleaned with distilled water, and then baked for an entire night at 60 °C.

3. Characterization

Chemical compositions were studied using the Bruker Alpha ATR FT-IR spectrometer, structural analysis was carried out using the Rigaku X-ray diffractometer (XRD), surface morphology was examined using Sigma Zeiss Field Emission Scanning Electron Microscopy (FESEM), and electrochemical performance was carried out using a PC-based CHI 660E electrochemical workstation.

4. Results and Discussion

4.1. FT-IR spectra and XRD analysis

Figure 1 (a) displays the FT-IR spectra of the CN/PPy composite and pure CN. (a) The breathing mode of the tri-s-triazine unit is represented by the peak in the CN spectra at 801 cm^{-1} [9]. The vibration stretching bands of C–N and C–N heterocycles are responsible for a number of peaks seen in the CN spectra, including peaks at 1201 cm^{-1}, 1404 cm^{-1}, and 1653 cm^{-1} [10,11]. The presence of PPy is confirmed by the characteristic peaks seen in the spectra of the CN/PPy composite at 692 cm^{-1}, 936 cm^{-1}, and 1089 cm^{-1}. In addition, a peak at 1201 cm^{-1} is seen, which suggests that the CN/PPy composite is in a deformed state as opposed to pure CN.

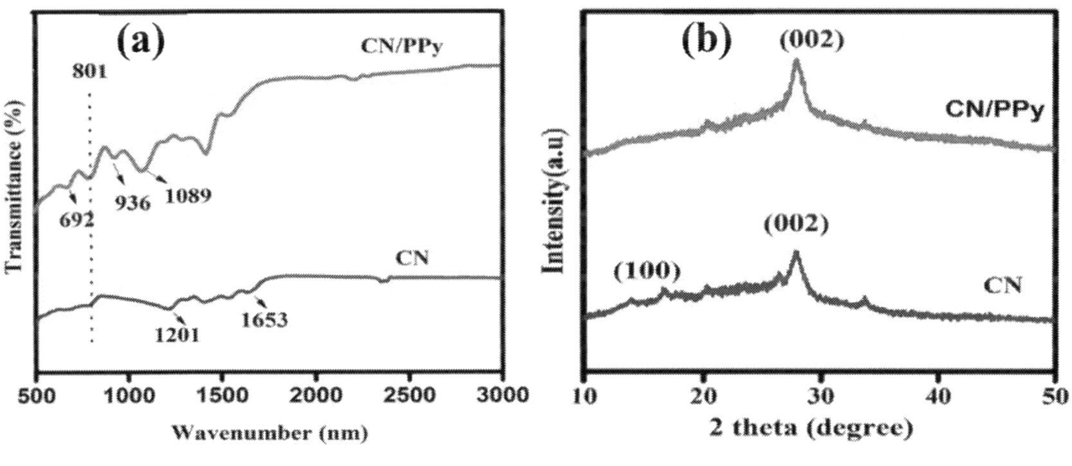

Figure 1: (a) FTIR spectra of CN and CN/PPy, (b)XRD Spectra of CN and CN/PPy

Figure 1 (b) shows the diffraction patterns of CN and CN/PPy composite XRD patterns of pure CN the CN sample exhibits a faint peak at 13.02◦, which is indexed as (100), [12] and represents the interlayer spacing in the in-plane direction of the repeating triazine units [13]. The dominant peak is a broad peak at approximately 27.8°. Similar XRD patterns can be seen in the CN/PPy composite, where a distinct peak can be seen around 27.8◦. The major diffraction peak at 27.8◦ corresponds to the (002) Bragg's reflection, indicating the interlayer spacing, and its presence validates the interlamellar stacking of the aromatic region in the composite [14]. Furthermore, it is clear from comparing the XRD patterns that the addition of PPy to the CN/PPy composite causes the broad peak to become a slightly sharp peak, indicating an improvement in electrochemical functionality.

4.2. FESEM Analysis

The FESEM images of pure CN that were used to analyze surface morphology are shown in Figure 2 (a) The prepared samples show plate-like structures instead of regular morphology [15]. The stacking morphology of CN is caused by van der Waals forces, which leads to a low specific surface area [16, 17].

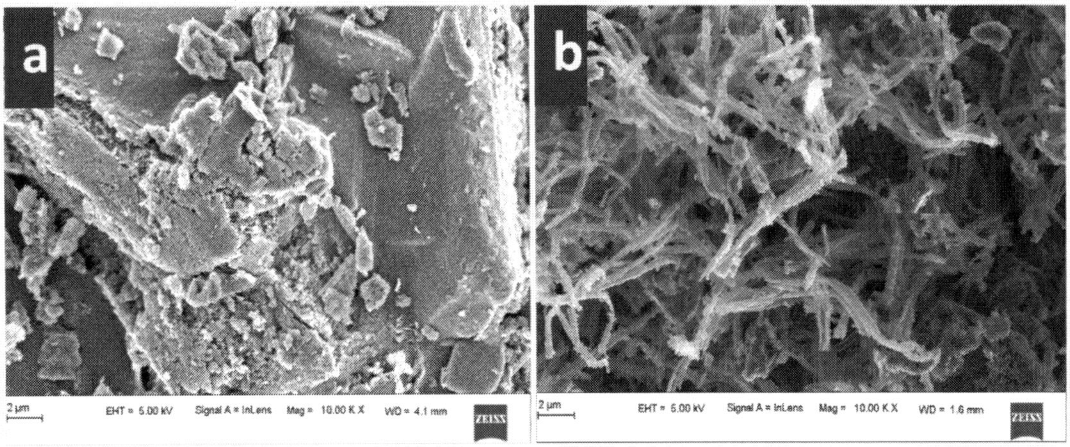

Figure 2: FESEM images of a) CN and b) CN/PPy

The agglomeration of CN particles on the PPy surface, which increases the composite material's specific capacitance, is depicted in Figure 2 (b) The PPy may increase specific capacitance because they have a larger surface area and use less mass. Another explanation for the decreased PPy could be that the CN particle clusters compress the nanotube by taking up space. Moreover, PPy may be compressed by the CN particles [18]. The observed morphological changes in the material may provide an opportunity to leverage its physical attributes, such as its large specific surface area, high conductivity, and good biocompatibility, of the CN/PPy composite. These characteristics aid in enhancing cyclic stability, raising the amount of energy stored, and speeding up the movement of hydrated ions [19].

4.3. Electrochemical performances

4.3.1. Cyclic voltammetry and Linear sweep voltammetry

The CHI 660E electrochemical workstation was used to perform the electrochemical performance. Two electrodes were used for the electrochemical measurements, and as Figure 3 illustrates, cyclic voltammetry can be used to determine the electrode's electrochemical performance in terms of reversible cathodic reduction and anodic oxidation in the voltage range of -0.8 to 0.8 V. (a) The reduction peak, which is connected to reduction in the cathodic zone and oxidation in the anode region, is clearly visible. The CN cyclic voltammetry displays a nearly rectangular-shaped area, but a smaller curve area is not beneficial for stable electrochemical performance. Greater specific capacitance and stable electrochemical behavior are caused by the larger curve integrated area, which is evident in the composite of CN/PPy Figure 3. (b) along with oxidation-reduction peaks. [20]

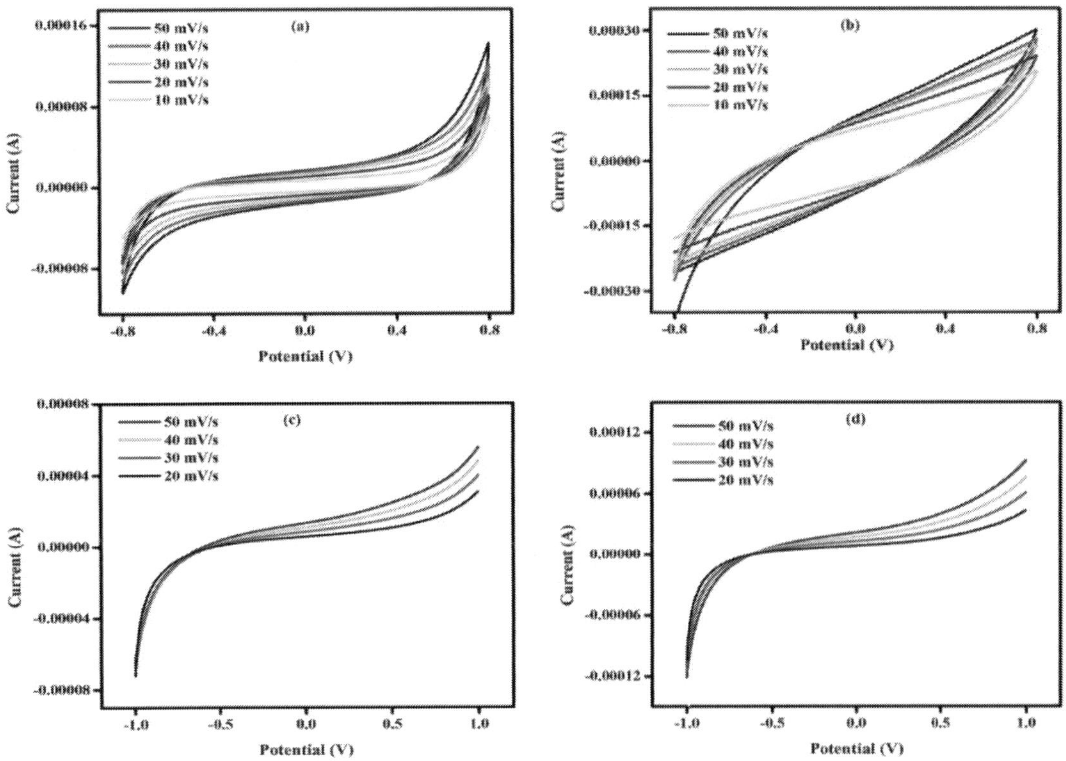

Figure 3: Cyclic voltammetry curves of (a) CN, (b) CN/PPy composite and linear sweep voltammetry of (c) CN (d) CN/PPy composite

The different scan rates applied to the CN, as shown in Figure 3, are analyzed by the LSV. (c) When comparing CN/PPy to CN, the I-V characteristic curve shows a linear increase in stability, which improves the superior I-V characteristics in the applied voltage range of -1 to 1 V. The LSV graph shows that the oxidation current increases in tandem with changing the sweep rate from 20 mV s-1 to 50 mV s-1. This suggests that the stability of the material is enhanced through electron diffusion, ions, and the charge transfer process, contributing to improved electrochemical performance.

5. Conclusion

In summary, this study successfully detailed the synthesis of pure carbon nitride (CN) and a composite with polypyrrole nanotubes using an in-situ polymerization method. Characterization through Fourier Transform Infrared Spectroscopy (FT-IR), Field Emission Scanning Electron Microscopy (FESEM), and X-ray Diffraction (XRD) affirmed the chemical bonding, structural, and morphological features of the prepared samples. The electrochemical measurements, conducted via cyclic voltammetry (CV) and Linear Sweep Voltammetry (LSV), demonstrated high current rates in both anodic and cathodic peaks, indicating superior electrochemical properties. This work contributes valuable insights into the synthesis, characterization, and electrochemical behavior of the fabricated materials, suggesting their potential for applications in diverse electrochemical devices.

Acknowledgment

The authors acknowledge Mangalore University for granting research fellowship and PURSE Lab., Mangalore University for providing FESEM characterization facility.

References

[1] Y. Fu, J. Zhu, C. Hu, X. Wu, and X. Wang. (2014). Nanoscale, vol. **6**, no. 21, pp.12555–12564.
[2] Y. Wang, Y. Song, and Y. Xia. (2016) Chem. Soc. Rev., vol. **45**, no. 21, pp. 5925–5950.
[3] A. Borenstein, O. Hanna, R. Attias, S. Luski, T. Brousse, and D. Aurbach. (2017) J. Mater. Chem. A, vol. **5**, no. 25, pp. 12653–12672.
[4] H. Shirakawa, E. J. Louis, A. G. MacDiarmid, C. K. Chiang, and A. J. Heeger. (1977). J. Chem. Soc. Chem. Commun., no. **16**, p. 578, 1977.
[5] K. Liao et al. (2016) J. Mater. Chem. A, vol. **4**, no. 15, pp. 5406–5409.
[6] C. Li, B. Xie, J. Chen, J. He, and Z. He. (2017) RSC Adv., vol. **7**, no. 22, pp. 13184–13190.
[7] T. Zhou and T. Zhang. (2021) Small Methods, vol**. 5**, no. 9, p. 2100515.
[8] Vijaykumar, S. P., Sapna, S., Suresh, D. S., Ganesha, H., Devendrappa, H. (2023). Diamond and Related Materials, **139**, 110257.
[9] M.B. Shekardasht, M.H. Givianrad, P. Gharbani, Z. Mirjafary, A. Mehrizad. (2020). Diam. Relat. Mater. **109,** 108008.
[10] R. Megha, Y.T. Ravikiran, S.V. Kumari, H.R. Prakash. (2018). Diam. Relat. Mater. **87,** 163–171.
[11] Q. Li, D. Xu, J. Guo, X. Ou, F. Yan. (2017). Carbon. **124,** 599–610.
[12] G. Palanisamy, K. Bhuvaneswari, M. Srinivasan. (2021). Diam. Relat. Mater. **118**,108540.
[13] L. Ye, S. Chen. (2016). Appl. Surf. Sci. **389,** 1076–1083.
[14] I. Rabani, R. Zafar, K. Subalakshmi, H.S. Kim, C. Bathula. (2021). J. Hazard. Mater. **407**,124360.
[15] X. Qu, S. Hu, P. Li, Z. Li, H. Wang, H. Ma, W. Li. (2018). Diam. Relat. Mater. 86159–166.
[16] S.-X. Zhou, X.-Y. Tao, J. Ma, L.-T. Guo, Y.-B. Zhu, H.-L. Fan. (2018). Vacuum. **149,** 175–179.
[17] J. Ma, X.-Y. Tao, S.-X. Zhou, X.-Z. Song, Yao-Wang Lin-Guo, (2019). J. Electroanal. Chem. **835,** 346–353.
[18] Z. Cai, H. Xiong, Z. Zhu, H. Huang, L. Li, Y. Huang, X. Yu. (2017) Synth. Met. **227,** 100–105.
[19] J. Liu, J. An, Y. Ma, M. Li, R. Ma, (2012). J. Electrochem. Soc. **159,** (6) A828–A833.
[20] Sharanappa S, Vijaykumar SP, Suresh DS, Shbil AB, Ganesha H, Devendrappa H. (2023). Journal of Energy Storage, **74**,109371.

IOP Conf. Series: Materials Science and Engineering 1300 (2024) 012010 doi:10.1088/1757-899X/1300/1/012010

Nilgiri tree peels-derived activated porous carbon for high performance supercapacitor applications

Abdullah Ba shbil *, Nagaraju Y S, Ganesha H, Veeresh S, Suresh D S, Vijaykumar S P, and Sapna Sharanappa.

Department of Physics, Mangalore University, Mangalagangothri, Mangalore-574199, India.
*Corresponding Author: aly72838@gmail.com

Abstract. In this study, the activated porous carbon (APC-x) was collected from Nilgiri tree peels, making it a sustainable and cost-effective green synthesis approach. The raw material undergoes carbonization and activating it using KOH as chemical activation agent. The chemical composition, crystallinity and morphology of the surface of ACP-x was examined by Fourier Transform infra-red (FTIR), X-ray diffraction and Field emission scanning electron microscope (FESEM). The cyclic voltammetry technique was used to measure the electrochemical performance of APC-x using PVA/KOH gel electrolyte in two-electrode system. The highest specific capacitance value 140 F/g was obtained for APC-0 at scan rate of 10 mV/s. The use of Nilgiri tree peels as precursor for activated porous carbon production presents a sustainable approach for developing high-performance supercapacitor materials.

1. Introduction

Activated porous carbon derived from biomass have gained significant attention for their potential use in supercapacitor applications due to its high surface area, regulated pore size distribution, and low cost in comparison to other carbon materials[1].Recently, biomass-derived activated carbon materials are considered a sustainable and cost-effective alternative to traditional carbon materials, which are typically derived from non-renewable sources such as coal or petroleum. The production of activated porous carbon from biomass involves the pyrolysis of biomass precursor at high temperatures, follwed by the activation of the resulting carbon material to create a highly porous structure with a large surface area. The activation process can be achieved through various methods, including chemical activation, physical activation, or a combination of both[2]. During the physical process, a significant amount of internal carbon is removed in order to impart well-developed carbon structures. In contrast, during the chemical activation process, dehydrating agents are utilized, which affects pyrolytic decomposition and causes the formation of tar, increasing the carbon yield[3]. In this paper, we prepared activated porous carbon by a simple process including carbonization of nilgiri tree peel with a chemical activation by KOH. It shown high-performance electrodes for supercapacitor applications. Nilgiri tree peels-derived carbon source is preferred due to their low cost, extensive availability, eco-friendly and renewable properties.

2. Experimental

2.1. Materials

The Nilgiri tree peels were obtained from nilgiri tree in Mangalore university (cumpus). Potassium hydroxide (KOH), charcoal, polyvinylidene fluoride (PVDF), were purchased from Sigma Aldrich, India. The solvents employed were distilled water (DW), hydrochloric acid (1M HCl), and N-Methyl-2-pyrrolidone (NMP).

2.2. Preparation of activated porous carbon derived from Nilgiri tree peels (APC-x)

Activated porous carbon was prepared by using Nilgiri tree peels as the carbon precursor, and chemical activation using KOH.The Nilgiri tree peels were first washed by using distilled water (DW) and dried at 90 °C for 24 h. Then, the peels were ground into a powder. After that 3 gm of Nilgiri tree peels powder were dissolved in 30 ml of 1 M KOH and heated at 100 °C for 15 h. The result was crushed and carbonization in a tube furnace at 450 °C in air atmosphere for 2 h. The product was filtered repeatedly with 1 M HCl and DW and dried in 100 °C for 10 h. Samples prepared by different mass ratio of Nilgiri tree peels powder to KOH (1:0.5 & 1:1) were labeled APC-0 and APC-1 respectively.

2.3. characterization

The surface morphology and the chemical change of the material were analyzed by Field Emission Scanning Electron Microscopy (Sigma Zeiss FESEM) and Fourier transformation infrared spectrometer (FTIR, ALPHA BURKUE model). The crystallization of the material was tested by the Rigaku Miniflex 500 (XRD). The CHI 660E model was used for electrochemical performance.

3. Results and duscussion

3.1. X-ray Diffraction Analysis

The XRD spectra of the activated porous carbon APC-0 and APC-1 samples are depicted in figure 1. All the samples show strong broad and weak peaks centered at 2θ values of 23.5° and 43° which ascribed to the reflection of (002) and (100) planes of graphitic carbon, respectively[4,5]. These strong peak suggest that the activated porous carbon contains some degree of graphite crystallite , which can contribute to its electrical conductivity and electrochemical properties. Furthermore, the APC-1 sample shows a new sharp peak at 2θ value of 29.5° which belongs to the graphite lattice of (110) plane[6]. The presence of sharp peak indicates to enhancement in crystallinity due to increase of KOH concentration.

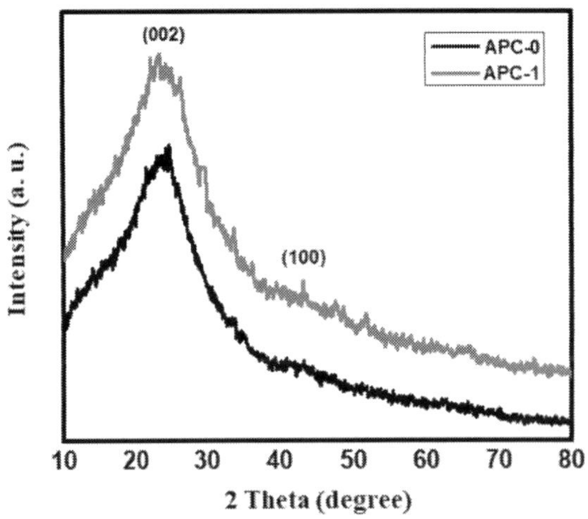

Figure 1. XRD patterns spectra of APC-0 and APC-1

3.2. FTIR Analysis

FTIR spectroscopy is a useful technique for analyzing the functional groups and chemical bonds present in the surface of activated porous carbon. Figure 2. shows the FTIR spectra of the APC-0 and APC-1 samles. It can be seen that the APC-0 and APC-1 samples display the FTIR characteristic bonds slightly similar. The peak around 3728 cm^{-1} represent the O-H stretching mode of hydroxyl groups. The strong peak centered around 1529 cm^{-1} associated with C-C or C-O stretching vibration in aromatic compounds[7]. The presence of strong band at 2342 cm^{-1} ascribed to the prsesent of CO2 from the atmosphere[8]. Additionally, the intensity of stretching peaks in the APC-1 sample is decreased with increase mass ratios of KOH. Moreover, the APC-0 sample shows new peak in the fingerprint region 1199 cm^{-1} represent C-O bonds[9]. The FTIR spectrum reveals the presence of hydroxyle groups, aromatic carbon, and various oxygen functinalities in the surface of activated porous carbon samples which endow it with suitable characteristics for supercapacitor electrode.

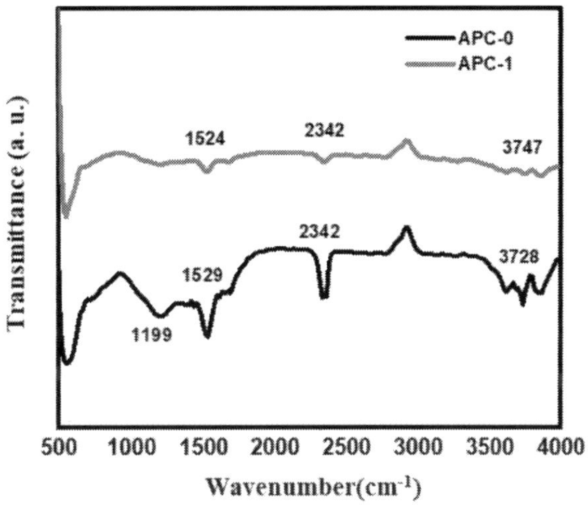

Figure 2. FTIR spectra of APC-0 and APC-1

3.3. FESEM Analysis

The surface structure and morphology of Nilgiri tree peels-derived activated porous carbon were characterized by FE-SEM. Both the activated porous carbon APC-0 and APC-1 samples displayed slight similar irregular porous structure with a smooth surface morphology as dipected in Fig. 3(a-b). Furthermore, the pore size of the APC-1 sample was increased as result to increase of mass ratio of KOH as shown in Fig. 3(b). The pore size can be controlled during the activation process, allowing for the optimization of the ion transport and charge storage capacity of the material[10].

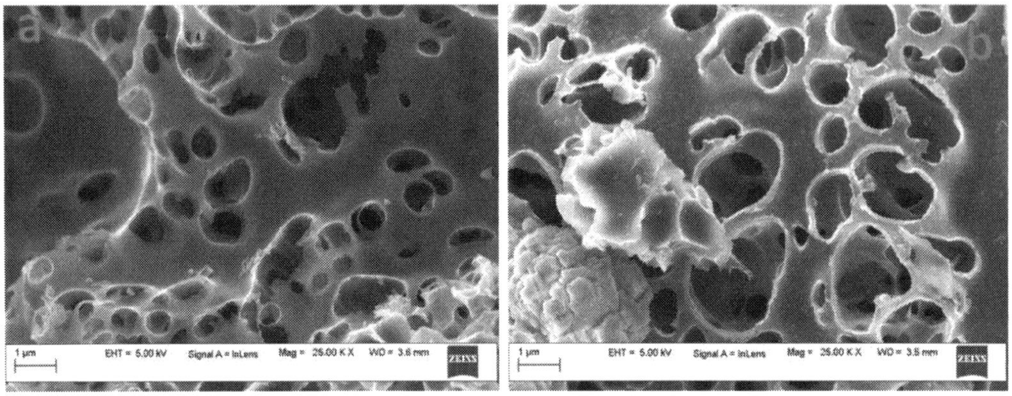

Figure 3. FESEM image of (a) APC-0 and (b) APC-1

3.4. Cyclic Voltammetry

The electrochemical properties of the activated porous carbon were tested in two-electrode system by using PVA/KOH gel electrolyte. Fig. 4(a&b) display the cyclic voltammetry (CV) curves of the APC-0 and APC-1 electrodes in the potential windows from -0.4 to 0.4 V at various scan rates from 10 to 50 mV/s. Both the electrodes display a quasi-rectangular shape with a rapid current response, indicating high performance of electrical double layer capacitor[11]. Additionally, the CV curves of the APC-0 demonstrate the highest current density area. The specific capacitance of APC-0 and APC-1 electrodes is 140 F/g and 110 F/g at scan rate of 10 mV/s respectively. The specific capacitance decreases with increasing the activation agent (KOH) ratio from 0.5 to 1.

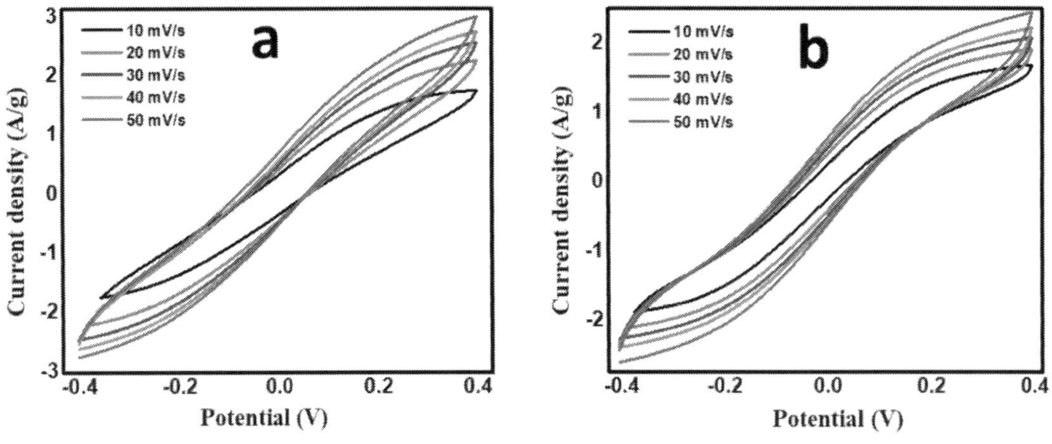

Figure 4. Cyclic voltammetry of (a) APC-0 and (b) APC-1

4. Conclusions

In this study, we used Nilgiri tree peels as precursor to produce activated porous carbon (APC-x) by activation process using KOH. The characterization techniques used for activated porous carbon include X-ray diffraction (XRD), Fourier Transform infra-red (FTIR), and Field Emission scanning electron microscopy (FE-SEM). These techniques provide information about the crystallinity, chemical composition and morphology of the APC-x surface. The APC-0 exhibits high specific capacitance of 140 F/g at scan rate of 10 mV/s. The results show that the activated porous carbon derived from Nilgiri tree peels is a promising material for supercapacitor applications due to its high specific capacitance, and environmentally friendly approach that can reduce waste and lower production costs.

Acknowledgement
Thanks to PURSE lab for providing characterization facility. And also acknowledged to University of Aden [(310) 2021] for providing fellowship to my Ph. D. program.

References

[1] P. Veerakumar, T. Maiyalagan, B. G. S. Raj, K. Guruprasad, Z. Jiang, and K. C. Lin, "Paper flower-derived porous carbons with high-capacitance by chemical and physical activation for sustainable applications," *Arab. J. Chem.*, vol. 13, no. 1, pp. 2995–3007, 2020, doi: 10.1016/j.arabjc.2018.08.009.

[2] C. K. Ranaweera, P. K. Kahol, M. Ghimire, S. R. Mishra, and R. K. Gupta, "Orange-Peel-Derived Carbon: Designing Sustainable and High-Performance Supercapacitor Electrodes," *C*, vol. 3, no. 4, p. 25, 2017, doi: 10.3390/c3030025.

[3] M. Natalia, Y. N. Sudhakar, and M. Selvakumar, "Activated carbon derived from natural sources and electrochemical capacitance of double layer capacitor," *Indian J. Chem. Technol.*, vol. 20, no. 6, pp. 392–399, 2013.

[4] H. Ma, J. Yang, X. Gao, Z. Liu, X. Liu, and Z. Xu, "Removal of chromium (VI) from water by porous carbon derived from corn straw: Influencing factors, regeneration and mechanism," *J. Hazard. Mater.*, vol. 369, no. October 2018, pp. 550–560, 2019, doi: 10.1016/j.jhazmat.2019.02.063.

[5] M. Usha Rani, K. Nanaji, T. N. Rao, and A. S. Deshpande, "Corn husk derived activated carbon with enhanced electrochemical performance for high-voltage supercapacitors," *J. Power Sources*, vol. 471, no. April, p. 228387, 2020, doi: 10.1016/j.jpowsour.2020.228387.

[6] N. S. Ammar, N. A. Fathy, H. S. Ibrahim, and S. M. Mousa, "Micro-mesoporous modified activated carbon from corn husks for removal of hexavalent chromium ions," *Appl. Water Sci.*, vol. 11, no. 9, pp. 1–12, 2021, doi: 10.1007/s13201-021-01487-1.

[7] Y. S. Nagaraju, H. Ganesh, S. Veeresh, H. Vijeth, and H. Devendrappa, "A strategy of making waste profitable: Self-templated synthesis of helical rod structured porous carbon derived from Ganoderma Lucidem for advanced supercapacitor electrode," *Diam. Relat. Mater.*, vol. 131, p. 109607, 2023, doi: 10.1016/j.diamond.2022.109607.

[8] Y. S. Nagaraju, H. Ganesh, S. Veeresha, H. Vijeth, and H. Devendrappa, "Synthesis of hierarchical ZnO/NiO nanocomposite Wurtz hexagonal nanorods via hydrothermal for high-performance symmetric supercapacitor application," *J. Energy Storage*, vol. 56, no. PA, p. 105924, 2022, doi: 10.1016/j.est.2022.105924.

[9] S. J. Rajasekaran and V. Raghavan, "Facile synthesis of activated carbon derived from Eucalyptus globulus seed as efficient electrode material for supercapacitors," *Diam. Relat. Mater.*, vol. 109, no. February, p. 108038, 2020, doi: 10.1016/j.diamond.2020.108038.

[10] S. Yang, S. Wang, X. Liu, and L. Li, "Biomass derived interconnected hierarchical micro-meso-macro- porous carbon with ultrahigh capacitance for supercapacitor," *Carbon N. Y.*, vol. 147, pp. 540–549, 2019, doi: 10.1016/j.carbon.2019.03.023.

[11] Y. Wang *et al.*, "Self-Templated Synthesis of Hierarchically Porous N-Doped Carbon Derived from Biomass for Supercapacitors," *ACS Sustain. Chem. Eng.*, vol. 6, no. 11, pp. 13932–13939, 2018, doi: 10.1021/acssuschemeng.8b02255.

IOP Conf. Series: Materials Science and Engineering 1300 (2024) 012011 doi:10.1088/1757-899X/1300/1/012011

Dielectric Analysis of Amine Modified Nano-graphite: Effects of Thickness, Temperature and Frequency

Bhanu Pratap Singh[1], Pushpendra Kumar[2] and S. P. Mahapatra[1]*
[1]Department of Chemistry, National Institute of Technology Raipur 492010 India
[2]Department of Chemistry, Guru Ghasidas Vishwavidyalaya, Bilaspur 497101 India
*Corresponding Author: spmahapatra.chy@nitrr.ac.in

Abstract. Amine modified nano-graphite (NH_2-NG) has been produced from nano-graphite (NG) powder via liquor ammonia (NH_3) aqueous solution by microwave, ultrasonic methods. The surface modified NG have been characterized by X-ray diffraction (XRD), scanning electron microscopy (SEM) and Fourier-transform infrared spectroscopy (FTIR) techniques which clearly demonstrate that synthesis of NH_2-NG was successful. The crystalline characteristics of amine modified nano-graphite powder have been studied through XRD technique. The morphology structure of synthesized surface modified nanoparticles has been determined by SEM analysis. FTIR spectra confirmed the existence of amine groups on the surface of produced nano-graphite powder. NH_2-NG pellet was prepared to analyse its dielectric properties as a function of thickness (1, 2, 3, 5 and 7 mm), frequency (range 10^{-2}-10^{5} Hz) and temperature (25, 50, 75 and 100 °C). As a function of frequency, the dielectric loss tangent and permittivity each exhibit a distinct relaxation peak that grows up to a thickness of 5 mm then decreasing. Because nanoparticles are capacitive, the capacitance values of NG and NH_2-NG increase with thicknesses up to 5 mm and drop with frequency. The drop in dielectric permittivity of NG and NH_2-NG pellets after 5 mm may be caused by an increase in resistivity. Positive temperature coefficient (PTC) of increased conductance is confirmed by the continuous increase in electrical conductivity of NG and NH_2-NG with temperature. The degree of temperature has a significant impact on the dielectric loss tangent, dielectric permittivity, and electrical conductivity of NG, and NH_2-NG.

1. Introduction

Different method has been reported for the surface modification of graphite oxide powder. Among them, introduction of surface modification of graphite oxide via reductive modification is considered as one of the most effective routes. in the literature, several reducing agents have been employed including ammonia solution (25 wt% in water) by one-pot solvothermal process [1], Graphite with an -NH_2 group was created using vacuum ultraviolet-induced photochemistry when ammonia was present [2]. However, due to their instrumental setup, some of the aforementioned approaches are costly and necessitate significant attention because multiple sensitive parameters must be maintained during the functionalization process. For this reason, it is thought that a straightforward wet chemical method offers a practical, affordable, and easy way to directly attach the -NH_2 group to the surface of nano-graphite. No such procedure has, as far as we are aware, been documented. Hydrazine-induced reduction of graphite and graphene oxides [3] Graphene oxide (GO) sheets, the layered structure of graphite oxide, are decorated on their basal planes and edges with a variety of oxygen functional groups, which significantly alter the van der Waals interactions between the layers and give GO its strong hydrophilicity. Thus, Surface modification of graphite oxide by hydrazine reduction [4] by simple ultrasonication and stirring. By adding improved NG to different kinds of polymers and other materials, composite materials can have their conductivity, tensile strength, and elasticity increased. They can be used as the conductor substantial for batteries, capacitors, double-layer capacitors, and improved microelectronic devices [5-8]. Capacitors use dielectric materials with low losses and high permittivity to store more electrical energy [9, 10]. Graphite is a well-known type of carbon that has a planar and layered structure. Because of its exceptional optical, thermal, electrical, and power properties as well as its distinct two-dimensional structure, graphene is now a viable substitute for many different applications, including photovoltaic, catalysis, nano-electronic, and energy conversion and storage [11–13]. Dielectric materials are used in the fabrication of microelectronics for large-scale power applications and energy storage devices [14]. Lyddane-Sachs-Taller states that lattice vibrations create a shift in the intrinsic dipole moment, which increases the permittivity to a very high value [15]. There aren't many studies on graphene oxide or graphene-based material composites' dielectric analysis in the literature [16, 17]. According to Verdejo et al.'s research, graphene oxide can be uniformly distributed throughout the polymer cross-section to achieve a satisfactory increase in dielectric permittivity [18]. The earliest techniques to synthesise graphene nanosheets from bulk graphite were the scotch-tape technique and epitaxial chemical vapours deposition [19, 20].

In the current work, graphite powder was used to create NG powder, and an aqueous liquid ammonia (NH_3) solution was used to modify the powder's surface. The morphological structure of NG and NH_2-NG was examined using SEM and XRD analysis. With the use of FTIR spectroscopy, the amine group of altered NG was identified. For use in various sectors, the dielectric relaxation spectra (DRS) of NG and NH_2-NG pellets was investigated. Capacitance, dielectric permittivity, dielectric loss tangent and electrical conductivity measurements were among the tasks included in this.

2. Experimental

2.1 Compounds

Graphite powder (size < 20 μm), liquor NH_3 (98%).

2.2 Preparation of nano-graphite

Synthesis of nano-graphite by graphite powder already discuss to our previous paper [21].

2.3 preparation of amine modified nano-graphite

A quantity of 2 gm of prepared nano graphite was reduced by liquor ammonia aqueous solution (20 ml liquor NH_3, 80 ml distilled water), accompanied by ultrasonic treatment for more than 24 h and then filtered many times and dried to 80 °C at 24 h. The resultant amine modified nano graphite. powder was finally dried in a vacuum oven at 120◦C.

3. Measurements

3.1 Characterization Techniques

Using a SEM (ZEISS EVO-50, Germany), the morphological properties of the synthesised NG and NH_2-NG powder were investigated. FTIR spectroscopy (Bruker Scientific Instrument, US) was used to analyse the amine groups in modified NG. X-ray diffraction (XPert Pro PANalytical, Netherlands) was used to confirm the crystal structures of the generated NG and NH_2-NG. In the 10-2-105 Hz frequency range, the dielectric characteristics of NG and NH_2-NG pellets were examined using the Hi-Tester LCR metre, Hioki 3533. Based on measurements of the dielectric, the following relationship has been used to evaluate electrical conductivity:

$$\sigma AC = \omega\varepsilon^0\varepsilon' \tan\delta \quad (1)$$

Where, $\omega = 2\pi f$ (f is frequency), ε' = dielectric constant (Cp/C_0), ε^0 = vacuum permittivity and $\tan\delta$ = dielectric loss tangent [22].

4. Results and Discussion

4.1 Scanning Electron Microscopy

SEM operating at 8.5 mm and 20 kV in high vacuum mode was used to examine the morphology of prepared NG and NH_2-NG. Figure 1(a) and (b) display SEM images of NG and NH_2-NG. As can be seen from Figure 1(a), NG has a smooth surface and a crumpled shrill sheet structure, whereas Figure

1(b) displays a wrinkled and rough surface because NH_2-NG's surface contains amine groups.

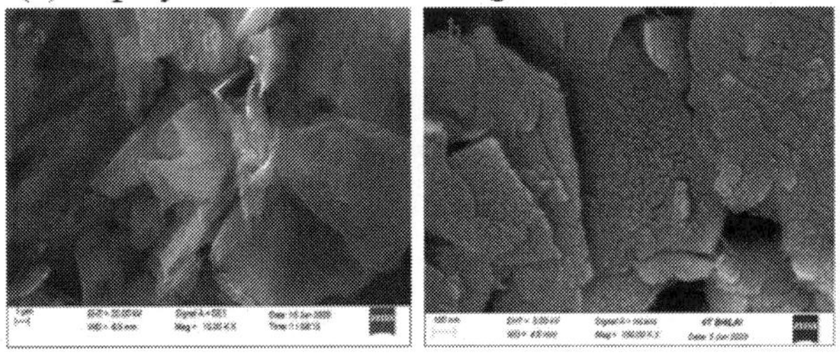

Figure 1 SEM image of (a) NG (b) NH_2-NG.

4.2 X-Ray Diffraction

Using the XRD method, the crystal structures of produced NG and NH_2-NG were identified. Figure 2 shown the XRD spectra of produced NG and NH_2-NG. NG and NH_2-NG display 2θ at 26.46° and FWHM of 1.9441 with 0.12 nm d-spacing [23]. NG and NH_2-NG have a crystalline size of 4.34 nm according to the Scherer and Bragg equations [24]. There are no structural changes, but the peaks have widened and become less intense, indicating a decrease in particle size.

$$Dp = (0.94\times \lambda) / (\beta\times \cos\theta) \quad (2)$$

Where, Dp is Average Crystallite size and β is Line broadening in radians,

Bragg's equation:

$$n\lambda = 2d\sin\theta \quad (3)$$

Where, n is Diffraction series, λ is X-Ray wavelength, d is Interlayer spacing and θ is Bragg angle,

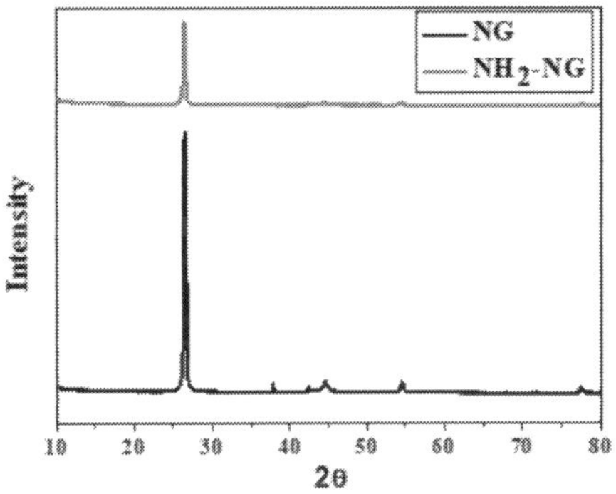

Figure 2 XRD of NG and NH_2-NG.

4.3 Fourier-Transform Infrared Spectroscopy

Using FTIR spectroscopy, the various functional groups of amines that changed NG were examined. Figure 3 shown the FTIR data of NH_2-NG. The results demonstrated that whereas NG exhibits no absorption peaks, NH_2-NG has those that are associated with particular functional groups. A portion of the absorption peaks at 3390.31 cm^{-1} are caused by the N-H stretching vibration. The FTIR spectrum of NH_2-NG demonstrates the effective amine modification of graphite powder. The C=C stretching vibration is attributed to the peak at 1625.57 cm^{-1}, whereas two new nitrogen-related peaks at 1019.18 cm^{-1} and 1619.56 cm^{-1} were seen. The amide's C–N stretching and N–H bending bonds can be roughly attributed to these peaks, respectively.

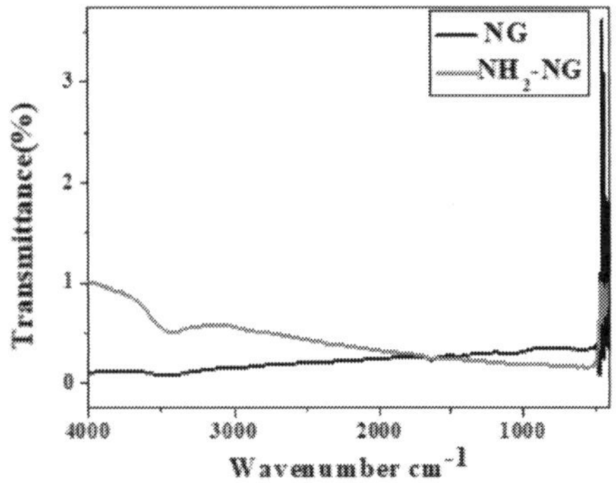

Figure 3 FTIR of NG and NH_2-NG

4.4 Dielectric Loss Tangent (tan δ)

The tan δ of synthesised NG and NH_2-NG has been calculated with frequency at different thicknesses (1 to 7 mm) at 25°C temperature and different temperatures (25°C, 50°C, 75°C, and 100°C) at 3mm thickness is shown in Figure 4 (a) and (b) Figure 5 (a) and (b) respectively. The tan δ rises with frequency at all thicknesses and temperatures of NG and NH_2-NG pellets; however, this rise does not continue after reaching a peak; rather, it decreases with frequency because polar groups are present in the prepared NH_2-NG. Additionally, it is noted that when amine functional groups are present, the peak of NH_2-NG shifts. The primary determinant of the relaxation behaviour of NG and NH_2-NG pellets at room temperature is their thickness. The maximum tan δ rises with NG and NH_2-NG pellet thickness up to 5 mm; this may be because of charge conduction and dipole relaxation that has occurred [25] and reductions for a thickness of 7 mm, which may be because of resistivity and dipole fluctuation that is increasing within NG and NH_2-NG pellets [26]. As temperature rises (Figure 5(a)), the maximum tan δ value increases towards lower to higher frequency; but, when temperature rises (Figure 5(b)), the maximum tan δ value increases towards higher to lower frequency. Temperature affects how frequently this maximum dielectric loss tangent manifests.

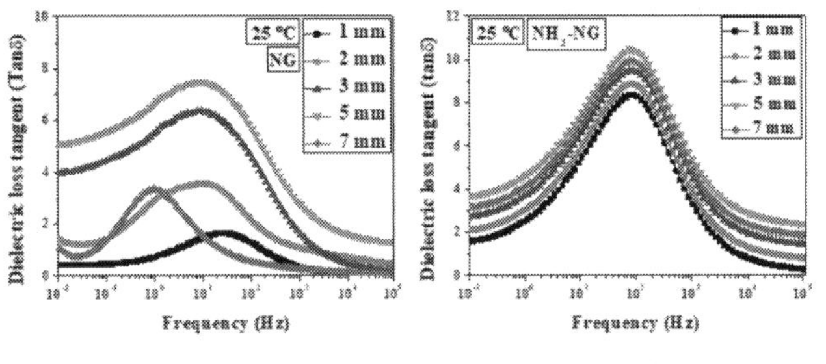

Figure 4 Dielectric loss tangant of (a) NG and (b) NH_2-NG effect of thickness.

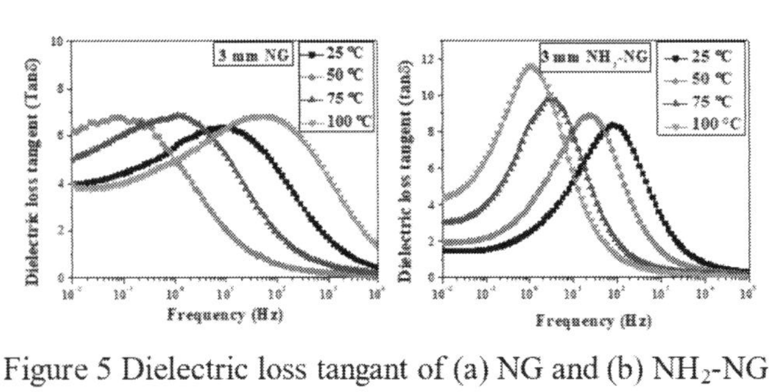

Figure 5 Dielectric loss tangant of (a) NG and (b) NH_2-NG effect of temperature.

4.5 Dielectric Permittivity (ε')

The ε' of a substance indicates the level of electrical polarisation. Synthesised NG and NH_2-NG nanoparticle thickness variations with frequency at 1, 2, 3, 5, and 7 mm are depicted in Figure 6(a) and (b) at room temperature and at various temperatures (25, 50, 75, and 100 °C) in Figure 7(a) and (b). The dielectric permittivity of modified NG grows with thickness and decreases with frequency as a result of the functional groups' ionic convergence and transmitting aerial change [27]. The graphic illustrates how the thickness of the NG and NH_2-NG pellets causes a rise in dielectric permittivity. The highly conductive character of the NG and NH_2-NG nanoparticles, as well as the positive temperature coefficient (PTC) impact [28], are the reasons behind the observation that dielectric permittivity rises with temperature (Figure 7(a) and (b)). The high electrical conductivity of NG and NH_2-NG and the ferroelectric character of the conductive particles may be the cause of the greater dielectric permittivity at low frequencies [29, 30].

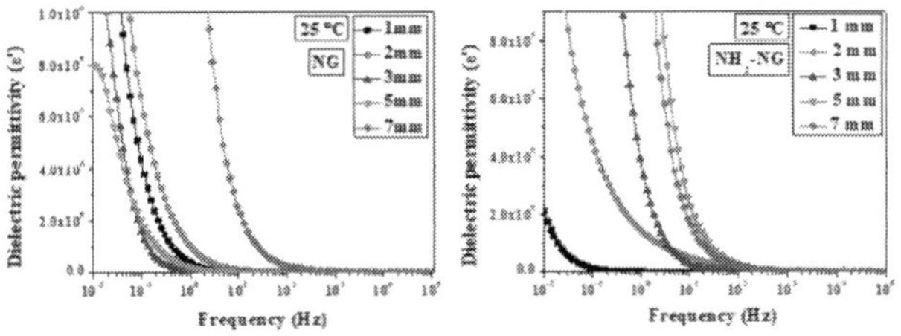

Figure 6 Dielectric permittivity of (a) NG and (b) NH_2-NG effect of thickness.

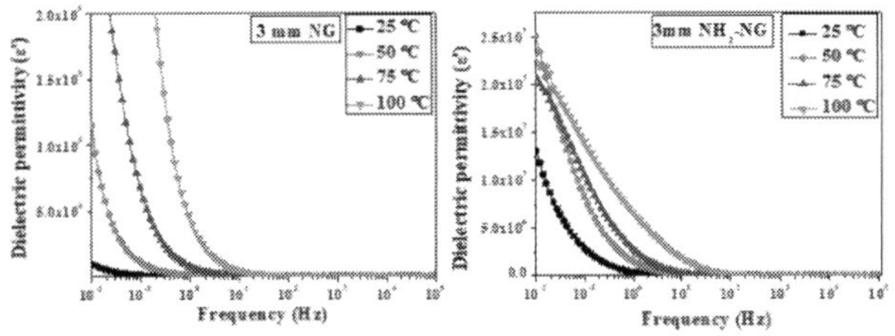

Figure 7 Dielectric permittivity of (a) NG and (b) NH_2-NG effect of temperature.

4.6 Electrical Conductivity (σ)

Electrical conductivity is the ability to carry electrical current. Figures 8(a) and (b) illustrate how the σ of NG and NH_2-NG pellets varies with frequency at various thicknesses. Both the thickness of NG and NH_2-NG pellets and their frequency cause a constant rise in σ. The image illustrates how the thickness of NG and NH_2-NG pellets up to 5 mm increases σ due to nanoparticle impacts creating a continuous conduit for electrons to traverse across the volume of the sample. The reduction in σ at a thickness of 7 mm could be attributed to the creation of agglomerates of nanoparticles and the deformation of continuous paths. Additionally, a drop in the dielectric permittivity and loss tangent may be the cause of the electrical conductivity decline [31]. It is also observed that σ of NG and NH_2-NG increases continuously with increasing temperature (Figure 9 (a) and (b)). This validates the greater conductance PTC effect. The curve displays both a DC and AC conductivity region up to 50 °C, but as temperature rises, it only displays AC conductivity. Thermal activation of charge carriers results in an increase in AC conductivity, and this rise in electrical conductivity with temperature is typically the result of the net effect [32].

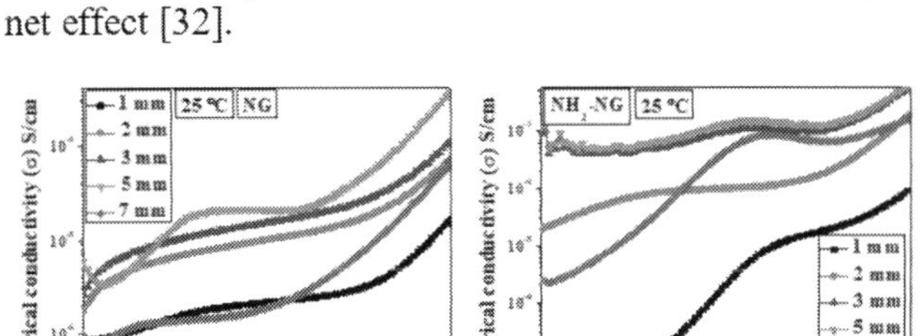

Figure 8 Electrical Conductivity of (a) NG and (b) NH_2-NG effect of thickness.

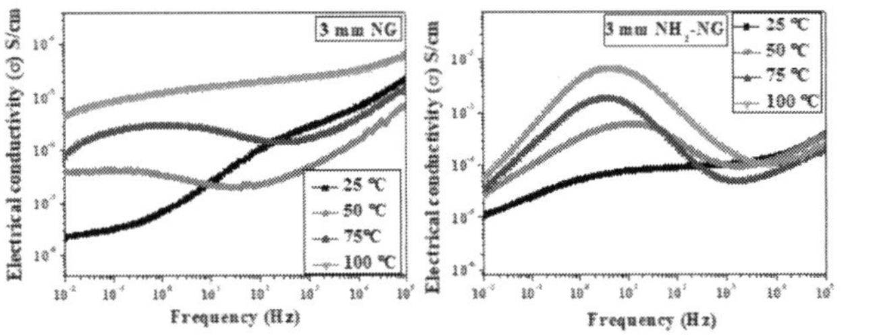

Figure 9 Electrical Conductivity of (a) NG and (b) NH_2-NG effect of temperature.

4.7 Percolation Behaviour

The percolation behaviour of the σ of NG and NH_2-NG pellets with thickness at various frequencies (1 to 100 kHz) is shown to vary in Figures 10(a) and (b). The σ of the NG and NH_2-NG pellets grows exponentially to a thickness of 5 mm, after which it decreases to a thickness of 7 mm. The agglomeration of nanoparticles may be the cause of this [33]. As shown in Figures 11(a) and (b), the σ of NG and NH_2-NG pellets can be divided into three zones. The first zone is inductive; electrical conductivity progressively increases with thickness as charge nanoparticle density falls. The third zone, known as the saturation region, is when the agglomerate's production of NG and NH_2-NG nanoparticles results in either a minor increase in σ or a decrease in it. The constant development of conductive paths results in a dramatic increase in σ in the second region, known as the percolation region. The percolation thresholds for NG and NH_2-NG pellets have been found to vary in thickness from 4 to 5 mm.

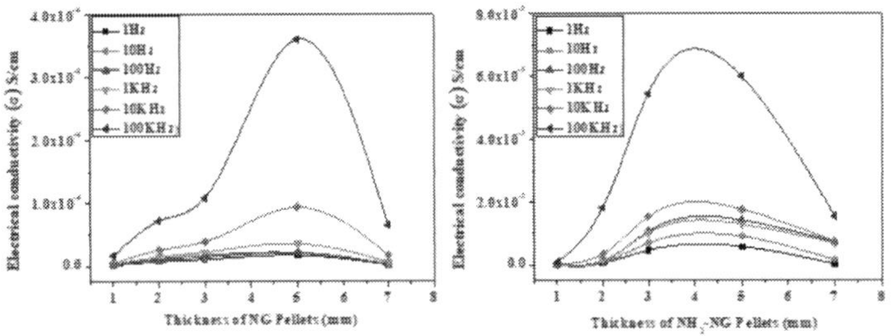

Figure 10 Percolation behaviour of (a) NG and (b) NH_2-NG.

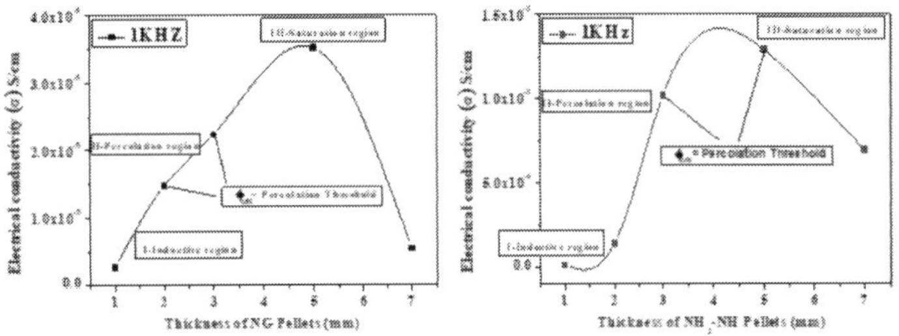

Figure 11 Electrical conductivity of (a) NG and (b) NH_2-NG.

5. Conclusions

Graphite powder was successfully converted into nano-graphite and amine modified nanoparticles using the oxidation process. The surface morphology of NG and NH_2-NG is confirmed by SEM pictures. The existence of amine groups on the surface of the manufactured NG powder was verified by FTIR measurements. The interlayer spacing of 0.12 nm is indicated by 2θ at 26.46° in the XRD data of NG and NH_2-NG. The inclusion of ionic groups in the modified NH_2-NG causes peak shifts and relaxation peaks with frequency in the tan δ spectra. Dielectric permittivity increases with thickness and temperature and decreases with frequency; these changes may be caused by the functional groups of modified NH_2-NG's charge orientation and dipole moment modification. Because conducting nanoparticles are ferroelectric, they have a larger dielectric permittivity at low frequencies. The thickness and temperature of NG and NH_2-NG pellets, along with frequency, all contribute to the ongoing rise in σ. The σ rises with NG and NH_2-NG particle thickness up to 5 mm, after which its reductions to 7 mm. This could be because the loss tangent and dielectric permittivity are decreasing. It has been determined that the percolation thresholds of NG and NH_2-NG pellets fall between 4-5 mm thicknesses. These findings show that NG and NH_2-NG nanoparticles have tremendous potential for use in energy storage devices, sensors, and electrical applications based on high dielectric.

Acknowledgements
The directors of the National Institute of Technology in Raipur are appreciated by the authors for their facilities and assistance.

References

[1] Zhang C, Hao R, Liao H and Hou Y 2013 *Nano Energy*. Jan **2** 88
[2] Girard-Lauriault PL, Illgen R, Ruiz JC, Wertheimer MR and Unger WE 2012 *Appl. Surf. Sci.* **258** 8448
[3] Park S, An J, Potts JR, Velamakanni A, Murali S and Ruoff RS 2011 *carbon*. **49** 3019
[4] Ren PG, Yan DX, Ji X, Chen T and Li ZM 2010 *Nanotechnology*. **22** 055705
[5] Dou L, Cui F, Yu Y, Khanarian G, Eaton SW, Yang Q, Resasco J, Schildknecht C, Schierle-Arndt K and Yang P 2016 *Acs Nano*. **10** 2600
[6] Hsu T, Epting WK, Mahbub R, Nuhfer NT, Bhattacharya S, Lei Y, Miller HM, Ohodnicki PR, Gerdes KR, Abernathy HW and Hackett GA 2018 *J. Power Sources*. **386** 1
[7] Schönhals A and Kremer F 2003 Analysis of dielectric spectra. In Broadband dielectric spectroscopy 59 Berlin, Heidelberg: *Springer Berlin Heidelberg*.
[8] Kumar P, Penta S and Mahapatra SP 2019 *Integr. Ferroelectr.* **202** 41
[9] Wu Y, Wang Z, Shen X, Liu X, Han NM, Zheng Q, Mai YW and Kim JK 2018 *ACS Appl. Mater. Interfaces*. **10** 26641
[10] Homes CC and Vogt T 2013 *Nat. Mater.* **12** 782
[11] Hao X 2013 *J. Adv. Dielectr.* **3** 1330001
[12] He L, Neaton JB, Cohen MH, Vanderbilt D and Homes CC 2002 *Phys. Rev. B*. **65** 214112
[13] Gao Y and Hao P 2009 *Physica E Low Dimens. Syst. Nanostruct.* **41** 1561
[14] Berger C, Song Z, Li X, Wu X, Brown N, Naud C, Mayou D, Li T, Hass J, Marchenkov AN and Conrad EH 2006 *Science*. **312** 1191
[15] Stankovich S, Dikin DA, Dommett GH, Kohlhaas KM, Zimney EJ, Stach EA, Piner RD, Nguyen ST and Ruoff RS 2006 *nature*. **442** 282
[16] De Wild J, Meijerink A, Rath JK, Van Sark WG and Schropp RE 2011 *Energy Environ. Sci.* **4** 4835
[17] Liu Y, Dong X and Chen P 2012 *Chem. Soc. Rev.* **41** 2283
[18] Hu W, Liu Y, Withers RL, Frankcombe TJ, Norén L, Snashall A, Kitchin M, Smith P, Gong B, Chen H and Schiemer J 2013 *Nat. Mater.* **12** 821
[19] Ramanathan T, Abdala AA, Stankovich S, Dikin DA, Herrera-Alonso M, Piner RD, Adamson DH, Schniepp HC, Chen XR, Ruoff RS and Nguyen ST 2008 *Nat. Nanotechnol.* **3** 327
[20] Dikin DA, Stankovich S, Zimney EJ, Piner RD, Dommett GH, Evmenenko G, Nguyen ST and Ruoff RS 2007 *Nature*. **448** 457
[21] Singh BP, Kumar P and Mahapatra SP, 2023 *Ferroelectrics*. **51** 617
[22] Kremer F and Schönhals A 2002 editors. Broadband dielectric spectroscopy. *Springer Science & Business Media*
[23] Shalaby A, Nihtianova D, Markov P, Staneva AD, Iordanova RS and Dimitriev YB 2015 *Bulg. Chem. Commun.* **47** 291
[24] Li J, Zeng X, Ren T and Van der Heide E 2014 *Lubricants*. **2** 137
[25] Sebastian MT, Silva MA and Sombra AS 2017 *Microwave materials and applications 2V set*. **5** 1
[26] Kumar P, Penta S and Mahapatra SP 2020 *J. Electron. Mater.* **49** 5801
[27] Tiwari SK, Choudhary RN and Mahapatra SP 2015 *High Perform. Polym.* **27** 274
[28] Saji J, Khare A and Mahapatra SP 2015 *Fibers Polym.* **16** 883

IOP Conf. Series: Materials Science and Engineering 1300 (2024) 012011 doi:10.1088/1757-899X/1300/1/012011

[29] Pastor M, Bajpai PK and Choudhary RN 2007 *J. Phys. Chem. Solids.* **68** 1914
[30] Kurchak AI, Eliseev EA, Kalinin SV, Strikha MV and Morozovska AN 2017 *Phys. Rev. Appl.* **8** 024027
[31] Das PS, Chakraborty PK, Behera B and Choudhary RN 2007 *Phys. B: Condens. Matter.* **395** 98
[32] Panda S, Goswami S and Acharya B 2019 *J. Electron. Mater.* **48** 2853
[33] Xu D, Sridhar V, Mahapatra SP and Kim JK 2009 *J. Appl. Polym. Sci.* **111** 1358

IOP Conf. Series: Materials Science and Engineering 1300 (2024) 012012 doi:10.1088/1757-899X/1300/1/012012

Synthesis and Characterization of Barium Hexagonal Ferrite ($BaFe_{12}O_{19}$)

Naveen C. S[1*] Anand Rohidas[1,2], Raghu S[2], B. N Havareddi[2], Ramanna R[2], S. Mattyappannavar[2]

[1]Department of Physics, Presidency University, Bengaluru

[2]Department of Physics, K.L.E Society's Basavaprabhu Kore Arts, Science and Commerce College, Chikodi

*Corresponding author: Email: naveen@presidencyuniversity.in

Abstract. Barium Ferrites is a good class of electrical material due to its high resistivity. It is mainly used in electronics and telecommunication industry due to their novel electrical properties. We have synthesized barium hexagonal ferrite ($BaFe_{12}O_{19}$) by solid state reaction method. The prepared precursor was calcined at temperature 750°C and characterized by FTIR, XRD and SEM. X-ray diffraction analysis confirmed the single-phase formation of $BaFe_{12}O_{19}$. Using the Debye Scherrer formula, the crystallite size of $BaFe_{12}O_{19}$ was calculated, and the average particle sizes were 26.52 nm, 27.17 nm, 19.17 nm and 23.67 nm. The result confirms that barium ferrite powder has a constant size of less than 30 nm. Four peaks at 2θ values of 35.52, 43.22, 57.21 and 62.84 degrees, corresponding to (110) (111) (210) and (211) planes of Barium ferrite have been observed. The FTIR spectra showed important signature absorption spectra of hexaferrite at 314 cm^{-1}, 499 cm^{-1} & 628 cm^{-1}. These peaks confirm the formation of barium hexaferrites. The SEM images of barium hexaferrite at 750°C showed large number of vertically grown barium hexaferrite nanorods with a grain size between 110 nm to 130 nm. Thus, barium ferrite is a good candidate in the use of electronics and communication sector.

1. Introduction

Barium ferrites ($BaFe_{12}O_{19}$), is an enduring magnet widely recognized for its magnetic importance in last few decades. $BaFe_{12}O_{19}$ is having magneto plumbite-type hexagonal structure crystal anisotropy along c-axis. Barium ferrite exhibits outstanding features like high saturation magnetization (72 emu/g), with good coercive force with minimum manufacture cost, high curie temperature (450°C) with large chemical constancy and resistance to corrosion [1]. They are alternatively mentioned as ferrite magnets and possess a different irreplaceability, finds wide-range application in magnetic recording media, computer data storage, fabrication of commercial permanent magnet, microwave devices, and electromagnetic shielding fields [2].

To obtain peak performance, it is critical to have enormously fine barium hexaferrite powder with a constantly maintained particle size distribution of approximately 0.1 micro meters, along with well-regulated magnetic properties. Hence it is required to make changes in the physical properties of component which sturdily dependent on the preparation technique. With this respect, several chemical processes carried out at low temperatures were explored to produce ultrafine $BaFe_{12}O_{19}$ Particles [3].

IOP Conf. Series: Materials Science and Engineering 1300 (2024) 012012 doi:10.1088/1757-899X/1300/1/012012

Some of the previous methods executed for synthesis were co-precipitation, hydrothermal synthesis, sol–gel process, microemulsion method, citrate-precursor, glass crystallization, sonochemical activation, and mechanochemical activation. The key drawback of these methods is that, the size of the particle is inadequate and uniformly distributed for utilization such as recording media [4].

The present research aimed to investigate the influence of the Fe^{3+}/Ba^{2+} mole ratio and the inclusion of surfactants in the synthesis of nanocrystalline barium hexaferrite powder via the solid-state reaction technique. This investigation holds significance as it aids in understanding the strategies for regulating the synthesis of single-phase barium hexaferrite powder, characterized by a uniform ultrafine size, relatively at low calcination temperatures. This research work presents the findings for magnetic and structural characteristics of synthesized $BaFe_{12}O_{19}$ powder.

2. Materials and Method

2.1. Sample preparation

The synthesis of Barium hexaferrite compounds followed the method described by (Asiri et al., 2018), via the conventional solid-state synthesis method, includes the mixing and combustion of high-purity reactants, viz. BaO (99.98% purity) and Fe_2O_3 (99.98% purity). [5]. The sample mixture was thoroughly combined and finely ground using an agate mortar and pestle. The sample was further placed into an alumina crucible and subjected to 3 hours heating process at 750°C. The sample was cooled at room temperature and once again blended and ground before it was subjected to an additional 2-hours heating cycle at a temperature of 750°C inside the furnace. The experiment was repeated for final 1 hour heating at 750°C [6]. The procedure was completed to facilitate material blending and enhance the thermal transport mechanisms within the mixture.

2.1.1. Characterization The X-ray Diffraction study was conducted for the synthesized sample of barium ferrite with goniometer (Ultima 3 theta-theta gonio, under 40kv/30mA- X-ray, 2θ/θ - scanning mode, fixed monochromator) by referring the method of Vinnik et al., 2020. Data was compiled within the angular range extended from 10 to 80 degrees (2θ), with increase of 0.02 degree. In the initial stage of indexing the powder diffraction pattern, Miller indices (h k l) were assigned to each analysed peak. This indexing process was achieved using two different methods and datasets. To achieve this, it is required to determine a division constant, so that the values in the third column can be converted into numbers at this point the constant is 43 (135.694 – 93.053). (The Scherrer equation, used in X-ray diffraction and crystallography, serves as a mathematical tool to determine the size of sub-micrometer particles or crystallites within a solid.) The four peaks 2θ values of 35.52, 43.22, 57.21 and 62.84 degrees. Corresponding to (110) (111) (210) (211) planes of Barium ferrite have been observed. To understand the morphology of the hexaferrite surface morphology was studied using SEM by the method describe by Mosleh et al., 2016. The SEM images of barium hexaferrites at 750°C showed large number of vertically grown barium hexaferrites developed in the powder sample. Fourier transform infrared (FTIR) analysis was carried out using the method of Carol Trudel et al., 2019. The infrared spectra of all samples were captured using a Perkin Elmer (Model 783) IR spectrometer in a KBr medium at ambient room temperature. To investigate the ferrite phase formation the synthesized $BaFe_{12}O_{19}$ hexaferrite particles were characterized by Fourier transform infrared (FTIR) analysis. Infrared spectra of the calcinated samples were recorded at room temperature, layer the mid-IR range from 4000 cm^{-1} to 400 cm^{-1}.

3. Results and Discussion

3.1 X-Ray Diffraction Studies

Figure 1 shows the structural characterization of $BaFe_{12}O_{19}$ studied by X-ray diffraction showed four peaks of 2θ values of 35.52, 43.22, 57.21 and 62.84 degrees. Corresponding to (110) (111) (210) (211) planes of Barium ferrite.

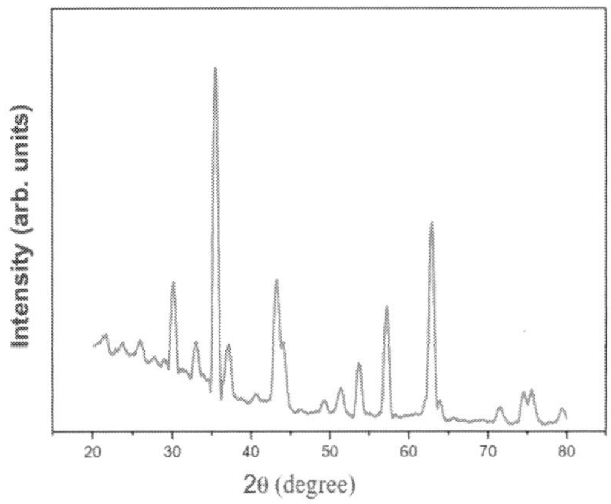

Figure 1. X-Ray Diffraction pattern of the $BaFe_{12}O_{19}$.

3.2 SEM

The microphotographs of Barium hexaferrite powder are presented in Figure 2. The SEM images of barium hexaferrites at 750°C showed large number of vertically grown barium hexaferrites developed in the powder sample. The grain size of the sample under investigation is observed and they are 116 nm, 127 nm. This indicates that the grain size ranges from 110 nm to 130 nm.

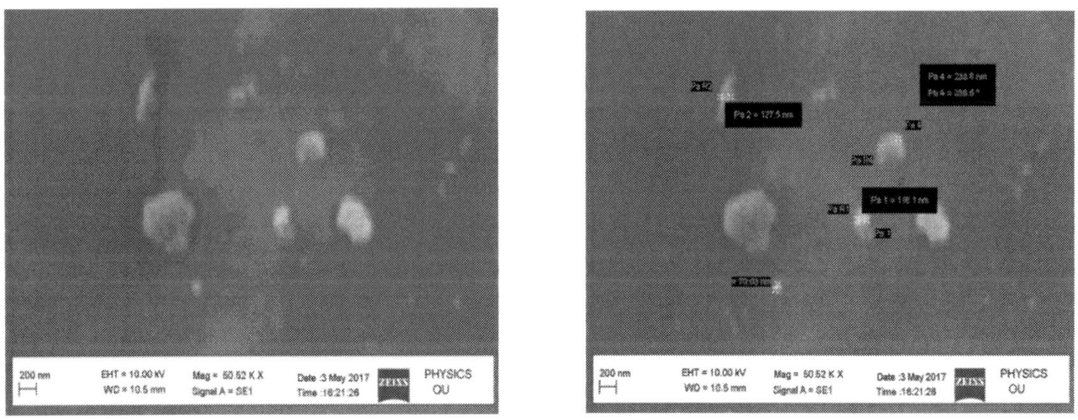

Figure 2: SEM image of Barium Hexaferrites.

3.3 FTIR

The infrared spectra for all the samples were obtained using a Perkin Elmer (Model 783) IR spectrometer in a KBr medium at standard room temperature. To investigate the ferrite phase formation the synthesized $BaFe_{12}O_{19}$ hexaferrite particles were characterized by Fourier transform infrared (FTIR) analysis. The infrared spectra of the calcinated samples were attained at room temperature within the mid-IR range of 4000 cm^{-1} to 400 cm^{-1}. Figure 3 shows FTIR spectra of $BaFe_{12}O_{19}$ hexaferrite particles calcinated at 750°C for this.

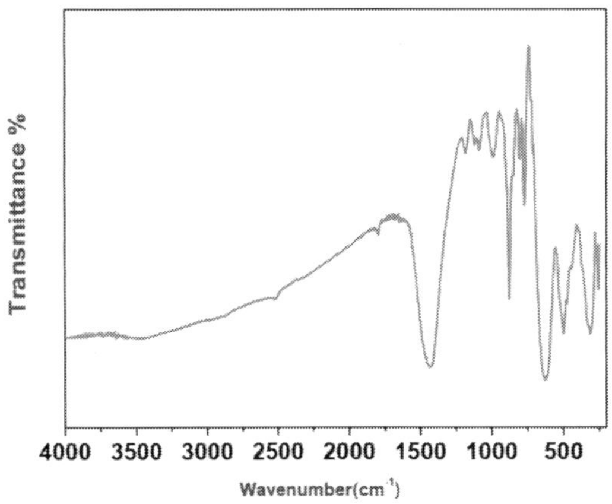

Figure 3: The FTIR Spectra of $BaFe_{12}O_{19}$

The absorption bands in the range of 350 cm^{-1} to 600 cm^{-1} are symbolic presence of hexaferrite. The peaks assign to magnetite, stretching vibration of FeO at tetrahedral and octahedral site confirms the synthesis of hexaferrite. Theoretical considerations suggests that there are three fundamental IR active modes to hexagonal structure [7]. The three characteristics absorption bands around 314 cm^{-1},499 cm^{-1} and 628 cm^{-1} are due to oxygen vibrations in perpendicular to octahedral cation oxygen axis and spinal block as a whole [8]. Typically, the vibration band specific to the spectral range from 430 cm^{-1} to 590 cm^{-1} shows the presence of hexagonal ferrite [9]. The frequency bands falling within the 550-580 cm^{-1} range parallel to the vibrations of the bond between the oxygen ion and the tetrahedral metal ion (O-M tetra), while those in the 430-470 cm^{-1} range are associated with the bond between the oxygen ion and the octahedral metal ion (O-M octa) [10]. The spectra showed a noticeable change in bond position, arising from alterations in the bond distance between Fe^{3+} and O^{2-} ions in both tetrahedral and octahedral complexes. [11].

4. Conclusions

X-ray diffraction analysis affirms the presence of a singular $BaFe_{12}O_{19}$ phase. The Debye Scherrer formula was applied to determine the crystalline size of $BaFe_{12}O_{19}$, resulting in average particle sizes as 26.52 nm, 27.17 nm, 19.17 nm, and 23.67 nm. The result confirms that barium ferrite powder uniformed size less than 30 nm. The FTIR spectra show the important signature absorption points of hexaferrites at 314 cm^{-1}, 499 cm^{-1} & 628 cm^{-1}. These points confirm the formation of barium hexaferrites. The SEM images of barium hexaferrite at 750°C showed large number of vertically grown barium hexaferrite nanorods. The grain size of the sample is lies between 110 nm to 130 nm.

5. References

[1] Radwan M, Rashad MM, Hessien MM. 2007 Synthesis and characterization of barium hexaferrite nanoparticles. *Journal of Materials Processing Technology*.; **vol 181,** pp 106-109.

[2] Huang K, Yu J, Zhang L, Xu J, Li P, Yang Z, Liu C, Wang W, Kan X. 2020 Synthesis and characterizations of magnesium and titanium doped M-type barium calcium hexaferrites by a solid state reaction method. *Journal of Alloys and Compounds*.; **vol 825,** pp 154072-10.

[3] Prithviraj Swamy PM, Basavaraja S, Havanoor V, Srinivas Rao NV, Nijagunappa R,

Venkataraman A. 2011 Barium ferrite nanoparticles prepared by self-propagating low-temperature combustion method and its characterization. *Bulletin of Materials Science.* **vol 34,** pp 1319-1323.

[4] Rashad MM, Ibrahim IA. 2011 Aug 1 Improvement of the magnetic properties of barium hexaferrite nanopowders using modified co-precipitation method. *journal of magnetism and magnetic materials.*; **vol 323,** pp 2158-2164.

[5] Asiri S, Güner S, Demir AY, Yildiz A, Manikandan A, Baykal A. 2018 Synthesis and magnetic characterization of Cu substituted barium hexaferrites. *Journal of Inorganic and Organometallic Polymers and Materials.*; **vol 28**, pp1065-1071.

[6] Park TJ. 2007 Multifunctional iron-based metal oxide nanostructured materials: Synthesis, characterization, and properties. *State University of New York at Stony Brook.*

[7] Kumar A, Yadava SS, Gautam P, Khare A, Mandal KD. 2019 Magnetic and dielectric studies of barium hexaferrite ($BaFe_{12}$ O_{19}) ceramic synthesized by chemical route. *Journal of Electroceramics.* **vol 42,** pp 47-56.

[8] Asiri S, Güner S, Demir AY, Yildiz A, Manikandan A, Baykal A. 2018 Synthesis and magnetic characterization of Cu substituted barium hexaferrites. *Journal of Inorganic and Organometallic Polymers and Materials.* **vol 28,** pp 1065-1071.

[9] Vinnik DA, Podgornov FV, Zabeivorota NS, Trofimov EA, Zhivulin VE, Chernukha AS, Gavrilyak MV, Gudkova SA, Zherebtsov DA, Ryabov AV, Trukhanov SV. 2020 Effect of treatment conditions on structure and magnetodielectric properties of barium hexaferrites. *Journal of Magnetism and Magnetic Materials.* **vol 498,** pp166190-20.

[10] Mosleh Z, Kameli P, Poorbaferani A, Ranjbar M, Salamati H. 2016 Structural, magnetic and microwave absorption properties of Ce-doped barium hexaferrite. *Journal of Magnetism and Magnetic Materials.* **vol 397,** pp101-107.

[11] Carol Trudel TT, Mohammed J, Hafeez HY, Bhat BH, Godara SK, Srivastava AK. 2019 Structural, dielectric, and magneto-optical properties of Al-Cr substituted M-type barium hexaferrite. P*hysica status solidi* **vol 216,** pp 1800928-9.

IOP Conf. Series: Materials Science and Engineering 1300 (2024) 012013 doi:10.1088/1757-899X/1300/1/012013

Determination of effective atomic number, electron density and Kerma of some ferroelectric materials using mass attenuation coefficients in the energy range 1 keV -100 GeV

Pushpa A.[1], Bhagyalaxmi I. B[1], Shivaleela B.[2] and Shivraj G. G[1]

[1]Department of Physics, Smt. V. G. Womens College, Kalaburagi-585102, Karnataka, India
[2]Department of Physics, Gulbarga University, Kalaburagi-585106, Karnataka, India

Abstract. The effective atomic number Z_{eff} and the corresponding effective electron density N_{el} of ferroelectric materials ($BaCoF_4$, $LiNbO_3$, $KNbO_3$, $BaTiO_3$, $PbZrO_3$, $PbTiO_3$, $KNbO_3$, $Bi_4Ti_3O_{12}$, $KHPO_4$, $PbBi_2Nb_2O_9$, $SrBi_2Ta_2O_9$, $Ba_{0.73}Sr_{0.27}TiO_3$ and $Mg_3B_7O_{13}Cl$) has been calculated in the extended energy region from 1 keV – 100 GeV using WinXCom program. The total photon interaction with coherent and incoherent one can distinguish three energy regions are approximately E<1 keV, 10 keV<E<1 GeV and E>10 GeV. The main photon interaction processes in these regions are photoelectric absorption, incoherent (Compton) scattering and pair production. Z_{eff} and N_{el} values also have been obtained for different processes such as coherent and incoherent scattering pair production in nuclear and electric field, photoelectric absorption and the variations are shown graphically. The variations are due to the presence of different elements in a compound. The behaviour of the energy dependence of electron density is similar to the energy dependence of effective atomic number. The photoelectric absorption is dominant for total coherent process for Kinetic energy released per unit mass (Kerma). Direct method is one of the method helpful to calculate effective atomic number for all types of materials, for all energies greater than 1 keV.

1. Introduction

Ferroelectric materials are utilized widely in a variety of applications due to their spontaneous electrical polarization [1]. Numerous beneficial features are available in ferroelectric materials. Ferroelectric hysteresis (used in nonvolatile memories), high permittivities (in capacitors), high piezoelectric effects (in sensors, actuators, and resonant wave devices like radio-frequency filters), high pyroelectric coefficients (in infrared detectors), strong electro-optic effects (in optical switches), and anomalous temperature coefficients of resistivity (in electric-motor overload-protection circuits) are a few of these. Ferroelectrics may also be produced in a wide range of materials, such as ceramics, single crystals, polymers, and thin films, which expands their use. The fundamental theories underlying the ferroelectric effect and the major kinds of ferroelectric materials are covered in this work, along with how their properties relate to their composition and various manufacturing processes. [2].

The total mass attenuation coefficient, which is the sum of absorption and scattering coefficients, is the most commonly used parameter in photon interaction with matter. The total mass attenuation coefficient is a measure of the probability of incident photons interacting with a matter of unit mass per unit area. The atomic number of the sample, the sample density, the incident photon energy and the chemical structure of the sample significantly affect this parameter. The total mass attenuation coefficient is an essential tool to obtain many other parameters like atomic, molecular and electronic cross sections, effective atomic number, electron density, kinetic energy released per unit mass (Kerma) [3] and molar extinction coefficient [4]. The measurement or calculation of these parameters is a pioneer attempt for many applied fields. These parameters are widely used in medical, radiation, nuclear, plasma and space physics, biological and agricultural industries, so accurate data of these parameters have a great importance [4, 5].

X-rays, Gamma rays and particle radiation such as electrons, protons or heavy ions are used in a variety of applications, which include radioisotope monitoring, cross-section studies of electromagnetic radiation absorption, scattering and attenuation, and testing multi-component, heterogeneous, and composite materials [6]. A thorough knowledge of the interaction of photons with ferroelectric materials is desirable. In radiation biology, as well as medical diagnosis and therapy, photons in the keV range are important [7]. Radiography and medical imaging require photons in the MeV range, while astrophysics and cosmology are interested in photons in the GeV range.

The incoming photon energy and the atomic number of the materials utilized determine how photons interact with matter. G. J. Hine [8] has pointed out that, unlike pure elements, a single number cannot uniquely represent the atomic number across the whole energy range in composite materials for photon interactions. The "effective atomic number" (Z_{eff}) for composite materials is a variable value that changes with energy. If certain constants are known, the energy absorption in a given medium can be computed. As a result, Z_{eff} is a parameter that varies with photon energy based on the interaction mechanism involved, rather than a genuine constant for a specific material. The electron density, which is defined as the number of electrons per unit mass of the absorber, is closely related to the effective atomic number [9].

Attenuation, absorption coefficients and photon interaction cross sections in materials, as well as effective atomic number and electron density, have been studied extensively in the field of radiation. X-ray and Gamma ray attenuation coefficients and interaction cross sections of various materials are typically calculated using the well-known XCOM program. Berger and Hubbell [10] published the first version of this program for estimating photon interaction cross sections and attenuation coefficients for each element or compound in the energy range 1 keV-100 GeV. Gerward et al [11] have released a new version of this program called WinXCOM, which runs on the Windows operating system. For incoherent and coherent scattering, photoelectric absorption, and pair production, the program offers total cross-sections and attenuation coefficients as well as partial cross sections. XmuDat, a similar program, calculates mass attenuation coefficients for elements, compounds and mixtures in the energy range of 1 to 50 MeV photons [12]. WinXCOM is another program that determines the removal and attenuation coefficients of transmitted fast neutrons, gamma rays, via mixtures and compounds [13]. The direct or interpolation approaches are commonly used to obtain the effective atomic number [9]. Z_{eff} is determined by the ratio of atomic or electronic cross sections in the direct method. Many writers have lately programmed this method to make computing Z_{eff} for any compound/material in a desired energy range. Taylor et al, on the other hand, used the interpolation method to report Auto-Z_{eff} soft quick determination of effective atomic numbers for any materials or compounds [14, 15].

In this study, the mass attenuation coefficients, effective atomic numbers and electron densities of all the selected ferroelectric materials have been selected in the energy range of 1 keV to 100 GeV were evaluated. In the energy range of 1 keV to 20 MeV, another crucial photon interaction feature known as Kerma values has also been investigated. We also explain the significance of the WinXCom program's single values for effective atomic numbers and electron densities for all photon interactions [coherent, incoherent, photoelectric, pair generation, total photon interaction (with coherent and without coherent)]. The acquired Z_{eff}, and N_{el} values are useful in medicine, research and dosimetry.

2. Theoretical background

The methods for calculating the Ferroelectric materials effective atomic number, electron density and Kerma ($BaCoF_4$, $LiNbO_3$, $KNbO_3$, $BaTiO_3$, $PbZrO_3$, $PbTiO_3$, $KNbO_3$, $Bi_4Ti_3O_{12}$, $KHPO_4$, $PbBi_2Nb_2O_9$, $SrBi_2Ta_2O_9$, $Ba_{0.73}Sr_{0.27}TiO_3$ and $Mg_3B_7O_{13}Cl$) [16, 17] are described in the following subsections. Using WinXCom [11, 18] we computed mass attenuation coefficients and photon interaction cross sections in the energy range of 1 keV to 100 GeV. This programme uses the same underlying cross-sectional database as Hubbell and Seltzer's well-known tabulation [19]. WinXCom allows you to export cross-sectional data to a predetermined MS Excel template, which makes future graphical and numerical data analysis, much easier.

2.1. Mass attenuation coefficients

When a beam of γ-photons experiences attenuation via scattering (coherent or incoherent scattering) and absorption (photoelectric effect, pair and triplet production) when passing through a material of given thickness [20]. The decrease in the intensity of photons is given by Beer-Lambert's law,

$$I = I_0 e^{-\mu x} = I_0 e^{-(\mu/\rho)x} \quad (1)$$

where I_0 and I are the incident and attenuated photon intensities respectively, t is the thickness of the absorber (cm) and μ is the linear attenuation coefficient (cm^{-1}). The total photon mass attenuation coefficient $(\mu/\rho)_{comp}$ has been estimated by the following 'mixture rule' with WinXCOM program for a chemical compound or mixture [21], [22].

$$\left(\frac{\mu}{\rho}\right)_{comp} = \sum_i \omega_i \left(\frac{\mu}{\rho}\right)_{comp} \quad (2)$$

where ω_i is the weight fraction of the ith constituent element present in the given compound and $(\mu/\rho)_i$ is the photon mass attenuation. For a material composed of multi elements the fraction by weight is given by

$$\omega_i = \frac{n_i A_i}{\sum_i n_i A_i} \quad (3)$$

where A_i is the atomic weight of the ith element and n_i is the number of formula units of the ith element.

2.2. Total atomic cross section

The total cross section (σ) and differential partial cross-sections are related by the relation

$$\sigma = \sigma_{coh} + \sigma_{incoh} + \tau + \kappa + \sigma_{ph,n} \quad (4)$$

where σ_{coh} is the coherent and σ_{incoh} is the incoherent scattering cross-sections, respectively. τ is the atomic photoelectric cross-section, κ is the positron electron pair production cross-section and $\sigma_{Ph,n}$ is the photonuclear cross-section [23].

2.3. Molecular cross section

The effective molecular cross section (σ_m) is estimated using the values of mass attenuation coefficients $(\mu/\rho)_{comp}$ by the following relation:

$$\sigma_m = \left(\frac{\mu}{\rho}\right)_{comp} \frac{\sum_i n_i A_I}{N} \quad (5)$$

where N is the Avogadro's number, n_i and A_i are the total number of atoms and atomic weight of the ith element in a molecule respectively [24].

2.4. Atomic and electronic cross-section

The effective atomic cross-section (σ_a) and effective molecular cross-section (σ_m) are related by the following equation [25, 26]:

$$\sigma_a = \frac{\sigma_m}{\sum_i n_i} \quad (6)$$

Similarly, electronic cross-section (σ_e) is given by the following equation [27]:

$$\sigma_e = \frac{1}{N}\sum_i \left(\frac{f_i A_i}{Z_i}\right)\left(\frac{\mu}{\rho}\right)_i \quad (7)$$

where and are the atomic number and fractional abundance of the constituent element

$f_i = n_i / \sum_i n_i$

2.5. Effective atomic number

Effective atomic number (Z_{eff}) is the ratio of the atomic and electronic cross-sections and it is given by

$$Z_{eff} = \frac{\sigma_a}{\sigma_e} \quad (8)$$

The definition of the effective atomic number can be found in equation (8). It is assumed that the molecule's actual atoms can be substituted by an equal number of identical (average) atoms, each with Z_{eff} electrons [28].

Effective atomic numbers Z_{eff} can also be calculated from the direct method of the low-Z materials for total photon interaction [29]. The formula is given below:

$$Z_{eff,pi} = \frac{\sum_i f_i A_i (\mu/\rho)_i}{\sum_j (f_j A_j / Z_j)(\mu/\rho)_j} \quad (9)$$

Where f_i is molar fraction in the mixture/compound with linear attenuation coefficient μ and mass attenuation coefficient (μ / ρ), A is atomic weight, Z is atomic number. The ratio (A/Z), of the atomic mass and the atomic number is approximately constant.

2.6. Electron density

Electron density (N_{el}) is the number of electrons per unit mass and is closely related to the effective atomic number and it is given by the following equation [30] ,

$$N_{el} = \frac{(\mu/\rho)_{comp} Z_{eff}}{\sigma_a} \quad (10)$$

2.7. Kerma relative to air

Kerma is the initial kinetic energy of all secondary charged particles liberated per unit mass by uncharged radiation at a site of interest (ICRU180, Attix 1986). It applies to photons (x-rays, gamma rays, bremsstrahlung, and neutrons) and uses the same measure as the absorbed dosage, J/kg = Gy. Photon fluence and the likelihood of an interacting medium are both closely related to Kerma [31] . To understand the relationship between mass and energy absorption coefficient (μ_{en} / ρ), where energy fluence of mono-energetic photon passing through area, A of an absorber is $\psi[Jm^{-2}]$ and is used to derive Kerma. Where $(\psi\mu_{en})Adx$ is the photon energy transmitted from uncharged ionizing gamma radiation to charge particles in a volume over a short distance dx behind the area, where μ_{en} is mass energy-absorption coefficient. Because the mass in a volume with density ρ is equals ρAdx, kerma equals

$$K = \frac{\psi \mu_{en} Adx}{\rho Adx} = \psi\left(\frac{\mu_{en}}{\rho}\right) \tag{11}$$

Therefore, the kerma is the product of the mass energy absorption coefficient and the energy fluence. The kerma of ferroelectric compounds with respect to air can be calculated as

$$K_a = \frac{K_{comp}}{K_{air}} = \frac{(\mu_{en}/\rho)_{comp}}{(\mu_{en}/\rho)_{air}} \tag{12}$$

Now the computation of Kerma is calculation of mass energy-absorption coefficient.

3. Results and Discussions

The variation of mass attenuation coefficient against optical energy of the compounds obtained by cross sectional data which were exported to MS Excel from. The molecular formula of crystalline materials is presented in Table 1.

Tabe 1 Name and molecular formula of ferroelectric materials.

S. No	Ferroelectric material	Molecular formula
1	Lithium titanate	$LiTaO_3$
2	Barrium tetrafuoride	$BaCoF_4$
3	Lithium niobate	$LiNbO_3$
4	Sodium niobate	$NaNbO_3$
5	Barium titanate	$BaTiO_3$
6	Lead zirconate	$PbZrO_3$
7	Lead titanate	$PbTiO_3$
8	Potassium niobate	$KNbO_3$
9	Bismuth titanate	$Bi_4Ti_3O_{12}$
10	Potassium dihydrogen phosphate	$KH2PO_4$
11	Lead bismuth titanate	$PbBi_2Nb_2O_9$
12	Strontium bismuth titanate	$SrBi_2Ta_2O_9$
13	Barrium strontium titanate	$Ba_{0.73}Sr_{0.27}TiO_3$
14	Boracite	$Mg_3B_7O_{13}Cl$

3.1. *Effective atomic number and electron density*

The calculation of Z_{eff} and N_{el} of ferroelectric materials were carried out using above mass attenuation coefficients in the energy region 1 keV-100 GeV. These data shows that both the parameters depend upon chemical composition of the given molecule or compound. The variations of N_{el} with photon energy in all the ferroelectric materials is same for partial and total interaction processes are similar to that of Z_{eff} shown in Figures and can be explained on a similar manner as that of Z_{eff}. The average values of Zeff and Nel are given in Table 2 and 3 for the selected ferroelectric materials.

Table 2. Average values of effective atomic number of ferroelectric materials.

Ferroelectric	Tot Coh	Tot Incoh	Coh	InCoh	PA	PN	PE
$LiTaO_3$	36.1	20.7	34.9	30.1	33.1	26.3	24.1
$BaCoF_4$	47.8	46.3	48.8	20.2	52.5	22.3	19.6
$LiNbO_3$	16.1	35.8	34.8	28.4	34.6	29.2	27.6
$NaNbO_3$	26.6	15.4	33.9	25.7	33.5	25.4	19.6
$BaTiO_3$	46.9	70.6	88.0	52.7	90.4	39.7	19.2
$PbZrO_3$	62.0	55.2	51.2	25.7	49.8	22.8	22.4
$PbTiO_3$	60.5	60.4	76.1	50.6	75.4	36.3	26.1
$KNbO_3$	27.8	27.8	39.0	23.7	39.4	21.2	14.9
$Bi4Ti3O_{12}$	48.9	70.6	92.1	50.8	94.0	41.2	18.1
$KHPO_4$	12.7	12.7	14.6	15.0	14.5	14.2	15.2
$PbBi_2Nb_2O_9$	51.9	50.9	55.3	40.7	55.4	34.5	32.8
$SrBi_2Ta_2O_9$	52.8	52.7	50.1	66.4	72.0	47.9	42.1
$Ba_{0.73}Sr_{0.27}TiO_3$	39.8	39.8	53.9	32.4	54.8	21.2	13.2
$Mg_3B_7O_{13}Cl$	14.2	14.1	25.4	24.2	26.3	24.0	22.9

*Photoelectric absorption
*Pair nuclear
*Pair electron

Table 3. Average values of electron density x10[24] (electrons/g) of ferroelectric materials.

Ferroelectric	Tot Coh	Tot Incoh	Coh	InCoh	PA	PN	PE
$LiTaO_3$	4.63	4.43	7.12	6.38	6.74	5.36	4.90
$BaCoF_4$	5.71	5.98	6.30	2.61	6.78	2.80	2.53
$LiNbO_3$	3.32	7.28	7.08	5.79	7.05	5.96	5.62
$NaNbO_3$	3.05	2.98	7.26	5.23	6.83	5.17	5.62
$BaTiO_3$	2.42	6.90	8.59	5.15	2.76	3.46	1.88
$PbZrO_3$	5.12	5.48	5.09	2.56	4.95	2.43	2.22
$PbTiO_3$	6.01	6.01	7.56	5.03	7.49	3.61	2.59
$KNbO_3$	4.65	4.65	6.53	3.96	6.59	3.54	2.50
$Bi4Ti3O_{12}$	6.90	6.90	9.02	4.96	2.76	4.02	1.76
$KHPO_4$	4.32	4.25	5.74	5.65	5.77	6.00	6.23
$PbBi_2Nb_2O_9$	8.17	8.02	8.81	6.41	8.74	5.44	5.17
$SrBi_2Ta_2O_9$	7.53	7.52	7.15	9.47	10.03	6.83	6.00
$Ba_{0.73}Sr_{0.27}TiO_3$	6.13	6.12	8.28	4.98	8.43	3.26	2.04
$Mg_3B_7O_{13}Cl$	5.29	5.27	9.48	9.65	9.83	8.98	8.53

*Photoelectric absorption
*Pair nuclear
*Pair electron

3.1.1. Total photon interaction *(with coherent)*

From Fig. 1 it can be easily seen that in the continuous energy range (1 keV– 100 GeV) the Z_{eff} and N_{el} are mainly dominated by different partial photon interaction processes [33-36]. Energy below 0.3 MeV, Z_{eff} has maximum values, where photoelectric interaction is dominant. Z_{eff} drops to a lower value typically for Compton scattering, in the energy region 0.3 MeV-2 MeV. Compton scattering or incoherent scattering is the main interaction process between 2-40 MeV energy region in this intermediate energy range and Z_{eff} is about constant and equal to mean (average) atomic number [37, 38]. In the transition region from 40 to 500 MeV, Z_{eff} increases with increasing energy as pair production gradually becomes dominant. At high energies above 500 MeV, Z_{eff} assumes an almost constant value determined by pair production.

The Z_{eff} varies from a higher value at lower energies to a lower value at higher energies, with a peak due to photoelectric effect near the K-edge of the high Z element present in the ferroelectric materials, then becoming constant with a minimum value at intermediate energies; further, there is an increasing trend in Z_{eff} values due to the relative dominance of photon interaction processes in various energy regions. The Z dependency of total atomic cross–sections explains all changes, leading to effective atomic numbers such as Z^{4-5} for photoelectric absorption, Z for Compton scattering, and Z^2 for pair creation.

As a result of the above, the photoelectric absorption cross-section, which is proportional to Z^{4-5}, gives larger weight to the high Z elements than the other processes. The greatest values of Zeff in the low energy zone are readily explained by this state. In contrast to photoelectric absorption and pair production processes, the Compton scattering cross-section is proportional to Z, giving high Z elements less weight [39].

Electron density is closely related to the effective atomic number, shown in Fig. 2. As a result, N_{el} qualitative energy dependence is similar to Z_{eff}. Furthermore, within the dominance region of any of the three major photon interaction processes, the effective atomic number values remain constant, but the values of effective atomic number vary in the regions where the dominance region for one process shifts to the other. The presence of absorption edges of the constituent parts creates an exception in the lower energy zone [20]. The N_{el} values where maximum upto 0.1 MeV, where photoelectric absorption is dominant. The values decrease from 0.1 MeV-1.5 MeV, where Compton scattering starts increasing attains minimum value between 1.5 MeV-18 MeV. N_{el} values gains minimum values between 18 MeV- 500 MeV and becomes independent of energy above 500 MeV. In this energy region Pair production is the dominant interaction process.

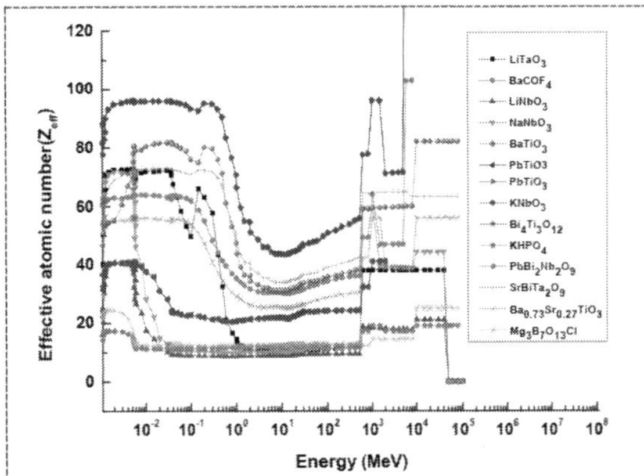

Figure 1. Plots of effective atomic number as a function of energy of ferroelectric materials (with coherent).

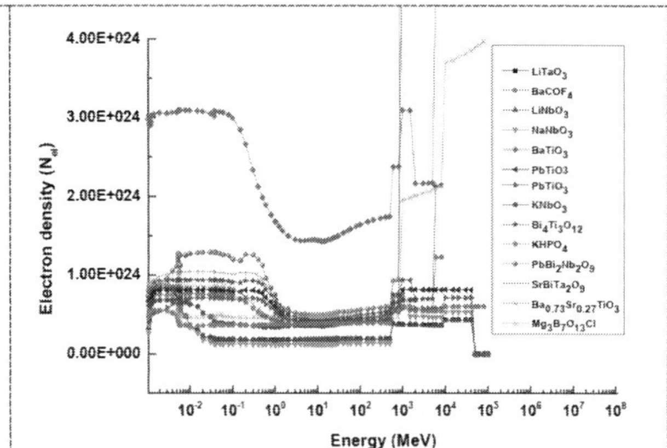

Figure 2. Plots of Electron density a function of energy of ferroelectric materials (with coherent).

3.1.2. Total photon interaction (with incoherent)

From Fig. 3, Z_{eff} values have maximum between 0.03 MeV-1 MeV, where photoelectric absorption is dominant. From 1 MeV-14 MeV it gradually decreases attains minimum values in the range 14 MeV-6 GeV, where Compton scattering is dominant interaction process. Above that pair production is dominant interaction process. Similarly, in Fig. 4 the N_{el} values have maximum values between 0.01 MeV- 1.5 MeV and after that gradually decreases and attains low values of Z_{eff} in the range 18 MeV-6 GeV. Above these values Z_{eff} values become independent of energy with small variation in energy.

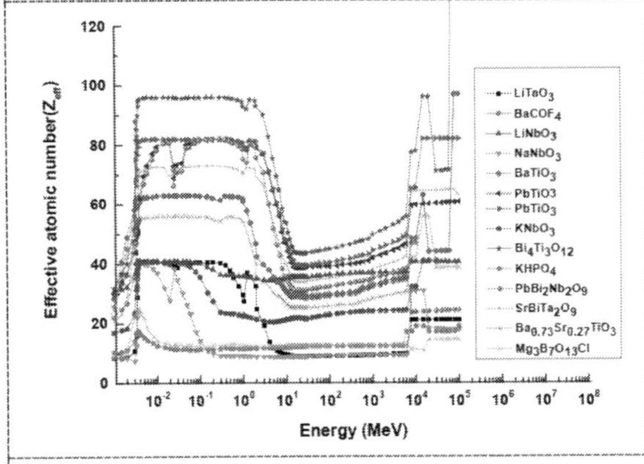

Figure 3. Plots of effective atomic number as a function of energy of ferroelectric materials (without coherent)

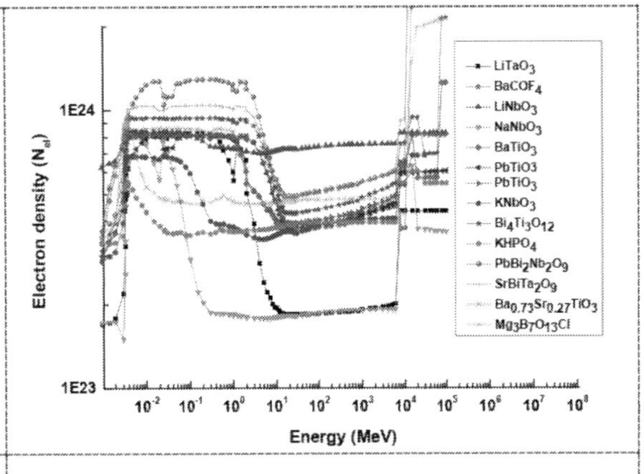

Figure 4. Plots of Electron density a function of energy of ferroelectric materials (without coherent).

3.1.3. *Coherent scattering*

The Fig. 5 shows the variation of Z_{eff} with photon energy for coherent scattering. It is apparent from the figure that Z_{eff} increases as energy increases from 1 keV to 0.005 MeV. After that it is independent of energy. There is drastic variation in Z_{eff} values for $BaCOF_4$, $PbTiO_3$ and $PbBi_2Nb_2O_9$ [40]. For the same energy range these compounds show maxima and minima values and above 600 MeV they attain constant values. Similary, the variation of N_{el} with photon energy for coherent scattering is shown in Fig 6 which indicates that N_{el} increases as energy increases from 1 keV-0.003 MeV and becomes independent of energy after that.

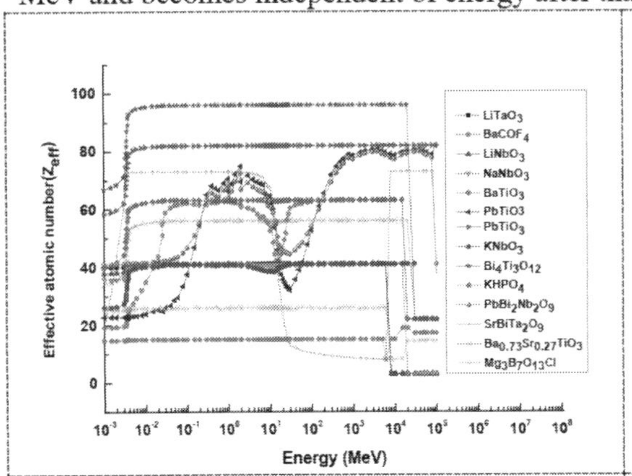

Figure 5. Plots of effective atomic number as a function of energy of ferroelectric materials (coherent).

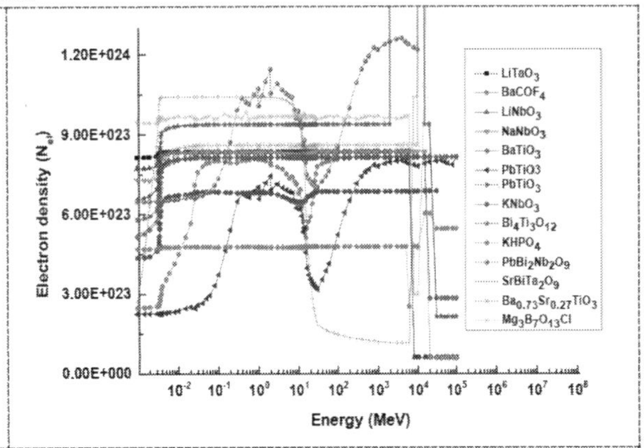

Figure 6. Plots of Electron density a function of energy of ferroelectric materials (coherent).

3.1.4. *Incoherent scattering*

The Fig. 7 shows the variation of Z_{eff} with photon energy for coherent scattering. It is apparent from the figure that Zeff increases as energy increases from 1 keV to 100 MeV. After that it is independent of energy. There is drastic variation in Z_{eff} values for ferroelectric materials from 1 keV-100 MeV due to elemental composition in them [40]. For the same energy range these compounds show maxima and minima values and below 600 MeV they attain constant values. Similary, the variation of N_{el} with photon energy for incoherent scattering is shown in Fig 8 which indicates that N_{el} increases as energy increases from 1 keV-100 MeV and becomes independent of energy.

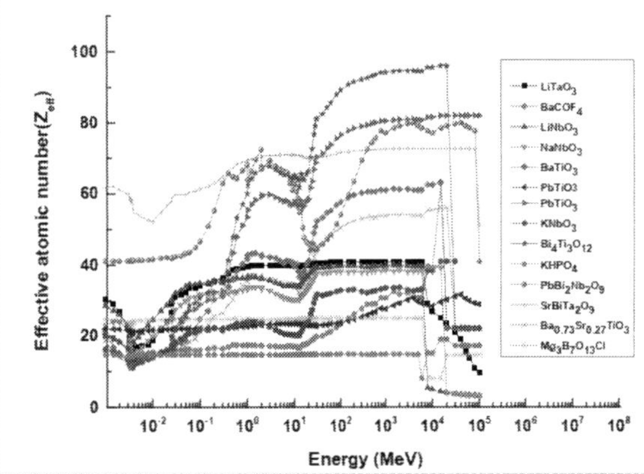

Figure 7. Plots of effective atomic number as a function of energy of ferroelectric materials (incoherent)

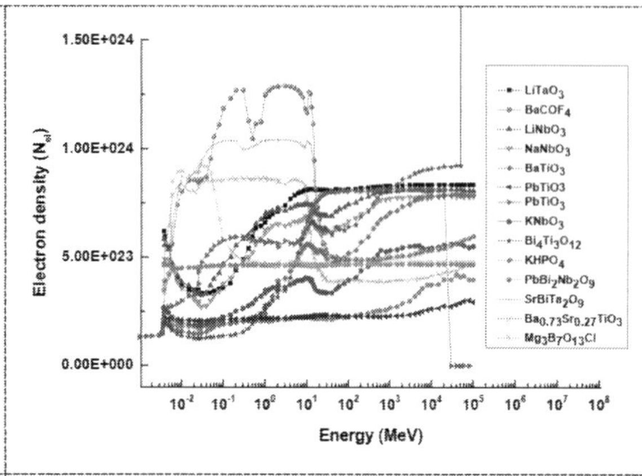

Figure 8. Plots of Electron density a function of energy of ferroelectric materials (incoherent).

3.1.5. Photoelectric absorption

A plot of Z_{eff} vs photon energy, especially for medium and high-Z materials, shows the distinctive absorption edges as the binding energy of each electron subshell is reached and a new pathway for photoexcitation is energetically enabled [17]. Fig 9 demonstrate the fluctuation of Z_{eff} with photon energy for photoelectric absorption, indicating that Z_{eff} is nearly independent of photon energy. There is drastic variation in Z_{eff} values for $BaCOF_4$, $PbTiO_3$ and $PbBi_2Nb_2O_9$. This is because; photoelectric process is predominant at low energies (1 MeV) and for materials of higher atomic numbers [42]. The N_{el} graph for photoelectric absorption process in Fig. 10 shows drastic variation for all the ferroelectric materials.

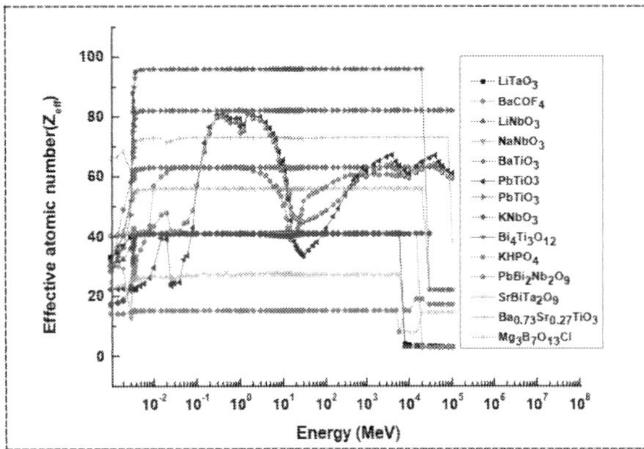

Figure 9. Plots of effective atomic number as a function of energy of ferroelectric materials (photoelectric absorption).

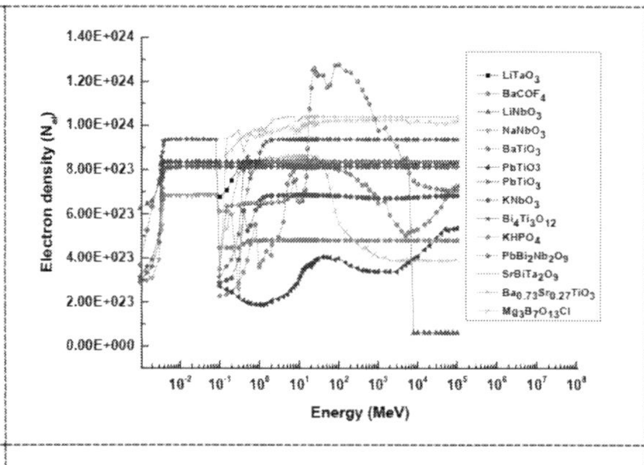

Figure 10. Plots of Electron density a function of energy of ferroelectric materials (photoelectric absorption)..

3.1.6. Pair production in nuclear field

Pair production in the nuclear field, the change of Z_{eff} with photon energy is shown in Fig. 11, which demonstrates that Z_{eff} increases significantly with increasing photon energy from 0.1 MeV to 10 MeV before becoming almost energy independent [41]. It could be because pair production in the nuclear field is Z^2 dependent. Due to the inclusion of elements with a wide range of atomic numbers, the increase in Z_{eff} with energy is more significant. Similarly, N_{el} values increases from 1 keV-5 MeV, after that they are independent of energy as shown in Fig. 12.

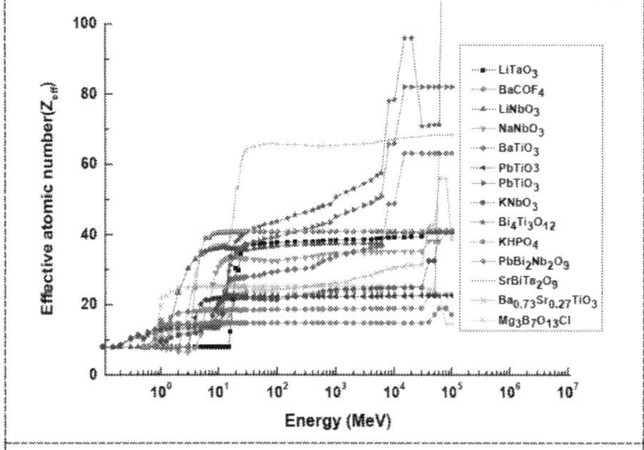

Figure 11. Plots of effective atomic number as a function of energy of ferroelectric materials (pair nuclear)

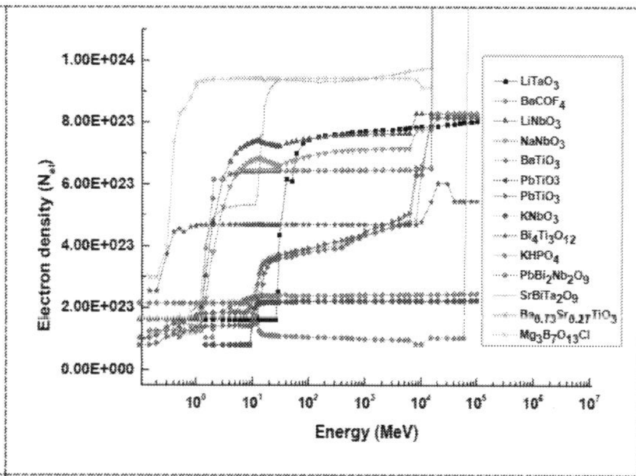

Figure 12. Plots of Electron density a function of energy of ferroelectric materials (pair nuclear).

3.1.7. *Pair production in electric field*

The Z_{eff} and N_{el} values increases with increasing photon energies up and then it is independent of energy thereafter for all ferroelectric materials. Similary, the variation of Z_{eff} and N_{el} with photon energy for electric field is shown in Fig.13 and 14.

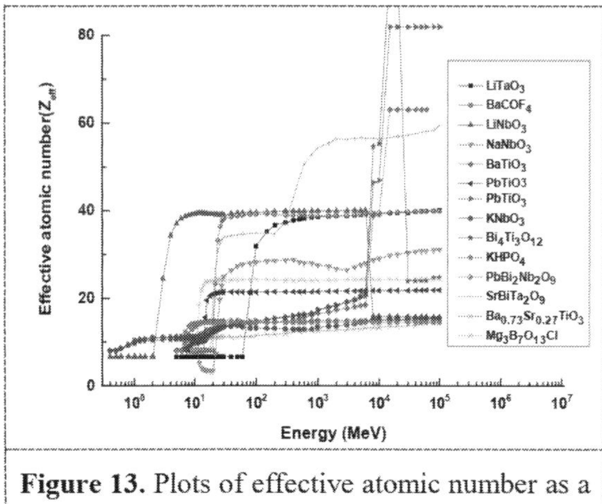

Figure 13. Plots of effective atomic number as a function of energy of ferroelectric materials (pair electron).

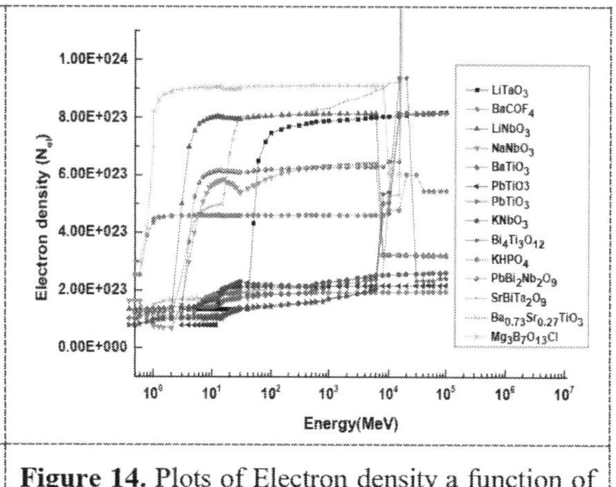

Figure 14. Plots of Electron density a function of energy of ferroelectric materials (pair electron).

3.1.8. *Kerma relative to air*

The variation of Kerma (K_a) values of the molecules for photon energy 1 keV to 20 MeV are shown in Fig. 15. The average of Kerma values are given in Table 4. It is to be noted that the K_a values show sharp peak due to photoelectric absorption of the high-Z constituent of the compound. From Fig. 15 one can observe that there is gradual increase in Kerma with photon energy, attains maximum value at 0.05 MeV. After that gradually decreases upto 20 MeV. The relative significance of the partial photon interaction processes.

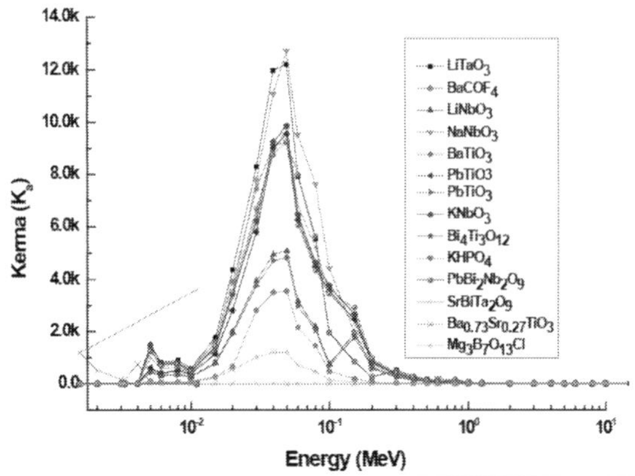

Figure 15. Plots of Kerma with energy for ferroelectric materials in the energy range 1 keV-20 MeV.

Table 4 Kerma values of ferroelectric materials for the energy range 1 keV-20 MeV.

Ferroelectric	Tot Cohx10^3	Ferroelectric	Tot Cohx10^3
$LiTaO_3$	1.67	$KNbO_3$	1.40
$BaCoF_4$	1.38	Bi4Ti3O_{12}	1.40
$LiNbO_3$	7.23	$KHPO_4$	4.09
$NaNbO_3$	6.48	$PbBi_2Nb_2O_9$	1.36
$BaTiO_3$	1.58	$SrBi_2Ta_2O_9$	1.37
$PbZrO_3$	1.28	$Ba_{0.73}Sr_{0.27}TiO_3$	1.74
$PbTiO_3$	1.28	$Mg_3B_7O_{13}Cl$	1.57

4. Conclusions

Study has been undertaken to get information of the mass attenuation coefficients, effective atomic number Z_{eff} and electron density Nel of the selected ferroelectric materials in the energy range 1 keV to 100 GeV using WinXCom program and its underlying database of atomic photon interaction cross sections. Variation of Z_{eff} and N_{el} can be distinguished in three energy regions as low, intermediate and high energies. The maximum value Z_{eff} and N_{el} are found at low- energy region where photoelectric absorption is the main interaction process and minimum values at intermediate energies than higher-energy region. At low-enrgy region (E<0.3 MeV), photoelectric absorption is dominant process. At intermediate energies (2 MeV<E> 40 MeV), the Compton scattering is dominant process. At high-energy region (E<500Mev), pair production is dominant process. We have also calculated Z_{eff} and N_{el} for individual photon interactions such as pair production in the nuclear and electric field, photoelectric absorption, coherent and incoherent scattering. Obtained value of Z_{eff} and N_{el} depends on the chemical composition of the sample we use. Therefore the calculated values of effective atomic number and electron densities of three molecules are the same. Set of formulas which are used for calculating the N_{el} and Z_{eff} in the present work can be used for any other materials. The kerma values shows peak due to photoelectric absorption at at 0.05 MeV. Calculated values of N_{el} and Z_{eff} will of ferroelectric materials hopefully helpful in the field of medical imaging, thin film industry, supercapacitors, memories, sensors, and actuators.

5. References

[1] Martin L W and Andrew M Rappe 2016 *Nat. Rev. Mater.* **2.2** 1.
[2] Whatmore R 2017 *Springer Handbook of Electronic and Photonic Materials* 1.
[3] Jayachandran C A 1971 *Phys Med Biol* **16** 617.
[4] Akman F, Kaçal M R, Akman F and Soylu M S 2017 *Can J Phys*, **95** 1005.
[5] Akkurt I 2009 *Ann Nucl Energy* **36** 1702. doi: 10.1016/j.anucene.2009.09.005.
[6] Kurudirek M and Onaran T 2015 *Radiat Phys Chem* **112** 125.
[7] Hubbell J H 1999 *Phys. Med. Biol.* **44** R1.
[8] Hine G J 1952 *Phys Rev,* **85** 725.
[9] Manohara S R, Hanagodimath S M and Gerward L, 2008 *Phys Med Biol*, **53** N377. doi: 10.1088/0031-9155/53/20/N01.
[10] Berger M J and Hubbell J H 1987 *Natl Bur Stand*, Washington, DC (USA).
[11] Gerward L, Guilbert N, Jensen K B, and Levring H 2004 *Radiat Phys Chem*, **71** 653. doi: 10.1016/j.radphyschem.2004.04.040.
[12] Nowotny R 1998 IAEA *Report IAEA-NDS*, **195** 52.
[13] El-Khayatt A M 2011 *Annals of nuclear energy*, **38** 128.
[14] Singh V P, Badiger N M, and Kucuk N 2014 *J Nucl Chem* 1 doi: 10.1155/2014/725629.
[15] Ali A M, El-Khayatt A M, and Akkurt I 2016 *Radiat Eff Defects Solids* **171** 202.
[16] Shivraj G G, Shivaleela B and Hanagodimath S M 2022 *J Mah Say Uni Bar* **55** 213.
[17] Shivraj G G, Shivaleela B and Hanagodimath S M, 2022 *J Mah Say Uni Bar*, **56** 169.
[18] Gerward L, Guilbert N, Bjørn J K and Levring H 2001 *Radiat Phys Chem* **60** 23, doi: 10.1016/S0969-806X(00)00324-8.
[19] Hubbell J H and Seltzer S M 1995 *J Res Natl Inst Stand Technol -PL.*
[20] Harish V, Nagaiah N, and Kumar H G 2012 *Indian J Pure Appl Phys* **50** 847.
[21] Jackson D F and Hawkes D J 1981 *Phy. Rep*, **70** 169.
[22] Chaiphaksa W, Limkitjaroenporn P, Kim H J and Kaewkhao J 2016 *Prog Nucl Energy* **92**, 48.
[23] Niranjan R S, Rudraswamy B, and Dhananjaya N 2012 *Pramana*, **78** 451.
[24] Manohara S R and Hanagodimath S M 2007 *Nucl Instrum Methods Phys Res B*, **264** 9. doi: 10.1016/j.nimb.2007.08.018.
[25] Manjunatha H C 2016 *J Radiat Res*, **7** 18.
[26] Limkitjaroenporn P, Kaewkhao J, Chewpraditkul W, Limsuwan P 2012 *Procedia Eng* **32**, 847.
[27] Gowda S, Krishnaveni S, Yashoda T, Umesh T K, and Gowda R 2004 *Pramana*, **63** 529.
[28] Issa S A 2016 *Radiat Phys Chem*, **120** 33.
[29] Manohara S R, Hanagodimath S M, Thind K S and Gerward L, *Nucl Instrum Methods Phys Res B*, **266** 3906.doi: 10.1016/j.nimb.2008.06.034.
[30] Sayyed M I, Akman F, Geçibesler H I and Tekin H O 2018 *Nucl Sci Tech* **29** 1.
[31] Demir D and Turşucu A 2022 *Ann Nucl Energy* 48 17.
[32] Bandal G S and Singh K 1993 *Appl. Radiat Isot*, **44** 505.
[33] SinghV P and Badiger N M 2013 *Radioprotection* **48** 431.
[34] Oto B, Gulebaglan S E, Madak Z and Kavaz E 2019 *Radiat Phys Chem* **159** 195.
[35] Manohara S R, Hanagodimath S M, Thind K S and Gerward L 2010 *Appl Radiat Isot* **68** 784. doi: 10.1016/j.apradiso.2009.09.047
[36] Kurudirek M, Büyükyıldız M and Özdemir Y 2010 *Nucl Instrum Methods Phys Res A* **613**, 251. doi: 10.1016/j.nima.2009.11.061.
[37] Al-Buriahi M S and Tonguc B T 2020 *Radiat Phys Chem* **166** 108507. doi: 10.1016/j.radphyschem.2019.108507.
[38] Manjunatha H C 2017 *Radiat Phys Chem* **137** 254 doi: 10.1016/j.radphyschem.2016.01.024.
[39] Ingalagondi P K, Omnath P, Mathapati G B, Shivraj G G, Sankarappa T and Hanagodimath S M 2017 *Int J Pure Appl Sci* **4** 42.
[40] Krishnaveni S, Gowda S, Yashoda T, Umesh T K and Gowda 2005 Radiat Phys Chem **74** 1. doi: 10.1016/j.radphyschem.2005.05.004.
[41] Shivraj G G, Shivaleela B and Hanagodimath S M 2022 *J Mah Say Uni Bar*, **56** 9.

[42] Singh V P and Badiger N M 2016 *Indian J Pure Appl Phys* **54**.
[43] Saniye T and Adnan K 2021 *J Biomed Res* **2** 024.

Synthesis and electrical properties of Reused-Graphite/Polyaniline composites

Basheerabegum Faniband[1], Sarvesha chandra S[1], Sreekanth R[2], Aravinda T[3], Jayadev Pattar[1,*], Sanna Kotrappanavar Nataraj[4], M Nagaraja[5]

[1]Department of Physics, School of Applied Sciences, REVA University, Bengaluru-560 064, Karnataka, India
[2]Department of Chemistry, School of Applied Science, REVA University, Bengaluru-560064, Karnataka, India
[3]Department of Chemistry, Nitte Meenakshi Institute of Technology, Bangalore 560064, Karnataka, India
[4]Centre for Nano and Material Sciences, Jain University, Bengaluru, Karnataka, India
[5]Department of Physics, Sri Siddhartha Institute of Technology, Tumakuru- 572105, Karnataka, India
***Corresponding author**: jayadev.pattar@gmil.com

Abstract

Polyaniline graphite composites have been synthesized using chemical oxidative polymerization method. The graphite powder has been reused which is obtained by used batteries. XRD patterns of graphite show (002) peak and no impurity peaks. Different wt. % (5, 10, 15, 20) of graphite is used for synthesis of composites and studied their electrical conductivity. XRD patterns of composites show increasing intensity with an increasing graphite content. SEM images show plate-like morphology of graphite whereas granular-morphology for PANI. Composite samples show both plate-like morphology and granular morphology indicating that graphite is embedded in the polymer matrix. Electrical characteristics of composite samples have shown a systematic increase in conductivity with an increase in the graphite concentration. Conductive fillers within the polymer matrix can increase the overall conductivity. Higher filler content provides more conductive pathways for electrons, resulting in enhanced conductivity. The detailed studies of these results are discussed and presented in this paper.

Keywords: Reused-Graphite; Polyaniline; Composite; Conductive polymer

1. Introduction

After the introduction of polymer (a Greek term meaning, a large chemical contains the combination of various small as well as alike substances) by Jons Jacob Berzelius in the year 1833, since then, it was observed that the polymers gained wide range of applications in various fields [1-3]. Generally, polymeric substances can be distinguished on bases of their source, structural characteristics, polymerization route, chemical bond strength, and conductivities [4, 5]. In the year 2000, a noble prize was awarded to scientists, MacDiarmid, Shirakawa and Heeger for their finding and development of conducting polymers [6]. Consecutively, various types of conducting polymers gained importance and hence, researchers synthesized polymers and their composites like polythiophene, polypyrrole, and polyaniline composites [7, 8]. During the process, electrons were delocalizing on the entire polymeric chain which ultimately offers flexibility to electrons especially in conducting polymers [9-11]. Such polymers are named as 4th generation of polymeric ingredients [12]. In current conditions, it is necessary to look for new ingredients having distinctive characteristics to improve existing technologies. For this reason, a new group of polymers (especially organic material) having the capacity of electricity conductivity was introduced and they were coined as Inherently conducting polymers (ICPs). These ICPs are extremely flexible and were being used in numerous research fields like electrical, electronics, electromechanical and so on [13-15]. Among these ICPs, polyaniline (PANI) was studied to a larger extent due to its various properties, having higher strength than other ICPs and has low cost [16-18]. However, in certain conditions, it was observed that PANI lose its elementary configuration. To tackle this problem, researchers were looked few ingredients like graphite, metallic carbon-based bases, metallic elements, and metallic sulphonated components to improve its mechanical characteristics as well as thermal strength [19-21]. However, not much work has been carried out on reused graphite. Hence, we have utilized graphite here to prepare PANI composite.

2. Materials and Methods

2.1. Preparation of Graphite

Graphite powder is prepared from a used battery cell. Briefly, graphite rod was dipped in concentrated hydrochloric acid for 2 h. Further, it was washed with double distilled water to remove acid and unwanted impurities. Later, fine powder of graphite was made using ball milling process. The powder was again washed with HCl and distilled water several times. Thus, obtained graphite powder was used in synthesis of PANI/graphite composites.

2.2. Synthesis of PANI/Graphite composite

Synthesis of PANI/Graphite was carried out as described in Ashwini et al., [22] with little modifications. Briefly, PANI/Graphite nanocomposites were prepared according to as described above. Initially, 5 g of graphite powder was dissolved in 100 ml of water for dispersion and kept in sonicator for half an hour. To this, 0.5 M aniline monomer was added with continuous stirring and was maintained in ice bath (less than 5 °C) for 15 min. Afterwards, 2 M of CH_3COOH solution was added and kept it for 15 min with continuous stirring. A 0.3 M APS (Ammonium persulphate) was added to the above solution drop wise and kept for continuous stirring for 24 hours.

A dark green solution of PANI/Graphite composite solution is stored in refrigerator for 24 hours. After this solution was filtered in vacuum filter and residue was washed with double distilled water and the washing process was repeated several times until the solution is purified. The obtained purified samples were dried at 50-60 °C. The dried samples were crushed in mortar to get fine powder of PANI/graphite nanocomposite.

2.3. Analytical techniques used for characterization of graphite/polyaniline composites.

Surface morphology and grain size distribution of different wt. % of composites (graphite/PANI) were analyzed using Field Emission Scanning Electron Microscopy (FESEM) (model JEOL, JSM-7100 F) [23]. The phase identification of different wt. % of composites (graphite/PANI) were determined by powder XRD (RIGAKU) functioning with Cu–Kα1 radiation (l = 1.54 A°) at scan of 3° min^{-1} and a 2-theta range of 5-80° [24, 25]. The PANI doped graphite samples were characterized with FTIR spectroscopy [24]. DC electric conductivity of different wt. % of composites (graphite/PANI) are measured by using two probe Keithley source meter.

3. Results and discussion

3.1. XRD analysis

XRD patterns of pure PANI, and PANI/graphite nanocomposites are shown in Figure 1.

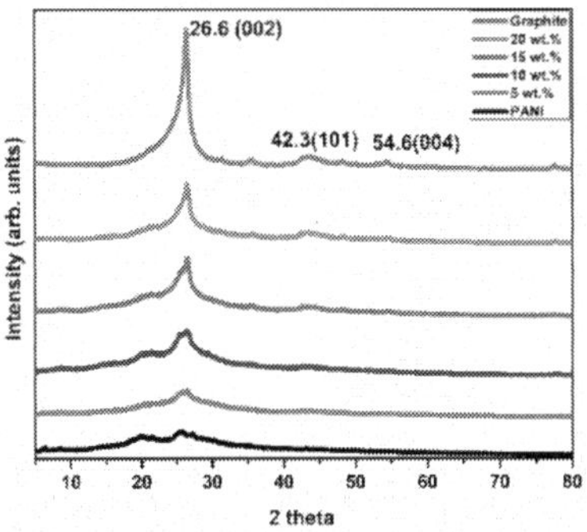

Figure 1. XRD pattern of PANI/graphite. (a) PANI, (b) 5% PANI/graphite composites, (c) 10% PANI/graphite composites, (d) 15% PANI/graphite composites, and (e) 20% PANI/graphite composites and (f) graphite.

The influence of PANI on the graphite crystal arrangement was displayed in the XRD patterns. The broad peak at 2θ = 25.5° was observed in pristine PANI. This peak corresponds to the orthorhombic structure of (200) plane. Pure graphite will be having a peak approximately at 2θ = 26.67° corresponding to (002) plane of carbon [26]. All composites show similar peaks comprising ICDD (Ref. code: 00-013-0148) [27, 28]. Similar kind of results were observed previously [29-31]. PANI/graphite composites show some typical peaks of both PANI as well as graphite. Further results reveal that the peak intensity increases with the increase in the content of graphite in PANI, it might be due to electronic interface at the atom. It is also noticed that there is a small shift in the peak positions which indicates the existence of structural changes. PANI in composites converts from partial amorphous to crystalline. The addition of a minute quantity of graphite into the PANI configuration by is responsible for the structural changes.

3.2. SEM analysis

SEM was used to determine surface morphology of the PANI/ graphite composites (Figure 2).

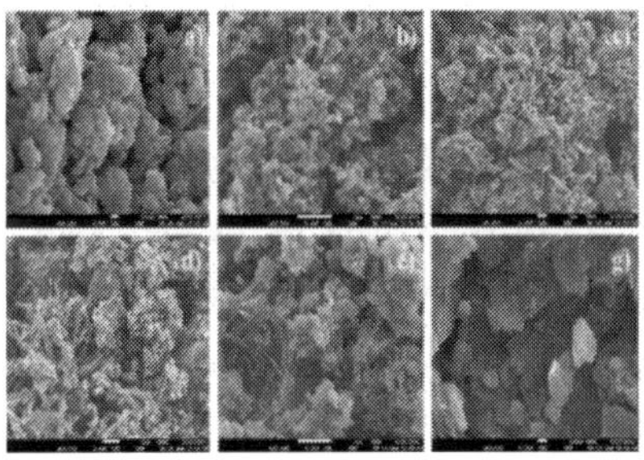

Figure 2. SEM images of PANI/graphite composite. (a) PANI, (b) 5% PANI/graphite composites, (c) 10% PANI/graphite composites, (d) 15% PANI/graphite composites, (e) 20% PANI/graphite composites, and g) Graphite.

The SEM image of graphite shows 2-D sheets or plate-like structure. PANI sample shows tubular structure because of acetic acid medium and synthesis conditions. The composite samples show the presence of both tubular structure and plate-like structure indicating the interconnection between PANI and graphite. Hence, the PANI-graphite composites synthesis under acidic condition (CH_3COOH, acetic acid) was successful. Similar kind of results were also observed previously [31, 32].

3.3. FTIR

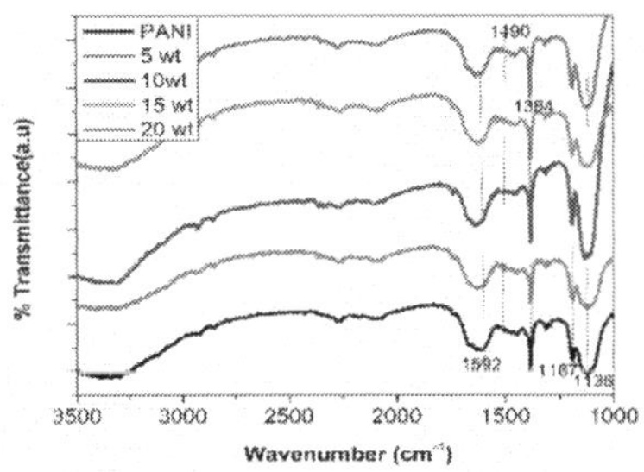

Figure 3. FTIR spectra of PANI, Graphite and PANI/ Graphite composites.

The presence of functional groups in PANI and PANI/ Graphite composites was studied by FT-IR spectrometer and results are shown in Figure 3. The FTIR spectrum of the PANI/Graphite composites exhibit many peaks that are representative of the spectrum obtained from composite. The FTIR spectrum of pure pristine PANI gave absorption bands at 1592, 1490,1384,1187, and 1136 cm^{-1} corresponding to benzenoid ring, C=C vibrations, C–N stretch of aniline, C–H in-plane bending mode and quinoid units respectively. Similar kinds of results were observed previously [33-36].

3.4. I-V Characteristic and Electrical Conductivity

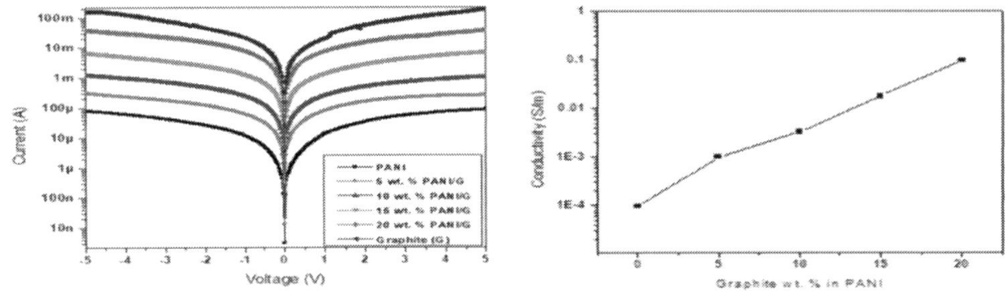

Figure 4a). I-V Characteristic and 4b) Electrical Conductivity of PANI/graphite composite.

I-V Characteristics and electrical conductivity of PANI/graphite composites were shown in Figure 4. The peak intensity increases with an increase in the content of graphite in PANI because of the electronic interaction at the atomic level. The symmetric behavior of I-V graph indicates that all samples have Ohmic contacts. In the Ohmic region at I-V electrical conductivity of the PANI is calculated and it found to be 9.6 $\times 10^{-5}$ S/m. The increase in conductivity with an increase in Graphite concentration is observed in composites shown in electrical conductivity image. The peak intensity increases with the increase in the content of graphite in PANI because of the electronic interaction at the atomic level. A similar kind of result was observed previously [37].

4. Conclusions

Synthesis of PANI/Graphite composite was carried out by chemical oxidative polymerization method using ammonium persulfate as an oxidizing agent and characterized for structural and electrical properties. The diffraction peaks of the composite differ considerably, the intensity of these peaks increases with graphite concentration implies the formation of PANI-graphite composite. Furthermore, it was found that the changes in surface morphology and functional groups were observed using SEM and FT-IR. These different linkages could give rise to different diffraction peaks in the composite.

Electrical conductivity increases with an increase in the concentration of graphite. Hence, PANI/graphite composite behaves more like a semiconductor than a semi-metal as a result of charge transfer between the graphite and PANI.

Conflict of Interest

None

Acknowledgments

All the authors are thankful to REVA University for providing research facilities.

References

[1] J.P. Coats, R. Cochereau, I.A. Dinu, D. Messmer, F. Sciortino, and C.G. Palivan, "Trends in the Synthesis of Polymer Nano- and Microscale Materials for Bio-Related Applications," *Macromol. Biosci.*, vol. 23, pp. 2200474, 2023, doi:10.1002/mabi.202200474.

[2] R. V. Lakshmi, P. Bera, S. T. Aruna, and H. C Barshilia, "XPS and SIMS studies of nanoscale polymer-based coatings." In: Polymer-Based Nanoscale Materials for Surface Coatings. Editors: Thomas S, George JS. 167-194, 2023, doi: 10.1016/B978-0-32-390778-1.00016-5.

[3] B. Wang, Y. Wang, S. Du, J. Zhu, and S. Ma, "Upcycling of thermosetting polymers into high-value materials," *Mater. Horiz.*, vol. 10, pp. 41, 2023, doi: 10.1039/D2MH01128J.

[4] A. Hayat, M. Sohail, A. El Jery, K. M. Al-Zaydi, S. Raza, H. Ali, Y. Al-Hadeethi, T. A. Taha, I. Ud Din, M. Ali Khan, M. A. Amin, E. Ghasali, Y. Orooji, Z. Ajmal, and M. Zahid Ansari, "Recent advances in ground-breaking conjugated microporous polymers-based materials, their synthesis, modification and potential applications," *Mater. Today*, vol. 64, pp. 180–208, 2023, doi: 10.1016/j.mattod.2023.02.025.

[5] Z. Zhao, L. Sun, Y. Li, and W. Feng, "Polymer-derived carbon materials for energy storage devices: A mini review," Carbon, vol. 210, pp. 118066, 2023, doi: 10.1016/j.carbon.2023.118066.

[6] S. C. Rasmussen, New Insight into the "Fortuitous Error" that Led to the 2000 Nobel Prize in Chemistry. *Substantia*, vol. 5, pp. 91-97, 2021, doi: 10.13128/Substantia-973.

[7] M. M. Kumar, P. K. Patil, K. Lakshmi, S. Reddy, S. Manjunatha, Y. T. Ravikiran, and M. Revanasiddappa, "Polythiophene/reduced graphene oxide nanocomposites for humidity sensing application," *Mater. Sci. Forum.*, vol. 1099, pp. 45-50, 2023, doi:10.4028/p-kid2pk.

[8] P. C. Himadri Reddy, J. Amalraj, S. Ranganatha, S. S. Patil, and S. Chandrasekaran, "A review on effect of conducting polymers on carbon-based electrode materials for electrochemical supercapacitors," *Synth. Met.*, vol. 299, pp. 117467, 2023, doi: 10.1016/j.synthmet.2023.117447.

[9] M. Alowakennu, A. O. Omoniyi, O. Ejeromedoghene, Y. A. Alli, E. Akor, and M. O. Nnyia, "Harnessing photo-induced processes for the fabrication and application of functional conjugated and conducting polymer-based materials," *J. Mol. Struct.*, vol. 1292, pp. 136149, 2023, doi: 10.1016/j.molstruc.2023.136149.

[10] J. M. Bone, and J. L. Jenkins, "Understanding polymer electrodeposition and conducting polymer modified electrodes using electrochemistry, spectroscopy, and scanning probe microscopy," *J. Chem. Educ.*, vol. 100, pp. 10, 4062–4071, 2023, doi: 10.1021/acs.jchemed.3c00656.

[11] S. Gu, J. Chen, I. Hussain, Z. Wang, X. Chen, M. Ahmad, S. P. Feng, Z. Lu, and K. Zhang, "Modulation of radical intermediates in rechargeable organic batteries," *Adv. Mater.*, online first, 2023, https://doi.org/10.1002/adma.202306491.

[12] V. Babel, and B. L. Hiran, "A review on polyaniline composites: Synthesis, characterization, and applications," *Polym. Compos.*, vol. 42, pp. 3142-3157, 2021, doi: 10.1002/pc.26048.

[13] C. Gao, X. Cui, C. Wang, M. Wang, S. Wu, Y. Quan, P. Wang, D. Zhao, S. Li, "3D-printed hierarchical porous and multidimensional conductive network based on conducting polymer/graphene oxide," *J. Materiomics.*, online first, 2023, doi: 10.1016/j.jmat.2023.05.012.

[14] S. Jose, Sariga, and A. Varghese, "Design and structural characteristics of conducting polymer-metal organic framework composites for energy storage devices," *Synth. Met.*, vol. 297, pp. 117421, 2023, doi: 10.1016/j.synthmet.2023.117421.

[15] F. Wei, T. Zhang, R. Dong, *et al.*, "Solution-based self-assembly synthesis of two-dimensional-ordered mesoporous conducting polymer nanosheets with versatile properties," *Nat. Protoc.*, vol. 18, pp. 2459–2484, 2023, doi: 10.1038/s41596-023-00845-4.

[16] A. Bibi, A. Shakoor, N. A. Niaz, *et al.*, "Electrical transport properties and thermoelectric power studies of polyaniline–$CaTiO_3$ composites," *Polym. Bull.*, vol. 80, pp. 5005–5021, 2023, doi: 10.1007/s00289-022-04299-8.

[17] G. Mandal, J. Bauri, D. Nayak, *et al.*, "Synthesis, structural study and various applications of Polyaniline and its nanocomposites," Polyaniline - From Synthesis to Practical Applications. IntechOpen, 2023, doi: 10.5772/intechopen.1002227.

[18] G. Sasikanth, B. V. R. S. Subramanyam, D. Maity, T. N. Narayanan, and T. P. Radhakrishnan, "Low conductivity polyaniline–gold nanocomposite via interfacial polymerization: efficient humidity sensing by thin films and mechanistic insights through *in situ* experiments," *J. Phys. Chem. C.*, vol. 127, pp. 18107-18116, 2023, doi: 10.1021/acs.jpcc.3c03866.

[19] R. Ikram, B. Mohamed Jan, P. B. Nagy, and T. Szabo, "Recycling waste sources into nanocomposites of graphene materials: Overview from an energy-focused perspective," *Nanotechnol. Rev.*, vol. 12, pp. 20220512, 2023, doi:10.1515/ntrev-2022-0512.

[20] M. Kour, S. Verma, B. Padha, P. Mahajan, A. Ahmed, and S. Arya, "An Introductory View About Supercapacitors," In: Hussain, C.M., Ahamed, M.B. (eds) Functionalized Nanomaterials Based Supercapacitor. Materials Horizons: From Nature to Nanomaterials. Springer, Singapore, 2024, doi:10.1007/978-981-99-3021-0_2.

[21] G. H. Tabhane, S. M. Giripunje, and S. B. Kondawar, "Fabrication and dielectric performance of RGO-PANI reinforced PVDF/$BaTiO_3$ composite for energy harvesting," *Synth. Met.*, vol. 279, pp. 116845. 2021, doi: 10.1016/j.synthmet.2021.116845.

[22] I. S. Ashwini, J. Pattar, P. Anjaneyulu, D. Prakash Babu, R. Sreekanth, S. R. Manohara, and M. Nagaraja, "Synthesis and electrical properties of polyaniline–cerium oxide composites," *Synth. Met.*, vol. 270, pp. 116588, 2020, doi: 10.1016/j.synthmet.2020.116588.

[23] N. C. Horti, M. D. Kamatagi, N. R. Patil, S. K. Nataraj, S. A. Patil, and S. R. Inamdar, "Synthesis and photoluminescence properties of polycarbazole/tin oxide (PCz/SnO_2) polymer nanocomposites," *Polym. Bull.*, vol. 78, pp. 6321–6336, 2021, https://doi.org/10.1007/s00289-020-03428-5.

[24] T. Aravinda, S. Rao, V. Kumbla, and J. Pattar, "Improved dielectric performance of polyvinylidene fluoride (PVDF) - Carbon dots composites," *Phys. E: Low-Dimens. Syst. Nanostructures.*, vol. 147, pp. 115589, 2023, doi: 10.1016/j.physe.2022.115589.

[25] V. T. Sharma, M. Manohara Halanur, S. V. Kamath, D. Mondal, and S. K. Nataraj, "Fe–Al based nanocomposite reinforced hydrothermal carbon: Efficient and robust absorbent for anionic dyes," *Chemosphere.*, vol. 259, pp. 127421, 2020, doi: 10.1016/j.chemosphere.2020.127421.

[26] Q. T. Ain, S. H. Haq, A. Alshammari, M. A. Al-Mutlaq, and M. N. Anjum, "The systemic effect of PEG-nGO-induced oxidative stress *in vivo* in a rodent model," *Beilstein J. Nanotechnol.*, vol. 10, pp. 901–911, 2019, doi: 10.3762/bjnano.10.91.

[27] T.N. Atiqah, S.J. Tan, K.L. Foo, *et al.*, "Effect of graphite loading on properties of polyaniline/graphite composites," *Polym. Bull.*, vol. 75, pp. 209–220, 2018, doi: 10.1007/s00289-017-2031-1.

[28] W. Xinming, Q. Shuhua, and D. Guocheng, "Polyaniline/graphite nanosheet, polyaniline/Ag/graphite nanosheet, polyaniline/Ni/Graphite nanosheet composites and their electromagnetic properties," *Synthetic Met.*, 162, 1609–1614, 2012, doi: 10.1016/j.synthmet.2012.07.012.

[29] S. E. Bourdo, B. A. Warford, and T. Viswanathan, Electrical and thermal properties of graphite/polyaniline composites. *J. Solid State Chem.*, vol. 196, pp. 309–313, 2012, doi: 10.1016/j.jssc.2012.06.038.

[30] Z.Q. Li, C.J. Lu, Z.P. Xia, Y. Zhou, and Z. Luo, "X-ray diffraction patterns of graphite and turbostratic carbon," *Carbon.*, vol. 45, pp. 1686–1695, 2007, doi: 10.1016/j.carbon.2007.03.038.

[31] A. Maddu, R. A. Nugroho, E. Rustami, S. Arjo, and M. Hidayat, "Synthesis of graphene/polyaniline nanocomposite for supercapacitor electrodes," *IOP Conf. Series: Journal of Physics: Conf. Series*, vol. 1171, pp. 012043, 2019, doi:10.1088/1742-6596/1171/1/012043.

[32] M. A. Gadyal, and K. S. Venkatesh, "Synthesis of polyaniline-graphite nano-composites," *Mater. Sci. Res. India*, vol. 12, pp. 85-88, 2015, doi: 10.13005/msri/120114.

[33] E. Akbarinezhad, "Synthesis of conductive polyaniline-graphite nanocomposite in supercritical CO_2 and its application in zinc-rich epoxy primer," *J. Supercrit. Fluids*, vol. 94, pp. 8-16, 2014, doi: 10.1016/j.supflu.2014.06.018.

[34] P. Bhattacharya, S. Dhibar, G. Hatui, A. Mandal, T. Das, and C.K. Das, "Graphene decorated with hexagonal shaped M-type ferrite and polyaniline wrapper: a potential candidate for electromagnetic wave absorbing and energy storage device applications," *RSC Adv.*, vol. 4, 17039-pp. 17053, 2014, doi: 10.1039/C4RA00448E.

[35] M. U. A. Prathap, B. Satpati, and R. Srivastava, "Facile preparation of polyaniline/MnO2 nanofibers and its electrochemical application in the simultaneous determination of catechol, hydroquinone, and resorcinol," *Sensor Actuat B-Chem.*, vol. 186, pp. 67-77, 2013, doi: 10.1016/j.snb.2013.05.076.

[36] T. Wang, and Y. J. Tan, "Understanding electrodeposition of polyaniline coatings for corrosion prevention applications using the wire beam electrode method," *Corros. Sci.*, 48, 2274-2290, 2006, doi: 10.1016/j.corsci.2005.07.012.

[37] X. S. Du, M. Xiao, and Y. Z. Meng, "Facile synthesis of highly conductive polyaniline/graphite nanocomposites," *Eur. Polym. J.*, vol. 40, pp. 1489–1493, 2004, doi: 10.1016/j.eurpolymj.2004.02.009.

IOP Conf. Series: Materials Science and Engineering 1300 (2024) 012015 doi:10.1088/1757-899X/1300/1/012015

Effect of temperature on electrical resistivity of EVA-LDPE nanocomposites for photovoltaic encapsulating application: An activation energy method

Moumita Naskar and Prabhat Kumar Maiti

Central Power Research Institute, Regional Testing Laboratory, Kolkata 700091, India

E-mail: moumita@cpri.in

Abstract. This article evaluates the effect of mixing strategy on thermo- electrical resistance of ethylene vinyl acetate (EVA), low-density polyethylene (LDPE) and fumed silica nanoparticles (FSN) based composite encapsulating materials. The nanocomposites were synthesized by solution compounding method with various sequence of ingredients addition. The quantity of ingredients is same for all nanocomposites where the ratio of EVA, LDPE and FSN addition is 10:1:0.11. The structural morphology of the samples prepared by different mixing sequence was studied under Transmission Electron Microscopy (TEM). DC volume resistivity (VR) were evaluated at 25 °C, 50 °C and 75 °C for all the compositions to check the insulation behaviour at extreme service temperature as VR is an assessing parameter for photovoltaic (PV) encapsulating materials. For obvious reason, the magnitude of VR decreases with increasing temperature. Activation energy (E_a) was estimated from Arrhenius equation and the value evaluated was very much dependent on the ingredients mixing strategy. Good electrical resistance with desired activation energy were obtained for the nanocomposites; prepared with a particular mixing strategy where percolation threshold achieved. Finally, the TEM structure and insulation property relationship has been addressed against the experimental findings.

Keywords: *EVA nanocomposites, Transmission Electron Microscopy, Volume resistivity; Arrhenius equation; Activation energy*

1. Introduction

The market for renewable energy has been increasing remarkably in the last decade due to limitless consumption of fossil fuels. Solar energy is a rich terrestrial renewable source of energy for the sunniest countries. PV module has prominent advancements and highest economical potential among all sustainable energy sources. Encapsulation of PV cell is a compulsory method for safeguarding the crystalline silicon wafer from unwanted environmental and mechanical stresses [1]. Encapsulants maintain the electrical insulation between photovoltaic cells and circuit elements and produces current & voltage during service time. It also combines the top glass and rear back sheet with the module and plays an important role for long-term stability. EVA is one of the most common encapsulant material. The encapsulating layer should eliminate the excessive heat of the silicon wafer [2]. But the uncontrolled service temperature has significant effect on the EVA encapsulation and it may oxidize in presence of heat, moisture and air. Finally, decomposition of EVA encapsulation leads to delamination, optical decoupling, metallic corrosion etc. with subsequent module efficiency loss [3]. Thus, it is essential to improve the thermal stability of the encapsulating material by blending with minimum quantity of more thermal resistant polymer and nanoparticle. Nanocomposite is the best option to achieve tailor made properties which is the aim of this research.

Blending of elastomeric EVA with very less amount of Low Density Polyethylene (LDPE) plastomer develops thermoplastic elastomer (TPE) with both flexible and rigid moieties in their phase structure which improves the thermal resistance of EVA. EVA is a copolymer of non-polar ethylene and polar vinyl acetate (VA) monomers and the structure of non-polar ethylene backbone is similar with LDPE. These two polymers have different structural polarity and degree of crystallinity and they are only miscible in amorphous zone [4]. Compounding of amorphous polymer with semi-crystalline polymer can provide immense possibilities in modifying desirable blend properties. Further, inclusion of nanoparticles can enhance the blend structure [5]. Generally, charge passes through the polymer-polymer and polymer-nanoparticle interfaces. Interface generation has strong influence on the temperature dependent electrical insulation properties of blend nanocomposites and which again depends on the sequence of components addition [6]. The purpose of this investigation is to study the effect of temperature on DC volume resistivity and calculation of activation energy as a function of sequential ingredients addition.

2. Sample preparation

Solution blending was carried out with 28% VA content EVA and LDPE with FNS in xylene medium at 90 ± 2°C with two various sequences of ingredients addition.

Sequence 1 – 40 g of EVA was initially dissolved in 100 ml of xylene followed by 0.44 g of FSN addition and the mixture was stirred vigorously. After 5 min of stirring 4 g of LDPE was incorporated to the mixture and blending continued for 12 min. Benzoyl peroxide (BPO) was added in to and the solution was stirred for final 5 min. 0.5 mm thin film was casted after solvent evaporation by 10 min compression moulding at 130±2°C and the sample is designated as ELS-1.

Sequence 2 – 40 g of EVA was initially dissolved in 100 ml of xylene followed by 4 g of LDPE addition and the mixture was well blended for 12 min. Thereafter, FSN was added to the solution by 1 wt% of total polymer content and mixed thoroughly for another 5 min. BPO was incorporated and the solution was stirred for final 5 min. After solvent evaporation, 0.5 mm thin film was casted by 10 min compression moulding at 130±2°C and the sample is designated as ELS-2.

Sequence 3 – 4 g of LDPE was initially dissolved in 100 ml of xylene followed by 0.44 g of FSN addition and the mixture was well blended for 4 min. Thereafter, 40 g EVA was added to the solution and mixed vigorously for 12 min. BPO was incorporated and the solution was stirred for final 5 min. After solvent evaporation 0.5 mm thin film was casted by 10 min compression moulding at 130±2°C and the sample is designated as ELS-3.

3. Characterization

3.1. *Transmission electron microscopy (TEM)*

Morphology of EVA-LDPE-FSN nanocomposites was investigated under Transmission electron microscope. 70 nm thin sample prepared by using an ultramicrotome equipped with 45° cutting edge diamond knife -50 °C in liquid nitrogen atmosphere. The micrographs were taken with JEOL JEM 2100 high resolution TEM (HRTEM) instrument by using LaB6 filament and the accelerated operating voltage was 200 keV having a resolution of 1.9 Å.

3.2. *Volume Resistivity (VR)*

Volume Resistivity (VR) was estimated by using a Sefelec Megohm Meter, model M 1500P and a Weiss technik climatic chamber. Three terminal resistivity probe ETS Model 803B was used to measure VR. 500V DC was applied for a duration of 60 sec to measure VR at 25±°C, 50±°C and 75±°C as per ASTM D257-21. The DC electrical resistance measurement set up was depicted in Figure 1.

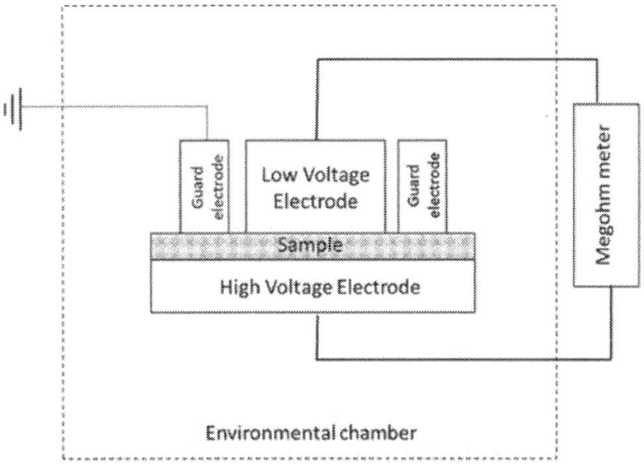

Figure 1. Volume Resistance measurement set up at different temperature

4. Results and Discussion

In order to study the effect of mixing strategy on the morphology of the EVA-LDPE-FNS nanocomposites TEM analysis has been carried out and the image presented in Figures 2 (a), (b) and (c). Figure 2(a) represents the bulk morphology of ELS-1 sample; where FNS was primarily mixed with EVA followed by blending with LDPE. EVA and LDPE are partially miscible and subsisted as two-phase systems. Noticeably it is found in ELS-1 sample that maximum FSN interacted with EVA due to initial high Vander-Wall interaction between EVA and FSN. These thermodynamic effect leads to the filler agglomeration in ELS-1 sample based on the affinity of FSN with EVA. A good homogenous FSN dispersion has been found for ELS-2 in Figure 2(b) as FSN was added to EVA-LDPE blend. This uniform distribution of FSN nano particles were observed in thermoplastic and elastomeric phases along with the interphase where FSN migration was fully controlled by kinetic effect [7]. Moderated agglomerated morphology has been observed for ELS-3 in Figure 2(c). The cluster structure of FSN broke in presence of solvent and could easily migrate into the dissolve LDPE phase. LDPE-FSN does not hold strong physical bonds. FSN nano particle had choice to migrate towards EVA phase through the interphase. FSN percolation rate is lower in ELS3 than ELS1. As a consequence, deagglomerated FSN distribution was developed in ELS3.

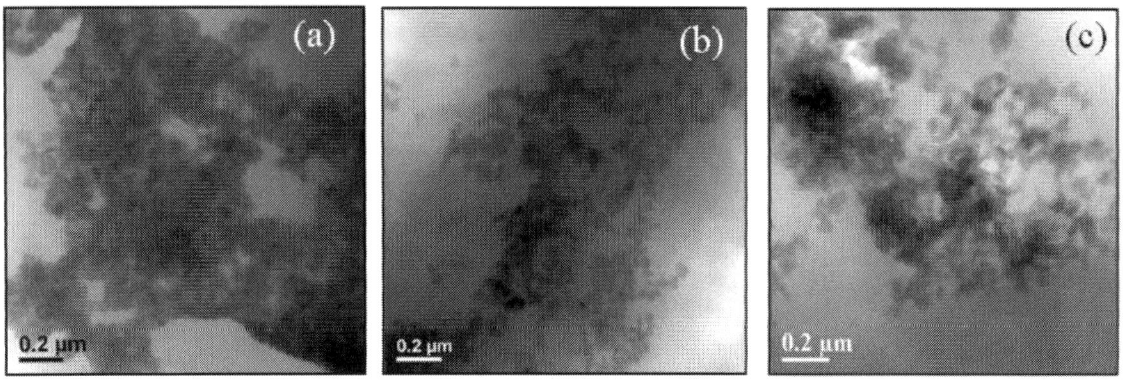

Figure 2. TEM micrographs of (a) ELS-1, (b) ELS-2 and (c) ELS-3

Temperature dependent VR of EVA-LDPE-FSN nano composites prepared with different mixing strategy are presented in Figure 3. Sufficient electrical insulation in a wide temperature span is essential for PV encapsulating material to prevent leakage current for maintaining module reliability and performance. Figure 3 reveals that all the prepared nanocomposite encapsulating materials have high VR, but with temperature it decreases significantly. The intensity of VR reduction with temperature is dependent on the ingredients mixing strategy. ELS2 sample provided higher electrical insulation than its other two counterparts. ELS1 presented a medium VR mangitude in the studied temperature range. The reason may be FSN is mainly confined in elastomeric EVA phase in ELS-1 sample and with temperature rise the mobility of VA part of EVA is restricted by the agglomerated nanoparticle as presented in schematic diagram Figure 4(a). The estimated retention values of VR are 61.6% and 20.0% at 50 °C and 75 °C respectively for ELS1. As per Figure 4(a) FSN agglomeration was prominent in EVA network phase which restrict the movement of EVA segment; but EVA starts flowing on melting at 75 °C caused less VR retention. Contrarily, FSN uniformly distributed in EVA-LDPE blend of ELS2 sample as depicted in schematic diagram Figure 4(b) due to good thermodynamic and kinetic effects. FSN are localized in the co-continuous EVA-LDPE matrixes and at the interphase and efficiently offered very good VR due to the stabilize morphology [5]. Viscoelastic response in more in homogeneous ELS2 system [6]. Simultaneously, VR retention are estimated as 21.6% at 50 °C and 3.13% at 75 °C.

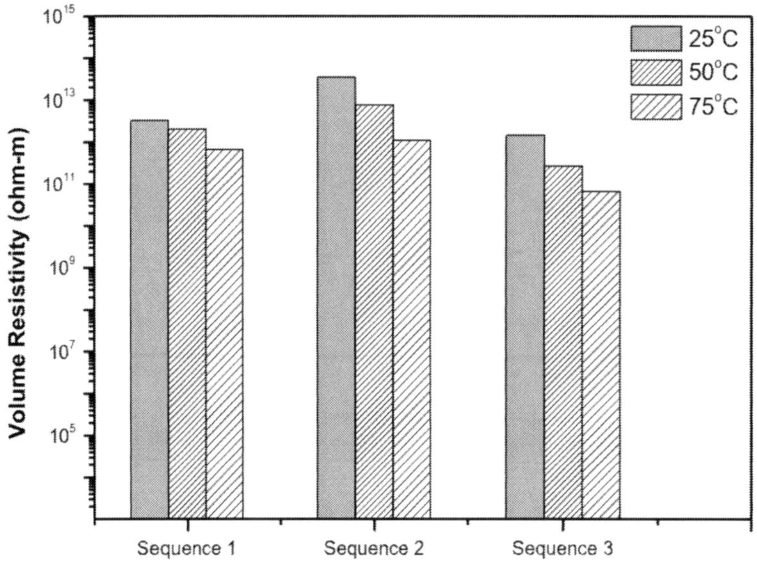

Figure 3. Variation of VR with temperature for ELS-1, ELS-2 and ELS-3

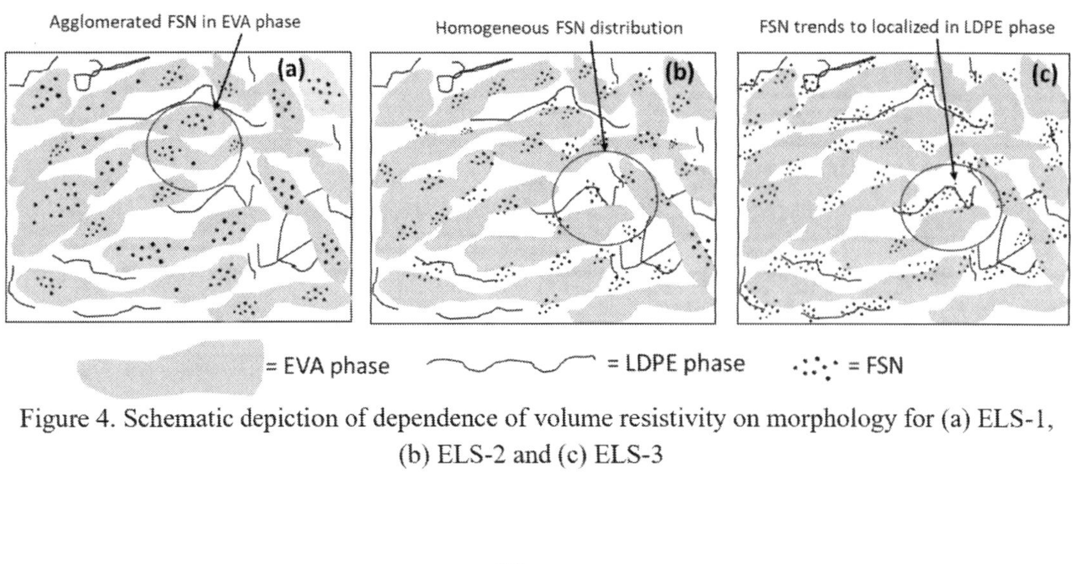

Figure 4. Schematic depiction of dependence of volume resistivity on morphology for (a) ELS-1, (b) ELS-2 and (c) ELS-3

Whereas, ELS3 sample could not afford good electrical insulation in all the temperatures under this study. This result is correlated with the TEM analysis in Figure 4(c) where FSN was premixed with the thermodynamically incompatible LDPE then blended with compatible EVA phase. This provoked the movement of FSN towards EVA through the interphase [8]. As a consequence, interphase is behaved as charge trap and the overall VR dropped [9]. The estimated magnitude of VR retentions are 18.5% at 50 °C and 4.51% at 75 °C for ELS-3.

Figure 5 confirmed that all the ELS nanocomposites have a strong inverse exponential relationship of volume resistivity (ρ_v) with temperature and the influence of nano particles addition strategy in the polymeric matrices and the chain movements of both polymers especially in EVA is an ionic conduction process and can be controlled by Arrhenius model. Thus, by using Arrhenius equation (Equation 1) the activation energy of ELS nanocomposites can be calculated.

$$\sigma = \sigma_0 exp(-E/kT) \quad (1)$$

Where, σ is the electrical conductivity which is equivalent to $1/\rho_v$, σ_0 is the pre-exponential factor, T is the absolute temperature, and k is Boltzmann constant. The activation energy, E_a can be obtained from the slope of the logσ versus 1000/T plot. In Figure 5, the ordinate is ln(1/ ρ_v), abscissa is 1000/T and the slope after linear approximation is equal to -E_a/k.

The estimated values of E_a for all the ELS samples are presented in Table1. It is confirmed that the energy acquired by the nano particle to free itself from the polymer matrix is less for ELS1 and maximum for ELS2. Agglomerated discrete nanoparticle phase in ELS1 has been found from TEM micrograph of ELS1 sample. In agglomerated structure particle-particle collision is more which decreases the energy barrier of the system [10]. Whereas, nanoparticles were distributed uniformly in ELS2 and moderately in ELS3. These aided in the increment of activation energy values.

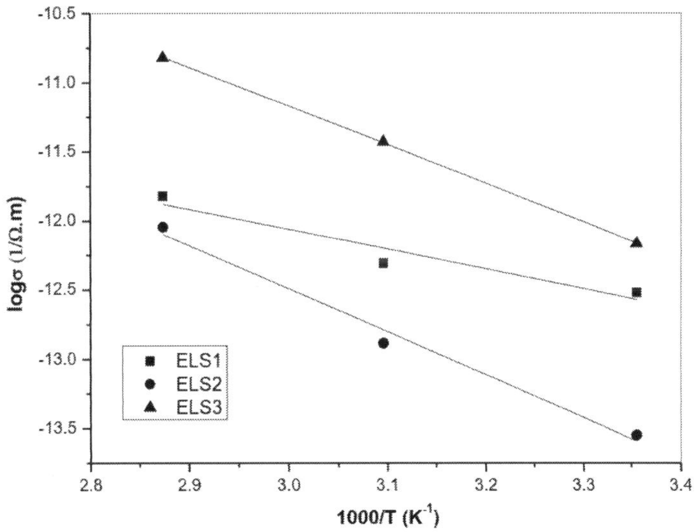

Figure 5. Arrhenius plot of E_a estimation

Table 1. Activation energy of various ELS samples

Sample Code	Activation Energy, E_a (meV)
ELS1	0.53594
ELS2	1.161248
ELS3	1.044673

5. Conclusion

Effects of FSN addition and sequence of mixing in fabrication of TPE nanocomposites were evaluated in this study. The sample films have been characterized by TEM and temperature dependent electrical bulk resistance followed by activation energies measurements. The magnitude of activation energy of ELS samples provided a strong structure-property relationship. FSN contributed a major role in tailoring the morphology of the nanocomposites besides its reinforcing ability. It was also perceived that ingredients mixing strategy is a strong function on the dispersion of FSN in the binary TPE system. Mixing of two polymers followed by FSN addition provided homogenous TPE/FSN nanocomposite as obtained in ELS2 sample. Therefore, ELS2 accorded with better electrical resistance and comparatively superior activation energy. A well balanced thermodynamic-kinetic effect has been established in ELS2. Finally, it can be concluded that the properties of EVA-LDPE blend nanocomposites are significantly ruled by the kinetics and thermodynamics effects by selective localization of nanoparticles in partially miscible blend.

Acknowledgements

The authors are thankful to the management of Central Power Research Institute, Bangalore for approving to publish this paper.

References

[1] Jiang S, Wang K, Zhang H, Ding Y, Yu Q 2015 *Macromol. React. Eng.* **9** 522
[2] Oliveira M C C, Diniz A S A C, Viana M M, Lins V F C 2018 *Renew. Sustain. Energy. Rev.* **81** 2299
[3] Li J, Shen Y C, Hacke P, Kempe M 2018 *Sol. Energy. Mater. Sol. Cell* **188** 273
[4] Elias L, Fenouillot F, Majesté J C, Alcouffe P, Cassagnau P 2008 *Polymer* **49** 4378
[5] De Luna M S, Filippone G. 2016 *Eur. Polym. J.* **79** 198
[6] Nofar M , Salehiyan R, Ray S S 2021 Compos. B. Eng. **215**, 108845.
[7] Khonakdar H A, 2015 *Compos. Part B Eng.* **76** 343
[8] Maani A and Carreau P J 2016 *J. Non-Newtonian Fluid Mech.* **233** 95
[9] Pandey J C and Sing M 2021 SPE Polymers **2** 236
[10] Loria H and Pereira-Almao P 2011 *Ind. Eng. Chem. Res.* **50** 8529

pH Responsive Properties of Methyl Red/PVA Polymers for Smart Food Packaging Applications

Swaroop K[1*, 2], Ashwini S G[3], Murari M S[4], Naveen C S[5], Vinayaka A K[6], Santhosh B M[7], Veerabhadraswamy B N[7] and Yashodhara I[2]

[1]Department of Physics, GM Institute of Technology, Davangere-577006, Karnataka, India.
[2]Center for Application of Radioisotopes and Radiation Technology, Mangalore University, Mangalore-574199, Karnataka, India.
[3]Department of Biotechnology, GM Institute of Technology, Davangere-577006, Karnataka, India.
[4]DST-PURSE, Mangalore University, Mangalore-574199, Karnataka, India.
[5]Department of Physics, School of Engineering, Presidency University, Bengaluru-560064, Karnataka, India.
[6]Department of Physics, Anjuman Arts, Science and Commerce College, Bhatkal-581320, Karnataka, India.
[7]Department of Chemistry, GM Institute of Technology, Davangere-577006, Karnataka, India.

*E-mail: swaroop.k.acharya@gmail.com

Abstract. Responsive polymers in particular have gained increasing interest during the last decades due to their unique properties. Their Ability to exhibit sharp and reversible changes in response to environmental pH condition have made them suitable for various applications. The pH responsive polymeric materials are being developed by the researcher in the recent year for the improvement of food safety issue in the packaging materials. In this study organic methyl red (MR) dye is used as a pH indicator induced with polyvinyl alcohol (PVA) films prepared by solution casting method. The prepared samples were characterized for microstructural, morphological, mechanical, water absorption and pH responsive properties using various analytical techniques. The XRD analysis has shown better crystallinity property which is in good agreement with the mechanical strength of the material. The morphological studies have shown homogeneous dispersion of MR on PVA matrix and change in the morphology of the MR/PVA polymer composites after acid treatment. The MR/PVA films were treated with rotten papaya fruit and change occurred in the samples were identified using UV-vis spectrometric analysis. The prepared MR/PVA pH-responsive polymers have shown good pH reactivity in the range of pH 4.5 to 6.5. Overall, the prepared samples have good pH responsive properties with additional qualities like good optical, structural, and mechanical characteristics.

Key words: Solution casting, pH response, powder XRD, crystallinity and mechanical strength.

1. Introduction

The pH responsive polymers, also known as smart polymers or stimuli-responsive polymers, are a class of materials that can undergo reversible changes in their properties in response to changes in the pH of their surrounding environment [1-3]. These polymers are designed to be sensitive to variations in the acidity or alkalinity of their surroundings and can exhibit significant changes in their structure, solubility, and other physical or chemical properties in response to pH changes [4]. This unique ability makes them valuable in a wide range of applications, including drug delivery, sensing, textiles, agriculture, environmental remediation, and food packaging applications [2].

Halochromic chemicals are introduced sparingly to the polymer matrix as pH indicators so that the colour changes caused by its reaction with bases or acids can be seen. The pH indicators are weak acids that are found as natural dyes that can change colour to show how many $H^+(H_3O^+)$ ions are present in a solution [5]. The pH value of the substance under evaluation, is determined using the logarithm of this concentration, serves as an indicator to its acidity, basicity, or neutrality. Indicators work on the basis that they react with water to produce the hydrogen cation H^+ or hydronium ion H_3O^+ [5]. The indicator molecule's colour changes as a result of the reaction. While some indicators alternate between coloured and colourless states, others change from one colour to another.

Polyvinyl alcohol (PVA) is a water-soluble synthetic polymer that has a wide range of applications in various fields, including the development of pH-responsive polymers [6]. The pH responsive polymers, when incorporated into food packaging materials, can help in extending the shelf life of perishable products. PVA-based films or coatings can be engineered to respond to changes in the pH of the packaged food, indicating the freshness of the product [7]. This application ensures that consumers are aware of the quality of the food they are purchasing. In the context of food packaging applications, MR could potentially be used as a pH indicator to monitor the acidity of food products. The change in pH could indicate spoilage. For instance, if a food product becomes more acidic due to spoilage, the colour change in MR can indicate this change.

In this context, a MR/PVA composite polymer was prepared using simple solution casting method, without the aid of any special equipment or any toxic compound involving in it. The MR/PVA polymers were characterized for structural, morphological, mechanical, swelling, and pH responsive properties using suitable analytic techniques.

2. Materials and Methods

The Polyvinyl alcohol (PVA) (~89,000 Mw, SDFCL, India) was used as base polymer material. Methyl Red (MR) (0.1% solution, indicator grade, Loba Chemicals, India) was used as a pH sensitive die for identifying the food spoilage. Glutaraldehyde (25% aqueous solution, Aldrich, USA) was used as crosslinking agent for the MR/PVA polymer composites. Double distilled water (DDW) was used throughout the experiments. All the chemicals were of analytical grade and used without further purification.

The MR/PVA polymer composites were synthesized using solution casting method. Initially 15 wt% of PVA aqueous solution were prepared by dissolving 15 gm of PVA into 100 ml of double distilled water at 70℃ under constant stirring and a small amount of glutaraldehyde was added to the prepared solution as a crosslinking agent and cooled to room temperature. Further, different concentration of MR namely, 0 ml, 1 ml 2 ml and 3 ml were added separately to the prepared PVA solution. Then the solutions were poured to the petri dishes separately and dried overnight at 50 ℃. The dried pure PVA polymer and 1 ml MR/PVA (MRPVA1), 2 ml MR/PVA (MRPVA2) and 3 ml MR/PVA (MRPVA3) polymer composites were used for further characterization. The pH response of the prepared samples were studied by treating them in the acidic environment of pH 4.5 to 6.5. The prepared polymer composites were also studied for pH responsive property with papaya fruit.

The prepared sample were tested for microstructural, morphological, mechanical, water absorbance and pH responsive properties using various analytical technique. The microstructural characteristics were carried out using powder X-ray diffractometer (powder XRD) (Rigaku; Miniflex, USA) within the diffraction angle (2Θ) range of 10-70° at a scan rate of 3°/min. The surface morphology of the prepared samples were characterized using Field Emission Scanning Electron Microscopy (FESEM) (ULTRA 55; Carl Zeiss, Germany). The mechcanical strength and elongation properties were studied using Universal Testing Machine (UTM) (LRX Plus; Lloyd Instruments, UK). The water absorption characteristics of the prepared samples were observed by immersing the samples in the DDW. The saturation swelling percentage were recorded after 24 hrs. The pH responsive properties of the prepared samples against different pH solution and with the papaya fruit were studied using UV-visible spectrophotometer (UV-vis) (UV-2600; Shimadzu, Japan).

3. Result and Discussion

The crystallinity of MR/PVA polymer were investigated using powder XRD analysis (figure 1). The pure PVA and the untreated MR/PVA have resembled same XRD pattern revealing the predominant diffraction angle at 2Θ=19.8°. This peak has a d spacing of 4.57, which corresponds to the crystalline PVA's (101) reflection plane [8, 9]. This pattern represents the semi-crystalline nature of PVA ascribing to the strong interaction between the PVA chains through intermolecular and intramolecular hydrogen bonding [10]. The acid treated MR/PVA polymer composite demonstrated slight increment in the crystallinity by showing increase in the intensity of the diffraction angle at 2Θ=19.8°. The powder XRD results shows that the acid treatment has marginally increased the crystallinity of the MR/PVA polymers. This result demonstrated that the interaction of acid with the MR/PVA polymer composite caused decrease in the amorphous nature of the films showing brittleness.

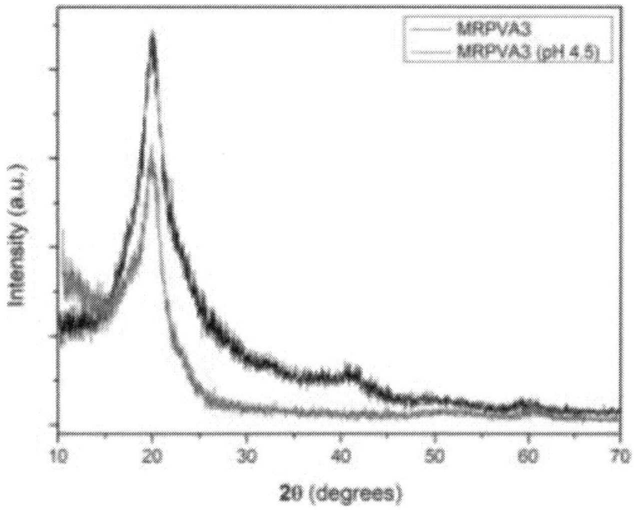

Figure 1: Powder XRD analysis of MR/PVA polymer composites.

To study the way molecules are arranged in the MR/PVA polymer composites, the surface morphology of the Pure PVA, MRPVA3, and MRPVA3 treated with buffer solution of pH 4.5 were investigated using FESEM analysis (figure 2). The morphology of pure PVA [figure 2(a)] has shown smooth morphology with no cracks of inhomogeneity on the surface confirming the amorphous nature complementing to the powder XRD analysis [8]. Whereas the morphology of the MRPVA3 [figure 2(b)] has also shown smooth morphology with small crystalline phases spread inhomogeneously throughout the surface ascribing to the presence of MR within the PVA matrix [11]. This indicates slight increment in the crystallinity after the addition of MR to PVA matrix. The acid treated MRPVA3 sample has shown major cracks throughout the surface and shown drastic increase in the crystallinity of the samples.

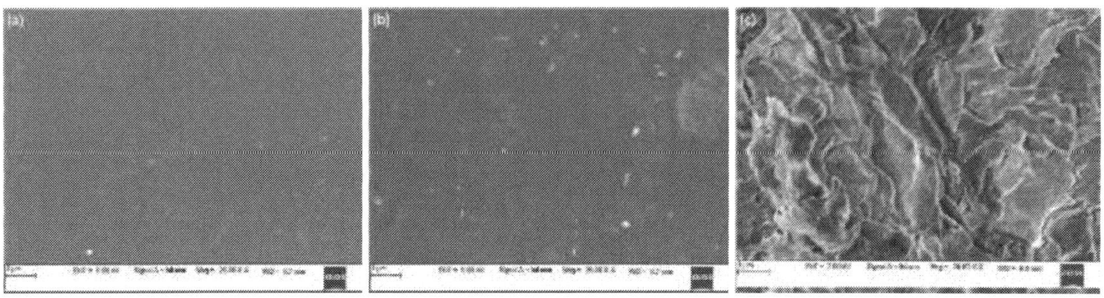

Figure 2: FESEM micrographs of a) Pure PVA, b) MRPVA3 and c) MRPVA3 polymer composites treated with 4.5 pH buffer solution.

The mechanical analysis of the MR/PVA polymer composites before and after the acid treatment were obtained from the stress vs strain curve and are provided in figure 3. The pure PVA polymer has shown the tensile strength of 49.89±1.2 MPa. The MR/PVA polymer composites have shown small decrement in the tensile strength with increase in the MR concentration as shown in the figure 3. Also, it was observed that the tensile strength has significantly decreased for MR/PVA polymer composites after the acid treatment. The % elongation of the samples has also shown decreasing trend with increase in the MR concentration. The % elongation of the MR/PVA polymer composites has considerably decreased after the acid treatment as shown in the figure 3. The mechanical studies of the prepared sample justifies increase in the crystallinity percentage with the acid treatment which can be correlated to the powder XRD and FESEM analysis results. The values of tensile strength, Young's modulus and % elongation were within the range of 40.5±0.5 to 49.89±1.2, 59.21±0.99 to 89.29±0.5 and 112.1±10.56 to 197.1±15.25 respectively. By considering the values of tensile strength, Young's modulus, and %elongation the developed MR/PVA polymer composites exhibit acceptable load bearing capacity, sufficient stiffness, and good elastic characteristics, which could make them suitable for the use in food packaging applications.

The swelling properties were analysed by studying the % swelling characteristics of prepared MR/PVA polymer composites. The samples were swollen to saturation and the equilibrium degree of swelling (%EDS) was recorded by soaking them in double distilled water for 24 hours. The %EDS of all the samples were shown in figure 3. The pure PVA has shown %EDS 192.5 ±21.2. The MR/PVA polymer composites have shown small decrement in the %EDS value with addition of MR concentration. The %EDS of MRPVA3 samples after treating with buffer solution of pH 4.5 and 6.5 were found to be decreasing significantly. The decrement in the %EDS is may be due to increase in the crystallinity of the sample after treating it with acids.

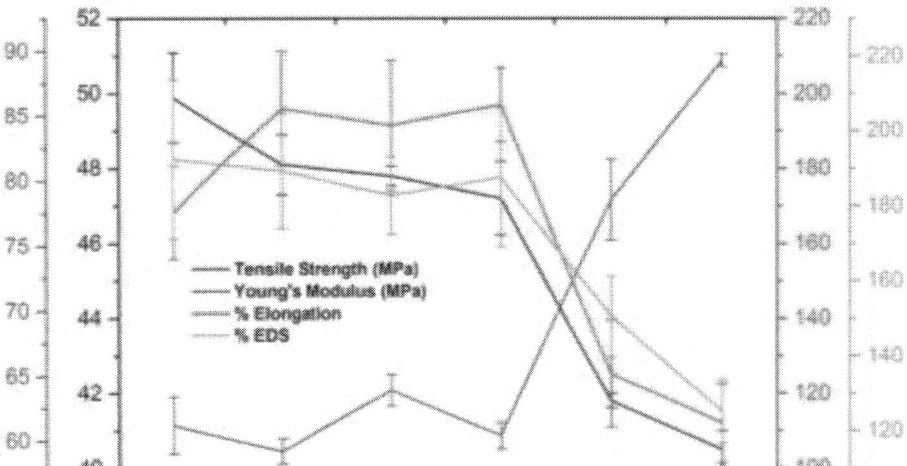

Figure 3: Mechanical and swelling analysis of MR/PVA polymer composites.

The UV-Vis spectroscopic analysis of the MR/PVA polymer composites were shown in figure 4 (a). The pure PVA polymer has shown an absorption peak at ~215 nm which can be attributed to the carbonyl group or the presence of residual acetate group of PVA. The MR/PVA polymers have shown a significant absorption peak at 283 nm and 408 nm along with the PVA characteristic peak [3, 12]. The MR/PVA polymers treated with buffer solution of pH 6.5 have shown maximum absorption peak at 435 nm confirming the yellowish appearance after the treatment. Whereas, MR/PVA polymers treated with buffer solution with pH 4.5 have shown maximum absorption peak at 525 nm confirming reddish appearance after treatment [12]. The intensity of the characteristic absorbance peak of the samples against both 6.5 and 4.5 pH buffer solutions found to be increased with increasing MR concentration confirming better pH response of MRPVA3 sample compared to others.

The pH response of the MR/PVA polymers were also tested against papaya fruit. The papaya fruit was allowed to rot for 9 days and the pH of the rotten papaya was recorded. The MRPVA3 sample was tested for pH response against rotten papaya at pH 4.3 and 6.6. The photographs representing the MRPVA3 sample after the treatment were shown in figure 4 (b). The MRPVA3 appeared to be pale yellowish and red in appearance with respect to pH 6.5 and 4.5 respectively. The results obtained are in good agreement with the UV-vis spectrometric analysis of MR/PVA polymer composites treated with the range of buffer solutions. The pH response trails showed that the MRPVA3, which has the highest MR concertation, demonstrated the most noticeable colour change that is easily visible to the human eye.

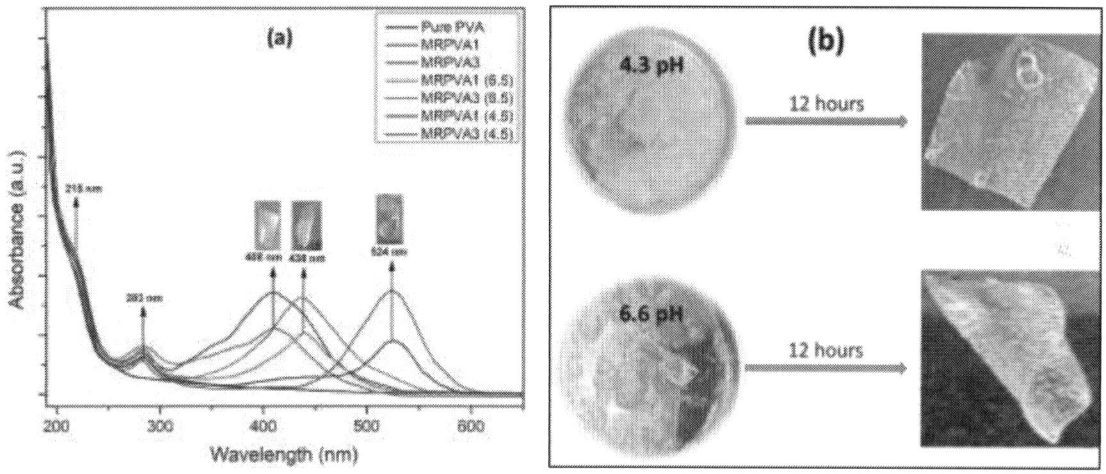

Figure 4: a) UV-vis spectroscopic studies of MR/PVA polymers treated with different pH buffer solutions, b) pH response study of MR/PVA polymer composites against papaya fruit.

4. Conclusion:

The pH responsive MR/PVA polymer composites were developed through solution casting method by physically blending MR in PVA matrix. The prepared samples were tested for pH responsiveness in the range from 4.5 to 6.5 pH values. The study demonstrates the feasibility of pH responsive MR/PVA polymer composites with the functionality of the colour change within the studied range of pH values. Change in microstructure and morphology of the MR/PVA polymers with addition of MR and acid treatment were studied using powder XRD and FESEM analysis. Both the results compliment the fact that the acidic environment has certainly increased the crystallinity of the polymers. Though, mechanical properties of the MR/PVA polymer composites has significantly varied after acid treatment, the material retained good mechanical strength, flexibility and ability to protect the food products. The colour changes occurred to the samples after the acid treatment were visually identified and also change in the wavelength were confirmed by the UV-vis spectroscopic analysis. The results of pH response of samples against papaya fruit during 9 days of rotting was in agreement with the UV-vis spectrometric analysis against range of acidic environment. The ability of the MR/PVA polymer composites to change colour in response to rotten papaya fruit allows consumers to determine the fruit's freshness using their naked eyes. Also, with better microstructural and mechanical stability the prepared MR/PVA polymer composites have the high potential to be used as a smart food packaging material.

Acknowledgement:
Authors are grateful to the Principal and Management, GMIT, Davangere for the financial support and laboratory facilities to carry out the research work. Authors are thankful to Dr. N Karunakara, Centre-in-Charge, CARRT, Mangalore University, Mangalore for extending the laboratory facilities to carry

out the research work. Authors are thankful to the Dr. Girish Bolakatti, Principal, GMIPSR, Davangere for providing the laboratory facilities and Mr. Pradeep H K, Assistant Professor, GMIPSR, Davangere for his support during the conduction of the experiments.

References:

1. El-Nasser H M 2011 *Phys. B Condens. Matter* **406** 1940.
2. Subramanian K 2022 *J. Nanomater.* **2022** 2022.
3. Chairunisa W and Imawan C 2015 *J. Phys. Conf. Ser.* **1321** 2019.
4. Khanifah S, Legowo A D K, Sholihun S and Nugraheni A D 2023 *Indones. J. Chem.* **23** 1021.
5. Alizadeh-Sani M, Tavassoli M, McClements D J and Hamishehkar H 2021 *Food Hydrocoll.* **111** 106237.
6. Gao L, Kong T and Huo Y 2016 *Macromolecules* **49** 7478.
7. Amaregouda Y and Kamanna K 2023 *Sustain. Food Technol.* **1** 738.
8. Kumaraswamy S, Patil S L and Mallaiah S H 2020 *J. Bioact. Compat. Polym.* **35** 435.
9. Gupta S, Pramanik A K, Kailath A, Mishra T, Guha A, Nayar S and Sinha A 2009 *Colloids Surfaces B Biointerfaces* **74** 186.
10. Nasar G, Khan M S and Khalil U 2009 *J Pak Mater Soc* **3** 67.
11. Jagadish K, Chandrashekar B N, Byrappa K, Rangappa K S and Srikantaswamy S 2016 Anal. Methods **8** 2408.
12. Păuşescu I, Dreavă D M, Bîtcan I, Argetoianu R, Dăescu D and Medeleanu M, *Polymers (Basel).* **14** 17.

IOP Conf. Series: Materials Science and Engineering 1300 (2024) 012017 doi:10.1088/1757-899X/1300/1/012017

Study the effect of cobalt chloride ($CoCl_2$) on the structural, optical, and electrical properties of polymer electrolyte (PVA - $CoCl_2$) films

Roopa K V[1], Udayaraj S[1], Annappa M[1], Gavisiddayya Mathad[1], Subramanya K[1], and Devendrappa H[2]
[1]*Department of PG Studies and Research in Physics, Kuvempu University, Shimoga-577451 India*
[2]*Department of Physics, Mangalore University, Mangalagangotri -574199 India*
**Email: subbuk17@gmail.com*

Abstract. A study of non-ionic polymer Polyvinyl Alcohol (PVA) in pure and transition metal salt Cobalt Chloride ($CoCl_2$) incorporated films with different weight percentages to change the optical, electrical, structural, and chemical structure also to make electrically conductive. To prepare the material in terms of film desired weight percentage of polymer PVA alone and for doped polymer different weight percentages of transition metal salts are dissolved in deionized water at a certain temperature and a solution cast technique is followed to get the desired thickness of film. Obtained films were transparent and achieved micro-thickness film. Experimentally confirmed for betterment in optical property to enhance in terms of absorbance, transmittance, direct and indirect energy band gap, extinction coefficient, reflectance property, refractive index, number of carbon clusters, and dielectric constant. The prepared film thickness is measured by using a Mittometer. Surface morphology studies are achieved by SEM analysis in terms of top view and sectional view also thickness of the sample is confirmed. Surface roughness and Maximum Height of the Profile are confirmed by profilometry studies. Electrical property is measured to find the desired range of IV value and to understand the sample behaviour as a resistor.

1. Introduction

Polyvinyl alcohol (PVA) is considered to be one of the most significant polymeric materials due to its wide range of applications in the industry and its relatively low manufacturing cost. Polymeric materials are a subject of great interest in scientific and technological research owing to their ability to be customized to meet specific requirements for a variety of applications. This is primarily due to their lightweight nature, excellent mechanical strength, and optical properties. Polymers have the potential to generate novel electronic levels within the forbidden gap of the electronic band structure. As a result, they transform from being insulating material to possessing electrical conductivity. This phenomenon presents various opportunities for the utilization of these materials in potential applications, such as the production of integrated circuits, optoelectronic devices, sensors, ion-detecting devices, solid-state batteries, which is capable of resisting water and thus inhibiting the entrance of moisture and other technological applications. [1234]

2. Experimental

2.1. Materials

(PVA) Polyvinyl Alcohol $(C_2H_4O)_n$ M_w=85000 to 125000, Cobalt Chloride hexahydrate $CoCl_2.6H_2O$ M_w= 183.32 was procured from SD Fine Chemicals, Double distilled DI water obtained at Kuvempu University.

2.2. Synthesis

To obtain the desired weight percentages of PVA-$CoCl_2$ sample films with concentrations of 0%, 5%, 10%, 15%, 20%, 25%, and 30% the technique of Solution cast was employed. The process of sample preparation involved taking 25 ml of deionized (DI) water in a beaker, which was then heated to 50°C using a hot plate. Subsequently, 4g of PVA was measured and added to the DI beaker, which was then stirred at 300rpm using a magnetic stirrer for approximately 8 hours, maintaining a temperature of around 50°C, just below its glass transition temperature. The resulting sample, with a specific consistency, was poured into a glass petri dish with a diameter of 95mm and left to dry for a period of 2-3 days at 50°C. Once the sample had solidified, it was peeled off from the petri dish and sealed in zip cover bags, after which it was subjected to desiccation for approximately 1 week. The film obtained was then cut into squares with a desired area of one square centimetre, it conduct further experimental characteristics.

3. Characterisation and Discussion

3.1. Sample thickness

The physical property, thickness and of the sample was found using Mittometer. The sample thickness obtained between the ranges 70μm to 100μm was measured using a Mittometer as shown in figure 1 and the values in Table 2.

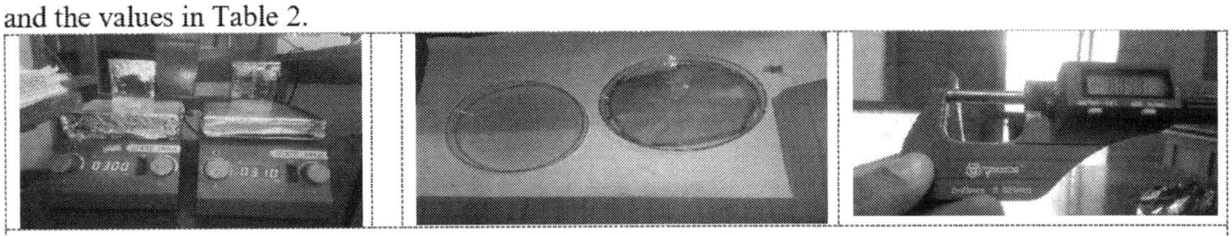

Figure 1a. Synthesis process flow. **Figure 1b.** Film obtained by Casting. **Figure 1c.** Measurement of film thickness using a Mittometer respectively.

3.2. Optical Properties

The optical properties were studied using UV–visible NIR spectroscopy using Quartz tungsten Halogen lamp as a source for pure and composites between 200 nm to 1500 nm wavelength range. The transmittance property has decreased from 92% to 45% evidencing an increase in absorbance property. Reflectance percentages decreased from 7 to 0.1% is found highest in 30% weight percentages as shown in figure 2a, 2b, 2c respectively. The optical band gap energy E_g is calculated using Tauc's equation, Absorption coefficient α, $\alpha h\nu = A(h\nu - E_g)^n$ $\alpha = 2.303.(t^{-1}).log(T)^{-1}$ where n=0.5 for Direct / n=2 for Indirect Energy, T is transmission, t is thickness of the sample, h is Plancks constant, ν frequency of source. Extinction Coefficient κ, refractive index η, Carbon Cluster Number N from [5 6] found tailored from pure to doped PVA-$CoCl_2$ films as in figure 3. [7] and values are found in Table 1. The refractive index was found to be increased, which helps in low scattering of the lattice materials, and the angle-dependent scattering profile results in a reduced translucent appearance. [8] Carbon cluster is found to know the elemental composition and excitation levels within the molecule using equation given by Robertson's and O'reilly [9].

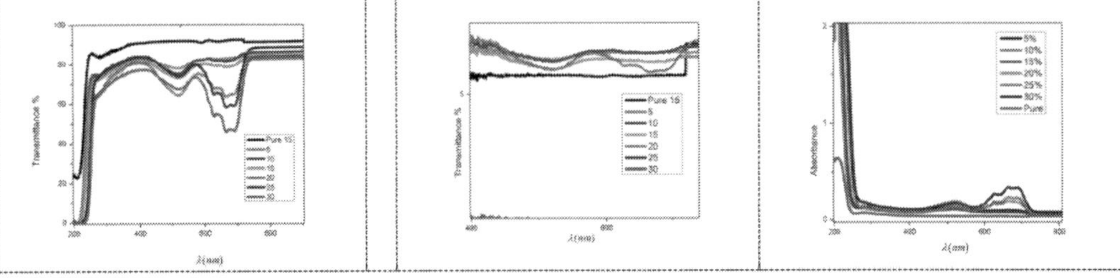

Figure 2a, 2b, 2c. The plot of Transmittance, reflectance and absorbance respectively for Pure PVA and different weight percentages of PVA-$CoCl_2$.

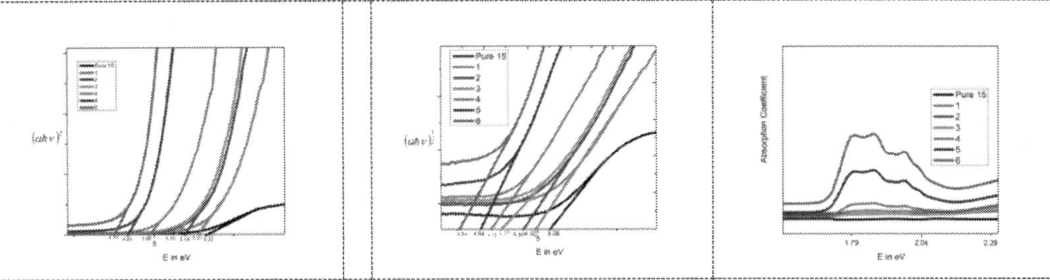

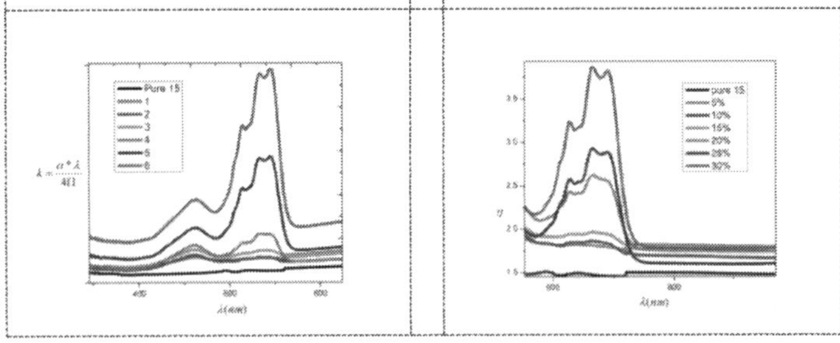

Figure 3. The graph for indirect - direct band gap, absorption coefficient, extinction co efficient, refractive index for Pure PVA and different weight percentages of PVA-$CoCl_2$.

Table 1. Summary of optical Property obtained from UV –NIR Spectroscopy for pure and different weight percentages of PVA doped films.

Sample	α x 10^{-5} eV	κ x 10^{-6} eV	η	Direct		Indirect	
				E_g eVcm^{-1}	N	E_g eVcm^{-1}	N
Pure PVA	0.09	0.45	1.47	5.08	3	5.32	3
5% $CoCl_2$	0.18	1.09	1.78	4.98	3	5.27	3
10% $CoCl_2$	0.21	1.24	1.86	4.90	3	5.19	3
15% $CoCl_2$	0.27	1.51	1.97	4.77	3	5.13	3
20% $CoCl_2$	0.42	2.28	2.63	4.70	3	4.96	3
25% $CoCl_2$	1.08	5.81	2.94	4.64	3	4.83	3
30% $CoCl_2$	1.81	9.78	3.85	4.54	4	4.74	3

3.3. *SEM*

Surface morphology studies and thickness of the sample were studies using Table top Scanning electron microscope (SEM) SNE4500M, with 5000V and magnification of 1000 times. Morphology analysis indicates that the amalgamation of the PVA – $CoCl_2$ sample is homogeneous, thereby minimizing agglomeration. Furthermore, it reveals a decrease in the amorphous characteristics with varying concentrations. Additionally, the thickness from a lateral perspective is assessed through scanning electron microscopy, which correlates with the aforementioned device measurement as in figure 4. [10] The surface roughness can be visualized and linked to surface profilometry. Table 2 Measured thickness using a Mittometer and SEM.

Figure 4. SEM images of Pure PVA and different weight percentages of PVA-$CoCl_2$ are illustrated in both the top and side views.

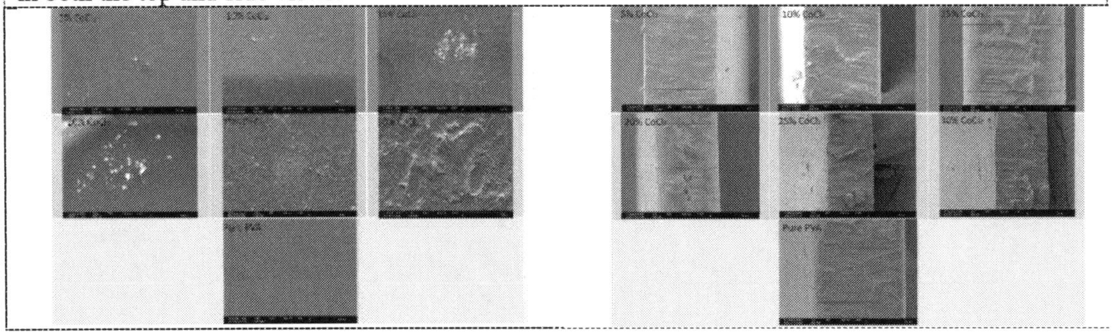

3.4. *Surface Profilometry*

Surface thickness was found using Surface Profilometry using Surface Profiler- Bruker 3D Non-contact Profiler Contour GT device.

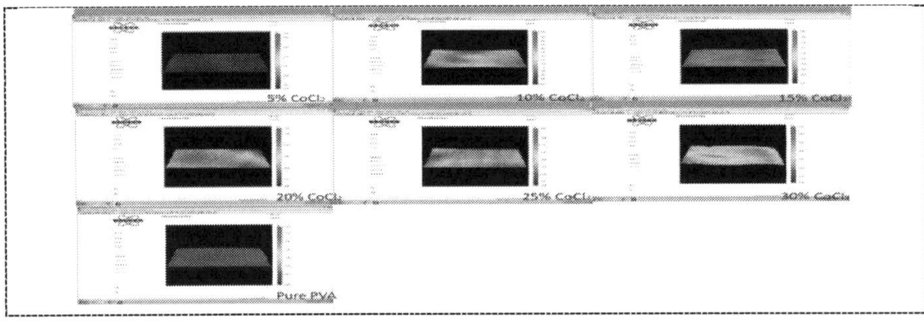

Figure 5. Surface profilometry for different doped percentages to know surface roughnes.

Table 2. Summary of the thickness measurement of PVA doped films was conducted using the Mittometer, scanning electron microscopy (SEM) image analysis, and the determination of average roughness values.

Sample	Thickness		Average Roughness
	Mittometer µm	SEM µm	R_a µm
Pure PVA	104	102	40.16
5% $CoCl_2$	93.5	83.7	46.21
10% $CoCl_2$	86.6	72	65.16
15% $CoCl_2$	81.6	120.5	88.57
20% $CoCl_2$	81.6	58.2	108.85
25% $CoCl_2$	50	43	159.14
30% $CoCl_2$	43	50.3	259.59

The sample roughness is understood by the surface profilometry image. This evidences the SEM morphology to know surface smoothness decreased and increase in average Roughness of the surface,

R_a to know amorphous nature of sample decreased with increased concentration [11] as shown in figure 5 and the values in Table 2.

3.5. *DC Electrical*

DC Electrical property studies were done using Probe Station - Cascade Microtech Summit 12000AP Current versus Voltage using probe station in vacuum for the temperature range -50^{0}C to +190C and voltage range -100V to +100V and checked for the current correlation. The electrical properties is studied using DC Conducting property. The current value has increased from nano to micro range with increased concentration of dopant. The graph nature of current I versus Voltage V is studies to know resistance and resistivity of the material. Hence conductivity has plotted corresponding to doped percentages in figure 6. The resistance R, resistivity ρ, conductance σ obtained values are tabulated in Table 3. [12]

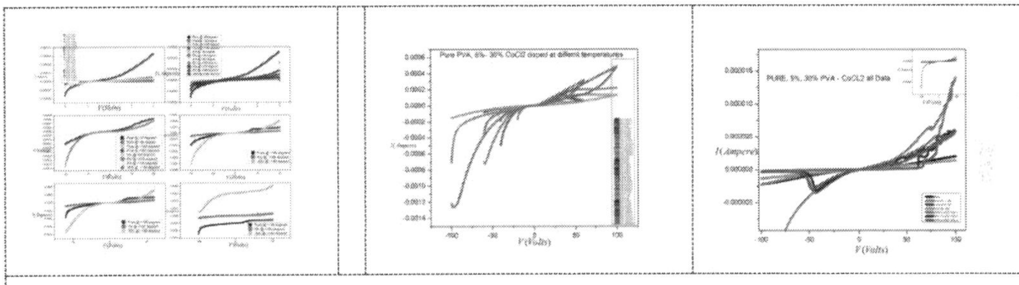

Figure 6. The graph for DC conductivity IV studies for Pure, 5% and 30% PVA-$CoCl_2$. Different sample at different Voltage range and temperature range at vaccum chamber, room chamber and two cycle multiple run respectively.

Table 3. Summary of electrical properties of PVA-$COCl_2$ films at different temperature and voltage between -20 to +20 V.

Samples	Average Resistance (Ω)	Resistivity, ρ X 10^{+6}(Ω.cm)	Conductivity, σ X 10^{-5} (Ω.cm)-1
Pure PVA	1.66 M	159	0.0627
5% PVA- $CoCl_2$	27.0 K	2.89	3.46
30% $CoCl_2$	32.6 K	7.58	1.32

4. Conclusion

The PVA – $CoCl_2$ sample with varying concentrations of dopant was economically, timely, and qualitatively, homogeneous film was obtained. UV studies proves band bap energy decreases resulting in enhancement of absorption for enhancing of photons jump within the level enhancing Π-Π* for transition metals. Decrease in the amorphous characteristics with varying concentrations proving the surface roughness using SEM morphology and Surface profilometry average roughness enhancement. Additionally, the thickness plays important role in calculating many parameter is confirmed from a lateral perspective through scanning electron and Mittometer. The significant value of conductivity, resistivity using DC electrical conductivity shows the material best suited for electrical devices. PVA-$CoCl_2$ can be used as coat resistant for preventing moisture penetration on the surface.

5. Acknowledgments

Authors has been acknowledged Kuvempu University for its provision of deionized water and space for the purpose of fabrication.
Dr. Praveen K C, a Scientist and Engineer in the Laboratory for Electro-Optics Systems (LEOS) within the Indian Space Research Organization (ISRO), has been recognized for conducting UV NIR Spectroscopy studies.
The Microelectronics and MEMS Laboratory, situated within the Department of Electrical Engineering at IIT Madras, has been commended for their involvement in the characterization of SEM, Profiler, and DC measurements as part of the INUP Program.

6. References

1. Jabbar, W. A., Habubi, N. F. & Chiad, S. S. *Optical Characterization of Silver Doped Poly (Vinyl Alcohol) Films. Journal of the Arkansas Academy of Science* vol. 64 http://scholarworks.uark.edu/jaashttp://scholarworks.uark.edu/jaas/vol64/iss1/21 (2010).
2. Kilarkaje, S., Manjunatha, V., Raghu, S., Prasad, M. V. N. A. & Devendrappa, H. Effect of 8 MeV electron irradiation on the optical properties of doped polymer electrolyte films. *Journal of Physics D: Applied Physics* **44**, (2011).
3. Heiba, Z. K., Mohamed, M. B. & Ahmed, S. I. Exploring the physical properties of PVA/PEG polymeric material upon doping with nano gadolinium oxide: Exploring the physical properties of PVA/PEG polymeric material. *Alexandria Engineering Journal* **61**, 3375–3383 (2022).
4. Abid, R. A. & Habeb, M. A. Effect of cobalt chloride (CoCl 2) on the electrical and optical Properties of (PVA-PVP-CoCl 2) films. vol. 18 at www.iiste.org (2013).
5. by B Zainab Falih Nazzal & ProfZainabAAl-Ramadhan, S. Physical properties of polymer(PVA) doped with Cobalt salts. *Sc. in Physics* at (2001).
6. Aziz, S. B. *et al.* Tuning the absorption of ultraviolet spectra and optical parameters of aluminum doped PVA based solid polymer composites. *Journal of Materials Science: Materials in Electronics* **26**, 8022–8028 (2015).
7. Ahmed, R., A. Ibrahiem, A. & A. El-sayd, E. Effect of Cobalt Chloride on the Optical Properties of PVA/PEG Blend. *Arab Journal of Nuclear Sciences and Applications* **0**, 0–0 (2018).
8. Zhang, X. A. *et al.* Ordered 3D Thin-Shell Nanolattice Materials with Near-Unity Refractive Indices. *Advanced Functional Materials* **25**, 6644–6649 (2015).
9. Gupta, D., Kumar, S., Kalsi, P., Manchanda, V. & Mittal, V. γ -Ray Modifications of Optical/Chemical Properties of Polycarbonate Polymer. *World Journal of Condensed Matter Physics* **05**, 129–137 (2015).
10. Elsaeedy, H. I., Ali, H. E., Algarni, H. & Yahia, I. S. Nonlinear behavior of the current–voltage characteristics for erbium-doped PVA polymeric composite films. *Applied Physics A: Materials Science and Processing* **125**, (2019).
11. Mei, L. & Guan, G. Profilometry and atomic force microscopy for surface characterization. *Nano TransMed* **2**, e9130017 (2023).
12. Abdul-Manaf, N. A., Salim, H. I., Madugu, M. L., Olusola, O. I. & Dharmadasa, I. M. Electro-plating and characterisation of CdTe thin films using CdCl2 as the cadmium source. *Energies* **8**, 10883–10903 (2015).

Alternating Current Response Study on PANI/V_2O_5 Composite at Room Temperature

M Annappa[1], B Chethan[2], S Udayaraj[1], M Gavisiddayya[1], K VRoopa[1], H Devendrappa[3], K Subramanya[1*]

[1]Department of PG Studies and Research in Physics, Kuvempu University, Shimoga-577 451
[2]Department of Physics, Indian Institute of Science, Bangalore-560012, India
[3]Department of Physics, Mangalore University, Mangalagangotri 574199

*Email: subbuk17@gmail.com

Abstract. In the present study, an Alternating Current (AC) response of Vanadium pentoxide (V_2O_5) doped Polyaniline (PANI) composite was studied. Pure PANI was synthesised by the In-situ polymerisation method. The PANI/Vanadium Pentoxide (PVO) composite was prepared by physical mixing of chemically synthesised PANI with dopant in a vibration mill. The structure and morphology of the samples were characterised by XRD and SEM studies. The Alternating Current response parameters of PANI, V_2O_5 and PVO composites were studied in the frequency scale of 40Hz to 110MHz at room temperature. The change in the AC conductivity of the composite as compared to PANI and V_2O_5 was observed and discussed based on the electron-hole exchange mechanism.

Keywords: Polyaniline; Vanadium pentoxide; Tangential loss; AC studies.

1. Introduction

In the present scenario, conducting polymers gained the immersion of so many researchers due to their advantageous properties, listed as easy synthesis, corrosion resistance, hygroscopic and room temperature operability. Despite this, these conducting polymers have some disadvantages, namely poor chemical stability and less mechanical strength. Thus, the conducting polymers are doped or composited with metal oxide, metal and bio-polymers to overcome these disadvantages and enhance their chemical and mechanical properties [1].

In the present work, we use the critical conducting polymer polyaniline (PANI) because of its essential properties: facile synthesis, high electrical conductivity and excellent bio-compatibility. PANI is considered to make a composite with vanadium pentoxide (V_2O_5) and vanadium pentoxide is taken because of its essential properties like good catalytic activity and thermal stability. The composite has been prepared by physical mixing method by taking equal amounts of PANI and V_2O_5 (50 Wt% of V_2O_5 in PANI matrix). The prepared polyaniline/vanadium pentoxide (PVO) composite has characterized structural and morphological studies like XRD and SEM techniques. The AC conductivity studies were carried out for the prepared PVO composite from40Hz to 110MHz range of frequency. The change in the conductivity, loss tangent and complex plane impedance are measured and recorded [2].

IOP Conf. Series: Materials Science and Engineering 1300 (2024) 012018 doi:10.1088/1757-899X/1300/1/012018

2. Material and Methods

2.1 *Chemicals*

The Aniline ($C_6H_5NH_2$), Ammonium persulfate ($(NH_4)S_2O_4$), Hydrochloric acid (HCL) analytical graded solvents and Inorganic compound V_2O_5 were purchased from Sigma-Aldrich chemicals, USA.

2.2 *Preparation of PANI*

The monomer aniline is polymerised by the In-Situ polymerisation method to prepare Polyaniline (PANI). The deionised water (200 ml) and a homogeneous mixture of HCL: aniline in a 1:6 ratio were mixed and stirred for 10m in a round bottom flask. To the above mixture, 4.8g of Ammonium persulfate (APS) dissolved in 60 ml deionised water was added drop wise using a burette to the above homogeneous solution was stirred constantly using magnetic stirrer about 4-5h. Then, the above resolution was allowed to polymerise for ten hours. After the complete polymerisation, through a suction filtration process, the filtrate was collected; the filtrate was washed using acetone,1M HCL and deionised water until it got a colourless residue, then the filtrate was kept to dry in a oven for 24h at 80°C. Finally, a green-coloured PANI powder was formed.

2.3 *PVO composite*

The PVO composites were prepared using a mechanical mixing technique with pure PANI with a weight ratio [0.50:0.50] in a one-hour vibration mill.

3. Experimental Techniques

3.1 *Characterization of the sample*

The FTIR spectra of powdered PANI and PVO composite were recorded by employing a NICOLET 750 FT-IR spectrometer, The XRD pattern of the sample was recorded using a Siemen D-500 powder X-Ray Difractometer.

3.2 *AC Conductivity Measurements*

The pellets were prepared by hydraulic press applying pressure of 4 to 5 tons. The electrodes were prepared using silver contact and coated on each sides of the pellet for better electrical connection. The Impedance analyser (Make: 4294A Precision Impedance Analyzer) accounted for AC electrical parameters with a 10 Hz to 110 MHz frequency range at room temperature. The electrical measuring apparatus is depicted in Fig. 1.

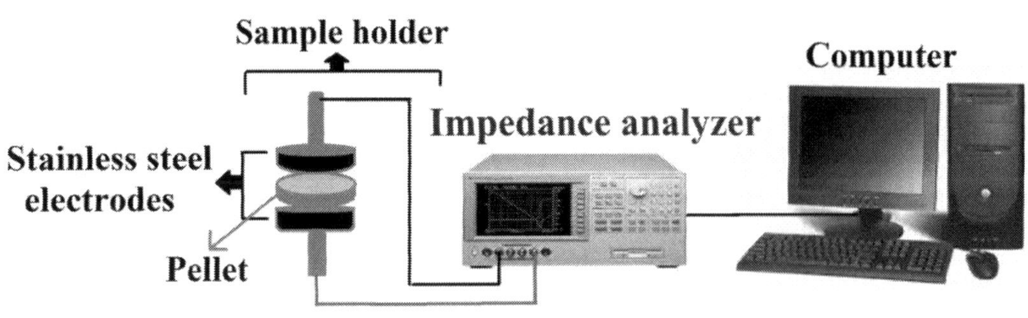

Figure 1: The AC conductivity measuring apparatus.

4 Results and discussion

4.1 XRD Analysis

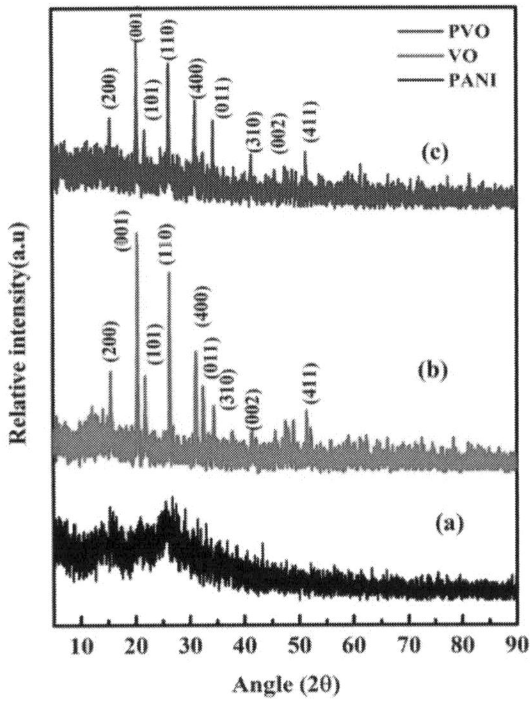

Figure 2: The XRD pattern of pure PANI, V_2O_5 and PVO Composite.

To criticise the structural behaviour of the pure PANI, V_2O_5 and PVO composite, XRD analysis was performed and depicted in Fig. 2. The typical diffraction peak of PANI showed in Fig.2 (a), and the diffraction peaks arriving at $2\Theta=15^{\circ}$, 19° and 26° confirms the semi-crystalline nature of the PANI. The sharp peaks arrived at the pure V_2O_5 [Fig. 2(b)] identifies the orthorhombic phase, which was well agreed with the JCPDS Card NO: 75-0457. The XRD pattern of PVO composite [Fig. 2(c)] clearly shows the appearance of both peaks of PANI and V_2O_5, exactly showing the formation of the composite [3].

4.2 SEM analysis

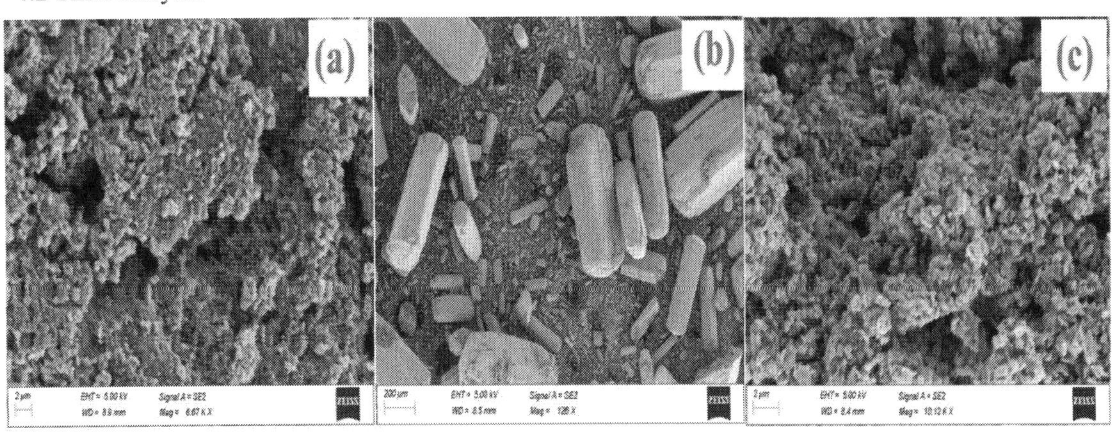

Figure 3: SEM images of (a) Pure PANI, (b) V_2O_5 and (c) PVO composite.

IOP Conf. Series: Materials Science and Engineering 1300 (2024) 012018 doi:10.1088/1757-899X/1300/1/012018

The scanning Electron Microscope (SEM) images report the surface morphology of the pure PANI, V_2O_5 and PVO composite, represented in Fig. 3. The fashioned SEM image of pure PANI [Fig. 3(a)] shows unusual spherical grains or granular shapes of conducting polymer material. Fig. 3(b) shows irregular oblong rod-like shapes of V_2O_5, with scattered orientation. Such type of morphologies was reported in the literature. The image of the PVO composite [Fig. 3(c)] shows agglomerated morphology. This emphasize the interest to notice that the PVO composite shows perfect incorporated non-linear dispersion of V_2O_5 in a polymer matrix, which was successfully achieved by mechanical mixing [4].

5. Dielectric Studies

5.1 AC Conductivity Response Studies

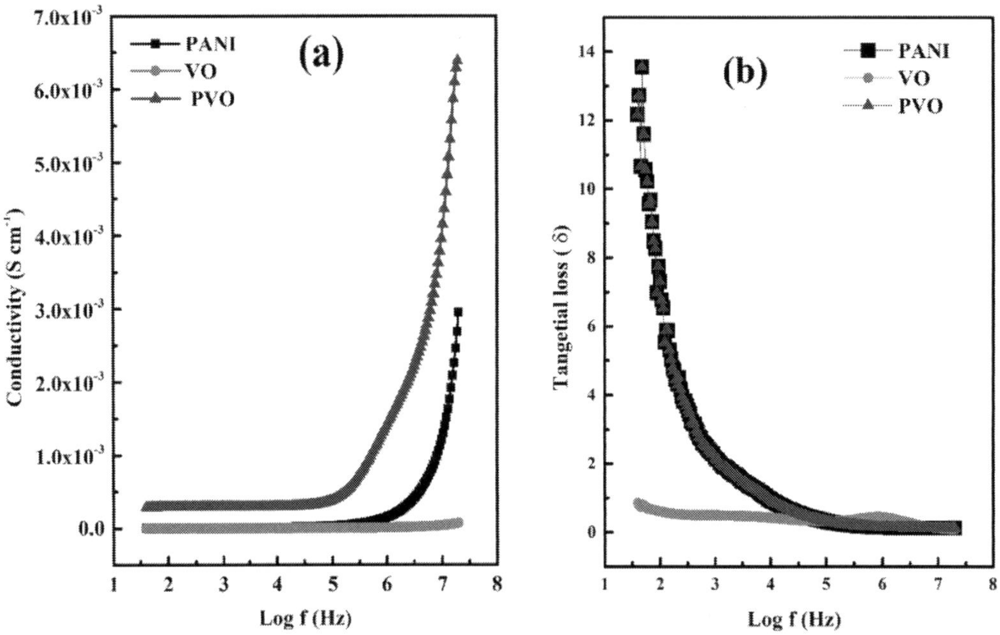

Figure 4: (a) Variation of AC Conductivity of PANI, V_2O_5 and PVO composite; (b) Frequency dependence of dielectric loss tangent of PANI, V_2O_5 and PVO Composite.

Fig. 4(a) shows the variation of AC conductivity concerning the frequency ranging from 40 Hz to 110 MHz for pure PANI, V_2O_5 and PVO composite. It is required to state that the enhanced conductivity of the composite has been noted as comparing with pure PANI and V_2O_5. The inclusion of V_2O_5 increases the AC conductivity; this was achieved due to extended hopping charge carriers in the composite. Mainly, V_2O_5 supplements in the PANI matrix enhance the conductivity of the PVO composite due to the charge polarisation effect. The V_2O_5 has helped most charge carriers to polarise at particular sites, which is the principal reason for enhanced conductivity [5]. The conductivity values are depicted in Table. 1.

The variation of the dielectric loss with varying frequency is represented in Fig. 4(b). The loss tangent for V_2O_5 is very less. The loss tangent for pure PANI, bearing higher loss at initial frequencies, decreases as the frequency higherd. A similar behaviour has been absorbed in the PVO composite—this higher loss at lower frequency is due to DC conductivity losses [6, 7].

5.2 Impedance Spectroscopy

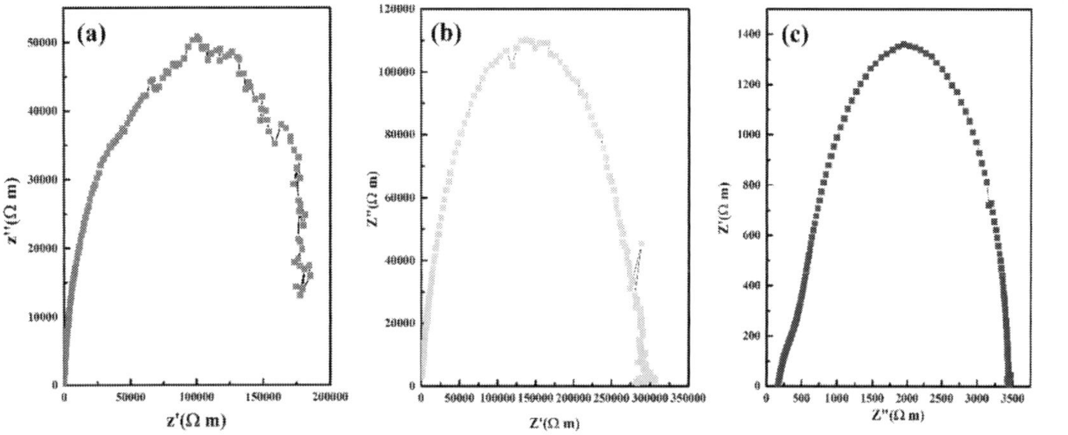

Figure 5: The plots of complex plane impedance for (a) Pure PANI, (b) V_2O_5 and (c) PVO Composite.

Fig. 5 shows the Cole –Cole plots for pure PANI, Pure V_2O_5 and PVO Composite. The pure PANI shown in Fig. 5(a) depicts a single semi-circle that implies the Debye-Model type. The impedance plot of V_2O_5 shown in Fig. 5(b) is a perfect single–semicircle, which agrees with the Debye model. However, a single semicircle in the PVO composite has been observed, which follows the non-Debye type. The diameter of the semi-circles decreases compared to PANI and$V_2O_5$with increased conductivity [8-10].

Table 1: The AC conductivity values for the Pure PANI, V_2O_5 and PVO Composite.

Sl. No.	Samples	AC Conductivity (σ_{ac}) at 1MHz in S cm^{-1}
01.	PANI	$9.4x10^{-5}$
02.	V_2O_5	$6.2x10^{-6}$
03.	PVO	$1.0x10^{-2}$

6.Conclusion

The PVO composite was prepared through the mechanical mixing method. Prepared samples have been characterized using XRD and SEM techniques. The AC conductivity was found to be $6.4x10^{-5}$S cm^{-1}at room temperature for the composite in the frequency range of 40 Hz to 110 MHz The conductivity was found to be increases with increase in applied frequency. The increase in conductivity in PVO composite attributes to the extended hopping charge carriers. Such frequency dependent AC responses of the composites at room temperature are very useful in the fabrication of memory devices, energy storage devices, sensing devices, super capacitors.

Acknowledgments

The authors thanks to CeNSE, Indian Institute of science, Bangalore for providing characterization facility. Also, authors thank the Chair, Department of Physics, Kuvempu University for supporting to carry out this Research work. Mr. Annappa M & Dr. Subramanya K thanks to Principal Sahyadri Science College for their unconditional support and encouragement.

REFERENCES

[1] Manjunatha S, Ravikirn Y T, Chethan B, Devendrappa H and Machappa T 2020 Alternating Current Response Studies on Polyaniline –Neodymiuim Oxide Composites, *AIP conference proccedings,* **2244** 080009.

[2] Chethan B, Raj Prakash H G,Vijayakumari S C,Ravikira Y T2018Structural and electrical properties of nickel substituted cadmium ferrite,*AIP Conf. Proc.* 1953 090028.

[3] Angadi B, Victor P, Jali V M, Lagare M T, Kumar R and Krupanidhi S B 2003 AC conductivity studies on the Li irradiated PZT and SBT ferroelectric thin films, *Mater. Sci. Eng. B,***100** 93–101

[4] Mane A T, Sartale S D and Patil V B 2015 Dodecyl benzene sulfonic acid (DBSA) doped polypyrrole (PPy) films: synthesis, structural, morphological, gas sensing and impedance study, *J. Mater. Sci. Mater. Electron,* **26** 8497–8506

[5] Elliott S R 1987 AC conduction in amorphous chalcogenide and pnictide semiconductors, *Adv.Phys***36** 135–217

[7] Ashis R, Anil R K and AmbikaPrasad M V N 2012 Studies of AC conductivity and dielectric relaxation behavior of CdO- doped nanometricpolyaniline, *Journal of Applied Polymer Science,***123.4** 1928-1934

[8] Patil R, Roy A S, Anilkumar K R, Jadhav K.M and Ekhelikar S 2012 Dielectric Relaxation and AC Conductivity of Polyaniline-Zinc ferrite composite, *Composites Part B: Engineering,***43.8** 3406- 3411

[9] Prakash R, Somani, Marimuthu R, Mandale A B 2001 Synthesis, characterization and charge transport mechanism of conducting polyaniline/V2O5 composites,*Polymer*, **42** 2991 - 3001

[10] Sasidhar N, Chandrashekar T, Chethan B, Ravikiran Y T and Megha R 2020 Room temperature AC electrical properties of polypyrrole/chromium oxide composite, *AIP conference proccedings,***2274** 020011

Ftir, uv-vis and conductivity studies in a copolymer of indole and carbazole

B Raghavendra, T Sankarappa*, Amarkumar Malge, Mohansingh Heerasingh, and Ashwini Devidas

Department of Physics, Gulbarga University, Kalaburgi, Karnataka

*Corresponding email: sankarappa@rediffmail.com

Abstract. The copolymer containing different wt% of indole and carbazole were synthesized by following in situ copolymerization method using ammonium persulfate as oxidizer, at room temperature. Crystalline and amorphous phases were noted from XRD patterns of the samples. Structural and optical studies were carried out by FTIR and UV-Vis spectroscopy. Different functional groups in the prepared copolymer have been identified and optical band gaps were determined to vary from 3.83 eV to 3.96 eV and 3.98 eV to 4.13 eV corresponding to direct and indirect transitions. Dc conductivity measured in the temperature range of 300 K to 423 K is observed to be semiconducting type. Two different regions of variation of conductivity with temperature have been observed and the corresponding activation energy was determined. The increase in conductivity and decrease in activation energy with increase of wt% of Cz content in PInCz copolymer is observed. This may be due to decrease of hopping distance of polarons with increase of Cz contents in the copolymer.

Key words: Indole/Carbazole co-polymer, FTIR, optical absorption, conductivity, activation energy

1. Introduction

Recently, researchers have been showing more interest on copolymers because they are helpful to overcome limitations of the monomers. They exhibit high optoelectronic properties, better thermal, electrochemical properties, and environmental stability etc., compared to their constituent monomers. Copolymer is a molecule made up of two different monomer units. The process of copolymerization takes place by the simultaneous polymerization of two different monomers. During the process, two different structures are incorporated into single polymer chain which in turn changes the properties of the copolymer molecule helpful for many applications. In the present work, our interest is with insertion of carbazole (Cz) as a comonomer into indole monomer (IN). Because, Polycarbazole has two fused benzene ring, good environmental stability, electrical and thermal conductivity, photoluminescence property and high mobility [1-3].

There are some reports on copolymers which were synthesized by different polymerization methods. Sipho E. M et al [4] have been prepared poly (aniline-co-pyrrole) copolymer using ferric chloride and ammonium per sulphate as oxidants separately. From XRD scans they have observed amorphous nature of copolymer and concluded that the conductivity of copolymer decreases by 2 orders compared to monomers it is due to unsuccessful complex of PANI and pyrrole. Sezai Sarac et al [5] have measured the conductivity of n-methyl pyrrole and carbazole with four probe method for different carbazole concentration in copolymer. They have found highest conductivity 9×10^{-5} S/cm for 1/3 ratio of monomers. Bhavana Guptha et al [6] reported fabrication and characterization of electrochemically synthesized Indole and Carbazole copolymer Schottky diode with combination of

Al/copolymer/indium. The conductivity of Indole is very low which is of the order of 10^{-5} to 10^{-7} Scm^{-1} and it was increased by incorporating Indole to thiophene by following the copolymerization process [7]. N. S. Wadatkar et al [8] prepared thiophene-indole copolymer by oxidative polymerization method, they have found optical band gap varying from 3.37 eV to 3.71 eV and observed increase of conductivity and decrease of activation energy for different Stoichiometry ratios of thiophene and indole. They concluded that the thiophene-indole copolymer is suitable for potential applications in optoelectronics devices.

2. Experimental

2.1 SAMPLE PREPARATION

Analytical grade chemicals such as Indole, Acetonitrile, Ammonium persulfate (APS), methonol, Carbazole and deionised water were used for synthesizing copolymers. The copolymer (PInCz) was prepared by following In situ copolymerization method. Indole and Carbazole monomer solutions were prepared by dissolving 2.672 gm (80 wt%) of Indole in 100 ml of acetonitrile and 0.668 gm (20 wt%) in 50 ml of acetonitrile separately and stirred. Carbazole solution was added to Indole solution stirred for few minutes. Solution of APS was prepared by dissolving 8.016 gm in 50 ml of water and added drop wise to combination of monomer solutions over a period of 30 minutes. The mixture of solutions was stirred for full day. The green colored solution at the end. The obtained solution was filtered and washed with deionised water and methonal successfully. The obtained product was the copolymer and it was dried at room temperature and labeled as PInCz-20. The same procedure was followed to prepare the copolymer of different ratios of Indole and Carbazole monomers (70:30, 60:40 and 50:50 wt%) and labeled the products as PInCz-30, PInCz-40 and PInCz-50 respectively.

2.2 MEASUREMENTS

XRD patterns of copolymer samples were collected from (RIGAKU ULTIMA MINIFLEX-600) in the 2θ range from 2° to 80°. The FTIR spectra of the samples were recorded using (SHIMADZUIR-Prestige-21) in the wave number range from 400 cm^{-1} to 4000 cm^{-1} and UV–Vis spectra of the samples recorded using (UV Visible1899 spectrometer) in the wavelength range from 200 nm to 900 nm. Hydraulic press was used to pelletize the powder sample. The two probe method was employed to carry out dc measurements of the samples in the temperature range from 300 K to 423 K. A constant voltage of 5 V applied across the pellet.

The electrical resistivity, ρ and Conductivity, σ of the samples has been determined using the following expressions.

$$\rho = R(A/t) \quad (1)$$

$$\sigma = 1/\rho \quad (2)$$

where, R = (V/I), A is the cross sectional area and t the thickness of the pellets. The obtained conductivity was observed to be within the accuracy range 2 to 3%.

3. Results and discussion

3.1 XRD

Figure 1. depicts XRD scans of PInCz copolymer and it reveals semi crystalline nature of the copolymer. XRD patterns show sharp peaks at 13.80° and 16.59°, which are the characteristics peaks of carbazole and therefore confirms the presence of carbazole in copolymer matrix [3]. In addition to that, the patterns contains the broad humps between 19° and 26° indicating amorphous nature of copolymer which further confirms the presence of indole in the copolymer matrix [9]. The intensity of the peak present at 16.59° increases with increase in the Cz concentration confirms copolymerization of indole and carbazole. Similar nature of patterns can be seen in the reports [8, 10]. The XRD patterns of Polyindole and Polycarbazole monomer presented in our earlier paper [11, 12].

3.2 FTIR

Figure 2. shows FTIR spectra of PInCz copolymer. The most characteristic difference is variation in the intensity of the band at 1051 cm^{-1}, appeared in the copolymer alone signifies the incorporation of ClO_4^- anions and confirms interaction between indole and carbazole monomers [6]. The band at 741 cm^{-1} is due to CH bond in the benzene ring of indole [6, 10, 13]. The band at 812 cm^{-1} attributed to C-H deformation in tri substituted benzene ring [6, 14]. The band at 3204 cm^{-1} indicates stretching vibration of N-H bond [11, 13]. The band at 1110 cm^{-1} originated from the C-N vibration [13] and at 1227 cm^{-1} stretching of C=N respectively [6, 14]. A sharp band around 1325 cm^{-1} refers to C-H out of the plane bending vibration of aromatic ring [14]. The presence of aromatic bands $-C_2=C_3-$ attributed to 1572 cm^{-1} [13] and the peak at 1613 cm^{-1} attributed to bending mode of aromatic alkene [6, 13]. The band at 1454 cm^{-1} can be ascribed to stretching of aromatic C-N bonds [13, 15, 16]. The peak at 621 cm^{-1} is due to NH wagging absorption [11, 17].

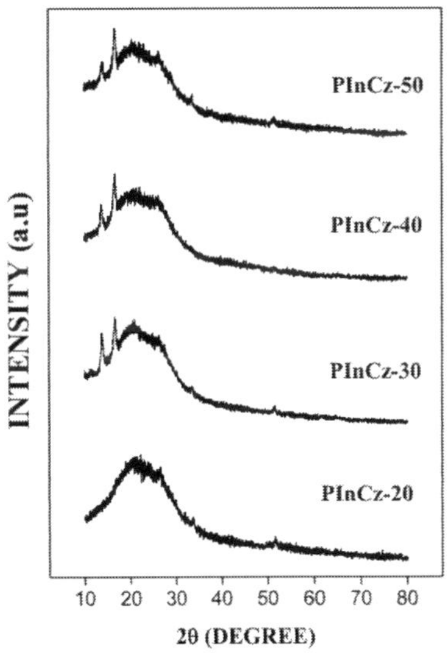

Figure 1. XRD patterns of PInCz copolymers of different composition.

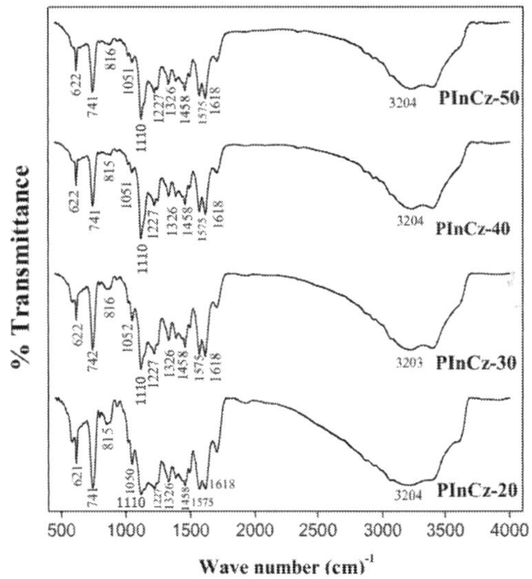

Figure 2. FTIR spectra for PInCz copolymers of different composition.

3.3 Uv-Vis

Figure 3. depicts UV–Vis absorption spectra of PInCz copolymers. In our previous work [11] we found two bands at 265 nm corresponds to π–π* transition of the benzene ring and 305 nm to conjugation and π–π* transitions of benzene ring of polyindole. And for polycarbazole we observed, broadband at 279 nm ascribed to bonding and anti bonding (π-π*) transition of the benzoid ring [12]. Similarly, PInCz copolymer shows two polaronic transitions at 260 nm and 285 nm and it is also observed that the peak at 285nm is shifted to 281nm as carbazole content increased. These results confirm the occurrence of copolymerization. The observed bands corresponding to Polyindole are comparable with thiophene-indole [8], Polyindole/Poly(vinyl acetate) [15].

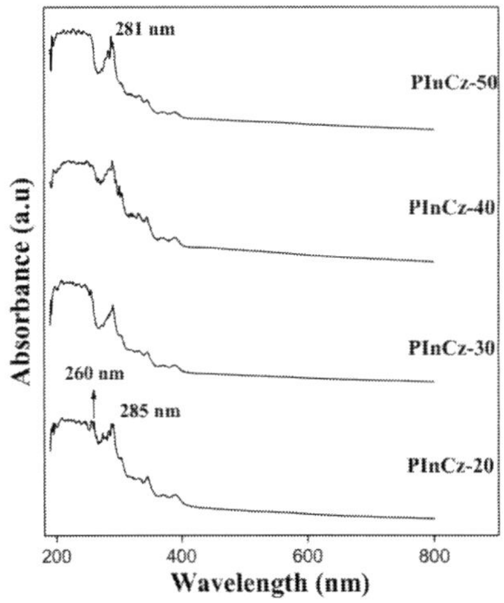

Figure 3. Optical Absorbance versus wavelength for PInCz copolymers of different composition

The direct and indirect optical energy gaps of the PInCz copolymer have been estimated using Mott-Davis-Tauc's, relation for linear absorption coefficient (α) [8],

$$(\alpha h\nu)^{1/n} = \frac{2.303}{d} = B\,(h\nu - E_g) \tag{3}$$

Where, α is the absorption coefficient, B the absorption constant, hν the energy of the photon, E_g the optical energy gap and d the sample thickness. In the above equation exponent (1/n) represents different electronic transitions between valance band and conduction band and it takes values 1/2, 2, 3/2 and 3 corresponding to allowed direct, indirect, forbidden direct and forbidden indirect transitions respectively.

Figures 4 & 5. shows the Tauc's plots for direct and indirect band gaps of the PInCz copolymer. Tauc's plots for all the samples are not shown in the Figure 4 & 5 to save space. The energy gaps were estimated by drawing tangents to the band edges and they were extrapolated on to the x-axis (hν-axis). The tangent where it intersects hν-axis gives value of energy gap. The direct and indirect band gap values of PInCz copolymer of different Indole and carbazole monomer ratios shows higher values compared to individual Polyindole and Polycarbazole and they tabulated in Table 1. Direct energy gaps of PInCz varies from 3.96 eV to 3.83 eV and where as indirect energy gaps from 4.13 eV to 3.98 eV. The obtained direct and indirect band gaps are found to be decreased with rise of Cz wt%. This confirms copolymerization of Indole and carbazole. The obtained band gap values shows that PInCz copolymer is acceptable for photovoltaic activities and suitable for constructing photovoltaic devices. The band gap values are in comparison with the thiophene-indole [8], polycarbazole-polyindole [6].

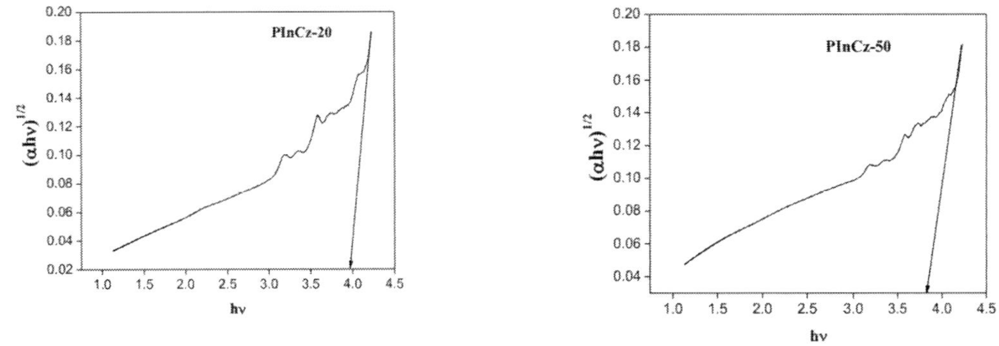

Figure 4. Tauc's plots of $(\alpha h\nu)^{1/2}$ versus hν for the determination of direct band gap for PInCz of different composition.

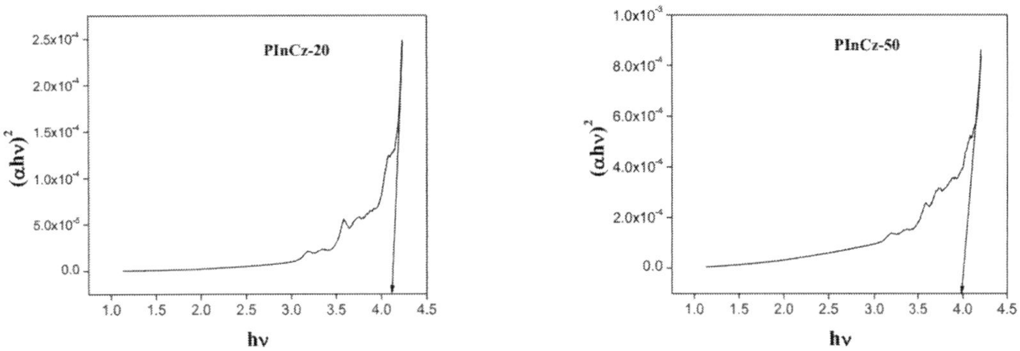

Figure.5. Tauc's plots of $(\alpha h\nu)^2$ versus hν for the determination of indirect band gap for PInCz of different composition.

Table 1. Optical Band gap energy (direct and indirect) values for PIn, PCz, PInCz copolymers of different ratios

Sl No	Sample	Direct band gap E_g (eV)	Indirect band gap E_g (eV)
1	PIn	2.08 [11]	2.75 [11]
2	PCz	3.32 [12]	3.42 [12]
3	PInCz-20	3.96	4.13
4	PInCz-30	3.88	4.03
5	PInCz-40	3.85	4.00
6	PInCz-50	3.83	3.98

3.4 Conductivity

Conductivity found to be of the order of 10^{-4} $(\Omega m)^{-1}$ and increased with increase of temperature. Conductivity is also observed to be increasing with Cz content in the copolymer. The increase of conductivity with temperature is due to increase of mobile ions. The copolymer with 50:50 wt% of Indole and Carbazole monomer ratio showed highest conductivity estimated to be 2.97×10^{-4} $(\Omega m)^{-1}$ at 413 K in region I and 2.86×10^{-4} $(\Omega m)^{-1}$ at 343 K in region II. Conductivity of the PInCz copolymer found to be higher than the Polyindole and Polycarbazole which are reported in our earlier wrok [11, 12] and this may be due to reduction in hopping distance and creation of network for transportation of charge carriers inside copolymer matrix. The order of estimated conductivity values are in comparison with the values reported in aniline-co-pyrrole, thiophene-indole [4, 8].

The temperature variation of electrical conductivity is analyzed using Arrhenius expression,

$$\sigma = \sigma_0 \exp (Ea/k_B T) \quad (4)$$

Where, σ_0 is pre exponential factor, Ea the activation energy and k_B the Boltzman constant. Figure 6. shows the plots of ln(σ) versus (1/T) for PInCz copolymer. To save space, the Arrhenius plots for all samples are not depicted in Figure 6. From the figure, we can observe that there exists at least two regions of variation of conductivity. Implying that there are two different sizes of absorption energy for conduction in region I and region II. Linear lines were fit in both the region to estimate activation energy. The conductivity increased because of the reduction in the scattering rate of polarons inside the copolymer matrix. The activation energy is found to be 5.50 meV for PInCz-40 in region I, and 2.53 meV for PInCz-50 in region II are the lowest values compared to the other concentration, indicates in this concentration electron, needs less amount of energy to overcome the potential barrier compared to other concentrations. The behavior of variation of conductivity with activation energy is in comparison with the report [8]. The conductivity and activation energy values of PInCz copolymer of different monomer concentration in both region I and II are tabulated in Table 2.

Table 2. DC conductivity and activation energy in region I and II for PInCz copolymers of different ratio

SL. NO	Sample	Activation energy E_a (meV)		Conductivity σ_{dc} $(\Omega m)^{-1}$ $(\times10^{-4})$			
		Region I	Region II	Region I		Region II	
				At 413 K	At 383 K	At 343 K	At 308 K
1	PInCz-20	17.63	3.22	2.22	2.18	2.13	2.11
2	PInCz-30	9.10	2.91	2.28	2.20	2.15	2.12
3	PInCz-40	5.50	1.81	2.36	2.33	2.30	2.29
4	PInCz-50	7.19	2.53	2.97	2.91	2.86	2.81

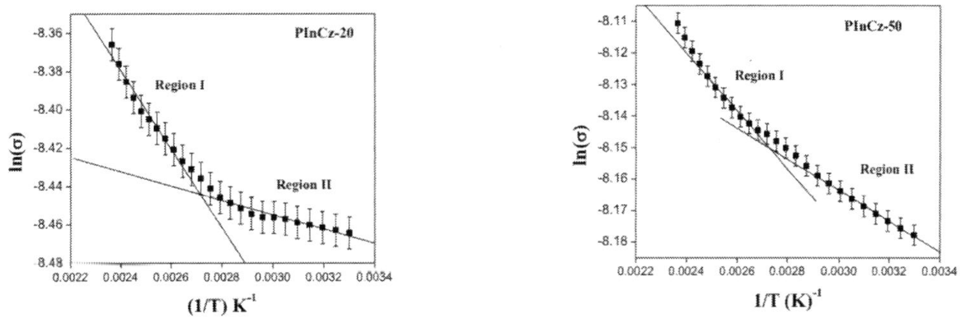

Figure 6. Plots of ln(σ) versus(1/T) in region I and II for PInCz copolymers of different ratios.

4. Conclusions

PInCz copolymer in different compositions has been synthesized by in situ copolymerization technique. The copolymer has been investigated for XRD, FTIR and Uv-Vis and conductivity. The XRD patterns reveals semi crystalline nature of copolymer and different functional groups in copolymer were identified by FTIR. The blue shift in the UV-Vis optical absorption gives information regarding intermolecular interaction between the bands of copolymer. Direct and indirect optical band gaps for allowed transitions were estimated by plotting Mott-Davis-Tauc's plots for the absorption edges and it is found that the band gaps of the PInCz copolymer are higher than the PIn and PCz. The Dc conductivity of the copolymer is increased with Corbazole concentration in copolymer and it is observed that (50:50) wt% of monomer has highest conductivity. Variation of conductivity with temperature indicates semiconducting nature of copolymer. Conductivity enhanced, activation energy reduced, and it is due to reduction of hopping distance of the charge carries in the copolymer matrix.

References

[1] N C Horti, M D Kamatagi, N R Patil, S K Nataraj, S A Patil, S R Inamdar, *Polym. Bull*, **78**, (2020) 6321-36.

[2] S. Madhurima Das, Roy, *Materials Today: Proceedings*, **18**, (2019) 5438–46.

[3] V Raj, D Madeswari, M Mubarak Ali, *Journal of advanced polymer science*, **116**, (2010) 147-154.

[4] Sipho E Mavundla, Gerald F Malgas, David E Motaung, Emmanuel I Iwuoha, *J Mater Sci*, **45**, (2010) 3325–30.

[5] A Sezai Sarac , Murat Ates & Elif Alturk Parlak, *International Journal of Polymeric Materials*, **53**, (2004) 785–798.

[6] Bhavana Gupta, Arun Kumar Singh, Ambrose A. Melvin, Rajiv Prakash, *Solid State Sciences*, **35**, (2014) 56-61.

[7] A Sezai Sarac, Serife Ozkara, and Esma Sezer, *International Journal of Polymer Anal. Charact*, 8, (2003) 395–409.

[8] N S Wadatkar, S A Waghuley, *Heliyon*, **5**, (2019) e01534.

[9] A Kumar, A C Pandey, R Prakash, *Catal. Sci. Technol*, **2**, (2012) 2533–38.

[10] N S Wadatkar, S A Waghuley, *Polymer Bulletin*, **77,** (2020) 4181-96.

[11] B Raghavendra, T Sankarappa, Amarkumar Malge, *J Inorganic and organometallic polymers*, **30**, (2020) 3586-94.

[12] B Raghavendra, T Sankarappa, Amarkumar Malge, *J Inorganic and organometallic polymers*, **32**, (2022) 2416-27.

[13] Madhu Tiwari, Ashish Kumar, Harshit Sunil Umre, Rajiv Prakash, *J. Appled Polymer Science*, 132, (2015) 1-9.

[14] A Kumar, M Tiwari, R Prakash, *Electro Chem*, **2,** (2015) 2001-10.

[15] Ozay Eraldemir , Bekir Sari , Ayşegul Gok, Halil, *Macromolecular Science Part A -- Pure and Applied Chemistry*, **45**, (2008) 205-211.

[16] Nursel sanli asik, Recep cas, Savaa sonmezoglu, Muzaffer can, Guven cankaya, *J. Non-Cryst Solids*, **356**, 1848–1853 (2010) 1848–53.

[17] R Pavul Raj, P Ragupathy, S Mohan, *J. Mater. Chem.* A, **3**, (2015) 24338–48.

Structural and Optical Properties of PMMA-MgO Nanocomposite Film

Shruthi K N[1], Ramaraja Varma V[1], Mohan Kumar[1], Sushma[1], Ganesh Sanjeev[1],*

[1]Microtron Centre, Department of Studies in Physics, Mangalore University, Mangalagangotri 574199, India

**Email:ganeshsanjeev@rediffmail.com*

Abstract: This study investigates the structural, morphological, and optical properties of poly(methyl methacrylate) (PMMA) films incorporated with magnesium oxide (MgO) nanoparticles. The PMMA-MgO polymer nanocomposite (PNC) films were fabricated via solution casting method using varying weight percentages (1-4 wt%) of MgO nanoparticles. X-ray diffraction analysis confirmed the integration of MgO nanofillers in the PMMA matrix. Fourier transform infrared spectroscopy revealed interactions between PMMA and MgO nanoparticles. Atomic force microscopy demonstrated increased surface roughness in PNC films with higher MgO loading. Optical characterization using UV-visible spectroscopy showed enhanced absorption in the UV region and a noticeable peak at 280 nm due to MgO nanoparticles. The refractive index of PMMA-MgO PNCs increased with rising MgO content while the optical bandgap marginally decreased. The study highlights the potential of PMMA-MgO PNC films for advanced optoelectronic applications requiring high optical transparency and tuneable refractive index.

Keywords. Polymer nanocomposites, XRD, FTIR, AFM, UV- Visible Spectroscopy.

1. Introduction

The capacity of transparent polymer nanocomposites to display a wide range of optical features, such as tailored emission/absorption characteristics, varied refractive indices (both high and low), and strong nonlinear optical properties, has attracted a lot of attention[1]. This heightened interest is rooted in the promising potential for optoelectronic applications[2]. Typically, these nanocomposites are crafted by integrating metal oxide nanoparticles (NPs) into a transparent polymer matrix, such as polymethyl methacrylate (PMMA). The polymeric component contributes essential attributes like processability, transparency and flexibility, while the inclusion of NPs imparts desired thermal, electrical and optical properties[3]–[6]. Demonstrating their technological and economic advantages, polymers have made significant strides in diverse fields, including automotive, chemical sensors and electronics[7]–[9]. The fusion of metal oxide NPs and polymers has become a focal point for scientists, offering a platform to seamlessly combine the unique properties of both components[2]. This convergence opens up new avenues for innovative applications and further reinforces the versatile role of transparent polymer nanocomposites in advancing modern materials and technologies[10].

Various studies have investigated the preparation and characterisation of Polymer Nanocomposites (PNCs), employing PMMA as the polymer matrix and integrating metal oxide fillers such as TiO_2, SnO, ZnO, CuO, and SiO_2[1], [11], [12]. Within the realm of transparent polymer nanocomposites, a research gap exists in exploring the preparation and characterization of PNC films using PMMA and MgO NPs. The inclusion of metal oxide nanoparticles, such as Magnesium Oxide (MgO), enhances the already impressive optical properties of these materials. Among other metal oxide nanomaterials, MgO stands out for its cost-effectiveness and eco-friendly nature, with lower toxicity than alternative metal oxides[13]–[15].

Blending PMMA with MgO nanoparticles introduces distinctive contributions to the optical characteristics of PNC films. This integration not only capitalizes on the inherent advantages of the polymeric component (processability, transparency and flexibility) but also utilises MgO's potential for tailored thermal, electrical, and optical properties. Polymers, known for their economic and technological benefits, especially in automotive, chemical sensors, and electronics, gain further advantages through the strategic integration of MgO nanoparticles.

The study recognizes these composites as particularly appealing for enhancing multifunctional optoelectronic devices and smart, flexible lightweight microelectronics. The synergistic relationship between polymer versatility and MgO's tailored properties positions these materials as promising solutions for those seeking cost-effective and eco-friendly advancements in advanced materials and nanocomposites. Given the mentioned details, this research aims to investigate the comprehensive structural behaviour through XRD data analysis, morphology via AFM, optical properties using UV–Vis spectroscopic measurements, and chemical interactions through FTIR. The focus is on PNC films prepared through a sonicated solution casting method with a PMMA matrix loaded with MgO nanoparticles.

2. Experimental Details

Polymer films were produced using the solvent casting method, involving the preparation of a poly(methyl methacrylate) (PMMA) stock solution by dissolving 1 gram of PMMA in 15 ml of chloroform. The PMMA solution was stirred until complete dissolution using a magnetic stirrer at room temperature. Simultaneously, magnesium oxide (MgO) nanoparticles were dispersed in chloroform at different weight percentages. The PMMA solution and MgO nanoparticle dispersions were combined and stirred for an hour to achieve proper blending and homogeneity, crucial for uniform distribution within the PMMA matrix. A separate pure PMMA film was prepared for comparative analysis, excluding MgO nanoparticles. The resulting doped nanocomposite and pure PMMA solutions were poured into Petri dishes and allowed to dry, forming MgO-doped PMMA polymer films. These films were systematically labelled based on the incorporated MgO weight percentages. The prepared PMMA-MgO PNC films exhibited a uniform thickness ranging from 0.12 to 0.17 mm and were devoid of air bubbles.

The X-ray diffraction (XRD) analysis of the prepared samples was conducted using a Rigaku Miniflex 600 X-ray powder diffractometer. Fourier-transform infrared spectroscopy (FTIR) spectra were obtained using an Alpha Bruker ATR-FTIR instrument. Surface morphology images were captured utilizing Atomic Force Microscopy (AFM). The optical properties of the films were assessed through UV-Visible microscopy.

3. Result And Discussion

3.1. XRD

Figure 1(a) showcases the X-ray diffraction pattern of MgO nanoparticles, with peaks aligning well with the typical MgO diffraction pattern [JCPDS No. 87-0652]. The grain size was calculated by employing Scherer's formula and it is found to be 22 nm. Figure 1(b) shows the XRD pattern of the PMMA-MgO sample. An amorphous peak at 13° confirms the presence of PMMA polymer in all the samples. Notably, MgO-doped PMMA polymer composite films exhibit robust peaks characteristic of

MgO nanoparticles, confirming their integration into the composite (16). The intensity of peaks varies nonlinearly with filler concentration. Furthermore, XRD analysis reveals that the positions of the original peaks of pristine MgO nanofillers remain unchanged upon dispersion in the host PMMA matrix. This indicates that MgO nanofillers maintain their original structure within the PMMA-MgO polymer nanocomposite (12). It's worth noting that no impurity peaks are detected, affirming the purity of the sample (4).

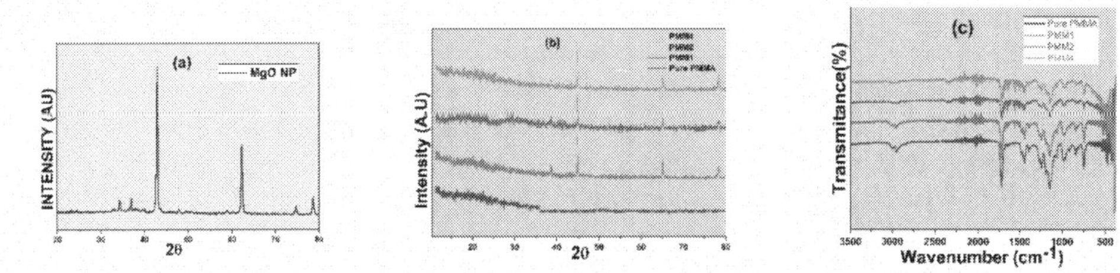

Figure 1.XRD Intensity Spectrum for (a) MgO NP, (b) PMMA-MgO PNC Films, and (c) FTIR Spectra of PMMA-MgO PNC Films.

3.2. FTIR

FTIR spectroscopy provides valuable insights into the interface affinity between PMMA-MgO. Figure 1(c) presents the FTIR spectra of pure PMMA, MgO nanoparticles and various compositions of PMMA-MgO nanocomposites. Notably, the intensity of transmission spectra demonstrates a linear reduction with an increase in filler concentration, suggesting the presence of MgO in the polymer composite [16]. All samples exhibit a consistent pattern of peaks without any deviation. Additionally, PMM2 and PMM4 samples reveal peaks below 400 cm^{-1}, attributed to the metal oxide Mg-O stretching. The band at 1719 cm^{-1} is associated with C=O stretching vibrations, while the band within the range of 1140 to 1270 cm^{-1} corresponds to C–O stretching vibrations. The band at 1067 cm^{-1} is observed due to -C-O-C stretching vibration. In the spectral range of 1492–1275 cm^{-1}, additional bands are linked to CH_3 bending vibrational modes. Additionally, asymmetric stretching of CH_3 vibrations was observed in the range of 2998-2950 cm^{-1}. The significant decrease in the intensities of the entire FTIR spectra may be attributed to the intermolecular bonding between the PMMA matrix and MgO NPs[17]. Thus, FTIR studies provide supporting evidence for the results obtained from XRD.

3.3. AFM

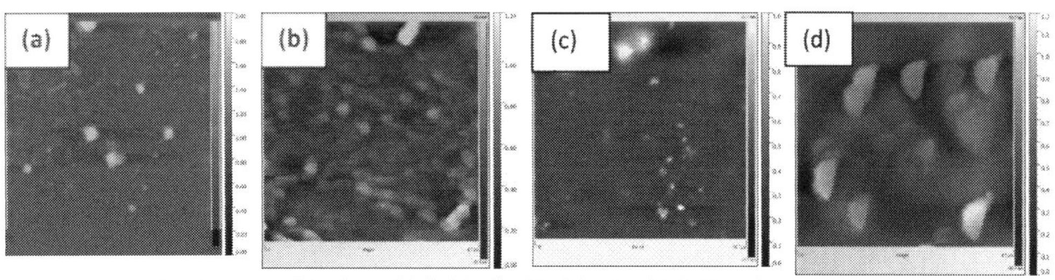

Figure 2. AFM images of (a) Pure PMMA, (b) PMM1, (c) PMM2 and (d) PMM4 samples

AFM was utilized to examine the morphology of PMMA-MgO nanocomposite films, providing insights into the surface roughness of the composite. Figure 2 displays the topological images of PNC films. The introduction of MgO nanoparticles has visibly altered the surface of PMMA-MgO nanocomposites, rendering it rougher. The root mean square (RMS) roughness has surged from 95 nm in the pristine PMMA film to 270 nm in the PMMA-MgO nanocomposite with a 4wt% MgO loading. The elevated RMS roughness values in PMMA-MgO nanocomposites, in contrast to neat PMMA, validate the presence of nanoparticles on the surface of the composite film[18]. This observation supports the effective dispersion of MgO nanoparticles within the PMMA matrix.

3.4. UV- VISIBLE SPECTROSCOPY

The optical absorption spectra analysis provides insights into the band structure and energy band gap of polymeric materials. Figure 3(a) illustrates variations in absorption across incident wavelengths (190-800 nm) for PMMA-MgO samples. In the UV spectrum of nanocomposite samples, absorption notably increases within the 190-250 nm range under UV radiation exposure. The pure PMMA film exhibits minimal absorbance across the visible region (λ = 400–800 nm) but shows increased absorbance in the UV region, confirming its ability to absorb UV radiation[19]. UV absorbance features are attributed to electronic transitions within the C=O group of the ester-linked to the repeat unit in the PMMA chain[11]. Figure 3(a) indicates that absorption in PMMA-MgO composite films consistently rises with a higher weight percentage of MgO particles integrated into the PMMA matrix, accompanied by a gradual red shift in the sharp absorbance edge. Additional absorbance peaks at 280 nm in doped PMMA samples, increasing with MgO concentration, confirm the presence of MgO nanofiller, supporting XRD results. The plateau in the visible region underscores the transparency of PMMA-MgO composites specifically for photons in this range[19].

The optical transmittance (T%) and reflectance (R%) of PMMA-MgO PNCs are illustrated in Figure 3(b). There was a decline in transmittance in the UV as well as the visible region as the MgO content increased. Within the wavelength range of 240-270nm, there is a sudden decrease in the value of extinction coefficient(k) as shown in Figure 3(d). Beyond 300 nm, k stabilizes at a constant value, suggesting low optical loss in this region[20]. Moreover, with an increase in MgO content in the PNCs, k also rises, indicating heightened light dissipation due to scattering and absorption by MgO nanoparticles[21]. This significant interaction between the MgO nanoparticles and the polymer blend induces changes in crystallinity, thereby influencing the band structure and absorption percentage[20].

Figure 3(e)presents variation of n, demonstrating a decrease in the refractive index of the films as the wavelength of the incident photon increases, eventually becoming constant at higher wavelengths. The findings further reveal that the refractive index of PMMA-MgO polymer composites rises with increasing MgO content in the visible region. Furthermore, the transparency of the composite samples is underscored by the small values of the extinction coefficient (k) compared to the refractive index (n)[21].

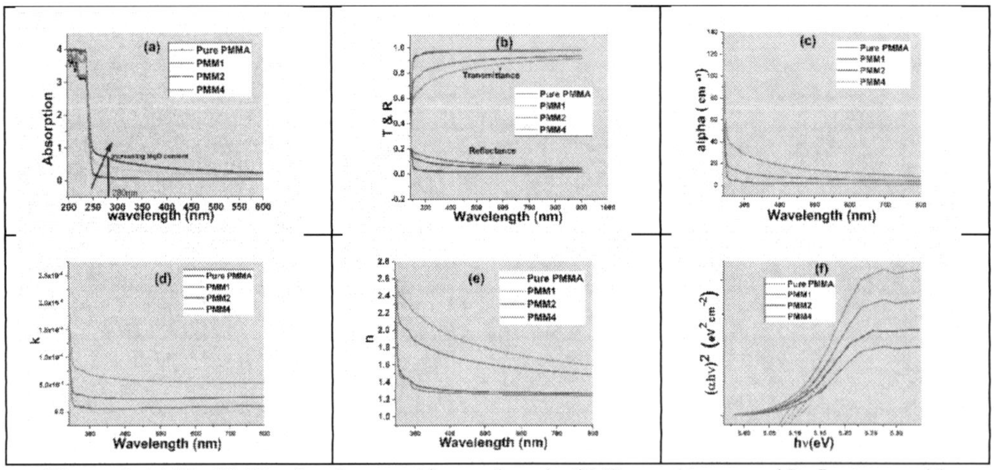

Figure 3. (a) Uv-Visible absorbance with wavelength, (b) Transmittance and Reflectance with wavelength, (c) Absorption coefficient(α) versus wavelength, (d) Extinction coefficient(k) versus wavelength, (e) Refractive index(n) versus wavelength and (f) $(\alpha h\nu)^2$ versus photon energy ($h\nu$) for PMMA-MgO PNC films.

The Tauc plot, featured in Figure 3(f), was used to examine the optical properties and distinguish the optical bandgap of the material. The observed linear segment confirms a direct bandgap in PMMA-MgO PNC film. The film's bandgap was established by extrapolating the linear section of the Tauc plots. The band gap of the pristine films slightly decreased from 5.089 eV to 5.069 eV as the filler concentration increased. PNCs characterized by transparency and a high refractive index hold significant potential across various applications in optical design and advanced optoelectronic devices. These applications encompass technologies like LEDs, image sensors, and waveguide systems[22].

4. Conclusion

This study comprehensively investigates the structural, morphological, and optical characteristics of PMMA-MgO nanocomposite films fabricated via a solution-casting approach. XRD analysis confirms the successful integration of MgO nanoparticles in the PMMA polymer matrix. FTIR spectroscopy reveals intermolecular bonding between PMMA and MgO. AFM studies demonstrate increased surface roughness with higher MgO loading. UV-visible spectroscopy highlights enhanced UV absorption and the presence of a distinctive MgO peak at 280 nm. The refractive index rises with increasing MgO content, while the optical bandgap shows a marginal reduction. The key findings underscore the potential of PMMA-MgO nanocomposites for diverse optoelectronic applications requiring customizable optical properties. The eco-friendly nature and cost-effectiveness of MgO make these nanocomposites economically viable. Overall, this study provides significant insights into the structure-property relationships in PMMA-MgO nanocomposites. Further tuning the properties through optimal selection of materials and methods can aid the development of these multifunctional nanocomposites for advanced technologies and devices.

Acknowledgement

The authors thank the DST-PURSE Lab at Mangalore University for providing characterization facilities.

References

[1] J. Ge *et al.*, "Preparation and Characterization of PS-PMMA / ZnO Nanocomposite Films with Novel Properties of High Transparency and UV-Shielding Capacity," pp. 1–6, 2010, doi: 10.1002/app.

[2] S. Hammani, A. Barhoum, and M. Bechelany, "Fabrication of PMMA/ZnO nanocomposite: effect of high nanoparticles loading on the optical and thermal properties," *J. Mater. Sci.*, vol. 53, no. 3, pp. 1911–1921, 2018, doi: 10.1007/s10853-017-1654-9.

[3] K. M. A. Hurayra–Lizu, M. W. Bari, F. Gulshan, and M. R. Islam, "GO based PVA nanocomposites: tailoring of optical and structural properties of PVA with low percentage of GO nanofillers," *Heliyon*, vol. 7, no. 5, May 2021, doi: 10.1016/J.HELIYON.2021.E06983.

[4] R. J. Sengwa, P. Dhatarwal, and S. Choudhary, "A comparative study of different metal oxide nanoparticles dispersed PVDF/PEO blend matrix-based advanced multifunctional nanodielectrics for flexible electronic devices," *Mater. Today Commun.*, vol. 25, no. June, p. 101380, 2020, doi: 10.1016/j.mtcomm.2020.101380.

[5] S. K. Marka, V. V. S. S. Srikanth, B. Sindam, B. K. Hazra, K. C. J. Raju, and S. Srinath, "Graphene-Wrapped MgO / Poly (vinyl alcohol) Composite Sheets : Dielectric and Electromagnetic Interference Shielding Properties at Elevated Temperatures," *ACS Appl. Mater. Interfaces*, vol. 11, pp. 23714–23730, 2019, doi: 10.1021/acsami.9b05137.

[6] J. K. Rao, A. Raizada, D. Ganguly, M. M. Mankad, S. V. Satayanarayana, and G. M. Madhu, "Investigation of structural and electrical properties of novel CuO–PVA nanocomposite films," *J. Mater. Sci.*, vol. 50, no. 21, pp. 7064–7074, 2015, doi: 10.1007/s10853-015-9261-0.

[7] A. Hashim and A. Hadi, "Synthesis and characterization of novel piezoelectric and energy storage nanocomposites: Biodegradable materials–magnesium oxide nanoparticles," *Ukr. J. Phys.*, vol. 62, no. 12, pp. 1050–1056, 2017, doi: 10.15407/ujpe62.12.1050.

[8] H. T. Vo and F. G. Shi, "Towards model-based engineering of optoelectronic packaging materials: Dielectric constant modelling," *Microelectronics J.*, vol. 33, no. 5–6, pp. 409–415,

2002, doi: 10.1016/S0026-2692(02)00010-1.

[9] M. Rashad, "Tuning optical properties of polyvinyl alcohol doped with different metal oxide nanoparticles," *Opt. Mater. (Amst).*, vol. 105, no. April, p. 109857, 2020, doi: 10.1016/j.optmat.2020.109857.

[10] Q. M. Al-Bataineh, A. A. Ahmad, A. M. Alsaad, and A. D. Telfah, "Optical characterizations of PMMA/metal oxide nanoparticles thin films: bandgap engineering using a novel derived model," *Heliyon*, vol. 7, no. 1, p. e05952, 2021, doi: 10.1016/j.heliyon.2021.e05952.

[11] P. Taylor, P. Liu, and Z. Su, "Journal of Macromolecular Science, Part B : Physics Preparation and Characterization of PMMA / ZnO Nanocomposites via In-Situ Polymerization Method Preparation and Characterization of PMMA / ZnO Nanocomposites via In-Situ Polymerization Method," no. January 2013, pp. 37–41, 2007, doi: 10.1080/00222340500408085.

[12] R. J. Sengwa and P. Dhatarwal, "Polymer nanocomposites comprising PMMA matrix and ZnO, SnO 2, and TiO 2 nanofillers : A comparative study of structural, optical , and dielectric properties for multifunctional technological applications," *Opt. Mater. (Amst).*, vol. 113, no. October 2020, p. 110837, 2021, doi 10.1016/j.optmat.2021.110837.

[13] V. A. Online, R. Verma, B. K. Gupta, and A. K. Srivastava, "RSC Advances," 2015, doi: 10.1039/C5RA21150F.

[14] H. Cui, X. Wu, Y. Chen, J. Zhang, and R. I. Boughton, "Influence of copper doping on chlorine adsorption and antibacterial behavior of MgO prepared by co-precipitation method," 2015, doi: 10.1016/j.materresbull.2014.10.067.

[15] M. R. Bindhu, M. Umadevi, M. Kavin Micheal, M. V. Arasu, and N. Abdullah Al-Dhabi, "Structural, morphological and optical properties of MgO nanoparticles for antibacterial applications," *Mater. Lett.*, vol. 166, pp. 19–22, 2016, doi: 10.1016/j.matlet.2015.12.020.

[16] C. Gavade, N. L. Singh, D. Singh, S. Shah, A. Tripathi, and D. K. Avasthi, "Study of dielectrical properties of swift heavy ion induced modifications in metal oxide/PMMA nanocomposites," *Integr. Ferroelectr.*, vol. 117, no. 1, pp. 76–84, 2010, doi: 10.1080/10584587.2010.489427.

[17] O. G. Abdullah and S. A. Saleem, "Effect of Copper Sulfide Nanoparticles on the Optical and Electrical Behavior of Poly(vinyl alcohol) Films," vol. 45, doi: 10.1007/s11664-016-4797-6.

[18] C. Backes and C. Backes, "Experimental Details," *Noncovalent Funct. Carbon Nanotub.*, pp. 191–198, 2012, doi: 10.1007/978-3-642-27582-1_5.

[19] D. Sun, N. Miyatake, and H. J. Sue, "Transparent PMMA/ZnO nanocomposite films based on colloidal ZnO quantum dots," *Nanotechnology*, vol. 18, no. 21, 2007, doi: 10.1088/0957-4484/18/21/215606.

[20] Q. M. Al-Bataineh, A. A. Ahmad, A. M. Alsaad, and A. D. Telfah, "Optical characterizations of PMMA/metal oxide nanoparticles thin films: bandgap engineering using a novel derived model," *Heliyon*, vol. 7, no. 1, p. e05952, Jan. 2021, doi: 10.1016/J.HELIYON.2021.E05952.

[21] O. G. Abdullah, S. B. Aziz, and M. A. Rasheed, "Effect of silicon powder on the optical characterization of Poly(methyl methacrylate) polymer composites," *J. Mater. Sci. Mater. Electron.*, vol. 28, no. 5, pp. 4513–4520, 2017, doi: 10.1007/s10854-016-6086-9.

[22] M. Nandimath, R. F. Bhajantri, J. Naik, and V. Hebbar, "Effect of Rhodamine 6G dye on chromaticity co-ordinates and photoluminescence properties of TiO2/PMMA polymer nanocomposites for LED applications," *J. Lumin.*, vol. 207, pp. 571–584, 2019, doi: 10.1016/j.jlumin.2018.11.048.

Structural, Physical and Optical studies of Sc^{3+} doped Lithium Aluminum Borate glasses

J Nagaraju[1] and B Eraiah[1*]

[1]Department of Physics, Bangalore University, Jnanabharathi campus, Bangalore 560056, Karnataka, India

*Corresponding author email: eraiah@rediffmail.com,

Abstract: Optically transparent lithium aluminum borate containing scandium glasses with composition $70B_2O_3 - 20Li_2O - (10 - x) Al_2O_3 - xSc_2O_3$ (where x= 0, 0.1, 0.2, 0.3, 0.4, 0.5, 0.6 and 0.7 in mol %) were fabricated by conventional melt quenching technique. The non-crystalline nature of the samples was monitored by X-ray diffraction and it confirms non-crystalline nature. Structure of the fabricated glasses are assessed using the Fourier transform infrared spectra. The absorption spectra of these glasses have been recorded in the wavelength range 200 nm – 1100 nm. Direct and indirect optical energy band gaps were calculated from Tauc's plots. The physical parameters such as refractive index (RI), optical basicity and interaction parameters were determined using proper formulas to know the characteristics of fabricated glass matrix. The results obtained indicate that 0.3 mol% Sc_2O_3-containing BLASc glass is more suitable for w-LEDs and lasers applications.

Keywords: Borate glass; Photoluminescence; Polarizability; Optical basicity; Energy band gaps.

1. INTRODUCTION

Due to its transparency, chemical inertness, and environmental friendliness, glass is one of the most fascinating materials and has drawn significant interest in both science and technology [1-3]. Among all glass hosts, borates have several special qualities that make them suited for use in fiberglass, including decreased thermal expansion, resilience to thermal shock, increased toughness and strength [3]. However, the negative qualities of borate glass include its hygroscopic nature, but it can be reduced by adding alkali oxides such as Li_2O, Na_2O, and K_2O. It can also increase several aspects of borate glasses. Among all alkali oxides, Li_2O has been used as a glass forming in a variety of compounds to increase physical property and viscosity while lowering melting point in the borate system. In addition to making borate glass an excellent network modifier, adding amphoteric oxide Al_2O_3 also enhances the host glass's chemical stability, mechanical strength, and emission qualities [4-6]. Al_2O_3- Li_2O_3-B_2O_3 glasses are of great interest to researchers because, depending on the concentration of Al_2O_3 in the glass matrix, Al_2O_3 acts as a glass former or modifier in the glass formations. In this research, the impact of Scandium on the lithium aluminum borate glass system is examined. Scandium is utilized as a lightweight material for high-performance applications in the military equipment, electrical, aviation, automotive, and transportation industries. X-ray diffraction (XRD) and Fourier transform infrared spectroscopy (FTIR) will be used to analyze the structural, optical, and morphological features. UV-Vis spectroscopy have been used to study the optical properties, such as absorbance and optical band gap. Sc_2O_3 doping to LAB glass matrix was limited in the current investigation to the range of 0 to 0.7 since this range has good glass forming capabilities.

2. EXPERIMENTAL

Boric acid, lithium carbonate, aluminum oxide, and scandium oxide were among the analar grade chemical agents that were added in the proper amounts to each glass composition. Each time, a porcelain crucible was used to melt the powdered substance at a high temperature (at 1040 °C) and produce a bubble-free, uniform liquid. The molten liquid was poured on to a brass moulds, cooled to ambient temperature (30 °C), and then annealed for two hours at 300 °C. The samples are polished using different grades of emery papers to produce well-polished and homogenous thickness samples. The chemical formulae used in the experiment was as follows: $70B_2O_3$ - $20Li_2O$ - (10 - x) Al_2O_3 - xSc_2O_3 (where x=0, 0.1, 0.2, 0.3, 0.4, 0.5, 0.6 and 0.7 in mol%). The glass samples were given the names BLA, $BLASc_{0.1}$, $BLASc_{0.2}$, $BLASc_{0.3}$, $BLASc_{0.4}$, $BLASc_{0.5}$, $BLASc_{0.6}$, and $BLASc_{0.7}$ depending on how much scandium was doped into them. The XRD-6100 SHIMADZU X-ray diffractometer's Cu-Ka radiation (a = 0.15406 nm) with a 2θ angle range of 10^0 to 80^0 was used to determine the present samples' amorphous nature. The Shimadzu UV-1800 spectrophotometer was used to record the absorbance spectra between 200 and 1100 nanometers. With KBr pellets working between 500 and 4000 per centimeter on a Brukar AT-IR spectrophotometer, FT-IR spectra of all glass samples have been recorded.

3. RESULTS AND DISCUSSION

3.1 XRD studies

The fabricated glass samples' X-Ray Diffractograms, which were captured at room temperature between 2θ values 10° and 90° at a rate of 2°/min as depicted in figure 1. It is evident that there is no sharp crystalline peaks, which clearly indicates the non-crystalline nature of glass. [7].

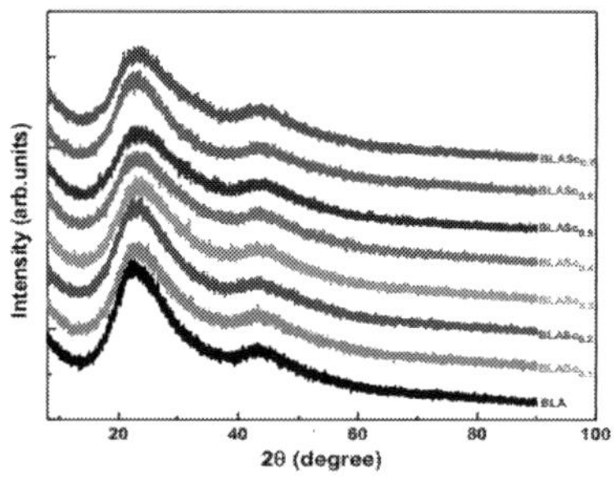

Figure 1: X-ray diffraction spectrum of fabricated glass samples

3.2, Physical Parameters

Table 1 displays several physical characteristics of the fabricated samples based on observed densities and computed molar volumes [7 - 9]. The physical characteristics of the glasses were identified, and they provided a detailed grasp of the atomic arrangements inside their network architectures [10].

3.2.1, Density

One of the most crucial analyses for comparing the investigated glass samples is density measurement. Toluene was used as the immersion solvent for the Archimedes principle density measurements on the created glass matrix. Glass matrix's molar volume can be calculated using the correct relation by knowing its density. As the percentage of scandium rises the density of the manufactured glass matrix falls where as its molar volume increases. This could be the result of NBO becoming bridging oxygen. The molecular volume grows as expected as the mol% of scandium in the glass matrix increases. This is because the amount of NBO in the glass samples changes when the scandium concentration rises, and the molecular volume rises as a result of an increase in bond length in the glass network. The negative relationship between the two values has been confirmed, figure 2 depicts how density and molar volume change as scandium concentration changes.

3.2.2, Optical basicity

Theoretical optical basicity (Λ_{th}), was estimated using the relevant formula [11] and the results are reported in Table 1. The creation of innovative functional materials with improved optical properties may benefit from an understanding of the optical basicity properties of glasses. As the concentration of scandium rises, the optical basicity values increase, indicating a higher capacity for oxide ions to donate electrons to neighboring cations and it is as depicted in figure 3 [12]

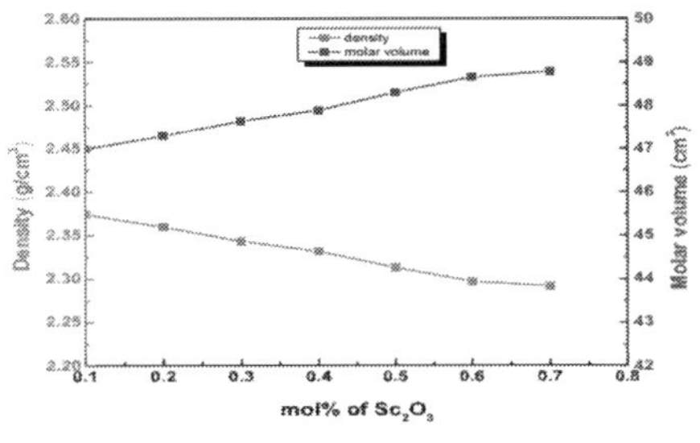

Figure 2: Variation of Density and molar volume with mol% of Sc_2O_3

The Clasius-Mosotti relation was used to determine the electronic polarizability [13], typically, polarization rises with the quantity of NBO atoms. As the concentration of scandium rises, the electronic polarizability for manufactured glass matrix rises which is depicted as in figure 3. Technology-wise, the majority of the physical and chemical properties of glasses that are governed by electronic polarizability are equally important to that of optical and electronic materials.

The charge overlap between adjacent ions is what Yamashita and Kuroswa's interaction parameter (A) (means physically) refers to [14]. Table 1 provides a summary of the A_{th} values that were acquired. Figure 3 depicts the relationship between optical basicity Λ_{th} and the interaction parameter A_{th} for current glasses, demonstrating a strong linear association between them.

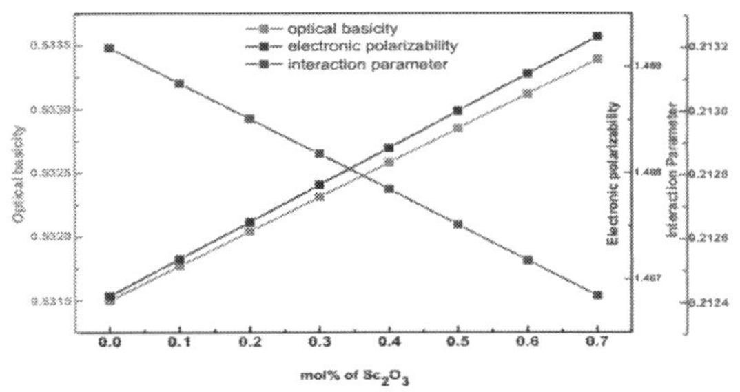

Figure 3: Variation of Optical basicity, electronic polarizability and interaction parameter with mol% of Sc_2O_3.

Table 1. Physical parameters of BLA glasses.

Glass sample	BLA	BLA $Sc_{0.1}$	BLA $Sc_{0.2}$	BLA $Sc_{0.3}$	BLA $Sc_{0.4}$	BLA $Sc_{0.5}$	BLA $Sc_{0.6}$	BLA $Sc_{0.7}$
Mol% of Scandium	0	0.1	0.2	0.3	0.4	0.5	0.6	0.7
Density ρ (g/cm³)	2.341	2.374	2.360	2.343	2.331	2.312	2.297	2.291
Molar volume Vm (cm³/mol)	47.636	46.990	47.295	47.641	47.895	48.300	48.649	48.783
Optical basicity	0.5315	0.5318	0.5320	0.5323	0.5326	0.5329	0.5331	0.5334
Electronic polarizability(Å³)	1.4668	1.4672	1.4675	1.4679	1.4682	1.4686	1.4689	1.4693
Interaction parameter (Å⁻³)	0.2132	0.2131	0.2130	0.2129	0.2128	0.2126	0.2125	0.2124

3.3, *FT-IR Spectroscopy*

FTIR spectroscopy is a method that uses mid-infrared radiation to analyse the molecular structure and composition of substances by creating a distinctive fingerprint for identification and study. The way FTIR operates is by identifying particular light frequencies that are related to the energy of the vibrational bonds inside molecules. The FT-IR spectra of fabricated glass matrix is as depicted in figure 4 and assignments for various bands as in table 2.

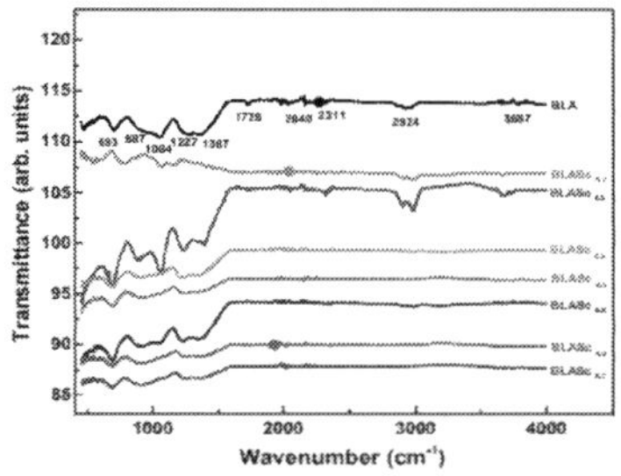

Figure 4: FTIR spectra of Scandium doped BLA glass system.

Table 2. FT-IR band assignments in Scandium doped BLA glasses

Band position(cm^{-1})	Assignments
693	Vibrations of B-O-B in the BO_4 and BO_3 groups [15]
887	The NBOs of the BO_4 groups [15]
1064	The BO_4 group vibration's of B-O bond stretching [16]
1227,1367	B-O bonds stretching vibration in the of meta-borate groups [16,17]
1726	Hydrogen bonding [17]
2040,2311,2924	Presence of OH group [17,18]
3687	OH group vibration of OH group[19]

3.4, *Optical absorption study*

The absorbance spectra of glass samples in the wavelength range 200 nm to 1200 nm are shown in figure 5. The notable absorption peak for the doped glasses is located at 277 nm. The strengths of the absorption bands varies as Sc_2O_3 concentration rises and it is maximum when mol% of Sc_2O_3 is 0.4. This peak match with absorption spectroscopy measurements satisfactorily [20]. For manufactured glasses, the optical absorption coefficient (α) has been computed using relevant relation

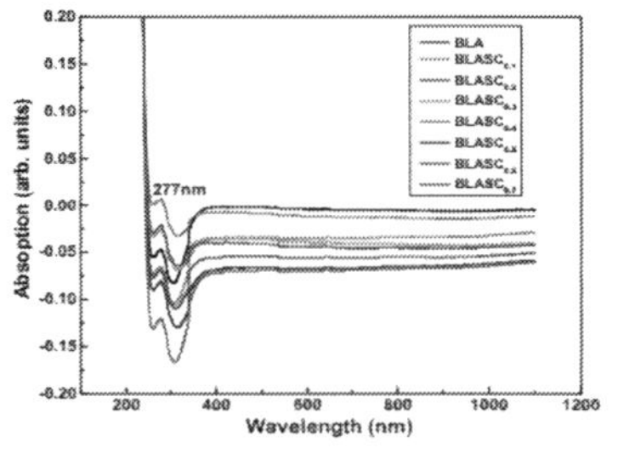

Figure 5: Absorption spectrum of Sc^{3+}doped BLA glass samples.

3.4.1, Direct and indirect optical energy band gaps

The optical band gaps (Eg), which are calculated using Davis and Mott's equation. Tauc's plots of these glass matrices are depicted in Figures 6 and 7. Using these plots, the optical energy gap for direct and indirect transitions was estimated. The evaluated energy gap values are shown in Table 3. According to Figure 10, a plot of Eg versus mol % of Sc_2O_3 content demonstrates that Eg declines as Sc_2O_3 content rises. The structural changes that are occurring in glasses is the main reason for Eg decrease as the Sc_2O_3 content rises.

In general higher negative charge is present on non-bridging oxygen atoms. The non-bridging oxygen atoms with bigger magnitude of negative charge makes it simpler for their electrons to be excited to the conduction band of cations, which reduces the optical band gap [22, 23]. Additionally, we assume that adding Sc^{3+} ions to the existing glass will cause them to occupy interstitial sites and function as modifiers, which will ultimately cause the breakdown of glass network. As a result, the band edge changes with increasing Sc_2O_3

content to longer wave lengths. Additionally, as NBO rises, the degree of electron localization also increases. This improves the glass network's donor centers and lowers the optical band gap [23].

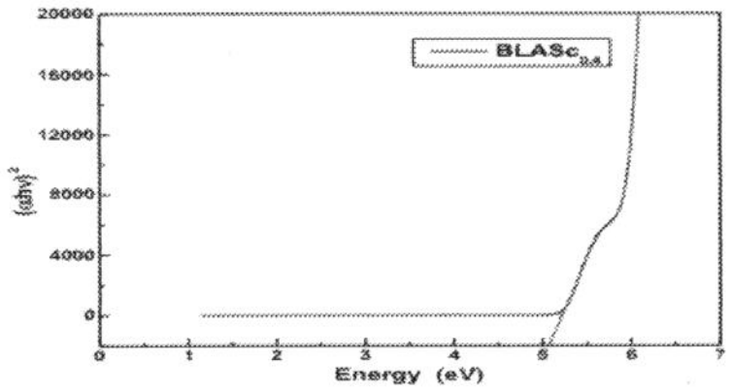

Figure 6 : $(\alpha h\nu)^2$ versus hv plot for 0.4 mol% Sc^{3+} ion BLA glass

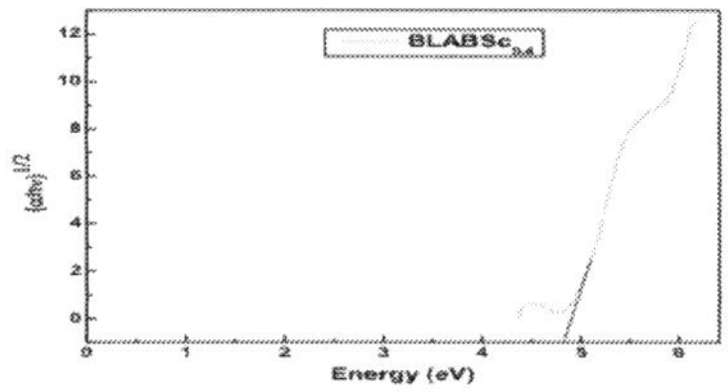

Figure 7. $(\alpha h\nu)^{1/2}$ versus hv plot for 0.4 doped mol% Sc^{3+} ion doped BLA glass

3.4.2 Urbach energy (E_u)

Glassy materials exhibit a tailing of state densities in the direction of the band gap. The breadth of these band tails is referred to as the urbach energy [24, 25].The urbach energy (E_u) has been calculated using the proper relation. The reciprocal of slope of linear portion obtained in the graph of logα verses energy of incident photon gives urbach energy. The urbach energy values obtained for fabricated glass matrix are tabulated in table 3 and variation of urbach energy with mol % of Sc_2O_3 is as in fig.8. It is observed that the Urbach energy values rise as the mol % of Sc_2O_3 increases. This behaviour is ascribed to the weak links turning into defects, which raises the level of disorder in the glass structure and increases the number of localised states in the band gap [26].

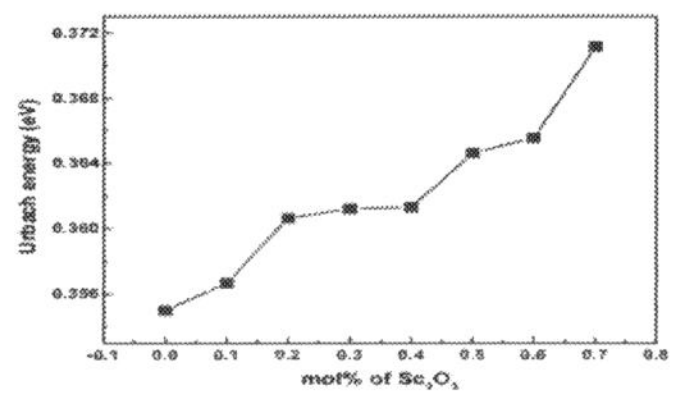

Figure 8: Variation of Urbach energy gap with mol% of Sc_2O_3

3.4.3 Index of refraction

Using the Dimitrov and Sakka relation, the refractive indices for the direct (n_d) and indirect (n_{ind}) energy gaps of fabricated glass samples was found [27]. With an increase in scandium concentration, the index of refraction for the current glass matrix rises. The results are as shown in table 3. Figures 9 show how the amount of scandium affects the optical energy band gap (for direct and indirect light) and index of refraction (n) .

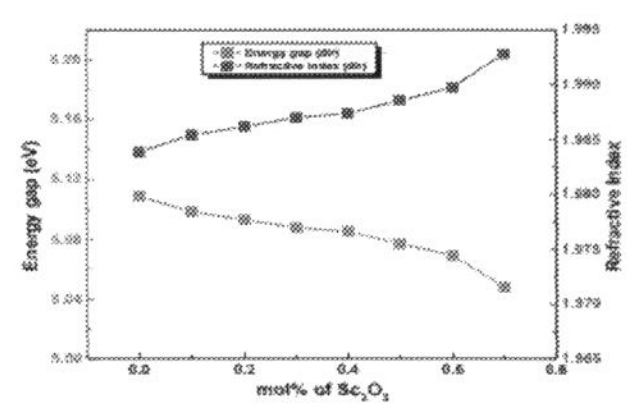

Figure 9: Direct optical energy band gap and indices of refraction with concentration of Sc_2O_3 plot for glass matrix

Table 3. Optical parameters of scandium doped BLA glass samples

Glass sample	BLA	BLA $Sc_{0.1}$	BLA $Sc_{0.2}$	BLA $Sc_{0.3}$	BLA $Sc_{0.4}$	BLA $Sc_{0.5}$	BLA $Sc_{0.6}$	BLA $Sc_{0.7}$
Energy gap E_g (eV)	5.109	5.098	5.093	5.088	5.085	5.077	5.069	5.048
R.I n	1.984	1.985	1.986	1.987	1.987	1.988	1.990	1.992
Molar mass M (g)	111.536	111.572	111.608	111.644	111.680	111.716	111.752	111.788
Urbach energy E_u(eV)	0.355	0.357	0.360	0.361	0.361	0.364	0.365	0.371

4 Conclusions

X-ray diffraction studies has been established the non-crystalline nature fabricated glass samples. Physical properties including density and molar volume, are clearly show that the addition of scandium to the glass sample decreases its compactness. The transition of borate units from $BO_3 \rightarrow BO_4$ with increase in scandium was confirmed by FTIR. It has been investigated that the production of non-bridging oxygen atoms in the glass network causes the structural stability of the glass system to decline. UV-visible spectroscopy indicated that this reduction in structural stability resulted in an improvement in the optical characteristics. The estimated urbach energy, which is brought about by the production of NBOs, ultimately causes the glass network to become more disordered and accelerates the rate of de polymerization. By enhancing the ionic character of the glasses, such disorder affects their optical qualities and makes them a good fit from an application point of view

5. REFERENCES

[1] J.E. Shelby, Introduction to Glass Science and Technology, second ed., The Royal Society of Chemistry, Cambridge, UK, 2005.
[2] R.S. Gedam, V.K. Deshpande, Solid State Ionics 177 (2006) 2589.
[3] F.H. ElBatal, M.S. Selim, S.Y. Marzouk, M.A. Azooz, Physica B 398 (2007) 126.
[4] N. Deopa, A.S. Rao, Opt. Mater. 72 (2017) 31-39.
[5] N. Deopa, A.S. Rao, J. Lumin. 196 (2017) 832-841.
[6] F. Ahmadi, R. Hussin, S.K. Ghoshal, J. Non-Cryst. Solids 499 (Nov. 2018) 131-141
[7] A. Rao, Y. Ahammed, R. Reddy, T.V. Rao, Opt. Mater. 10 (4) (Sep. 1998) 245-252.
[8] A. Ichoja, S. Hashim, S.K. Ghoshal, I.H. Hashim, R.S. Omar, J. R. Earths 36 (12) (2018) 1264-1271.
[9] Y.A. Yamusa, R. Hussin, W.N.N.W. Shamsuri, Y.A. Tanko, S.A. Jupri, Optik 164 (2018)324-334.
[10] R. Laopaiboon, C. Bootjomchai, Mater. Des. 80 (Sep. 2015) 20-27.
[11] Algradee, M. A., Sultan, M., Samir, O. M., & Alwany, A. E. B. (2017). Applied Physics A, 123(8) (2017)524
[12] E.S. Yousef, M.M. Elokr, Y.M. AbouDeif. J. Mol. Struct. 1108, 257-262 (2016)
[13] M. Mariyappan, K. Marimuthu, M.I. Sayyed, M.G. Dong. U. Kara, J. Non-Cryst. Solids 499(2018) 75-85.
[14] V. Dimitrov, T. Komatsu, J. Solid Stat. Chem. 163, 100-112 (2002).
[15] Y. Gandhi, K.S.V. Sudhakar, M. Nagarjuna, N. Veeraiah, J. Alloy. Compd. 485 (2009) 876.
[16] G. Pal Singh, D.P. Singh, Physica B 406 (3) (2011) 640–644.
[17] G. Lakshminarayana, S. Buddhudu, Spectrochim. Acta 63 (2006) 295.
[18] Frank. Berkemeier, Stephan. Voss, Árpád.W. Imre, Helmut. Mehrer, J. NonCryst. Solids 351 (2005) 3816–3825.
[19] L. Baia, S. Simon, Phys. Chem. Glas. 46 (2005) 279-283.
[20]. L.R. Moorthy, T.S. Rao, M. Jayasimhadri, A. Radhapathy, D.V.R. Murthy. Spectrochim. Acta Part A 60, (2004) 2449-2458
[21] M.AAlgradee, AElwhab, B. Alwany, A.AHigazy. J Adv. Phys. (2017) 6163-170
[22] M.A. Algradee, A. Elwhab, B. Alwany, M. Sultan, M. Elgo- shimy, Q. Almoraisy. Optik 142, (2017) 13-22.
[23] M. Farouk, A. Abd El-Maboud, M. Ibrahim, A. Ratep, I. Kashif, Spectrochim. Acta A 149, (2015) 338-342.
[24] F. Ahmad, E. Hassan Aly, M. Atef, M.M. ElOkr, J. Alloys Compd. 593, (2014) 250-255
[25] FUrbach, Phys. Rev. 92, (1953) 1324.
[26] Y. Chen, Q. Nie, T. Xu, S. Dai, X. Wang, X. Shen. J. Non Cryst. Solids 354, (2008) 3468-3472.
[27] V. Dimitrov, S. Sakka. J. Appl. Phys. 79, 1741-1745 (1996)

Situ synthesis and characterization of polypyrrole/ZnO Nanocomposites for optical and photocatalytic activity

Mahadevi V[1], Veeresh S[2], B T Vijaykumr[1], Reshma[1], Priyanka Kolhar[1], Bharati Basavaraj[1], Basavaraja Sannakki[1*]

[1]Department of physics, Gulbarga University, Kalaburagi-585106
[2]Department of physics, Mangalore University, Mangalagangothri-574199
Author's for correspondence : madhupreddy6553@gamil.com & sannakki.phy@gmail.com

Abstract. In situ polymerisation process used for manufacturing polypyrrole (PPy) nanoparticle-reinforced various doping concentration of zinc oxide (ZnO NPs) nanocomposites. Fourier transform infrared(FTIR), Scanning electron microscopy(SEM), Photo luminescence (PL) and UV-vis spectroscopy was used to characterise the polymer nanocomposite. The ZnO nanoparticles are distributed throughout PPy matrix as shown in the SEM pictures of the nanocomposites. Distinctive peaks of Pure PPy and metal oxide bond of ZnO are as displayed in the FTIR spectra. Optical property of PPy, ZnO and PPy/ZnO was studied by using PL and UV-DRs spectrophotometer. These outcomes demonstrated that the synthesised PZ30 nanoparticles could potentially use in the treatment of wastewater.

Keywords: Situ polymerization; polymer nanocomposite; methylene blue.

1.Introduction

Pollutants in water cause serious ecological issues for every living organism. The main contributors to pollution are synthetic organic compounds, which include dyes produced by a variety of industries, including pigment, cosmetics, paper, plastic, skin, leather, photography, food, and medicines, among others. Highly poisonous and possibly cancer-causing dyes have been connected to a number of human and animal disorders [1-3]. Associated with textile dyes are allergenic to skin conditions and problems with the central neurological system. These problems may originate from the inactivation of the enzymes due to cofactor replacement [4, 5]. When ingested or inhaled, textile dyes can irritate the skin and eyes, especially if they have been exposed to dust [6, 7]. The aromatic structure of dyes increases their resistance to deterioration in the presence of the environment. Because pollution has significantly increased, worldwide environmental standards are growing stricter to cut pollution, and technology to remove dyes has been progressing [8, 9]. One of the popular methods to remove contaminants in water is photocatalytic degradation (PCD) [10, 11]. Photocatalysts are the best materials for converting solar energy for use in oxidation and reduction processes. Photocatalysts are used in a variety of processes, including as removing impurities from water and the air, splitting water to create H_2, controlling Oduors, inactivating bacteria and cancer cells, and managing odors [12-14].

Due to its potential to remove hazardous and harmful substances from the environment, photocatalyst materials have gained significant research during the past 10 years [15, 16]. There is a great deal of promise for the removal of hazardous substances with semiconductor photocatalysts. Because of recent developments in its synthesis and its distinct optoelectronic, catalytic, and photochemical capabilities, zinc oxide (ZnO), that has a broad band gap of 3.2 eV, has attracted a large amount of interest [17]. Photocatalysts cause the generation of reactive oxygen, which helps in the decomposition of contaminants. Incorporating two photocatalysts with various bandgaps leads to a greater separation of electron/hole pairs, where more species would exist for reactions to degrade contaminants.

In this study, the PPy/ZnO NPs were synthesized by insitu polymerisation method. Synthesized nanocomposites were characterized by FTIR, SEM with EDX, UV-Vis and PL. This study evaluated visible light using PPy/ZnO nanocomposites as a catalyst.

2.Experimental section

2.1 *Materials*

Zinc Nitrate Hexahydrate [$Zn(NO_3)_2 6H_2O$], Ethylenediaminetetraacetic acid disodium salt (EDTA-Na- salt), Pyrrole monomer from SRL Pvt Ltd. Methylene blue (MB), Ammonium persulphate (APS) ($(NH_4)_2S_2O_8$) and 25% Aqueous Ammonia from SD Fine-Chem Limited. Tetra butyl alcohol (TBA), ascorbic acid (AA) from

Sigma-Aldrich, potassium dichromate ($K_2Cr_2O_7$) from Analytical CS Reagent.

2.2 Synthesis of PPy

Chemical oxidative polymerization was employed to synthesize PPy using the pyrrole (C_4H_5N) monomer. As an oxidising agent, ammonium per-sulphate of analytical grade (APS, $(NH4)_2S_2O_8$) was used. A 1:1 volume ratio of 0.1M aqueous solution of pyrrole and 0.1M of APS was used to carry out the chemical polymerization in a beaker. The polymerization process took place at room temperature for 4 hours. Black precipitate was produced after the polymerization process ended. To get rid of any potential oligomers, the precipitate was filtered, repeatedly washed with distilled water, and then dried for six hours at 60°C.

2.3 Synthesis of ZnO NPs

ZnO nanoparticles (NPs) were created using the co-precipitation technique. To prepare ZnO, 0.04M solution of $Zn(NO_3)_26H_2O$ (2.97g/250ml) was stirred continuously while sodium hydroxide solution (4M) was gradually added. White precipitate was obtained at pH 11, then filtered and was dried at room temperature after being rinsed with ethanol/DI water. In order to create ZnO NPs, the precipitate was ultimately heated to 450°C for three hours in a muffle furnace (heating rate: 5°C min-1).

2.4 Synthesis of PPy/ZnO Nanocomposite

Pyrrole is dissolved in water at 0.1M was mixed with the necessary weight percent of ZnO NPs (15, 30, and 45%), and this mixture was added to the monomer solution to begin the chemical polymerization process. Drop wise addition of 0.1M of APS in a volume-to-volume ratio of 1:1. The polymerization took place for 4 hours at the ambient temperature. The result of the polymerization process was the formation of the black precipitate. To create the PPy/ZnO nanocomposite and remove any possible oligomers, the precipitate was filtered and washed multiple times in distilled water. It was dried for six hours at 60 °C.

2.5 Characterization

Using a JASCO FTIR 6800, FTIR spectrometer in the range from 400 to 4000 cm^{-1} was used measuring the functional group of the PPy/Zno. SEM analysis was used to analyse the morphological investigations (Neo-Scope JCM-6000 PLUS system). UV spectroscopy was used to analyse the absorption spectra for the presence of functional groups (DS5Graphic_RGB-1024x667), and double excitation and emission monochromator photoluminescence spectrometer (FLS1000) was used to detect the PPy/Zno photoluminescence.

3 Result and discussion

3.1 Fourier Transform Infrared Spectrometer (FTIR)

The FTIR spectra of pure PPy, ZnO nanoparticles, and PPy/ZnO nanoparticle composites loaded with varying ZnO contents are displayed in Figure 2. From spectra it is observed that, the FTIR spectrum of synthesized nanoparticle in the range of 400 – 4000 cm^{-1} the stretching mode of ZnO appears at around 433 cm^{-1} (Figure 2) . The synthesis of PPy was confirmed by the spectra of bulk PPy, as illustrated in Figure 2. Bands at 1556 cm^{-1} and 1480 cm^{-1}, respectively, are responsible for the weak band's and C=C and C-C's stretching vibration in the pyrrole ring [18]. At 1182 cm^{-1} and 1051 cm^{-1} in the IR spectrum, respectively, PPy displays the usual C-N and C-H stretching vibration of pyrrole. The peak at 1030 cm^{-1} is the result of PPy being absorbed by N-H in-plane deformation. The band found at 917 cm^{-1} and 602 cm^{-1} may be caused by the N-H vibration in polymers and the out-of-plane ring deformation.

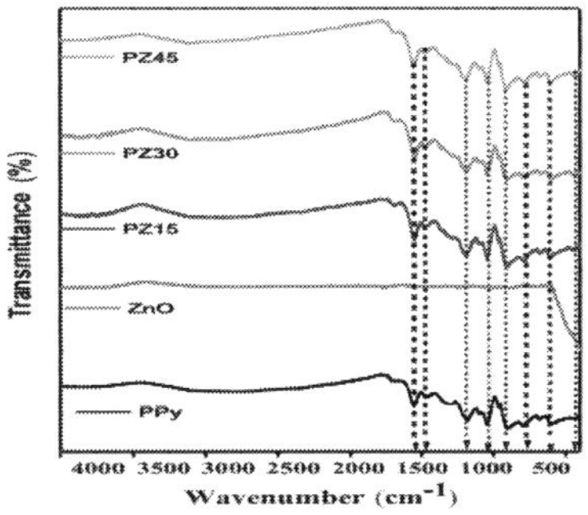

Figure 1. FTIR spectrum of prepared PPy,ZnO and PPy/ZnO nanocomposites

3. UV-DRS Analysis

UV-DRS spectra was used to determine the band-gap of synthesized nanoparticles. As shown in Figure 3, maximum reflectance of PPy, ZnO, 15PZ, 30PZ, and 45PZ nanocomposites was 395nm, 380 nm, 375 nm 345nm and 366 nm respectivly. The equation (3) was used to determine the energy band gap (Eg) of the PPy, ZnO and nanocomposites of PPy/ ZnO from the Tauc's plot, to get $(\alpha hv)^2$ *vs* hv graphs,

$$\alpha hv = A[hv - E_g]^n, \qquad (1)$$

In this case, the index n relies on the kind of electronic transitions, the photon's energy is hv, the absorption coefficient is, the band gap energy is Eg, the absorption constants for indirect transitions are A. For PPy, ZnO, 15PZ, 30PZ, and 45PZ nanocomposites, the observed Eg values were 1.9 eV, 3.2 eV, 3.1 eV, 2.82 eV, and 2.86 eV, respectively. This suggests that increasing PPy can decrease bandgap, which in turn can enhance photocatalytic activity in the presence of sunlight.

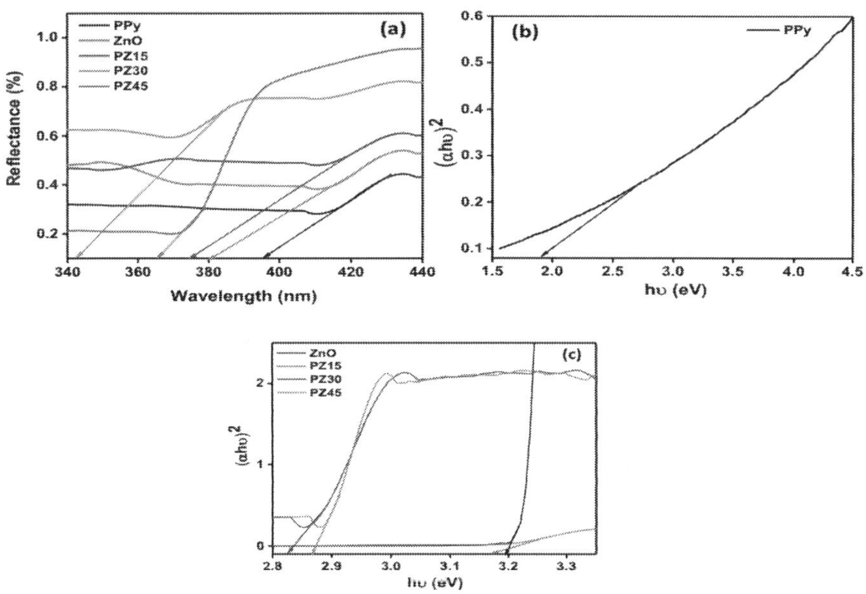

Figure 2. (a) UV-Vis diffused reflectance spectrum of of PPy, ZnO and PPy/ZnO nanocomposites, (b)&(c) Energy band gap of PPy, ZnO and PPy/ZnO nanocomposites.

3.3 Photoluminescence (PL) studies

Figure 4 displays the PL spectra for PPy, ZnO, and PPy/ZnO nanocomposites. At room temperature, excitation wavelength of PPy was 269 nm, while ZnO, PZ15, PZ30, and PZ45 had excitation wavelengths of 330 nm. At the same time, PPy emission peak was 293 nm, and the emission wavelengths of ZnO, PZ15, PZ30, and PZ45 was 492 nm. One of the best methods for determining a photocatalyst's efficiency is the PL study. High rates of electron and hole recombination decrease photocatalytic activity, as electrons and holes are essential to photocatalysis. Consequently, reduced photoluminescence intensity would result in higher photocatalytic activity [19]. According to photoluminescence spectra, PZ30 is a highly effective photocatalyst. Due to the fact that the photoluminescence intensity is lower than in other nanocomposites.

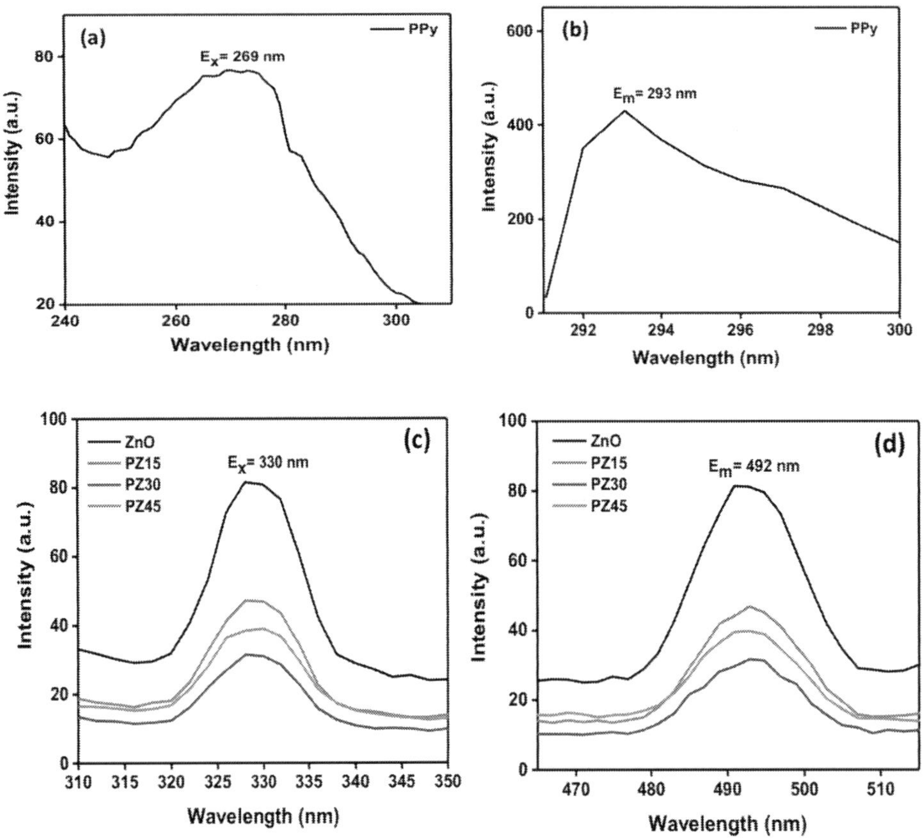

Figure 3. (a)& (b) Excitation peak and emission peak of PPy,
(c)&(d) Excitation peak and emission peak ZnO and PPy/ZnO nanocomposites

3.4 Scanning electron microscopy (SEM) and energy dispersive X-ray analysis (EDX) analysis

SEM and EDX analysis are two possible techniques for analysing the morphology and chemical composition of prepared samples. SEM picture in Figure 5 (a &c) shows that, PPy spherical grains structures are loosely aggregated and ZnO nanoparticles have structure that resembles a flower like structure with sharp edges. The surface morphology of 30PZ is displayed in Figure 5 (e). SEM pictures make it abundantly evident that the PZ30 nanocomposite has an agglomeration- sponge-like heterostructure, a porous surface, and erratically formed clusters of spherical grains (Figure 5(e)). With this heterostructure, the surface area, active sites, and photocatalytic activity are all boosted. Figure 5(b, d &f) displays the EDX spectrum for the PPy, ZnO and PZ30 nanocomposite, It demonstrates that synthetic samples contain C, N, O, and Zn.

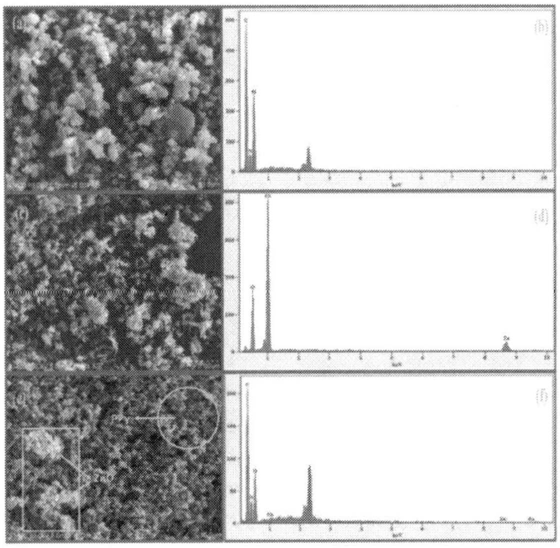

Figure 4. (a, c & e) SEM images of prepared PPy, ZnO and PZ30 and (b, d & f) EDX pattern of PPy, ZnO and PZ30.

4. Conclusion

This study effectively made PPy/ZnO NPs synthesized by eco- friendly coprecipitation and in situ polymaerisation method. Morphological research SEM pictures display the PZ30 nanocomposite has an agglomeration- sponge-like heterostructure, a porous surface, and erratically formed clusters of spherical grains. The optical energy bandgap increases as ZnO doping increases. ZnO, PZ15, PZ30 and PZ45. These outcomes demonstrated that the synthesised PZ30 nanoparticles could potentially use in the treatment of wastewater.

Reference

[1] Ardila-Leal L D, Poutou-Piñales R A, Pedroza-Rodríguez A M, Quevedo-Hidalgo B E, 2021 Abrief history of colour, the environmental impact of synthetic dyes and removal by using laccases, *Molecules* (Basel, Switzerland) **26**(13) 3813.

[2] Alsukaibi A K, 2022 Various approaches for the detoxification of toxic dyes in wastewater, *Processes* **10**(10) 1968.

[3] Mohan D, Shukla S P, 2022 Hazardous consequences of textile mill effluents on soil and their remediation approaches, *Cleaner Engineering and Technology* **7** 100434.

[4] Briffa J, Sinagra E, Blundell R, 2020 Heavy metal pollution in the environment and their toxicological effects on humans, *Heliyon* **6**(9).

[5] Lansdown A B, 1995 Physiological and toxicological changes in the skin resulting from the action and interaction of metal ions, *Critical reviews in toxicology* **25**(5) 397-462.

[6] Akter T, Protity A T, Shaha M, Al Mamun M, Hashem A, 2023 The Impact of Textile Dyes on the Environment, Nanohybrid Materials for Treatment of Textiles Dyes, *Springer*, pp. 401-431.

[7] Smith C, 2023 Allergic contact dermatitis: chemical and metabolic mechanisms, *CRC Press.*

[8] Pereira L, Alves M, 2012 Dyes—environmental impact and remediation, *Environmental protection strategies for sustainable development,* 111-162.

[9] Bhat A P, Gogate P R, 2021 Degradation of nitrogen-containing hazardous compounds using advanced oxidation processes: a review on aliphatic and aromatic amines, dyes, and pesticides, *Journal of Hazardous Materials* 403 23657.

[10] Chakrabarti S, Dutta B K, 2004 Photocatalytic degradation of model textile dyes in wastewater using ZnO as semiconductor catalyst, *Journal of hazardous materials* ***112***(3) 269-278.
[11] Nasr O, Mohamed O, Al-Shirbini A S, Abdel-Wahab A M, 2019 Photocatalytic degradation of Zacetaminophen over Ag, Au and Pt loaded TiO2 using solar light, *Journal of Photochemistry and Photobiology A*: Chemistry **374** 185-193.
[12] Al-Nuaim M A, Alwasiti A A, Shnain Z Y, 2023 The photocatalytic process in the treatment of polluted water, *Chemical Papers* **77**(2) 677-701.
[13] Yaqoob A A, Parveen T, Umar K, Mohamad Ibrahim M N, 2020 Role of nanomaterials in the treatment of wastewater: A review, *Water* **12**(2) 495.
[14] Mills A, Le Hunte S, 1997 An overview of semiconductor photocatalysis, *Journal of photochemistry and photobiology A*: Chemistry **108**(1) 1-35.
[15] Long Z, Li Q, Wei T, Zhang G, Ren Z, 2020 Historical development and prospects of photocatalysts for pollutant removal in water, *Journal of hazardous materials* **395** 122599.
[16] Di Paola A, García-López E, Marcì G, Palmisano L, 2012 A survey of photocatalytic materials for environmental remediation, *Journal of hazardous materials* **211** 3-29.
[17] Ali A, Phull A R, Zia M, 2018 Elemental zinc to zinc nanoparticles: Is ZnO NPs crucial for life? Synthesis, toxicological, and environmental concerns, *Nanotechnology Reviews* **7**(5) 413-441.
[18] Batool A, Kanwal F, Imran M, Jamil J, Siddiqi S A, 2012 Synthesis of polypyrrole/zinc oxide composites and study of their structural, thermal and electrical properties, Synthetic Metals **161**(23-24) 2753-2758.
[19] Lu M, Javed M, Javed K, Tan S, Iqbal S, Liu G, Khalid W B, Qamar M A, Alrbyawi H, Pashameah, 2022 Construction of a Well-Defined S-Scheme Heterojunction Based on Bi-ZnFe2O4/Sg-C3N4 Nanocomposite Photocatalyst to Support Photocatalytic Pollutant Degradation Driven by Sunlight, *Catalysts* **12**(10) 1175.

IOP Conf. Series: Materials Science and Engineering 1300 (2024) 012023 doi:10.1088/1757-899X/1300/1/012023

Investigations on Impact of YSZ Nanoparticles on Morphology & Electrical conductivity of PEDOT: PSS

C J Vijaykumar[1], S S Bulla[1], C K Chavan[2] and R F Bajantri[3]

[1]Department of Studies in Physics, Davangere University, Shivagangotri, Davangere, 577 007, Karnataka, India
[2]Department of Physics, V.V. Sangha's Kottureshwara Degree College, Kottur 583134, Karnataka, India
[3]Department of Studies in Physics, Karnatak University, Dharwad - 580 003 Karnataka, India
soumyabulla@gmail.com
Mr.physicist@rediffmail.Com

Abstract: Conductive polymer nanocomposites, which combine the flexibility and conductivity of polymers with the unique properties of nanofillers, have generated interest in various fields such as energy storage, sensors, coatings, and corrosion protection. This study discusses the effect of Yttria-stabilized zirconium nanoparticles (YSZ NPs) on the morphology and electrical conductivity of poly(3,4-ethylenedioxythiophene): poly(4-styrenesulfonate) (PEDOT: PSS). Thin films of PEDOT: PSS and PEDOT: PSS: YSZ were fabricated using spin coating technique. FTIR and XRD characterizations confirmed the interaction between YSZ nanoparticles and the PEDOT: PSS matrix, leading to changes in chemical morphology and film crystallinity. The investigation of the current-voltage (I-V) relationship showed improved electrical conductivity in PEDOT: PSS films with the addition of YSZ nanoparticles, with respective conductivities of 0.028×10^{-6} S cm^{-1} and 0.0885×10^{-4} S cm^{-1} for pristine and YSZ-containing films. Additionally, the sensing properties of the PEDOT: PSS: YSZ film in detecting organic vapours were studied. These findings suggest that these conducting polymer nanocomposite thin films could potentially be used as electrolyte components in battery cells, supercapacitors, and fuel cells, as well as serve as sensors for certain organic vapours.

1. Introduction

In the field of polymeric materials research, conducting polymers have attracted a lot of attention since their discovery in 1978. [1]. This is mostly justified by their numerous uses, including in secondary batteries and biosensors. Nevertheless, conducting polymers' restricted processibility prevents their prospective applications from getting completely fulfilled. Because of their favourable processability, polymers that can dissolve in water or other solvents that are polar are therefore widely known and highly desirable. Among the wide range of conductive polymers that are readily accessible for purchase, because of its exceptional ability to dissolve in water, poly(3,4-ethylenedioxythiophene): poly(4-styrenesulfonate), or PEDOT: PSS, is frequently referred to as the most efficient conducting polymer in the field of practical applications. [2]. In addition, it should be emphasized that PEDOT: PSS exhibits notable characteristics such as high transparency, favourable mechanical flexibility, and exceptional thermal stability [3–5]. It's also important to note that popular solution-based techniques like spin coating, spray pyrolysis, solution casting, and printing may be employed to create high-quality PEDOT: PSS films on a variety of substrates. It is crucial to emphasize that most other conducting polymers cannot be effectively treated with these methods. There are several uses for poly(3,4-ethylenedioxythiophene): polystyrene sulfonate (PEDOT: PSS), ranging from its use as an electrode constituent to its use as a photographic substance. Its intrinsic flexibility, transparency, and oxidative stability are responsible for its adaptability It is widely used in many different sectors, including transistors, biosensors, electrochemical displays, and flexible electrodes. However, a noteworthy disadvantage of PEDOT: PSS is its comparatively low electrical conductivity, with a value below 1 S-cm^{-1}.

PEDOT: PSS consists of two components: Poly(3,4-ethylenedioxythiophene) (PEDOT), which is insoluble in water, and poly(4-styrenesulfonate) (PSS), which is soluble in water. The insulating feature shown by the PSS component is responsible for the low conductivity seen in virgin PEDOT: PSS. Therefore, it is essential to enhance the conductivity of PEDOT: PSS films. To enhance the connection between PEDOT: PEDOT chains and facilitate the efficient passage of electric charges, it is necessary to mitigate the columbic interaction between the positively charged PEDOT polyions and the negatively charged PSS polyions [6–8].
Numerous studies have been conducted on the improvement of conductivity in PEDOT: PSS using various ways and procedures. Among these approaches, the inclusion of different additives, particularly metallic nanoparticles in the PEDOT: PSS matrix, has shown superior outcomes. In a study conducted by Bhowal et al. (2019), gold (Au) and silver (Ag) nanoparticles were introduced as dopants into a matrix of PEDOT: PSS. The researchers observed changes in the morphology of the PEDOT: PSS and an improvement in its electrical conductivity [6]. In a study conducted by Vidyashree Hebbar et al. (2018), the authors investigated the impact of incorporating graphene oxide particles on the conductivity of PEDOT: PSS films [9]. In a study conducted by Zhuang Zhang et al. (2015), a coating of multi-walled carbon nanotubes (CNTs) was applied over PEDOT: PSS in order to investigate the relationship between the concentration of CNTs and the resulting increase in conductivity [10]. According to the study conducted by Melendez et al. (2010), the authors observed the impact of silver nanoparticles on the electrical conductivity of PEDOT: PSS[11].
Yttrium-stabilized zirconia's remarkable thermal expansion coefficient, low thermal conductivity, high ionic conductivity, and desirable thermal and chemical stability make it a valuable material for a wide range of applications, such as high-temperature structural materials [12], thermal barrier coatings [13, 14], catalysis [15], and the electrolyte of solid oxide fuel cells [16]. Furthermore, compared to bulk materials, nanostructured materials have better or distinctive characteristics and a wide range of capabilities. As an illustration, consider nanocrystalline rare-earth-stabilized zirconia, which, in comparison to its traditional microcrystalline form, has shown improved radiation resistance, decreased thermal conductivity, and increased ionic conductivity [17,18]. Taking into account all of these aspects, we looked at how YSZ nanoparticles, a flexible metal ceramic material, affected the electrical conductivity and morphology of pure PEDOT: PSS in the current study. Additional research was done on the sensing capabilities of the PEDOT: PSS: YSZ thin film in the detection of various organic vapours.

2. Experimental Methods

2.1 Materials used

The water soluble & conductive poly(3,4-ethylenedioxythiophene): poly(4-styrenesulfonate) (PEDOT: PSS) was purchased from Sigma Aldrich, India. Yttria Stabilized Zirconium oxide (3YSZ) with a molecular weight of 349.03 g mol^{-1} is acquired from Get Nanomaterials, France.

2.2 Preparation of pristine PEDOT: PSS & PEDOT: PSS-YSZ nanocomposite thin films

By employing the spin coating technique, PEDOT: PSS and PEDOT: PSS: YSZ nanocomposite thin films were fabricated through the combination of Pure Yttria Stabilized Zirconium oxide (3YSZ) with liquid PEDOT: PSS. The process involved taking 5ml of Pristine PEDOT: PSS and another mixture of 5wt% YSZ nanoparticles dispersed in 5ml of PEDOT: PSS, which were separately subjected to stirring for an hour at 500 RPM and subsequent ultrasonication to ensure even distribution of the components. The resulting samples were then spin coated onto dry and pristine glass plates to produce transparent thin films suitable for further analysis and measurements.

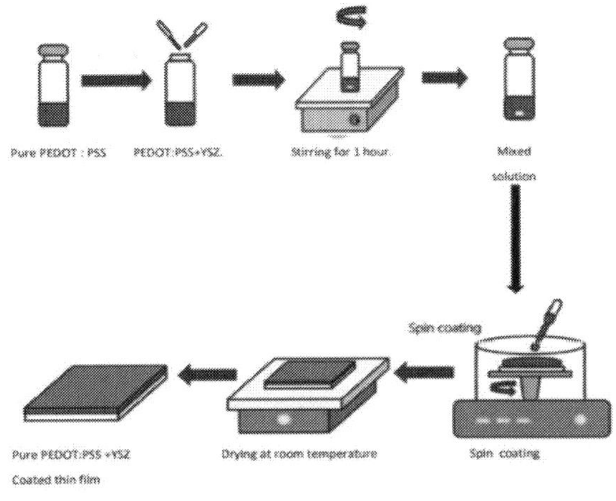

Figure 1: Schematic diagram of synthesis of PEDOT: PSS-YSZ thin films by spin coating

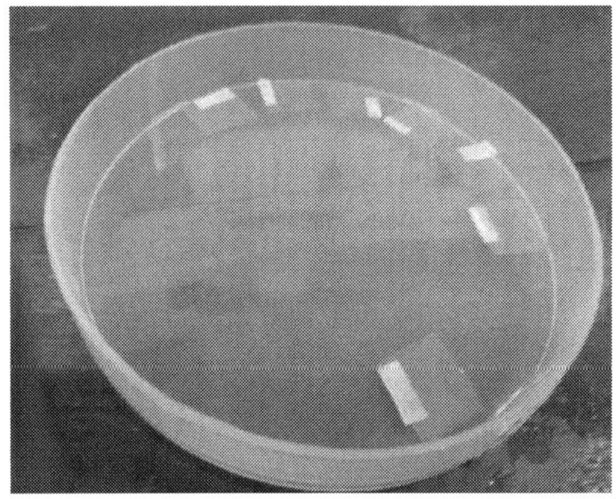

Figure 2: PEDOT: PSS-YSZ thin films

3. Characterization techniques

3.1 FTIR spectroscopy

FTIR was used to confirm the interaction of YSZ nanoparticles with PEDOT: PSS polymers mixture. FTIR spectra was recorded between 400 to 4000 cm^{-1} (wavelength) in transmittance mode using Shimadzu IR Prestige-21 ATR-FTIR with resolution of 4 cm^{-1}.

3.2 X-ray diffraction

To check the effect of additive & to know the variation of dopant induced crystallinity, X-ray diffraction measurements were done. A powder X-ray diffractometer (model Miniflex-II; M/S Japan) was used to assess the samples utilizing Cukα X-ray radiation (λ=1.5418Å). To get the pattern the scanning range was set to a diffraction (2Θ) angle of 10° to 80°.

3.3 Electrical measurements

I-V graph yields valuable information about resistance and conductivity. I-V curves were plotted to know the electrical conductivities of both the samples, by the four-probe method. I-V characteristics was measured using 4-point probe (Napson-3000).

4. Results and Discussion

4.1 X-ray Diffraction Studies

X-ray diffraction (XRD) investigation was conducted in order to examine the influence of an additive and examine the modifications to crystallinity accomplished by the dopant. The X-ray diffraction patterns acquired for the pristine PEDOT: PSS and YSZ-doped PEDOT: PSS samples are shown in Figure 4, with the diffraction angle ranging from 10 to 80 degrees. A prominent diffraction peak was seen at around 22.61° in the X-ray diffraction pattern of virgin PEDOT: PSS. This peak corresponds to the alternating lamella stacking arrangement of PEDOT and PSS, characterized by a lattice spacing of approximately 2.55Å. In the case of PEDOT: PSS, a wider peak was seen at an angle of about 21.86° (2θ), indicating a lattice spacing of around 2.63 Å for YSZ. The presence of faint crystallinity in the materials is indicated by the Border peaks seen in the X-ray diffraction (XRD) patterns. The X-ray diffraction (XRD) examines reveals corroboration of the amorphous characteristics of both virgin PEDOT: PSS and PEDOT: PSS: YSZ thin films.

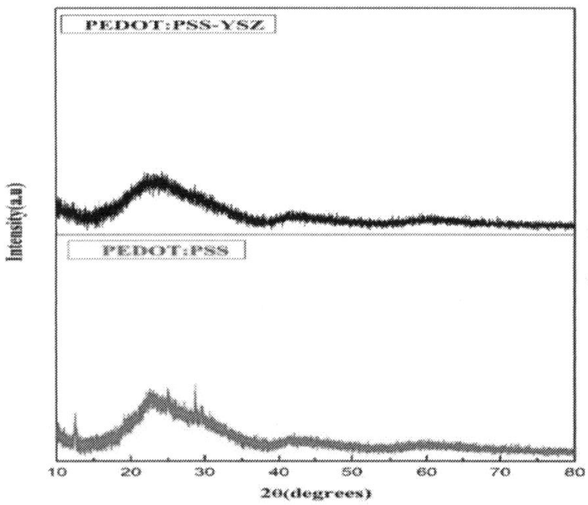

Figure 3: X-ray diffraction Spectrograph

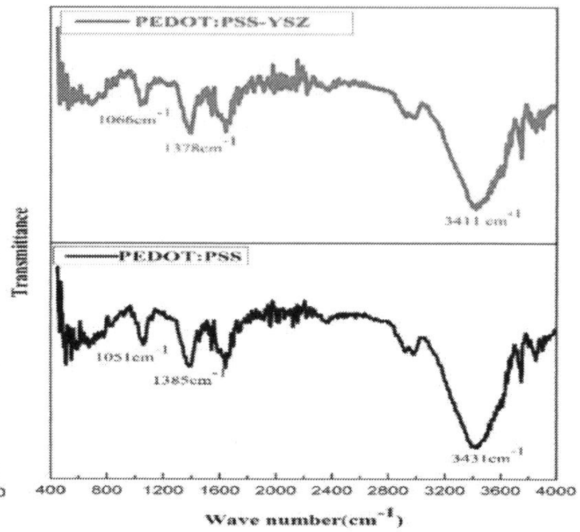

Figure 4: FTIR Spectrograph

4.2 FTIR Studies

FTIR spectroscopy provided molecular-level data to validate changes in the chemical morphology of PEDOT: PSS films when YSZ nanoparticles were added. A comparative analysis was conducted between PEDOT: PSS-YSZ nanocomposite thin films and pristine PEDOT: PSS films using FTIR spectrographs. Wave numbers at 785 and 830 cm^{-1} indicated the presence of a C-S stretching bond in PEDOT: PSS films, while 1051 cm^{-1} indicated stretching of C-O-C bonds. The infrared absorption bands at 1385 cm^{-1} and 1648 cm^{-1} were due to stretching vibrations of C-C bonds, C=C bonds, and sulfonic acid groups in Poly styrene sulfonate. Other peaks corresponding to the presence of C=S, C-N, C=O, C-H stretching, and O-H stretching were also observed. Additionally, a peak at 3753 cm^{-1} was attributed to stretching vibrations of O-H bonds in Sulfoxide functional groups.

YSZ nanoparticles treatment of PEDOT: PSS results in observed peak shifting and the appearance of new peaks. The C-O-C bond stretching peak at 1051 cm^{-1} shifts to 1066 cm^{-1}. The C-S stretching peaks at 785 cm^{-1} and 830 cm^{-1} shift to 683 cm-1 and 812 cm^{-1}, respectively. The shifts in these peaks indicate specific molecular vibrations. The stretching of C-C bonds causes the initial peak at 1385 cm^{-1} to shift to 1378 cm^{-1}, while stretching of C=C bonds causes the initial peak at 1648 cm^{-1} to shift to 1652 cm^{-1}. The stretching of C=S bonds leads to the shift of the initial peak at 1969 cm^{-1} to 1913 cm^{-1}. The presence of a C-N bond causes the initial peak at 2213 cm^{-1} to shift to 2910 cm^{-1}, and stretching of C=O bonds causes the initial peak at 2976 cm^{-1} to shift to 2970 cm^{-1}. The stretching of C-H bonds results in the change of the absorption peak from 3431 cm^{-1} to 3411 cm^{-1}. Finally, the stretching of O–H bonds causes the initial peak at 3753 cm^{-1} to shift to 3739 cm^{-1}. These observed shifts indicate conformational changes in PEDOT and PSS chains, resulting from electrostatic interactions between YSZ nanoparticles and PSS. FTIR measurement confirms the alteration in the chemical structure of PEDOT: PSS chains with the inclusion of YSZ nanoparticles.

4.3 I-V Characteristics

The investigation of the current-voltage characteristics of produced polymer nanocomposite thin films and the subsequent analysis of their corresponding I-V graphs provide significant insights into the resistance and conductivity properties of these materials. The Figures 5 and 6 displays the I-V graphs of acquired for pristine PEDOT: PSS and PEDOT: PSS-YSZ thin films respectively. For PEDOT: PSS: YSZ thin films with a thickness of 96 micron, we observed high electrical conductivity and low electrical resistance. However, while considering the virgin PEDOT: PSS with a thickness of 33μm, it was observed that the electric current flow was significantly reduced and exhibited a high electrical resistance. Therefore, it can be deduced that the electrical resistance of PEDOT: PSS is reduced with the incorporation of YSZ nanoparticles. The observed phenomenon may likely be attributed to the electrostatic contact between YSZ nanoparticles and PSS chains. This interaction facilitates enhanced connectivity between PEDOT chains, hence promoting the unhindered motion of electrons along the PEDOT chain. It was determined that the electrical conductivity of the pure form of PEDOT: PSS is 0.028×10^{-6} S cm^{-1}, whereas the conductivity of PEDOT: PSS: YSZ was found to be 0.0885×10^{-4} Scm^{-1}. The inclusion of YSZ nanoparticles resulted in a threefold improvement in the conductivity of pure PEDOT: PSS. The investigation of current-voltage (I-V) characteristics provides empirical evidence supporting the enhanced electrical conductivity seen in virgin PEDOT: PSS thin films as a result of the incorporation of YSZ nanoparticles.

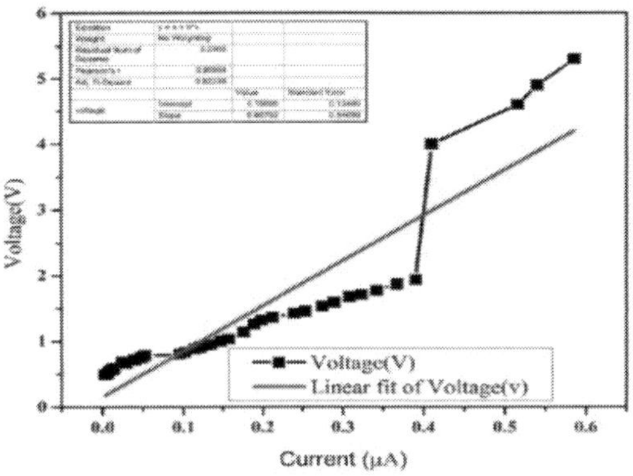

Figure 5: I-V characteristic graph of PEDOT: PSS films

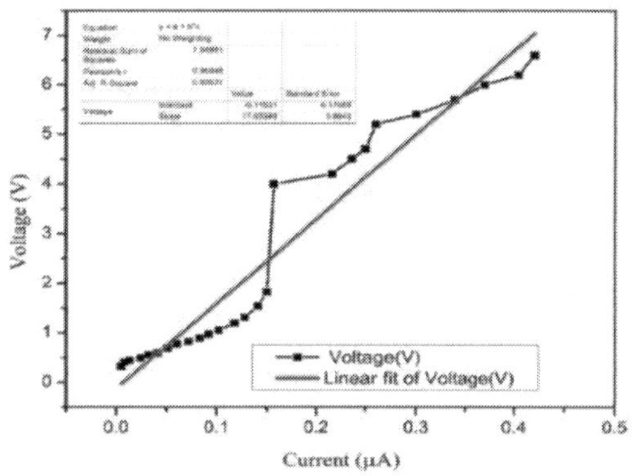

Figure 6: I-V characteristic graph of PEDOT: PSS: YSZ films

4.4 Sensing characteristics

The sensor has an extensive range of functional features linked to their particularity, reaction time, sensibility, and functional life and may be characterized as a system that can identify, measure, and send a clear output to a physical quantity[19]. Percent sensitivity (S), which is shown in the equation below, was used to measure the sensor's performance.

$$S(\%) = \frac{\Delta R}{Ra} \times 100$$

$$\Delta R = R_g \sim R_a$$

Where, R_a and R_g represent the resistance of the sensor measurements in the presence and absence of an organic solvent gas, accordingly.

Experiments were conducted utilizing low-boiling range organic vapor (OV) such as acetone, ethanol, hydrogen peroxide, and tetrahydrofuran (THF) to assess the sensing properties of PEDOT: PSS composite. The % sensing response, response time, and recovery time of PEDOT: PSS sensor thin films were used to evaluate their sensing capabilities. The chosen thin films (prepared composites) were first carefully set into the sample container. A 250 mL beaker containing 3 mL of testing sample OV (acetone, ethanol, hydrogen peroxide, and tetrahydrofuran) was then used to hold it. The thin films were exposed to the test sample's vapor for 250 s. It was then separated off the beaker and left out in the open for a further 250 seconds. Each sensor's resistance was seen to abruptly and steadily drop in the testing sample's vapor until eventually becoming constant. A sensor's resistance increased while it was in the air and finally reached the saturation point, returning to its previous value.

4.4.1 Sensing response

Fundamentally CP based nanoparticles are one of the most significant families of inexpensive sensing materials, outperforming other sensing materials in terms of sensitivity, response time, and reversibility at room temperature. Dopant and its composition may change the resistance of the CPs and their nanocomposites. Using Origin software, the prepared sensor (thin film) PEDOT: PSS collected the data after being exposed to four organic solvents (analytes) using a comparable technique.

It may be expected that PEDOT: PSS and its composite materials exhibit a strong sensitivity response (sensitivity%) to Hydrogen per oxide more than other gasses based on the chemical sensor readings (Table 1) acquired from this study.

4.4.2 Response and recovery time

The time it takes for a thin film's resistance to fall by 90% off its starting point following gas contact is referred to as the reaction time. The amount of time required for a thin film's resistance to return to 90% of its original level once the gas (air environment) has been eliminated is determined. Good reaction and recovery times are necessary to enhance sensing capabilities. In this work, liquid reducing analytes at ambient temperature (30 °C) were utilized rather than gas vapor. A liquid can be heated to almost boiling point at PPM value to get the same result. The evaporation rate is thus lower at room temperature. As a result, while the sensor's reaction time is quick, the response and recovery times are longer. Thus, the desirable gas-sensing properties are obtained. The measurements of the sensing response, response time, and recovery time of the PEDOT: PSS: YSZ sensor for various OVs are also shown in Table T1, and the same data is presented in a separate graph in Figure 7. From the reaction and recovery times, it can be seen that the time intervals range widely, from mere seconds to many minutes. PEDOT: PSS: YSZ sensor reaction time for acetone medium was 82 seconds, while tetrahydrofuran medium response time was 210 seconds. Figure 7 illustrates the PEDOT: PSS: YSZ sensor's dynamic recovery behaviour with acetone, ethanol, hydrogen peroxide, and THF OVs

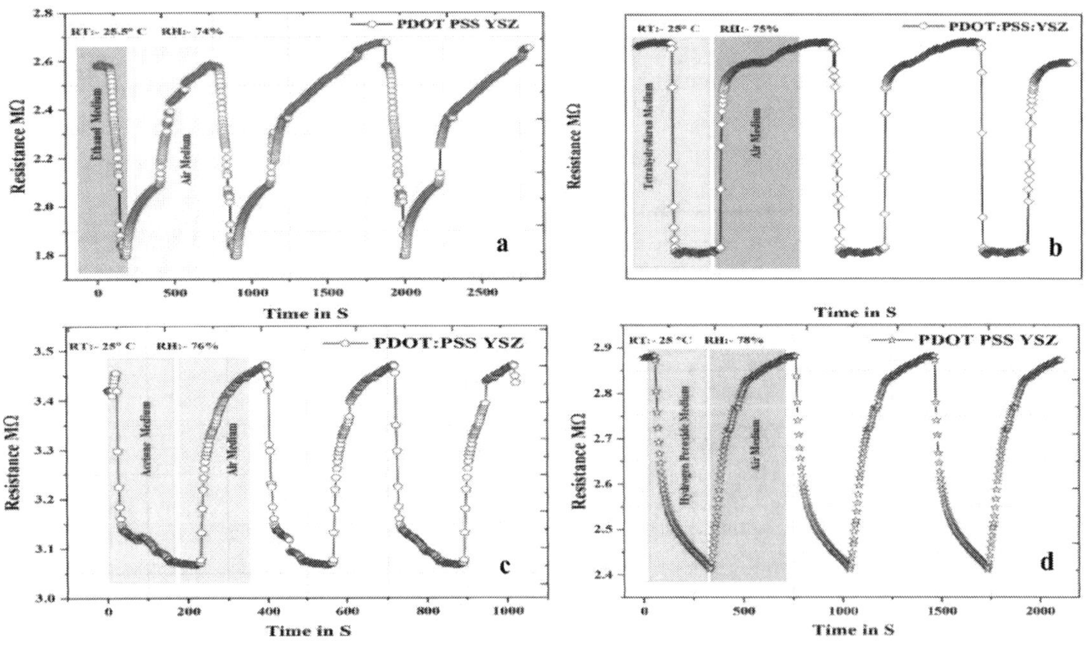

Figure 7: Dynamic response curve of PEDOT PSS YSZ in a) Ethanol b) Tetrahydrofuran c) Acetone d) Hydrogen per Oxide OVs

4.4.3 Reversibility test

Superior sensitivity and reversibility, the two features of a commercial gas sensor that must be present for a practical use. The ability of the sensor to switch back and forth between the analyte and the surroundings without losing any detecting effectiveness is known as reversibility. Under various OVs, the reversibility test was run on synthetic thin film. PEDOT: PSS: YSZ were subjected to acetone, ethanol, hydrogen peroxide, and THF OVs for the first 250 s before being exposed to air for the next

250 s, for an overall of four cycles, to test for reversibility. Figure 7 depicts the behaviour that was noticed.
For PEDOT: PSS: YSZ, the percentage sensitivity for ethanol, THF, acetone and hydrogen per-oxide OVs was found to be 0.44%, 2.81%, 4.32%, 7.11% respectively. Thus, the prepared PEDOT: PSS: YSZ sensor material shows good sensitivity to hydrogen per oxide compared to other OVs.

Table 1: Sensing characteristics of PEDOT: PSS: YSZ thin films for different Organic Solvents

Composite Name	Solvent Used	Response Time(S)	Recovery Time(s)	Resistance In Air(MΩ)	Resistance In Gas (MΩ)	Percentage sensitivity
PEDOT PSS YSZ	Ethanol	100	477	2.291775	2.301865	0.4402700
	Tetra Hydrofuran	210	524	2.781416	2.703222	2.8113018
	Acetone	82	277	3.305175	3.162104	4.3286966
	Hydrogen per oxide	164	548	2.776211	2.578744	7.1128239

5.Conclusion

Pure PEDOT: PSS and PEDOT: PSS: YSZ nanocomposite thin films were prepared using a spin coating technique, and the effect of YSZ nanoparticles on pristine PEDOT: PSS was investigated using FTIR, XRD, and I-V characterization techniques. The analysis of FTIR and XRD confirms the interaction between YSZ nanoparticles and the PEDOT: PSS chain. The expansion of XRD peaks and the noted changes in chemical morphology support this finding. The thin films' amorphous structure is shown by the X-ray diffraction (XRD) analysis. The inclusion of YSZ nanoparticles boosted the electrical conductivity of pristine PEDOT: PSS film from 0.028×10^{-6} S cm^{-1} to 0.0885×10^{-4} S cm^{-1}, according to examinations of I-V characteristics. Several metrics, including sensing response, response, and recovery time, were investigated for the produced thin films of PEDOT: PSS: YSZ for various OVs. Of these, PEDOT: PSS: YSZ shown a good sensitivity for hydrogen peroxide gas vapours.

Referenes:

[1] Ouyang J 2013 “Secondary doping” methods to significantly enhance the conductivity of PEDOT: PSS for its application as transparent electrode of optoelectronic devices *Displays* **34** 423–36
[2] Nardes A M, Kemerink M, De Kok M M, Vinken E, Maturova K and Janssen R A J 2008 Conductivity, work function, and environmental stability of PEDOT: PSS thin films treated with sorbitol *Org. Electron.* **9** 727–34
[3] Groenendaal L, Jonas F, Freitag D, Pielartzik H and Reynolds J R 2000 Poly (3, 4-ethylenedioxythiophene) and its derivatives: past, present, and future *Adv. Mater.* **12** 481–94
[4] Xia Y and Ouyang J 2012 Significant different conductivities of the two grades of poly (3, 4-ethylenedioxythiophene): poly (styrenesulfonate), Clevios P and Clevios PH1000, arising from different molecular weights *ACS Appl. Mater. Interfaces* **4** 4131–40
[5] Chen Y, Xu J, Yang Y, Zhao Y, Yang W, Mao X, He X and Li S 2016 The preparation and electrochemical properties of PEDOT: PSS/MnO2/PEDOT ternary film and its application in flexible micro-supercapacitor *Electrochim. Acta* **193** 199–205
[6] Bhowal A C, Talukdar H and Kundu S 2019 Preparation, characterization and electrical behaviors of PEDOT: PSS-Au/Ag nanocomposite thin films: An ecofriendly approach *Polym. Bull.* **76** 5233–51
[7] Xia Y and Ouyang J 2010 Anion effect on salt-induced conductivity enhancement of poly (3, 4-ethylenedioxythiophene): poly (styrenesulfonate) films *Org. Electron.* **11** 1129–35

[8] Wang Q and Eslamian M 2016 Improving uniformity and nanostructure of solution-processed thin films using ultrasonic substrate vibration post treatment (SVPT) *Ultrasonics* **67** 55–64

[9] Hebbar V, Bhajantri R F, Ravikumar H B and Ningaraju S 2019 Role of free volumes in conducting properties of GO and rGO filled PVA-PEDOT: PSS composite free standing films: A positron annihilation lifetime study *J. Phys. Chem. Solids* **126** 242–56

[10] Zhang Z, Chen G, Wang H and Li X 2015 Template-directed in situ polymerization preparation of nanocomposites of PEDOT: PSS-coated multi-walled carbon nanotubes with enhanced thermoelectric property *Chem. Asian J.* **10** 149–53

[11] Melendez R G, Moreno K J, Moggio I, Arias E, Ponce A, Llanera I and Moya S E 2010 On the influence of silver nanoparticles size in the electrical conductivity of PEDOT: PSS *Materials Science Forum* vol 644 (Trans Tech Publ) pp 85–90

[12] Mebrahitom Asmelash G, Mamat O, Ahmad F and Prasada Rao A K 2015 Thermal shock and fatigue behavior of pressureless sintered Al 2 O 3–SiO 2–ZrO 2 composites *J. Adv. Ceram.* **4** 190–8

[13] Wang L, Shao F, Zhong X H, Ni J X, Yang K, Tao S Y and Wang Y 2018 Tailoring of self-healing thermal barrier coatings via finite element method *Appl. Surf. Sci.* **431** 60–74

[14] Wang J, Sun J, Zhang H, Dong S, Jiang J, Deng L, Zhou X and Cao X 2018 Effect of spraying power on microstructure and property of nanostructured YSZ thermal barrier coatings *J. Alloys Compd.* **730** 471–82

[15] Białas A, Kondratowicz T, Drozdek M and Kuśtrowski P 2015 Catalytic combustion of toluene over copper oxide deposited on two types of yttria-stabilized zirconia *Catal. Today* **257** 144–9

[16] Carneiro J S A, Brocca R A, Lucena M L R S and Nikolla E 2017 Optimizing cathode materials for intermediate-temperature solid oxide fuel cells (SOFCs): Oxygen reduction on nanostructured lanthanum nickelate oxides *Appl. Catal. B Environ.* **200** 106–13

[17] Huang Z, Qi J, Zhou M, Gong Y, Shi Q, Yang M, Han W, Cao Z, Peng S and Lu T 2018 A facile solvothermal method for high-quality Gd2Zr2O7 nanopowder preparation *Ceram. Int.* **44** 1334–42

[18] Huang Z, Han W, Feng Z, Qi J, Wu D, Wei N, Tang Z, Zhang Y, Duan J and Lu T 2019 The effects of precipitants on co-precipitation synthesis of yttria-stabilized zirconia nanocrystalline powders *J. Sol-Gel Sci. Technol.* **90** 359–68

[19] Bulla S S, Bhajantri R F, Chavan C and Sakthipandi K 2021 Synthesis and characterization of polythiophene/zinc oxide nanocomposites for chemiresistor organic vapor-sensing application *J. Polym. Res.* **28** 251

IOP Conf. Series: Materials Science and Engineering 1300 (2024) 012024 doi:10.1088/1757-899X/1300/1/012024

Optical and Structural Properties of TiO2 doped Calcium Borate Glasses

Pruthviraja B and Eraiah B*

Department of P.G. Studies and Research in Physics, Bangalore University, Jnana Bharathi campus, Bangalore 560056, Karnataka, India

*Corresponding author email: eraiah@rediffmail.com

Abstract: The novel glasses were created using the melt quenching technique and the glass composition xTiO2-(50-x) CaO-50B2O3, where x= 0, 1, 3, 5, 7, 9 in mol%, which was characterized by XRD spectra and confirmed the glass samples' amorphous nature. Using the Archimedes principle, the density of the glasses was determined, and the associated molar volume was estimated. Density increases with increasing mol% TiO2, whereas molar volume decreases, as anticipated. The absorption spectra from the wavelength region of 200nm–1100nm were obtained using a UV-visible spectrometer. Tauc's plot was used to evaluate the direct and indirect energy band gap, and the fluctuation of the energy band gap with mol% TiO2 has been discussed. The refractive index has also been estimated using energy band gap data. In the 400-4000 cm-1 range, the FT-IR functional groups were confirmed.

Keywords: Borate glass; TiO2; Amorphous materials; Density; Energy band gap.

1. Introduction

Glasses are particular resources due to their exceptional characteristics. These possessions are used in virtually every field of engineering and technology. They exhibit a wide range of features, including as insulators and conductors, magnetic, linear, and NLO (nonlinear optical) properties, piezoelectricity, zero thermal expansion, and excellent mechanical and chemical properties [1, 2]. Because of the spacious range of potential production methods, researchers are still investigating unsullied aspects of glasses. Glass formers such as Borate, Silicate, and Phosphate are habitually working in oxide glasses. These are also used in bio applications and a diversity of optical applications [3, 4]. Borate-based glass have received a lot of attention recently for NLO device applications due to their extensive broadcast window, strong laser smash up threshold, and NLO coefficients [5, 6]. The material's physical properties [7] are prejudiced by the module oxides present in the borate network. Because of their high refractive indices and capacity to crystallize NLO vigorous phases such as $BaTi(BO_3)_2$, and $Ba_3Ti_3O_6(BO_3)_2$, glasses in the BaO-TiO_2-B_2O_3 system have piqued the interest of a number of researchers [8]. BTBO glasses were previously investigated in order to produce $BaTiO_3$-based glass for capacitor and ferroelectric applications [9, 10]. In the BTBO system, heterogeneous hugeness nanocrystallization of TiO_2 (anatase phase) was discovered recently [11]. Transparent TiO_2 nanocrystal glasses have high refractive indices. Hydrophobic conditions (contact angle 90°) were likewise applied to the BTBO glasses [12]. Even though B_2O_3 oxide is an immaculate glass, the manufacture of such glasses requires an assortment of additives to widen their physicochemical properties, such as Na_2O, which helps squash the melting point of the glass; CaO, which promotes the homogeneity of the glass; and other additives based on the attributes desired. Because of their ability to alter the magnetic, electrical, attenuation, and optical properties of such glasses, multi-valence cations, such as transition metals and rare earth cations, are frequently utilized as additions in borate glasses **[13–16].** TiO_2 is considered an advantageous additive since it improves and alters the properties of glass, predominantly its electrical and optical properties. Trivalent and tetravalent titanium cations, due to their low coordination statistics, can participate in the glass matrix not only as a glass modulator, TiO_6, but also as a glass architect, TiO_4 **[17, 18].** According to certain experts, adding TiO_2 to the Na_2O-Bi_2O_3-B_2O_3 glass permutation boosts its optical absorption **[19, 20].** TiO_2 is also known as a potentially affordable **[21-23]** and non-toxic n-type semiconductor. The objective of this study is to inspect the optical characteristics of TiO2-doped CaO-borate glasses comprehensively. The optical band gap, Urbach energy, and refractive index have all been revealed.

2. Materials and Methods:

The melt quenching procedure was used to successfully make borate-based glasses with the following composition: 45 xTiO2-(50-x) CaO-50B2O3, where x=0, 1, 3, 5, 7, and 9 in mol%. In a porcelain mortar, the correct amounts (in mol%) of high purity H3BO3, crude CaCO3, and TiO2 were properly mixed. An electrical furnace was used to heat the mixture to 1000°C in porcelain over the period of 40 minutes. The molten specimens were spun to guarantee maximum homogeneity. They were then heated to room temperature and poured upon a brass plate.

Table-1 The sample were prepared by varying concentrations of TiO_2 and CaO as mentioned in the below table.

Sample code	TiO_2mol%	CaOmol%	B_2O_3mol%
S1	0	50	50
S2	1	49	50
S3	3	47	50
S4	5	45	50
S5	7	43	50
S6	9	41	50

3. Results Analysis

3.1. XRD Studies.

At room temperature, X-ray diffraction (XRD) examinations were performed on the resources using a Bruker AXS D8 Advance diffractometer (Germany) (CuK radiation). The XRD patterns of calcium borate glass as prepared samples are shown in **Figure 1** as a function of TiO_2 concentration. The lack of identifiable peaks in the XRD pattern, along with the large XRD hump, suggests that the glass samples produced are amorphous. Because of oscillations in interatomic distances, glass atoms are neither evenly spaced nor homogeneous, as in crystals **[24].** The generated glass samples all have a glassy appearance.

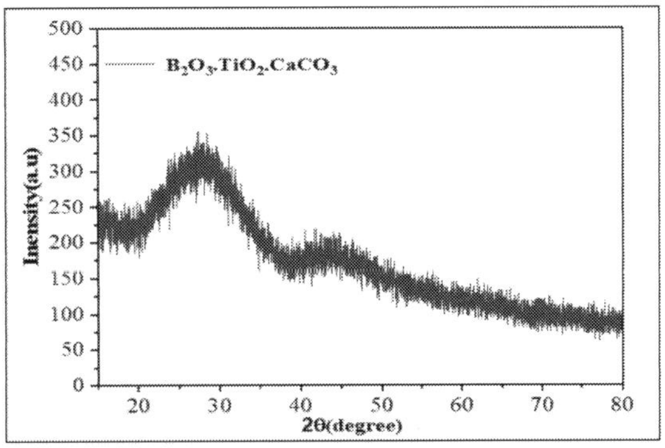

Figure 1: XRD of TiO_2-CaO-B_2O_3 Glasses

3.2. Physical Parameters.

The well-known Archimedes rule is used to estimate material densities and molar volumes in this research. The weights of the shaped glass specimens were subsequently estimated using a 4-digit vulnerable microbalance in air and toluene (At normal temperature, liquid density is 0.866 g/cm^3). **Figure 2** depicts the density and molar volume of TiO_2-doped calcium borate contents. **Table 2** illustrates the practical values of density and molar volume for various TiO_2 concentrations. The thickness increased with increasing the amount of TiO_2 in $CaCO_3$ from 2.5710 to 3.4620 g/cm^{-3}, up to 9 mol% TiO_2.

Table 2. The physical properties of the glass samples tested.

Glass Series	TiO_2 (mol%)	Density ρ (g/Cm^3)	Molar Volume V_m (cm^3/mol)	Concentration of TiO_2 ion N_{Tb}x 10^{23} (ions/Cm^3)	Polaron radius r_p x 10^{10} (Cm)	TiO_2–TiO_2 distance r_i x 10^{10} (Cm)
S1	0	3.0833	36.2841	0	0	0
S 2	1	3.71	36.51	0.0273	2	6
S 3	3	3.4620	34.45	0.1110	1	3
S 4	5	2.7331	20.4549	0.1493	1	2
S 5	7	2.8255	38.3940	0.1097	1	3
S6	9	2.8870	38.2540	0.1416	1	3

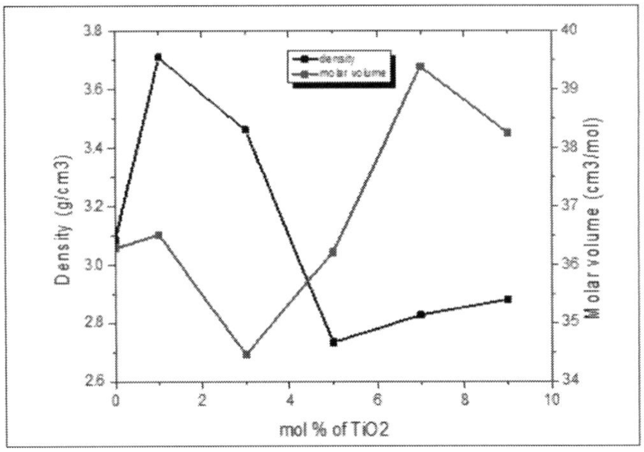

Figure 2: TiO_2-CaO-B_2O_3 Glasses Molar Volume and Density

3.3. Raman Spectral Studies.

Raman spectroscopy was working in this study to inspect the CaO-B_2O_3-TiO_2 glass mixture. The impacts of TiO_2 concentration and CaO-B_2O_3 molar ratio on CaO-B_2O_3-TiO_2 glass arrangement were explored. The preliminary results show that TiO_2 exists in the form of [TiO_2] or B-O-Ti linkage, [BTi_2O_{10}] structure units, and causes some caustic effects on big borate groups while inducing the development of a large number of relatively small structure units. The enhancement in CaO indemnity is the B-O-Ti connection and big borate groups, which are critical in forming relatively tiny units. As a result, TiO_2 and CaO both decrease the borate network structure and the capacity of the CaO-B_2O_3-TiO_2 glass system to generate borate glasses **[25]**.

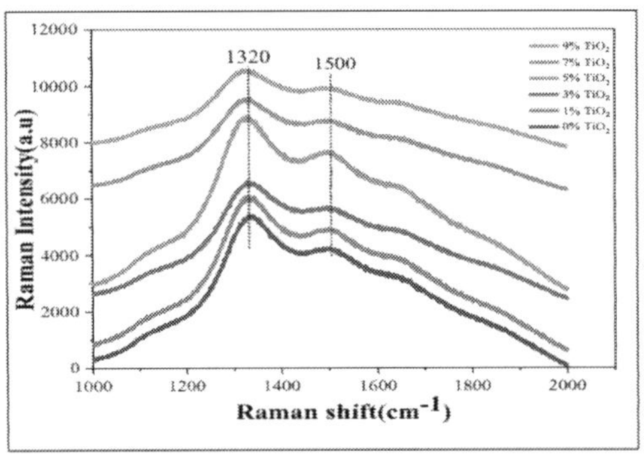

Figure 3: Raman spectra of TiO_2-CaO-B_2O_3 Glass

3.4 FT-IR Spectral Studies.

Figure 4 depicts the Fourier infrared spectra of TiO_2-doped calcium borate glasses. The FT-IR spectra, which exhibit a lot of bending or stretched vibrational links, can be used to learn about the local structure of a manufactured glass matrix. **Table 3** offers information regarding bands and their roles.

Table 3. TiO_2 doped glasses with FT-IR band allocations.

Band width (cm^{-1})	Assignments
650	Ring-type metaborate groups. **[26]**
900	B-O-B bonds, stretching vibration.
1076	BO_4 symmetric stretching vibration **[27]**.
2980	B-O-H functionality.
3045	O-H bending vibration has a weak band. **[28]**

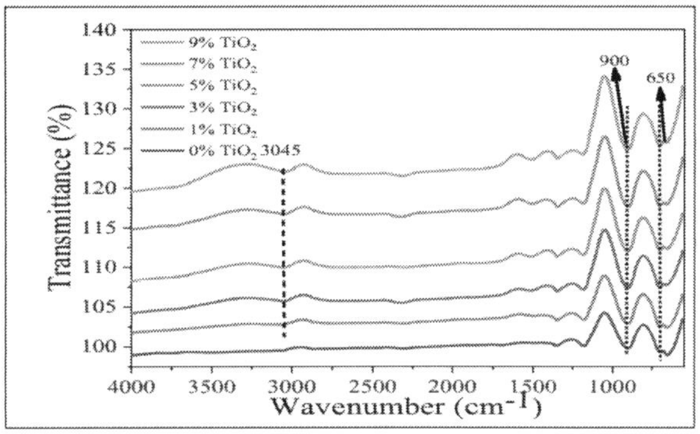

Figure 4. FT-IR spectra of TiO_2-CaO-B_2O_3 Glasses

3.5. Optical absorption spectra.

Figure 5 depicts the absorbance and wavelength correlation. The absorbance of the sample is chosen to be between 200 and 1200 nm. The absorbance sample falls from 200-400 nm and the leftovers remain constant at 300-400 nm, however the sample width reduces as sample meditation increases. The absorbance in the visible region remains constant, and the width increases as the molar concentration of titanium ions increases. **Figure 5** clearly demonstrates that the glass samples are in charge of both the UV and visible ranges. Using the glass samples' preliminary data (A) and thickness (t), the Beer Lambert's law was used to compute the absorbance coefficient (λ) **[29, 30]**.

Figures **6a** and **6b** illustrate Tauc's plots, which are used to determine direct and indirect band gap energies.

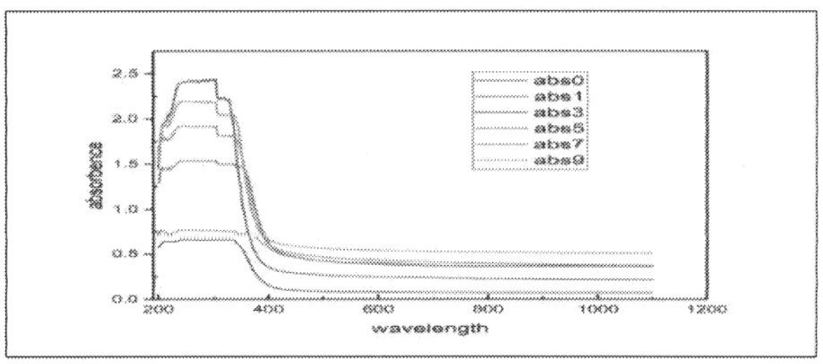

Figure 5: UV-Visible Spectra of TiO_2-CaO-B_2O_3 Glasses

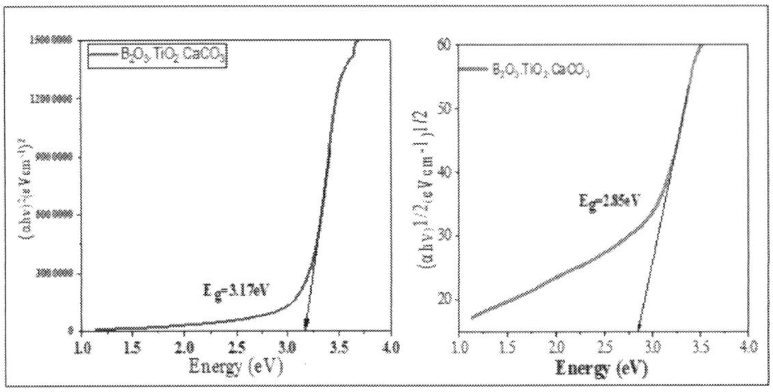

Figure 6: (a) & (b) Direct and Indirect band gap

4. Conclusion:

Titanium-doped borate glass samples were made by melt quenching according to their chemical composition. X-ray diffraction corroborated this. A number of physical tests, such as density, molar volume, borate-borate separation, ion concentration, and inter-ionic distance, show that when titanium is introduced, the glass sample becomes less compact. Raman and FT-IR investigations reveal the vital groups present in the glass as structural alterations in the glass matrix. According to UV-visible spectroscopy, this loss in structural stability improves optical individuality, specifically a significant reduction in the optical band gap.

References:

1. Manzani, D., Junior, J. B. S., Reyna, A. S., Neto, M. L. S., Bautista, J. E., Ribeiro, S. J., & De Araújo, C. B. (2019). *Dalton trans,* 48(18), 6261-6272.
2. Pavić, L., Graca, M. P., Skoko, Ž., Moguš-Milanković, A., & Valente, M. A. (2014). *J. Am. Ceram,* 97(8), 2517-2524.
3. Abdelghany, A. M., ElBatal, F. H., & ElBatal, H. A. (2014). *Process. Appl. Ceram,* 8(4), 185-193.
4. Baino, F., Tomalino, M., & Tulyaganov, D. (2021). Ceramics, glass and glass-ceramics. Springer International Publishing.
5. Feitosa, C. A. C., Mastelaro, V. R., Zanatta, A. R., Hernandes, A. C., & Zanotto, E. D. (2006). *Opt. Mater,* 28(8-9), 935-943.
6. Maia, L. J. Q., Mastelaro, V. R., Schneider, J. F., Parent, P., & Laffon, C. (2005). *J. Solid State Chem,* 178(5), 1452-1463.
7. Abdelghany, A. M., & ElBatal, H. A. (2013). *J Non Cryst Solids,* 379, 214-219.
8. Bamford, C. (1977). Colour generation and control in glass. https://lccn.loc.gov/77024505.
9. Naresh, P., Raju, G. N., Kumar, V. R., Piasecki, M., Kiytyk, I. V., & Veeraiah, N. (2014).*Ceram. Int,* 40(1), 2249-2260.
10. Biswas, K., Sontakke, A. D., & Annapurna, K. (2010). *Alloys Compd,* 489(2), 493-498.
11. Agarwal, A., Seth, V. P., Gahlot, P., Goyal, D. R., Arora, M., & Gupta, S. K. (2004). *Spectrochim. Acta B: At. Spectrosc,* 60(13), 3161-3167.
12. Komatsu, R., Sugawara, T., Sassa, K., Sarukura, N., Liu, Z., Izumida, S., & Yamanouchi, K. (1997). *Appl. Phys,* 70(26), 3492-3494.
13. Farouk, M., Samir, A., Metawe, F., & Elokr, M. (2013). *J Non Cryst Solids,* 371, 14-21.
14. Sadeq, M. S., & Morshidy, H. Y. (2019). *Ceram. Int,* 45(15), 18327-18332..
15. Murawski, L., Chung, C. H., & Mackenzie, J. D. (1979). *J Non Cryst Solids,* 32(1-3), 91-104.
16. Naresh, V., & Buddhudu, S. (2012). *Ferroelectr. Lett. Sect,* 437(1), 110-125.

17. Heidari, S., Hooshmand, T., Yekta, B. E., Tarlani, A., Noshiri, N., & Tahriri, M. (2018). *Ceram. Int,* 44(10), 11682-11692.
18. Abdelghany, A. M., & ElBatal, H. A. (2013). *J Non Cryst Solids,* 379, 214-219.
19. Abouhaswa, A. S., Zakaly, H. M., Issa, S. A., Rashad, M., Pyshkina, M., Tekin, H. O., & Mostafa, M. Y. (2021). *Ceram. Int,* 47(1), 185-204.
20. Sadeq, M. S., & Morshidy, H. Y. (2019). Ceramics International, 45(15), 18327-18332.
21. Thakur, S., Thakur, V., Kaur, A., & Singh, L. (2019). *J Non Cryst Solids,* 512, 60-71.
22. Wright, A. C., & Vedishcheva, N. M. (2013). *Phys. Chem. Glas.: Eur*, 54(4), 147-156.
23. Elkelany, E. A., Hassan, M. A., Samir, A., Abdel-Ghany, A. M., El-Bahnasawy, H. H., & Farouk, M. (2021). *Opt. Mater,* 112, 110744.
24. Al-Hazmi, F., Mansour, S. F., AlHammad, M. S., Abdo, M. A., & Sadeq, M. S. (2021). *Ceram. Int,* 47(6), 8566-8572..
25. dos Santos, V. R., Campos, T. M. B., Thim, G. P., Borges, A. L. S., & de Sousa Trichês, E. (2023). *Ceram. Int*, 49(7), 11236-11248.
26. Rajyasree, C., Krishna, S. B. M., Babu, A. R., & Rao, D. K. (2013). *J. Mol. Struct,* 1033, 200-207.
27. Moustafa, Y. M., & El-Egili, K. (1998). *J Non Cryst Solids,* 240(1-3), 144-153.
28. Boudlich, D., Bih, L., Archidi, M. E. H., Haddad, M., Yacoubi, A., Nadiri, A., & Elouadi, B. (2002). *J. Am. Ceram,* 85(3), 623-630.
29. Eraiah, B. (2006). *Bull. Mater. Sci,* 29, 375-378.
30. Kumar, M. S., & Eraiah, B. (2023). *Mater. Today.*

IOP Conf. Series: Materials Science and Engineering 1300 (2024) 012025 doi:10.1088/1757-899X/1300/1/012025

Synthesis, Characterization, Dielectric, and Electrical Conductivity Studies of Zinc Oxide Nanoparticles

A Sharma, P kumar and S P Mahapatra
Department of Chemistry, National Institute of Technology Raipur 492010 India
*Corresponding Author: spmahapatra.chy@nitrr.ac.in

Abstract A co-precipitation technique is used to synthesize of ZnO nanoparticles. The synthesized ZnO nanoparticles were characterized by UV-visible spectroscopy, scanning electron microscopy (SEM), high-resolution transmission electron microscopy (HR-TEM), photoluminescence, Fourier-Transform Infrared Spectroscopy (FTIR), and X-ray diffraction (XRD). The size of the nanoparticles was wide-ranging from 1 nm to 100 nm, corresponding to the HR-TEM analysis. The photoluminescence study of ZnO nanoparticle shows emission in the UV region. The particle dimension of ZnO nanoparticle has also been studied through XRD. Dielectric spectroscopy of synthesized ZnO NPs pellet has been studied at a wide frequency range 10^{-1} to 10^{5} Hz. The capacitance and dielectric permittivity of ZnO nanoparticles drop continuously with frequency as dipoles have less time to align in the field. Dielectric permittivity of ZnO pellets increase up to 5 mm thickness and subsequently drop, perhaps due to raise in resistivity. The dielectric loss of ZnO pellet has been examined as a function of frequency. The electrical conductivity of ZnO nanoparticles rise exponentially with frequency. Based on the dielectric studies, the dielectric permittivity and electrical conductivity of ZnO are highly depending on thickness and frequency range. The percolation threshold of ZnO pellets has been found between 4-5 mm thicknesses.
Keywords: Zinc oxide, capacitance, dielectric, permittivity, electrical conductivity, percolation.

1. Introduction

The bulk of studies in this area of nanomaterial synthesis methodology are focused on dealing with their shape, size, and content. Every one of such elements plays crucial role in shaping the characteristics of materials that give rise to various technological applications [1]. ZnO nanoparticles (ZnO NPs) are a versatile substance with exceptional chemical along with physical properties such as excellent chemical stability, higher electrical coupling coefficient, in addition to broad radiation absorption spectrum and excellent photo stability [2-3]. We must alter the nanometer-scale surfaces of these materials to create goods with novel features [4]. Based on its relatively small dimensions as well as highly specialized surface shape, synthesized metal/metal oxide nanostructures exhibit significantly better activity of the catalyst compared to the bulk state. Centrifugation can successfully separate them as a result of reactions mixture and they can be recycled with no effect any of their effectiveness. Several noble metal also metal oxide nanostructures work catalytically well in the breakdown of dyes [5]. ZnO is a semiconductor of the n-type with a band gap of 3.27 eV at room temperature along with activation binding energies of 60 meV. For the reason of their bulky surface-to-volume percentage, ZnO nanostructures offer appealing chemical and physical aspects. Zinc oxide nanoparticles surface needs to be changed in order to increase dispersion. Vapor-liquid-solid (VLS) synthesis is one method for producing ZnO nanostructures [6]. The composition and structure of ZnO have a significant impact on its electrical characteristics [7]. Furthermore, there has lately been extensive research into metal

oxide nanoparticles such as ZnO, In_2O_3, SnO_2, as well as dope oxides. ZnO is one of these metal oxide nanomaterials that is attracting greater interest since, compared to other materials, it is plentiful, inexpensive, as well as environmentally friendly [8]. Among NPs fillers, metal oxides NPs are essential for improving the dependability of electrical insulation and achieving a compact design in electric power apparatuses through polymer nano dielectrics [9]. Numerous businesses, including those in research, engineering, medicine, and other fields are using nanotechnology today. Nanoparticles are subatomic particles that have properties dependent on their size, quantum confinement, and surface plasma resonance. Researcher's focus during the last 20 years has been on metal oxide nanoparticles because of their distinct electrical, optical, mechanical, magnetic, also chemical capabilities [10]. Due to their innovative optical, transport characteristics, semiconductor compounds have grabbed significant a lot of interest since over the last few decades and hold tremendous promise for a many applications in optics. In the near ultra violet and visible ranges, the wide band gap semiconductor ZnO exhibits great optical transparency and luminous characteristics [11]. Although the nanoparticles were created using a variety of physicochemical techniques, the environment is however negatively impacted by them. Now a day, most academics focus their attention on eco-friendly methods to sustain the planet as a result of environmental worries [12]. ZnO NPs were used to decorate the surface of bacterial nano wires in an effort to increase their electrical conductivity due to their distinct physical, electrical, and structural features [13]. The exceptional ability of these methods to produce metal and metal oxide based NPs to serve as catalysts and aid in several industrial, electrical and physical processes is the driving force behind the growing interest in doing so. These methods' oxidation-reduction processes which are extremely sensitive but also very particular, describe their formational mechanisms [14]. Han et al. have investigated the ZnO NPs single crystal's transverse optical phonon mode and low frequency dielectric characteristics [15].
Zinc oxide nanoparticle was synthesized via the co-precipitation technique in present investigation. The morphological properties of ZnO nanoparticle were studied using high-Resolution transmission electron microscopy (HR-TEM) scanning electron microscopy (SEM). Its optical characteristics were investigated with UV-Vis spectroscopy. FTIR were used to explore the functional groups attached to the surface of ZnO nanoparticles. Dielectric characteristics of ZnO pellets have been measured in form of capacitance, dielectric permittivity, electrical conductivity, percolation behaviour for use in various fields.

2. Experimental

2.1. Chemicals

$Zn(NO_3)_2$, distilled H_2O, sodium hydroxide (NaOH), and ethanol (C_2H_5OH) were purchase from standard supplier (Thermo-Fisher Scientific) and use for synthesis of ZnO nanoparticles.

2.2. Synthesis of Zinc Oxide (ZnO) nanoparticles

ZnO nanoparticles were synthesized by a co-precipitation procedure using $Zn(NO_3)_2$ and NaOH as precursors. Throughout the experiment, a 0.1M aqueous solution of zinc nitrate ($Zn(NO_3)_2.6H_2O$) was maintained during constantly stirring for one hour using a magnetic stirrer, and a 0.8M aqueous solution of sodium hydroxide was also prepared using the same method with one hour of stirring. After the zinc nitrate was completely dissolved, 0.8M NaOH aqueous solution was added drop by drop (slowly for 45 minutes) while constantly shaking at high speed. After adding all of the sodium hydroxide, the reaction was left to run for 2 hours. In this state, the beaker was sealed for 4 hours. The solution was permitted to rest overnight until the supernatant solution had been removed appropriately. Residual solution was centrifuged for 10 minutes for removing the precipitation. Precipitated ZnO NPs were rinsed three times with distilled water along with ethanol to remove impurities prior drying in an air environment at 60°C. $Zn(OH)_2$ is completely converted into ZnO NPs after dry. The optical and nano structured substances characteristics of the produced ZnO nanoparticles were investigated [16].

3. Measurements

3.1 Characterization Techniques

XRD spectrum shows the crystalline structure of the generated substance ZnO nanoparticle, whereas UV-Visible spectroscopy reveals optical properties of ZnO nanoparticles. A UV-Visible spectrophotometer was used to get the UV-Visible spectrum between 250 to 550 nm (Perkin Elmer, USA). At the surrounding temperature, the photoluminescence spectrum of zinc oxide nanoparticles has been studied with wavelengths of excitation ranging from 440 to 600 nm. FTIR spectroscopy was used to study the functional categories of zinc oxide nanoparticle pellets prepare with KBr. Zinc oxide spectrum were investigate in the 400–4000 cm^{-1} region. The morphology of ZnO was investigated using SEM and HR-TEM.

3.2. Dielectric Relaxation Spectra

The dielectric relaxation spectra (DRS) of ZnO NP pellets were measured using a Hi Tester LCR meter and a Hioki 3533, with silver foil acting as a blocking electrode. The item under investigation was round, with a diameter of 1 cm and a thickness of 2.0 mm. Using the connection, electrical conductivity (σ) was determine from dielectric data by following equation.

$$\sigma\ AC = \omega \varepsilon 0 \acute{\varepsilon} \tan\delta \quad (2)$$

Where ω is 2πf (f is frequency), ε_0 is vacuum permittivity and $\acute{\varepsilon}$ is dielectric permittivity or dielectric constant

$$\acute{\varepsilon} = C_p/C_0 \quad (3)$$

The electrical conductivity (σ) was calculated by dielectric data [17].

4. Results and Discussion

4.1. Optical Properties

The optical absorption spectra of zinc oxide nanoparticles were recorded using UV-Visible spectrophotometer Model LT-2900 (Labtronics). The spectrophotometer's spectral bandwidth is 0.0, 0.1, 0.2, 0.3, 0.4, 0.5, 0.6, and 0.7 absorbances, and in the wavelength ranges from 200 to 600 nm. The UV-Vis absorption spectra of zinc oxide nanoparticles can be seen in Figure 1. Its absorption spectra have been measured between 250 and 550 nm. The absorbance peak at 400 nm corresponded to the typical band of zinc oxide nanoparticles in the spectra [18].

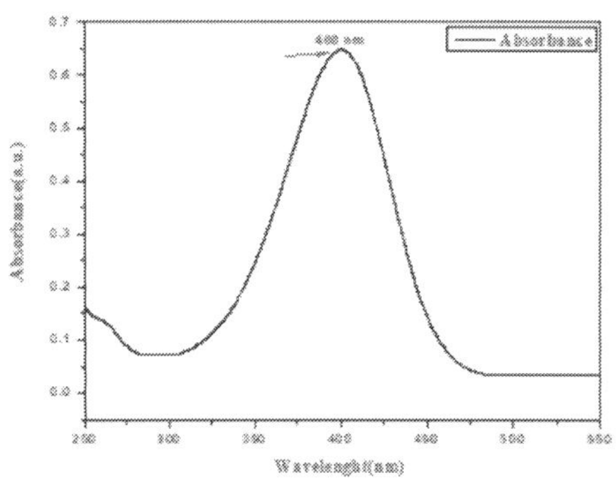

Figure 1. UV–Visible spectrum of ZnO nanoparticles.

4.2 Photoluminescence Spectra

It is essential to note that the physical properties of semiconducting materials alter when their size reduces to the nanoscale, a phenomenon known as "quantum size effects." Photoluminescence shows that quantum confinement raises the band gap energy of ZnO. [19]. Figure 2 represents the photoluminescence spectrum of ZnO nano powder at room temperature. The wavelength range displays two emission peaks: one at 480 nm (UV region) due to close band gap excitonic emissions, and another at 550 nm due to the presence of single ionized oxygen vacancies. The atomic interactions of a photogenerated hole that an electron filling the oxygen vacancy causes the emission [20-21]. Additionally, the scale indicates the powder's narrow size allocation of nanoparticles with the luminescence peak full width half-maximum (FWHM) just a small number of nanometers wide [22].

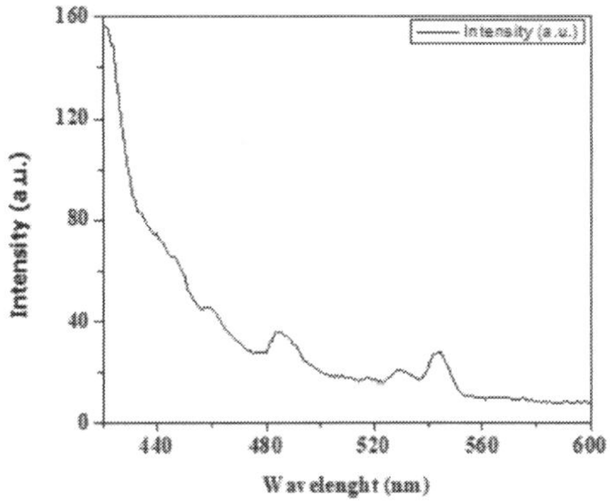

Figure 2. Photoluminescence spectra of ZnO nanoparticles.

4.3 Fourier-Transform Infrared Spectroscopy (FTIR)

FT-IR experiments were carried out to validate the bond structure and identify related functional groups of synthesized ZnO nanoparticles utilizing optimal conditions. Figure 3 shows the infrared absorption spectra of ZnO nanoparticles in the 3500-500 cm^{-1} wave number region. The band at 437.71 cm^{-1} corresponds to the Zn-O stretching mode of the ZnO lattice. The bands at 3434.61 cm^{-1} correspond to the O-H mode of vibration. C=O exhibits a significant asymmetric mode of vibration between 1635.48 and 1635.48 cm^{-1}. Symmetric stretching arises between 1510.34 and 1337.66 cm^{-1} due to the existence of C-O. There is also a C-O-C peak there [23]. The presence of multiple hydroxyl group-related vibration modes in this example is most likely due to ambient moisture absorption by the pellets sample during FTIR analysis.

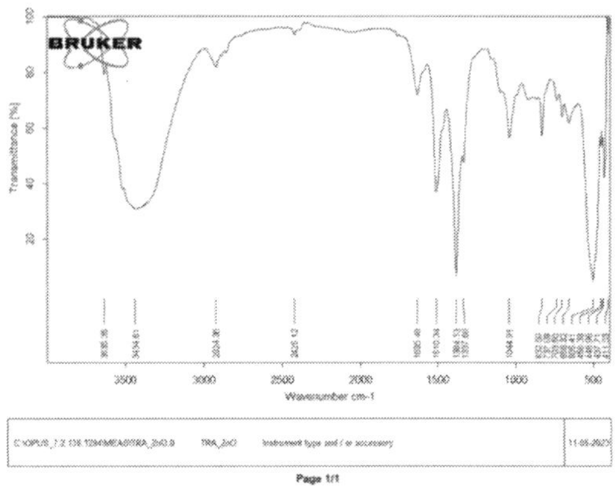

Figure 3. FTIR spectra of ZnO nanoparticles.

4.4. Scanning Electron Microscopy (SEM)

A scanning electron microscope is a category of electron microscope that uses a high-intensity electron beam to generate features on the surface. Scanning is also used to agree on the external morphology (structure) along with size of crystalline particles. SEM was used to study the variance in particle size within the samples. The sample is composed up of electrons that combine through atoms to generate signals that give data regarding the material's constituent as well as surface shape. As shown in Figure 4, a considerable number of ZnO nanoparticles have a spherical form, and agglomerates are also changed into bigger particles. The SEM photomicrograph revealed individual ZnO nanoparticles as well as various aggregates. At a magnification of 25.00 KX, observable morphologies ZnO NPs are expected to have spherical forms and aggregated nanoparticle states. In SEM , a focused beam is swept over a large sample, and pixel by pixel, backscattered or secondary electrons are counted to generate an image of the material's surface. Variances in the process of synthesis of ZnO nanoparticles could lead to variations in particle shape, size, and dispersion [24].

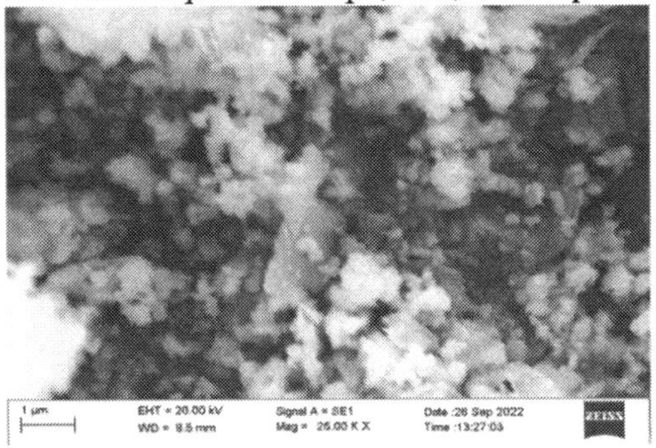
Figure 4. SEM image of ZnO nanoparticles

Figure 5. TEM image of ZnO nanoparticles

4.5. High Resolution Transmission Electron Microscopy (HR-TEM)

HR-TEM has been utilized to evaluate the shape and size of the produced NPs. The minuscule powder of ZnO NPs was disseminated in methanol over a carbon-coated copper grid, as well as ultra high resolution HR-TEM pictures were obtain at an increasing current of 200 kV. Figure 5 represents the HR-TEM study of ZnO NPs, which are spherical in shape as well as evenly scattered with no agglomeration. The particle size distributions in the figure found in the range of 20-100 nm. The physical parameters (diameter, dispersion and morphology) of the nanostructures and crystallinity have been seen in Figure 5 [25]. ZnO NPs possessed a spherical morphology as desired by the relative quantity fractions used in the synthesis as well as an exceptionally size distribution has limits with diameters ranging 20-50 nm [26].

4.6. X-Ray Diffraction (XRD)

X-ray diffraction was utilized to determine the segment purity and crystallinity of the produced zinc oxide nanoparticles. Figure 6 shows the XRD patterns of the zinc oxide nanoparticles. The XRD spectra of zinc oxide nanoparticles exhibit 2θ values at 31.84, 34.52, 36.33, 47.63, 56.64, 62.91, 66.54, 68.05, and 69.12 degrees, which correspond to (100), (002), (101), (102), (110), (103), (200), (112), and (201) the hexagonal wurtzite planes [27]. The presence of nanoparticles in the product has been clearly demonstrated by the widening of the diffraction peaks. In this analysis, an unusually large intensity peak is detected at $2\theta = 34.52^0$ (101), which is deviant when compare to the standard and previous experiments. The average crystallite size (D) of the synthesized ZnO nanoparticles was determined to be 25.97 nm using the Debye-Scherer formula (eqn. (1)) [28].

$$D = [0.9\,\lambda / \beta cos\theta] \quad (1)$$

D is the crystallite size (nm), λ is the wavelength of Cu-Kα radiation 0.15406 Å, and β is the full width at half maximum of the diffraction peak (radian).

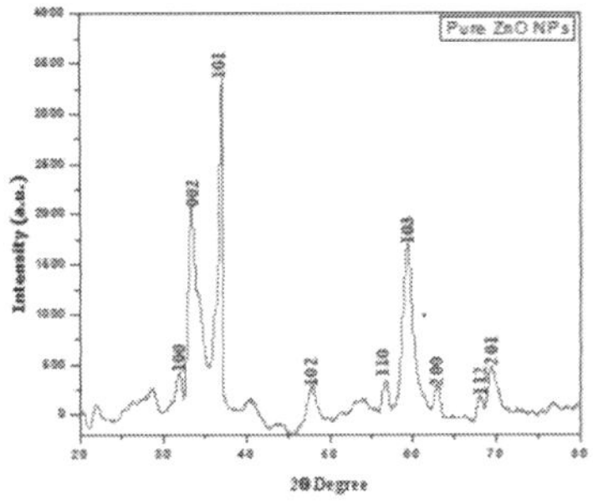

Figure 6. XRD spectra of ZnO nanoparticles.

4.7. Capacitance (Cp)

Capacitance (Cp) is an important property for a new class of single-electron devices developed by utilizing nanoparticle quantum dot arrays. Based on the concept of few bodies, nanoparticle quantum dot capacitance theoretical model is proposed. Figure 7 demonstrates the change in Cp as a function of frequency such as different thicknesses (1 to 5 mm) of produced ZnO nanoparticles pellets. Capacitance increases with frequency as a result of charge orientation along with changes in the function group dipole moment of ZnO NPs. Every type of polarizations (dipole, interface, electric, and ionic, for example) are present in the compounds results in a greater value of capacitance at low frequency. The figure also indicates which values rises as the thickness of the ZnO pellets that are used grows up to 4 mm, after then, it decreases, which might be due to the increased resistance of the ZnO pellets [29-30].

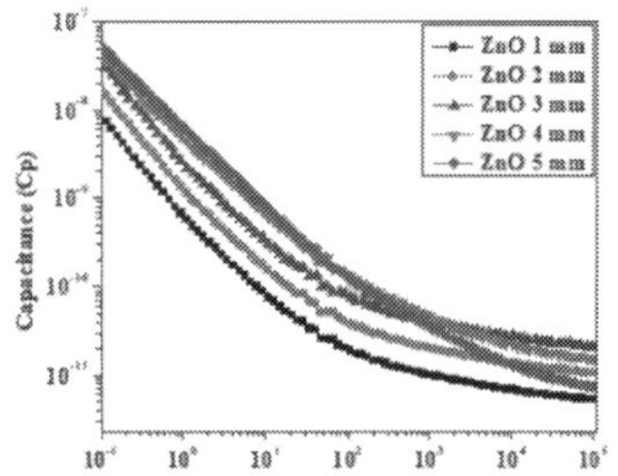

Figure 7. Capacitance plots of ZnO nanoparticles pellets measured as a function of frequency.

4.8. Dielectric Permittivity

The Dielectric permittivity (ϵ') of a material reveals its electrical polarization. Figure 8 shows the variations in dielectric permittivity as a function of frequency for various thicknesses (1 mm to 5 mm) of synthesized ZnO nanoparticles. Dielectric permittivity decreases with frequency due to the charge orientation and rotation of dipole moments. The presence of dipoles, interfaces, electric, and ions in the compounds allows in an increased dielectric permittivity value at low frequencies. Figure also shows that dielectric permittivity increases with increasing thickness of the ZnO pellets up to 4 mm beyond which it might be related to the complex confrontation due to decrease in capacitance [31].

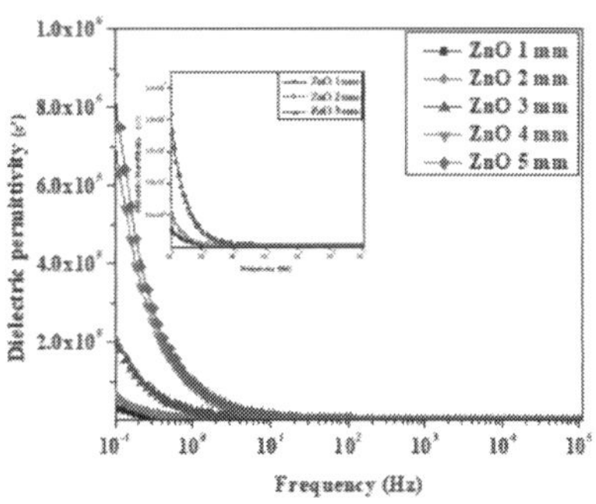

Figure 8. Dielectric permittivity plots of ZnO measured as a function of frequency.

4.9. Electrical Conductivity

Electrical conductivity (σ) is a measurement of a material's electrical current carrying capacity, as seen in Figure 9. The change in electrical conductivity with frequency for changing thicknesses and rising frequency of ZnO NPs pellets can be observed from the figure. The electrical conductivity of ZnO nanoparticles rises constantly with rising thickness of ZnO pellets up to 4 mm, after which the improvement or increase is very negligible due to increase in resistance. The increase in electrical conductivity confirms the conductive nature of zinc oxide nanoparticles. This is most likely owing to the information that zinc oxide nanoparticles have a higher stability as a function of frequency [32-34].

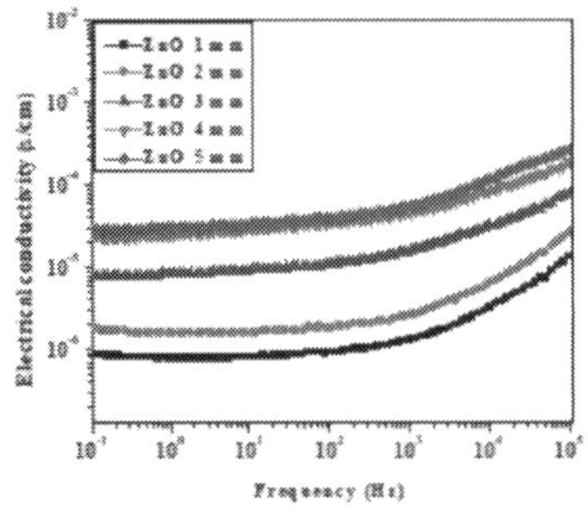

Figure 9. Electrical conductivity plots of ZnO measured as a function of frequency.

4.10. Percolation

Figure 10 represents the percolation behaviour of the electrical conductivity of zinc oxide nanoparticles at various thicknesses. At all frequencies (1, 10, 100 Hz and 1, 10, 100 kHz), electrical conductivity grows exponentially until 4 mm thickness, indicating the presence of a percolation limit, before decreasing at 5 mm thickness of ZnO pellet. It could be because the nanoparticles contact each other and an ongoing pathway is formed throughout the capacity of sample for electrons to pass through up to 4 mm thickness, but at 5 mm thickness, the nanoparticles contact every single different as well as agglomerates form, and persistent path is reduced by the volume of sample for the movement of electron [35].

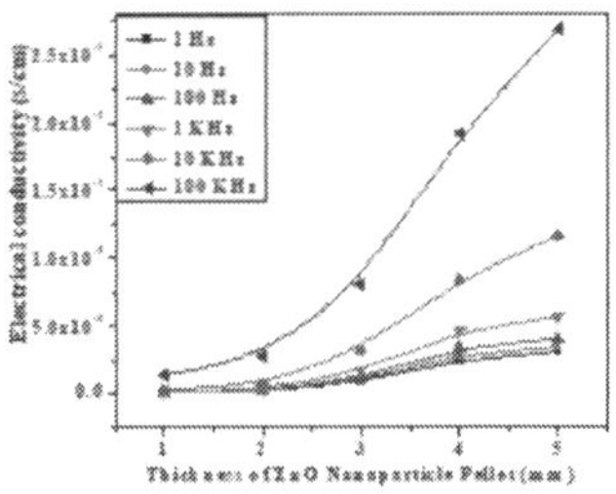

Figure 10. Percolation behavior of ZnO nanoparticles pellets at different frequencies.

As highlighted in Figure 10, the fluctuation of electrical conductivity of ZnO pellets because of rising thickness can be divided into three zones. The following are the regions: The conductivity of zinc oxide pellet increases gradually with thickness due to a small amount of charge particle passing through the system without the formation of continue conductive path; (ii) percolation region: an abrupt rise in conductivity of ZnO NPs thus a continuous conductive path is formed; and (iii) saturation region: the minor effect or electrical conductivity decreases precisely because of the formation of agglomerates or non-continuous conductive path [36]. According to the Figure 11, the percolation threshold occurs at around 4 mm thickness based on the dielectric data and electrical conductivity of ZnO nanoparticles.

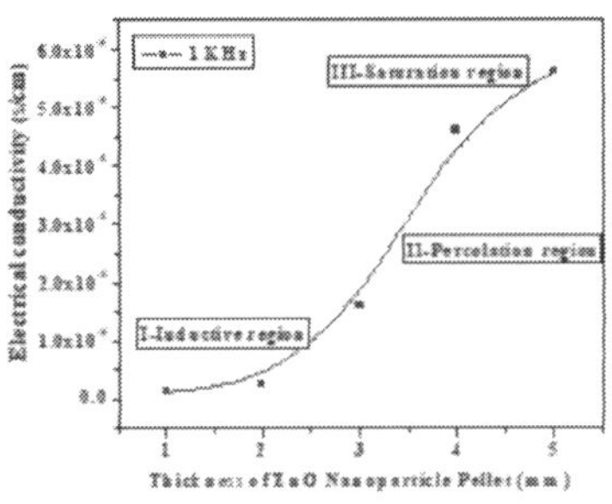

Figure 11. Electrical conductivity of ZnO pellet as a function of thickness at 1 KHz frequency.

5. Conclusions

ZnO nanoparticles were successfully synthesized by co-precipitation technique. UV-visible spectrum of ZnO has a maximum absorption peak of around 400 nm. Based on the sharp and strong peaks, ZnO is crystalline. At normal temperature, the photoluminescence spectra of ZnO nanoparticles show excitation wavelength at 480 nm. The spectrum exhibits two emission peaks, one at 550 nm (UV region) associated with close band gap excitonic emission with the other at 560 nm associated with the occurrence of single ions in spaces in oxygen. X-ray diffraction (XRD) higher peak values for diffraction suggest that it is showing crystalline nature of the nanomaterial. The SEM image indicates thin sheets aggregated as well as spherical in shape, and the HR-TEM picture reveals that the particle size of ZnO can range from 1-100 nanometers, due the contribution of oxygen-containing functional groups with several other spherical structures. Capacitance value decreases with frequency due to charge orientation and variations in the coefficient of the group dipole moments of ZnO NPs. The capacitance value shows separate plot with increasing thickness of ZnO pellet and the maximum value is continuous up to 4 mm thickness. Because of charge orientation and dipole rotation, dielectric permittivity decreases as a function frequency. Dielectric permittivity values of ZnO nanoparticles increase with thickness and decrease with frequency as like capacitance. Electrical conductivity of ZnO nanoparticles pellets rises exponentially, indicating higher alternating current conductivity. The percolation behaviour based on electrical conductivity values of ZnO nanoparticles have been observed for various thicknesses of ZnO pellets. At all frequencies, electrical conductivity grows exponentially until 4 mm thickness, indicating the presence of a percolation limit, before saturation at 5 mm thickness of ZnO pellet. The characteristics of ZnO at the nanoscale enhance its applications in industrial, integration with other nanomaterials, electrochemical and in composite materials. Specifically, ZnO can also be used as accelerator and filler in elastomer composites to enhance its curing and strength. Also ZnO nanoparticles can be used as conducting materials in electrical and thermal applications. Hence, the features and characterizations of ZnO nanoparticles may be used in a variety of engineering applications.

ACKNOWLEDGMENTS

The author would like to express their gratitude to the Director of NIT Raipur, India for giving them laboratory resources and encouragement.

References

[1] Ghorbani HR, Mehr FP, Pazoki H, Rahmani BM. *Orient. J. Chem.* 2015 Jun **31** 1219-21.
[2] Segets D, Gradl J, Taylor RK, Vassilev V, Peukert W. *ACS nano.* 2009 Jul 28;**3**:1703-10.
[3] Segets D, Martinez Tomalino L, Gradl J, Peukert W.*J. Phys. Chem. C.* 2009 Jul 16;**113**:11995-2001.
[4] Das D, Nath BC, Phukon P, Dolui SK.*Colloids Surf. B.* 2013 Nov **1**;111:556-60.
[5] Singh KB, Gautam N, Upadhyay DD, Abbas G, Rizvi M, Pandey G *Inorg. Chem. Commun.* 2022 Aug **6**:109855.
[6] Anbuvannan M, Ramesh M, Viruthagiri G, Shanmugam N, Kannadasan N. *Spectrochim. Acta A Mol. Biomol. Spectrosc.* 2015 May15;**143**:3048.
[7] Wang XS, Wu ZC, Webb JF, Liu ZG. *Appl. Phys. A* . 2003 Aug;**77**:561-5.
[8] Hassan MM, Ahmed AS, Chaman M, Khan W, Naqvi AH, Azam A. *Mater. Res. Bull.*2012 Dec 1;**47**:3952-8.
[9] Mansour SA, Elsad RA, Izzularab MA. *J. Polym. Res.*2016 May;**23**:1-8.
[10] Sagadevan S, Vennila S, Lett JA, Marlinda AR, Hamizi NA, Johan MR. *Results Phys.* 2019 Dec 1;**15**:102543.
[11] Madhumitha G, Elango G, Roopan SM. *Appl. Microbiol. Biotechnol.*2016 Jan;**100**:571-81.
[12] Azizi S, Ahmad MB, Namvar F, Mohamad R. *Mater. Lett.*. 2014 Feb 1;**116**:275-77.

[13] Maruthupandy M, Anand M, Maduraiveeran G, Suresh S, Beevi AS, Priya RJ. *Adv. Nat. Sci.: Nanosci. Nanotechnol.* 2016 Oct 13;**7** 045011.
[14] Malik P, Shankar R, Malik V, Sharma N, Mukherjee TK. *J Nanopart Res.* **2014**;2014.
[15] Tripathi R, Kumar A, Bharti C, Sinha TP. *Curr. Appl. Phys.* 2010 Mar 1;**10:**676-81.
[16] Ahamed AJ, Kumar PV. *J. Chem. Pharm. Res.*2016;**8**:624-8.
[17] Schönhals A, Kremer F. Analysis of dielectric spectra. 2003 (pp. 59-98). Berlin, Heidelberg: Springer Berlin Heidelberg.
[18] Yedurkar S, Maurya C, Mahanwar P.*Open j. synth. theory appl.*2016 Jan 31;**5**(1):1-4.
[19] Mendoza DV. *Orient. J. Chem.*. 2022 Dec 1;**38**(6).
[20] Huang MH, Wu Y, Feick H, Tran N, Weber E, Yang P. *Adv Mater.* 2001 Jan;**13**:113-6.
[21] Williams G, Kamat PV. Langmuir. *J. Am. Chem.* 2009 Dec15;**25**:13869-73.
[22] Rao BS, Kumar BR, Reddy VR, Rao TS, Chalapathi GJ. *Chalcogenide Lett.* 2011 Mar 1;**8**:177-85.
[23] Jayarambabu N, Kumari BS, Rao KV, Prabhu YT. *Int. J Curr. Eng. Technol.* 2014 Oct;**4**:3411-6.
[24] Romadhan MF, Suyatma NE, Taqi FM.. *Indones. J. Chem.*. 2016 Jul **1;**16:117 23
[25] PP V. *J. Bionanosci.*2020 Mar;**10**:112-21.
[26] Bai DP, Zhang XF, Zhang GL, Huang YF, Gurunathan S. *Int. J. Nanomedicine.* 2017 Sep **5**:6521-35.
[27] Kalpana VN, Kataru BA, Sravani N, Vigneshwari T, Panneerselvam A, Rajeswari VD. *Opennano.* 2018 Jan **1;**3:48-55.
[28] Mendes CR, Dilarri G, Forsan CF, Sapata VD, Lopes PR, de Moraes PB, Montagnolli RN, Ferreira H, Bidoia ED. *Sci. Rep.* 2022 Feb 16;**12**:2658.
[29] Beyazay T, Oztuna FE, Unal O, Acar HY, Unal U.*ChemElectroChem.* 2019 Jul 15;**6**:3774 81.
[30] Kumar P, Penta S, Mahapatra SP. *Integr. Ferroelectr.* 2019 Oct **13;**202:41-5.
[31] Saji J, Khare A, Mahapatra SP. 2015 Apr;**16**:883-93.
[32] Sahu G, Gaba VK, Panda S, Acharya B. *High Perform. Polym.* 2018 Apr;**30:**365-74.
[33] Saji J, Khare A,. *High Perform. Polym.* 2016 Feb;**28**:3-13.
[34] Mahmood A, Maqsood A. *Appl. Nanosci.*. 2021Aug;**11**:2311-36.
[35] Guo Z, Wei S, Shedd B, Scaffaro R, Pereira T, Hahn HT. *J. Mater. Chem.*. 2007;**17**:806-13.
[36] Zhu J, Wei S, Ryu J, Sun L, Luo Z, Guo Z. *ACS Appl. Mater. Interfaces.* 2010 Jul 28;**2**:2100-7.

Wide band gap tuning of Mg doped ZnO thin films for optoelectronic applications.

K. Mohammed Salman[1], Mohamed Zikriya[1], C.G. Renuka[1*]

[1] Department of Physics, Bangalore University, Bengaluru 560056, India

*Corresponding author: renubub@gmail.com

Abstract: Zinc oxide (ZnO) thin films with magnesium (Mg) doping have demonstrated significant potential due to their wide band gap, which enables efficient absorption of ultraviolet (UV) radiation while maintaining transparency to visible light. Sol-gel spin coating methods were used to deposit Mg-doped ZnO thin films. The solution was made with methanol as the solvent, and the starting materials were magnesium acetate tetrahydrate and zinc acetate dihydrate. The integration of Mg content in deposited ZnO films was investigated for structural, morphological, and optical properties using an X-ray diffractometer (XRD), scanning electron microscope (SEM), energy dispersive X-ray analysis (EDAX), and UV-VIS spectrophotometer. Mg-doped ZnO films were produced on quartz substrates and post-annealed at 400°C. According to absorption coefficient calculations, the optical band gap of Mg-doped ZnO films is in the range of 5.4 eV which is ideal for a variety of applications in the field of optoelectronic devices.

Keywords: MgZnO, Wide band gap, Optoelectronic applications.

1. Introduction

The group II-VI n-type semiconductor zinc oxide is extensively utilized. It demonstrates a 3.3eV direct wide band gap and possessing a substantial 60 meV exciton binding. Because of its high electron mobility, ability to produce light at room temperature, and well-defined hexagonal crystal structure, ZnO plays a significant role as a potential material for optoelectronics, photonics, and spintronics applications. However, considering its band gap, there are limitations on how effectively it can be applied to integrated optical devices. Additionally, to accomplish barrier multiple layers, quantum well structures, etc. wide band gap tailoring is required for short wavelength range in heterostructure optoelectronic and photonic devices [1-2]. When a semiconductor is doped, impurity states are generated which alters the fermi energy levels resulting in modification of the energy band gap [3]. Altering the bandgap of zinc oxide is therefore possible by alloying it with elements from groups II and III, such as Mg, Be and Ca. Because Zn (0.06 nm) and Mg (0.057 nm) have almost equal ionic radii, magnesium doping in ZnO does result in limited lattice distortion [4,5]. The energy of the optical bandgap in thin films of ZnO doped with Mg could be effectively modulated between 3.4 eV to 7.8 eV. When alloyed with ZnO, MgO's high bandgap allows deep UV bandgap expansion, allowing for bandgap tuning by modification of Mg content [6,7]. MgZnO thin films with wide optical band gap characteristics are suitable for electro-optical applications such as buffer and TCO layers for thin film solar cells and layers of light-emitting diodes. Additionally, MgZnO is an ideal semiconductor for UV photodetectors due to its distinct qualities, such as robust radiation hardness and a wide range of tunable band gaps [8]. Considering that MgO and ZnO have distinct structures of lattice are influenced by the electronegativity and the small variation in ionic radius between Mg and Zn ions, doping also modifies the optical bandgap. Phase separation is caused by doping large Mg content. Due to this, obtaining a hexagonal pure wurtzite MgZnO in one phase is challenging whenever the Mg content would be raised above a specific threshold. Moreover, the synthesis method and other growth variables that impact the crystallinity of the MgZnO alloy affect MgO's solubility in MgZnO [9,10].

IOP Conf. Series: Materials Science and Engineering 1300 (2024) 012026 doi:10.1088/1757-899X/1300/1/012026

ZnO thin films doped with magnesium have been synthesized by variety of deposition techniques, including pulse laser deposition, RF magnetron sputtering, sol-gel spin coating, and spray pyrolysis. Sol-gel spin coating technology stands out from the rest of them due to its ease of use, energy efficiency, cost-effectiveness, and ability to produce MgZnO thin films at lower temperatures [11-15].

In this study we have synthesized the films using the sol-gel spin-coating process to study Mg doped ZnO thin films structural, morphological and optical properties.

2. Experimental

2.1 Materials and Methods

The precursor materials used were magnesium acetate and zinc acetate dihydrate. In the process of preparing the ZnO solution, 1 gram of zinc acetate is dissolved in 10 ml of isopropanol with continuous stirring. For the MgO solution, 0.2, 0.4, 0.6, 0.8, and 1.0 grams of magnesium acetate were individually dissolved in 10 ml of methanol. Nitric acid and monoethanolamine were added as stabilizers to the ZnO and MgO solutions until they achieved a homogeneous and transparent consistency. The prepared solutions were subjected to sonication at 50°C for 10 minutes. Subsequently, they were constantly mixed at a temperature of 60°C for a period of 2 hours at a rate of 200 rpm. The prepared Mg solution was added dropwise to the ZnO solution under the same conditions, and the resulting mixtures were allowed to age for 24 hours to form a gel-like structure.

For the deposition of thin films onto a quartz substrate, a spin coater was employed. The Mg-doped ZnO solution was carefully dispensed onto the substrate, and the rotation speed was set at 1500 rpm for 60 seconds. This process was repeated for various concentrations of magnesium at a drying temperature of 120°C for 10 minutes in a furnace. After achieving the desired film thickness, in a muffle furnace, the thin films were annealed for two hours at 400°C.

2.2 Characterization

Using a theta-theta arrangement and Cu Kα radiation, we used Bruker AXS D8 advanced device for X-ray diffraction (XRD) to analyse the structural characteristics of the thin films. Using CARL ZEISS scanning electron microscopy (SEM) and EDS (Energy Dispersive Spectroscopy), surface morphology and elemental composition was evaluated. Cary 60 UV-Visible spectrophotometer is used for the transmission and absorption spectra. Using an excitation wavelength of 328 nm, photoluminescence (PL) spectra were captured using a Agilent Technology-cary-60 Eclipse spectrophotometer.

3. RESULTS AND DISCUSSION

3.1 XRD Analysis

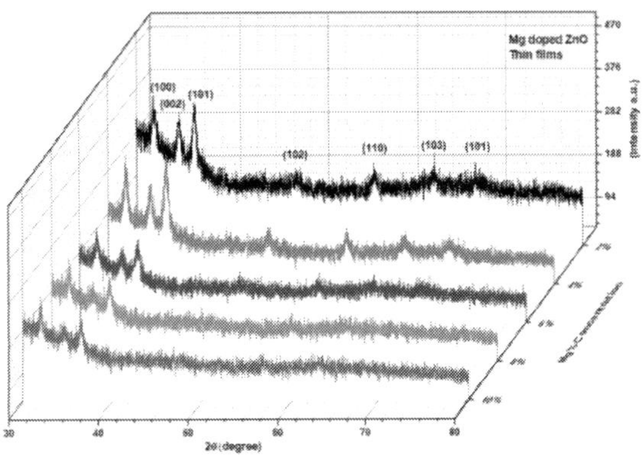

Figure 1. XRD pattern of Mg doped ZnO Thin Films

Table 1. Structural parameters of Mg-doped ZnO from XRD patterns

Sample	FWHM (degree)	2θ (degree)	Crystallite Size(nm)
2%MgZnO	0.4537	31.79	31.81
4%MgZnO	0.4572	31.83	31.57
6%MgZnO	0.4293	31.85	33.61
8%MgZnO	0.4188	31.94	34.45
10%MgZnO	0.3961	31.99	38.09

Using the x-ray diffraction method, the crystalline structures of the produced thin films were investigated. depicts the XRD patterns of MgZnO thin films that were deposited with different Mg concentrations on quartz substrates 2, 4, 6, 8, and 10%, respectively and annealed at 400 °C for 2 hours as shown in Figure 1. The films are all polycrystalline in structure with hexagonal wurtzite structure of ZnO. It has been identified that three primary individual XRD peaks for each sample were derived from (1 0 0), (0 0 2), and (1 0 1) planes, respectively, at 31.8, 34.6, and 36.4 degrees (JCPDS card no.: 00-036-1451). MgO and ZnO phase separation is due to Mg atoms being more chemically reactive and favoring oxygen reactions. [19]. Despite having noticed that the diffracted peaks reduce as concentration of Mg increases. The addition of Mg into the ZnO lattice might be responsible for the moderate shifting of the diffraction peaks towards lower angles. This reveals a slight modification in the nucleation and growth of preferred planes, which may be related to the film deposition process. The average crystallite size is calculated using Debye-Scherrer's formula [11]:

$$D = \frac{0.9\lambda}{B\ Cos\theta} \qquad (1)$$

where B is the FWHM (full width at half-maximum) of the angle of diffraction, θ is the Bragg diffraction angle, and λ is the x-ray wavelength of 1.54 Å. The average size of crystallites was measured to be in the range of 34 nanometers for 2% to 48nm for 10% MgZnO thin films.

3.2 SEM Micrographs

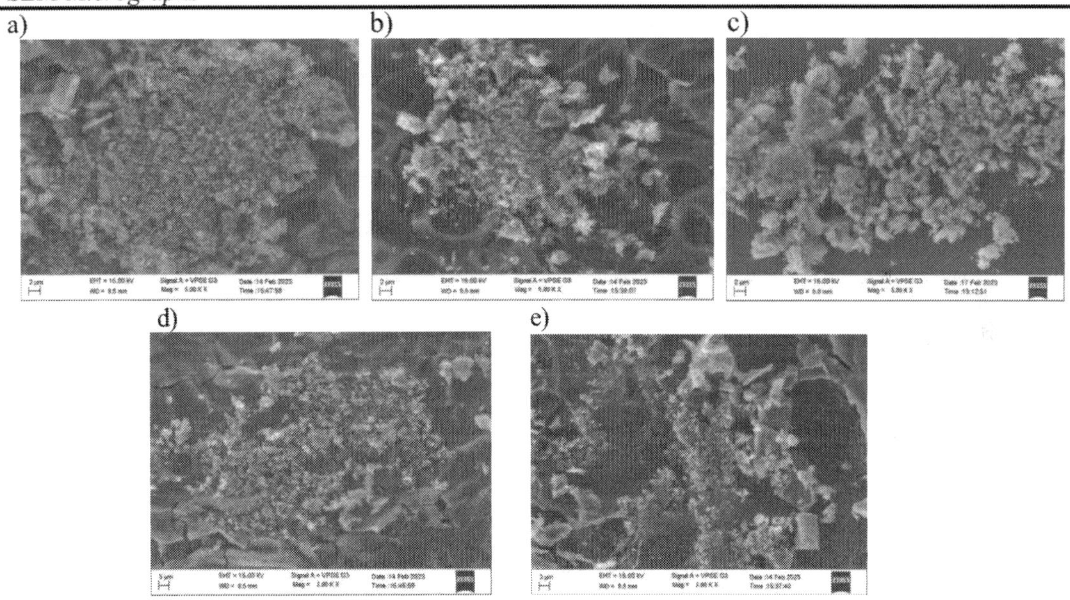

Figure 2. SEM images depicting surface morphologies of MgZnO thin films

Properties of the ZnO thin film surface at various Mg doping levels were studied using SEM results, it is seen in the Figure 2, that the thin films surface is composed of nanosized inhomogeneous particles spherical in shape. As the Mg concentration is increased, the grain size of substituted MgO structures has grown, leading to gaps and agglomerations on the surface to occur as well [8]. The average particle size was measured to be within a range of 46 nanometers for 2% to 87nm for 10% Mg incorporated ZnO Thin films.

The thin films only had a small number of nanopores. These nanoscale holes were connected to microstructural flaws that developed as a result of the coating precursors' thermal breakdown and the persistence of organic residues in the oxide films [19].

3.3 EDAX Analysis

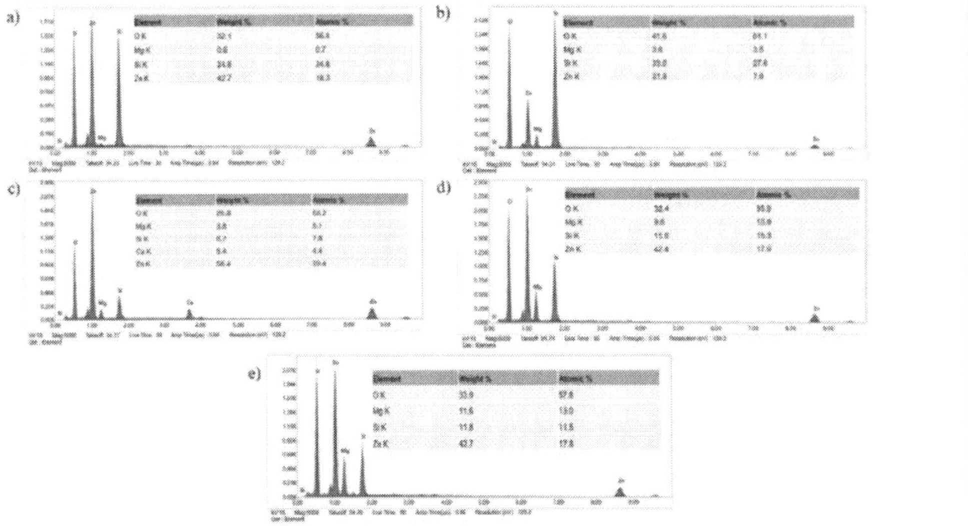

Figure 3. The EDAX of Mg-ZnO thin films a) 2%, b) 4%, c) 6%, d) 8%, e) 10% Mg doped ZnO Thin films.

The EDS spectra revealed that the sol gel method's effective incorporation of Mg into ZnO thin film. The deposited film's weight % of Mg incorporation in ZnO is 0.6, 3.6, 3.8, 9.6, and 11.6% at 400°C as shown in the Figure 3. It has been reported that the maximum amount of Mg that may be dissolved in ZnO relies on both the growth techniques and the growth conditions employed. Mg incorporation can be observed in each film at the 1.25 KeV peak. Peaks with energies of 1.012 keV and 0.525 keV, respectively, indicating the existence of zinc and oxygen. Due to the substrate being composed of silica and containing calcium, as in the case of soda-lime glass, it is also depicted [15].

3.4 *The Uv-Vis absorption spectrum*

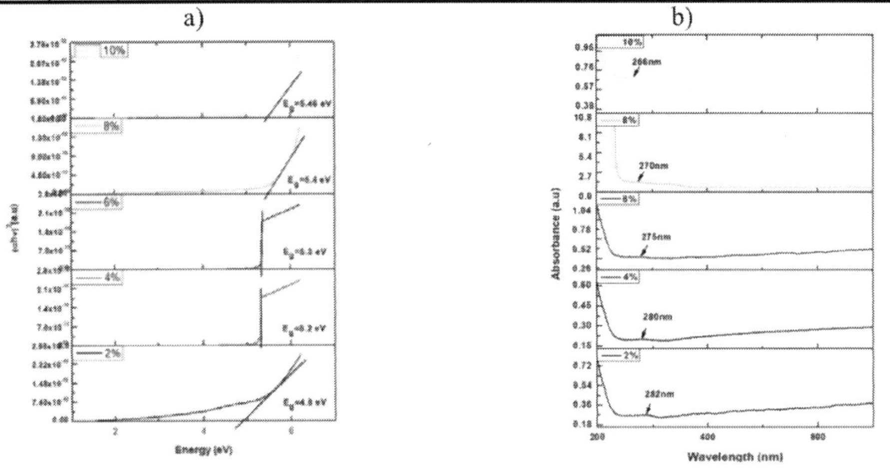

Figure 4. a) Uv-Vis absorption peaks b) Band gap values

The thin films optical bandgap has been calculated using Tauc's relationship given by [22],

$$(\alpha h\nu)^{\frac{1}{n}} = A(h\nu\text{-}E_g) \qquad (2)$$

where E_g, α, h, and ν represents optical energy band gap, absorption coefficient, Planck's constant and frequency of radiation respectively; n and A are constants. Figure 4 represents a) Uv-Vis absorption peaks, b) Band gap values, c) Urbach energy plots for MgZnO thin films.

Taking MgZnO as a direct transition material to consideration, n's value has been defined to be 2 in order to construct the graph of $(\alpha h\nu)^2$ vs $h\nu$. After that, the optical bandgap of thin films synthesized from varying Mg concentrations of 2, 4, 6, 8, and 10% were estimated to be 4.8, 5.2, 5.3, 5.4 and 5.46eV, Individually, by projecting the linear segment of the non-linear curve until it reaches the x-axis [16,17,21]. It is observed that a higher percentage of MgO atoms get substituted into the ZnO lattice the wider the E_g. These influences can alter how the spectrum shifts, depending on the ratio of magnesium to zinc atoms. As the Mg doping level increased, the absorption region peaks shifted towards shorter wavelengths. The observed blue shift in the optical band gap is directly related to this result [17]. Due to variations in electronegativity and ionic radius differences between Mg and Zn, structural defects that may be created by the substitution of Mg atoms can be linked to an increase in band gap [18].

3.5 Photoluminescence Spectrum

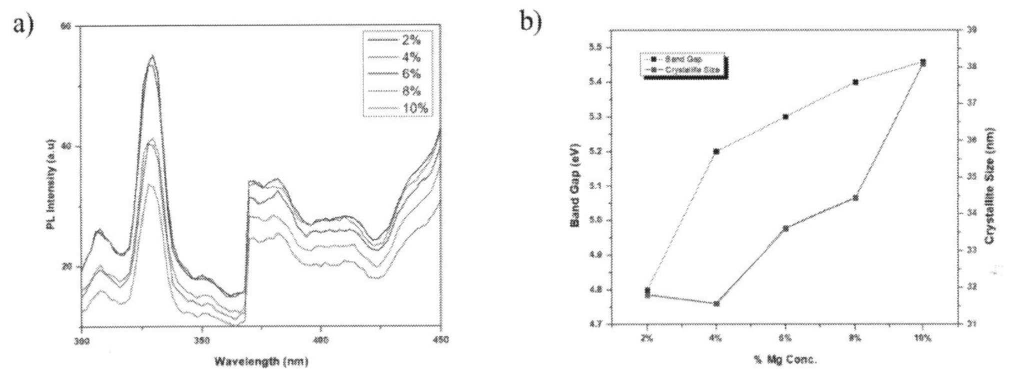

Figure 5. a) PL peaks of Mg-ZnO Thin Films b) relation between % Mg content , Band Gap and Crystallite Size.

The photoluminescence spectrum as shown in Figure 5a), suggests information on excited state relaxation mechanisms and Defects that exhibit optical activity. An intense PL emission (300–350 nm) can be noticed. The source of the PL emission peak is commonly recognized as typically arising from a large bandgap to a near-band-edge (NBE) transition, primarily associated with exciton annihilation [8].

The findings show that the PL emission intensity increases with increasing Mg content and reaches its maximum at 2 and 6% Mg concentration before decreasing. Due to Mg substitution altering the band-gap, the PL emission peak simultaneously blue shifts, demonstrating that optical band gap may be controlled by altering Mg concentrations. The bandgap widening and the amplitude of the peak's blueshift are strongly correlated, and both of these factors could have impacted by surface recombination, impurities, and defects during high-energy excitation [20]. Figure 5b) represents relation between Mg % concentration, optical band gap and crystallite size. The bandgap increases nonlinearly and steadily as Mg content increases. Substituted Mg at the sites and grain borders could account for the non-monotonic connection [23]. Mg doping initially decreases crystallite size, but at higher doping concentrations, it tends to increase [24].

4. CONCLUSION

In our study, we explored ZnO thin films doped with magnesium (Mg) deposited via sol-gel spin coating, we deposited these films, then investigated their morphological, structural, and optical characteristics. X-ray diffraction revealed structural characteristics, with increasing Mg content leading to more prominent phase separation. SEM analysis showed spherical shaped particles with larger grain sizes and surface imperfections with Mg addition, and Mg integration was confirmed by energy-dispersive X-ray spectroscopy. The optical bandgap of these films was approximately 5.4 eV, making them promising for optoelectronic applications. Additionally, photoluminescence spectra exhibited strong PL emission, with peak intensity at 2% Mg content. These findings underscore the versatility incorporating ZnO thin films doped with magnesium for a range of optoelectronic applications.

6. Acknowledgements

The author C G Renuka gratefully acknowledges the Bangalore University, Bengaluru of the financial support provided with a project number UNI.ORDER.NO.DEV:D2a:BU_RP.2020-21 and my sincere gratitude Directorate of Minorities, Government of Karnataka.

5. References:

[1] Baig, F., Ashraf, M.W., Asif, A. and Imran, M., 2020. *Optik*, **208**, p.164534.

[2] Mia, M.N.H., Pervez, M.F., Hossain, M.K., Rahman, M.R., Uddin, M.J., Al Mashud, M.A., Ghosh, H.K. and Hoq, M., 2017. *Results in physics*, **7**, pp.2683-2691.

[3] Andriotis, Antonis N.; Menon, Madhu, 2015. *Journal of Applied Physics,* **117**(12), 125708.

[4] Devi, Vanita; Joshi, B C; Kumar, Manish; Choudhary, R J. 2014. *Journal of Physics: Conference Series,* **534,** 012047.

[5] Jiang, D.Y., Shen, D.Z., Liu, K.W., Shan, C.X., Zhao, Y.M., Yang, T., Yao, B., Lu, Y.M. and Zhang, J.Y., 2008. *Semiconductor science and technology*, **23**(3), p.035002.

[6] Hu Y, Cai B, Hu Z, Liu Y, Zhang S, Zeng H. 2015, *Curr Appl Phys*; **15**:423–8.

[7] Kara, R., Mentar, L. and Azizi, A., 2020. *RSC advances*, **10**(66), pp.40467-40479.

[8] Temel, Sinan & Nebi, M & Peker, D. 2018. *Optoelectronics and advanced materials-rapid communications*. **12**, No. 1-2 p. 76 – 79.

[9] Kaushal A, Kaur D. *Sol Energy Mater Sol Cells*. 2009. **93**:193–8.

[10] Reddy KTR, Prathap P, Revathi N, Reddy ASN, Miles RW.2009, *Thin Solid Films*; **518**:1275–8.

[11] A.A. Hussain, K.A. Aadim, H.M. Slman, 2014, *Iraqi J. Phys*. **12** ,56–61.

[12] K. Vijayalakshmi, A. Renitta, K. Karthick, 2014 *Ceram. Int*. **40**, 6171–6177.

[13] Mujdat Caglar, Yasemin Caglar, Saliha Ilican, 2016, *Phys. B* **485**,6–13.

[14] K. Karthick, K. Vijayalakshmi, 2014, *Super Lattice Microstructure* **67**,172–180.

[15] B. K. Sonawane; M. P. Bhole; D. S. Patil, 2009. *Opt Quant Electron* **41**(1),17–26.

[16] Kaushal A, Pathak D, Bedi RK, Kaur D. 2009, *Thin Solid Films* ;**518**:1394–8.

[17] Bouachiba, Y., Mammeri, A., Bouabellou, A. et al. 2022, *J Mater Sci: Mater Electron* **33**, 6689–6699

[18] Hussain, Fayyaz; Imran, Muhammad; Khalil, R.M. Arif; Niaz, Niaz Ahmad; Rana, Anwar Manzoor; Sattar, M. Atif; Ismail, Muhammad; Majid, Abdul; Kim, Sungjun; Iqbal, Faisal; et.al, 2019. *Physica E: Low-dimensional Systems and Nanostructures,* **115,** 113658.

[19] Kai Huang, Zhen Tang, Li Zhang, Jiangyin Yu, Jianguo Lv, Xiansong Liu, Feng Liu, 2012 *Applied Surface Science*, **258**, Issue 8, Pages 3710-3713.

[20] Bang, Jungsik; Yang, Heesun; Holloway, Paul H., 2006, **17**(4), 973–978.

[21] Shizuo Fujita, Hiroshi Tanaka, Shigeo Fujita, 2005, *Journal of Crystal Growth*, **278**, 1–4, Pages 264- 267.

[22] Mohamed Zikriya, Y.F. Nadaf, P. Vijai Bharathy, C.G. Renuka, *Journal of Rare Earths*, 2019, **37**, 1, Pages 24-31.

[23] Hu Y, Zeng H, Du J, Hu Z, Zhang S.2016, *Mater Chem Phys*;**182**:15–21.

[24] Varghese, J., Saji, S.K., Aswathy, N.R. et al.2021, *Eur. Phys. J. Plus* **136**, 1206.

Effect of aspect ratio on the nonlinear optical properties of ZnO nanorods

M Ramya[1,2], Swathy Satheesh[3], Sreelakshmi Sreedhar[3], Jitha P Marydasan[3] and M Kailasnath[1]

[1] International School of Photonics, CUSAT, Kochi, Kerala, India.
[2] Sree Vyasa NSS College, Wadakkanchery, Thrissur, Kerala, India.
[3] Ansar Womens College, Perumpilavu, Thrissur, Kerala, India.
Email: swathysatheesh527@gmail.com

Abstract. In this work, we synthesized ZnO nanorods with two different aspect ratios using the solution method. Morphological and crystalline structures were investigated using TEM and XRD analysis. Also investigated the effect of aspect ratio on the optical properties of ZnO nanorods. With an increasing aspect ratio of the nanorods, the absorption peak is red-shifted and near-band edge emission is suppressed. The nonlinear absorption is enhanced with the increasing aspect ratio of ZnO nanorods.

1. Introduction

Zinc Oxide (ZnO) is a widely used semiconductor with a wide direct bandgap, high mobility, large exciton binding energy, and high chemical and thermal stability[1,2]. ZnO is used in numerous scientific fields due to its excellent electro-optical, chemical, and mechanical properties[1,3]. By changing the synthesis conditions, controlled synthesis of ZnO various morphologies are possible such as nanoparticles, nanotubes, nanoneedles, nanoflower, nanorods, nanoplate, nanowire, nanopellets, etc[1,2]. Among these nanostructures, ZnO nanorods have attracted great attention because of their wide applications in the light detector, LED, lasers, sensors, solar cells, and photocatalysis[2,4]. In this paper, we synthesized ZnO nanorods with two different aspect ratios and examined their linear and nonlinear optical properties.

2. Materials and Methods

2.1. Materials for ZnO nanorod synthesis

Zinc acetate dihydrate (Zn $(CH_3COO)_2.2H_2O$) and Sodium hydroxide (NaOH) were used as chemical reagents. 1-Butanol ($C_4H_{10}O$) and 1-Hexanol ($C_6H_{14}O$) used as a solvent.

2.2. Preparation method

Ultrasound sonication-assisted simple solution method was employed in the preparation of ZnO nanorods. 0.1M $Zn(CH_3COO)_2.2H_2O$ and 0.5M NaOH were dissolved in selected solvents. A transparent white solution was formed and heated at 80°C for 2 hours and kept at room temperature for 24 hours. The resulting white precipitate was centrifuged and dried to get in powder form. Figure 1 shows the schematics of the synthesis procedure.

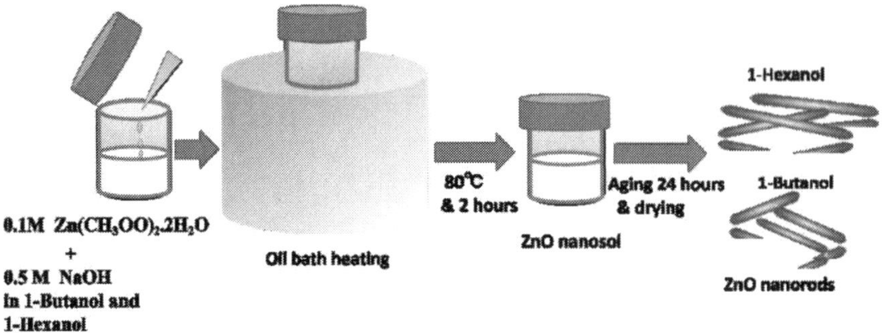

Figure 1. Synthesis procedure for ZnO nanorod in 1-Butanol and 1-Hexanol.

2.3. Material Characterizations

A transmission electron microscope (TEM; JEM 2100) was used to analyze the morphology. The crystallinity of the samples was characterized by XRD (Bruker AXS D8 Advance). The absorption of prepared ZnO nanorods was measured using a JASCO V-570 UV-vis spectrometer and emission spectra was analyzed using a Cary Eclipse Varian spectrofluorometer.

2.4. Nonlinear optical studies

First, we prepared a stable colloidal ZnO nanofluid. For this synthesized nanorods in the powder form were dispersed in water at the concentration of 0.1mg/ml under ultrasonication. These colloids were used as a sample for the nonlinear studies. The nonlinear optical studies were carried out using the Z-scan method, in which a Q-switched Nd-YAG laser (532nm, 7ns, 10Hz) was used as an irradiation source. Z-scan is a highly sensitive and simple technique for estimating the nonlinear absorption of the sample. The sample was fixed on the translational stage and moved along the beam focus, self-focusing/ defocusing effects caused the change in transmittance value. When the sample is irradiated with the beam, the transmittance at focus either decreases or increases depending on the property of the material. More details of Z scan techniques have been described in our previous publications[3].

3. Results and Discussion

TEM images of the ZnO nanorods synthesized in 1-Butanol and 1-Hexanol are shown in Figure 2. The nanorod synthesized using 1-Butanol and 1-Hexanol was named as Sample A and Sample B. The TEM image revealed the formation of nanorod-like morphology. Longer nanorods are formed in 1-Hexanol (Sample B) compared with 1-Butanol (Sample A) with an aspect ratio of 5 and 2 respectively. The aspect ratio of nanorod increases with increasing alkyl group length of the selected solvent. More details for the growth mechanism were explained in our previous publication[2].

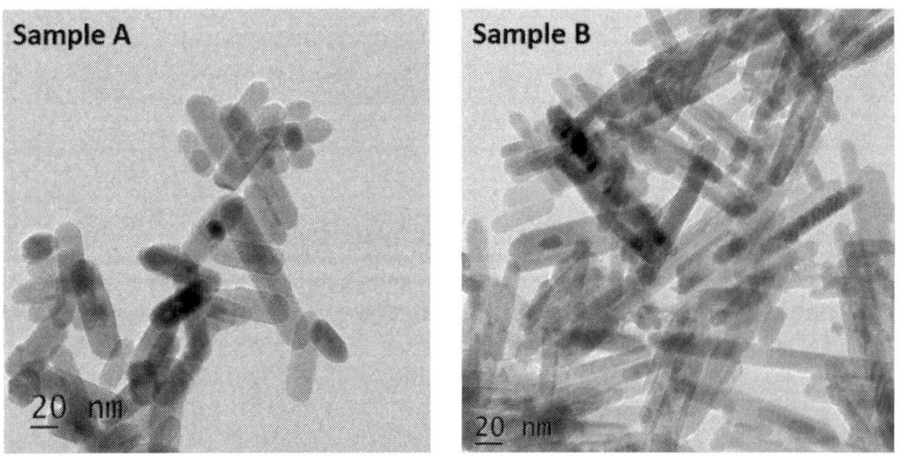

Figure 2. TEM image of synthesized ZnO nanorods in 1-Butanol and 1-Hexanol as solvent.

The crystal structure of nanorods was studied using XRD analysis shown in Figure 3. Two samples exhibited a hexagonal wurtzite structure of ZnO in agreement with JCPDS file number 01-089-1397. The peaks corresponding to the planes are identical and no peak shift or other crystalline phases are detected.

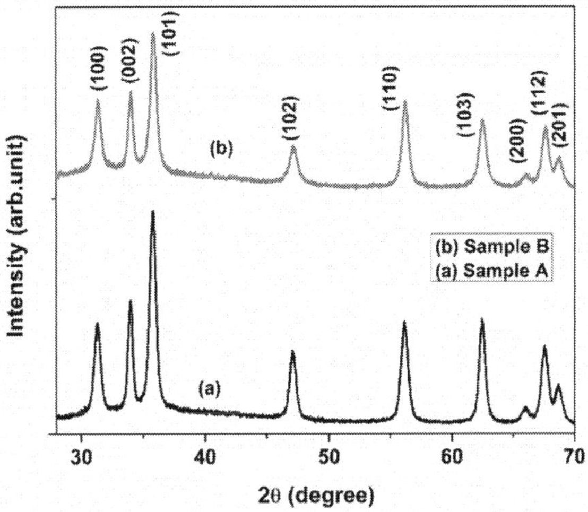

Figure 3. XRD pattern of ZnO nanorods synthesized in 1-Butanol and 1-Hexanol.

The UV-vis absorption spectra and emission spectra are shown in Figure 4. As the aspect ratio increases absorption peak is shifted to a higher wavelength region. The absorption spectra exhibited blue shift and band edge absorption[2]. Generally, the photoluminescence spectra include two bands. The first band represents the UV emission and the second band is a broad visible state emission. In Figure 4(B), only near-band edge emission is observed for both samples. The intensity of emission intensity increases with decreasing the aspect ratio. This increment is due to the presence of minimum lattice surface defects. Lattice surface defects act as a nonradiative recombination trapping center, which increases the surface recombination and decreases the near band edge emission[5].

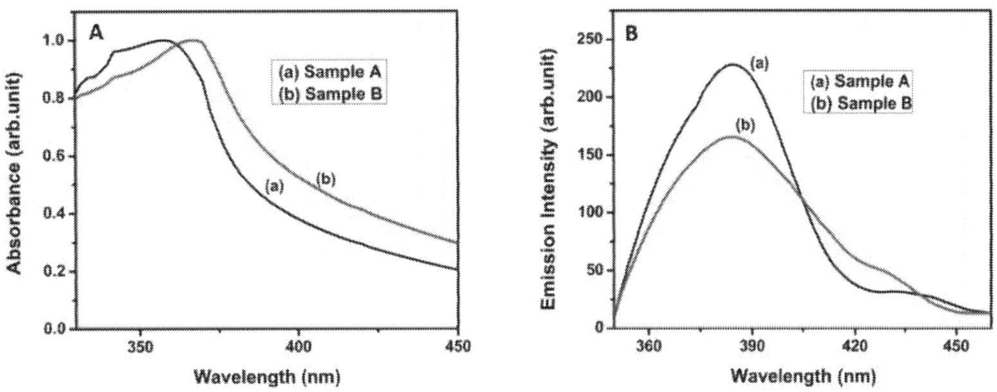

Figure 4. (A) Absorption and (B) emission spectra of ZnO nanorods synthesized in 1-Butanol and 1-Hexanol.

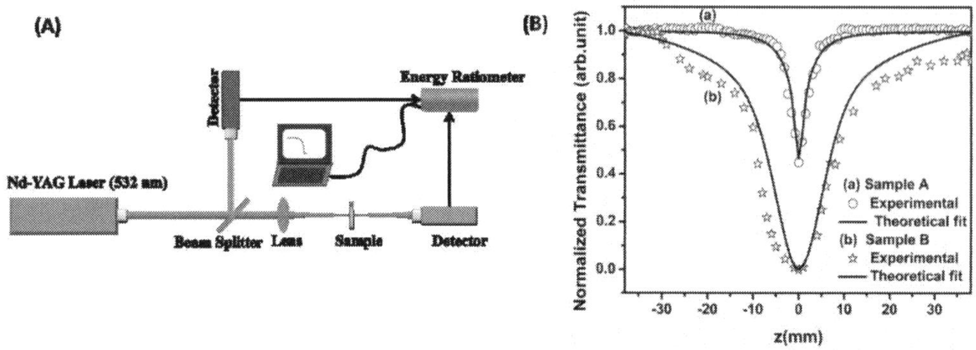

Figure 5. (A) Z scan experimental setup and (B) open aperture Z scan trace of ZnO nanocolloids.

Nonlinear optical properties of prepared ZnO nanocolloids are studied using an open aperture Z scan experiment at the input intensity of 25 MW/cm^2. Figure 5 shows the Z-scan experiment setup and open aperture curve. The open aperture curve exhibited an induced absorption and transmittance at focus decreases. The obtained RSA curve was fitted with two-photon absorption process[3,6].

$$T(z, s=1) = \sum_{m=0}^{\infty} \frac{[-q_0\ (z,t)]^m}{(m+1)^{3/2}}$$

Here q_0 is the fitting parameter. As the aspect ratio of ZnO nanorods increases, the depth of the valley increases. This increment in depth indicates the enhancement in nonlinear behavior. Here larger nanorods (Sample B) possess higher nonlinear optical properties. Calculated nonlinear absorption (β) is 600 cm/GW and 760 cm/GW for Sample A and Sample B respectively. In this work, synthesized nanorods lie in a weak confinement regime, and the exciton oscillation strength increases with an increase in particle size. Thus, the enhancement in nonlinear absorption (β) is related to the quantum confinement effect as well as the exciton oscillator strength[3].

4. Conclusion

The present study revealed the aspect ratio-dependent linear and nonlinear optical properties of ZnO nanorods. According to the increase in aspect ratio, the UV absorption peak is shifted towards the higher wavelength region and increases the surface defects which suppressed the UV emission. In the case of nonlinear optical studies increase in aspect ratio enhances the nonlinear absorption. This study confirmed that the optical properties of ZnO nanostructures are controlled by changing their aspect ratio.

5. References

[1] Ramya M, Nideep T K, Nampoori V P N and Kailasnath M 2021 Solvent assisted evolution and growth mechanism of zero to three dimensional ZnO nanostructures for dye sensitized solar cell applications *Sci. Rep.* **11** 1–14

[2] Ramya M, Nideep T K, Nampoori V P N and Kailasnath M 2019 Understanding the role of alcohols in the growth behaviour of ZnO nanostructures prepared by solution based synthesis and their application in solar cells *New J. Chem.* **43** 17980–90

[3] Ramya, M., Nideep, T. K., Nampoori, V. P. N. , Kailasnath M 2021 Shape dependent heat transfer and nonlinear optical limiting characteristics of water stable ZnO nanofluid *Surfaces and Interfaces* **26** 1–10

[4] Fiedler S, Lee Cheong Lem L O, Ton-That C, Hoffmann A and Phillips M R 2020 Enhancement of the UV emission from gold/ZnO nanorods exhibiting no green luminescence

Opt. Mater. Express **10** 1476
[5] Sharma R, Khanuja M, Islam S S, Singhal U and Varma A 2017 Aspect-ratio-dependent photoinduced antimicrobial and photocatalytic organic pollutant degradation efficiency of ZnO nanorods *Res. Chem. Intermed.* **43** 5345–64
[6] Ramya M, Nideep T K, Vijesh K R, Nampoori V P N and Kailasnath M 2018 Synthesis of stable ZnO nanocolloids with enhanced optical limiting properties via simple solution method *Opt. Mater. (Amst).* **81** 30–6

Acknowledgments
This work was financially supported by DST-SERB (grant number EEQ/2018/000468, EMR/2016/003614) and Cochin University of Science and Technology, India.

The mechanical and tribological properties of composites of nanographene and an aluminium-metal matrix were studied, as was their corrosion behaviour, in order to develop materials suitable for use as shielding materials

Romala rajesh [1], V.V.S. Prasad[2], Maduthuri Venkatesh[3]*, P Surya nagendra [4], Gopalapalli Madhav Pavan Surya[5]

[1,3,5] Research Scholar, Department of Marine Engineering, Andhra University, Visakhapatnam, India
[2, 4] Professor, Department of Marine Engineering, Andhra University, Visakhapatnam, India
Email: venkymaduthuri1114@gmail.com

Abstract
In this investigation, pitting corrosion methods were used to investigate the corrosion behaviour of nanographene oxide-reinforced aluminium-copper metal matrix composites. S1 (AL 100% + 0% Cu + 0% Nano Graphene), S2 (AL 91% + 7% Cu + 2% Nano Graphene), S3 (AL 82% + 14% Cu + 4% Nano Graphene), and S4 (AL 73% + 21% Cu + 6% Nano Graphene) were the sample compositions. Ultrasonic stir casting was utilised to produce the aluminium-copper-nanographene composites. Nanographene was analysed with BET to determine its surface area, porosity volume, etc. The effects of nanographene oxide weight fraction on the corrosion behaviour of aluminium-copper composites were investigated. The surface texture of the corroded material was evaluated with an atomic force microscope, and the hardness was measured with a Vicker's microhardness tester. The results of the experiment revealed that the corrosion rate decreased as the percentage of nanoparticles in the matrix increased up to S3 composition. The microhardness values increased as the weight fractions of reinforcements in the matrix increased. And additional mechanical tests, such as tensile, wear, and flexural, to ascertain the composites' mechanical behaviour.

1. Introduction
Aluminium metal matrix composites (AMMCs) are a group of materials that combine the lightweight properties of aluminium with advanced mechanical and thermal properties through added strength The most commonly hardened component is copper, which provides mechanical strength. Another emerging reinforcement is nanographene, which offers unique properties such as high strength and thermal stability. The mechanical behaviour of materials is an important factor in determining their suitability for different architectural applications. In recent years, aluminium-metal matrix composites (AMMCs) labelled with copper nanographene have received increasing attention due to their improved mechanical properties Understanding the mechanical behaviour of these materials is important and to improve efficiency and expand their potential applications.

The incorporation of copper and nanographene into AMMCs offers several mechanical advantages. Copper with high strength and good ductility, acts as a reinforcing material to provide tensile strength, stiffness and toughness in composites while nanographene, with unique mechanical properties at the nanoscale, contributes to strength, stiffness and wear resistance. Several studies have been conducted to investigate the performance of AMMCs reinforced with copper nanographene. This study uses various testing methods such as tensile testing, hardness testing and wear testing to evaluate the mechanical properties of the composites.

The addition of copper nanographene significantly improved the tensile properties of AMMCs, enhancing their tensile strength and parameters. This addition also enhanced machinability and toughness, making the composites more suitable for applications requiring high strength and resistance[1]. In addition, numerical modelling techniques such as finite element analysis (FEA) have also been used to predict and understand the mechanical behaviour of AMMCs These models provide valuable insights into stress distribution, deformation behaviour and failure mechanisms in composites, contributing to they design and optimize their technology assets. The mechanical behaviour of aluminium-metal matrix composites with copper nanographene filler enhances tensile strength, corrosion resistance, and toughness, enhancing their application in construction, aerospace, automotive, and electronics industries.

Pitting corrosion is a localized form of corrosion that can lead to the formation of small pits or holes on the metal surface, potentially compromising the structural integrity of the material. Understanding the pitting corrosion behaviour of AMMCs is crucial to ensure their long-term durability and reliability[2]. Research studies have investigated the pitting corrosion resistance of AMMCs with copper and nanographene, considering factors such as the corrosion potential, corrosion current density, and morphological analysis of the pits formed. These studies have focused on evaluating the corrosion behaviour of the composites in different corrosive environments, such as acidic, alkaline, and chloride-containing solutions[4].

The characterization of nano-scale materials is important to understand their properties and optimize their performance in various applications. For aluminium-metal matrix composites (AMMCs) with copper nanographene reinforcements, growth characteristics such as atomic force microscopy (AFM) and Brunauer-Emmett-Taylor (BET) analysis in analyzing their surface topography porosity play an important role. AFM is a powerful imaging technique that allows high-resolution imaging and measurement of surface topography at the nanoscale. It uses a sharp edge that scans the sample, detects forces between edge and sample and provides a detailed map This method provides valuable information about AMMC surface morphology, roughness and particle distribution, and empowers researchers investigate the effects of copper and nanographene reinforcement on composite microstructure.

BET analysis measures surface area and porosity of materials, providing insight into pore size distribution, surface area, and pore volume. It helps researchers understand the effect of copper and nanographene on porosity and surface area in materials like AMMCs. AFM-BET analysis was used to evaluate surface roughness in AFM composites, showing that nanographene increased smoothness and decreased roughness, highlighting its significant role in 19th-century AMMC surface appearance[4]. In another study, BET analysis was used to investigate the porosity and specific surface area of different AMMCs of copper and nanographene materials. The results showed that the addition of nanographene increased the specific surface area of the composites, while copper affected the pore size distribution. These data show the importance of copper and nanographene on the porosity and surface of AMMCs the combination of attributes[5].

In conclusion, AFM and BET characterization techniques provide valuable insights into the surface area, roughness, and porosity of AMMCs incorporated with copper-nanographene fillers These techniques allow researchers to understand the effects of these fillers under the microstructure and surface properties in composites The combination of analysis contributes towards a broader understanding of AMMC and contributes to the development of improved materials with improved performance.

2.Materials and methods

Aluminium alloy was used as the matrix material in this investigation. Aluminium was used as the primary material for this. Copper and nanographene reinforcements were used. Four categories of Al-based samples were characterised:

S1 (100 % AL, 0% Cu, 0% Nano Graphene), S2 (AL 91% + 7% Cu + 2% Nano Graphene), S3 (AL 82% + 14% Cu + 4% Nano Graphene) and S4 (AL 73% + 21% Cu + 6% Nano Graphene).

Aluminium alloy was chosen due to its excellent ductility and castability.The Al-Cu-Gr particle composite is manufactured in a pit furnace (model: INDFUR suppliers, Chennai), which is essentially a three-phase resistance furnace that is electrically heated and has a 12.5 kilowatt capacity. It also contains four 14 gauge, A-1-grade heating coils. The tank's capacity is 200 litres. The electrical furnace utilised in this project was a PLC automatic temperature control with electrical resistance type, with a maximum operating temperature of 1,200°C and a control precision of 5 °C for molten aluminium. The electrical stirrer can stir at a maximum speed of 2000 rpm. The process involves melting around 0.450 kg of Al metal matrix alloy in a graphite crucible at 1150°C. Magnesium ribbons are used to improve wettability and ensure uniform distribution of strength. When the temperature reaches 1150°C, nanographene powder is gradually added to the molten metal (2% of the base metal's weight) after preheating it to 600°C. A mechanical stirrer covered with alumina and equipped with four coated blades is used to blend the reinforcement with the molten metal, creating a vortex. The molten metal is stirred forcefully at 500 rpm in an organogas atmosphere to prevent oxidation. To create cast specimens measuring 100 x 16 x 5 mm, the high-temperature molten metal is poured into a preheated cast iron mold at 800°C.

3.Mechanical Testings

The tensile strength of nanocomposites was tested using an electro-mechanical UTM 'Instron Make' at room temperature and a strain rate of 1×10^{-4} s^{-1}. Four specimens of Al (base metal), unreinforced S1, and each reinforced composite were fabricated for tensile testing. The results showed that friction stir processing (FSP) of the base matrix led to a 5.94% decrease in ultimate tensile strength, while smaller particle size increased tensile fracture strength[6]. The addition of nanoscale second-phase reinforcement particles increased the composites' elastic modulus, yield strengths, ultimate tensile strengths, and tensile failure stresses compared to an S1 matrix without reinforcement. The gradual improvement of properties from the unreinforced S1 matrix to the hybrid composite suggests equal refinement of grains and uniform distribution of dislocations in the matrix. The tensile properties of the unreinforced S1 specimen decreased from 53 m to 18 m while the average particle size decreased. The addition of nanographene to composite samples increases Vickers hardness from Hv 85.44 to Hv 167. However, the hardness decreases with higher amounts of graphene. Composite samples containing 4 weight % of nanographene have lower hardness than those containing 2 weight %. Al/GO composite samples have higher microhardness values than unreinforced Al alloy. The weight percentage of nanoparticles in the Al matrix can be raised by up to 1% without affecting hardness values. The hardness value of metal matrix composites is influenced by reinforcement amount and surface characteristics. The wear rate of hybrid composites made of aluminium-copper-nano graphene (MMCs) was studied using microns as a measure of wear. Sample 3 had the least wear, measuring 78.13 microns. The study found that higher graphene content percentages significantly reduced wear loss in MMCs. Hybrid composites made of Al + 4% Gr exhibited greater wear resistance than unreinforced S1 and other composites, regardless of load and sliding velocity circumstances. The graph depicting the wear rate over time demonstrates how nanoparticles affect the wear rate throughout periodic intervals. The flexural test examines the reaction of composite materials to deflection, using test samples created in accordance with ASTM A: 370 standards. Sample 2 exhibited higher flexural strength due to copper and nanographene content, while its deflection was smaller due to aluminium distribution. The result values of tests are shown in table 1 and represented graphically in figure 1.

SAMPLE	TENSILE STRENGTH	HARDNESS	AVERAGE WEAR	FLEXURAL STRENGTH
S1	99.74	85.44	201.49	213.21
S2	106.04	128.5	254.52	319.42
S3	130.16	90.5	78.13	281.32
S4	142.94	167	99.84	285.52

Table1: Samples and their respective test values

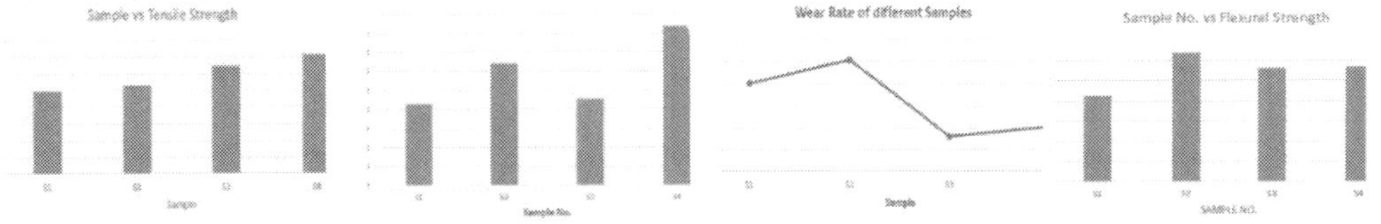

Sample vs Tensile Strength **Sample vs Hardness Number** **Sample Vs Wear rate** **Sample Vs Flexural strength**

Figure 1: Graphs showing samples vs Mechanical tests

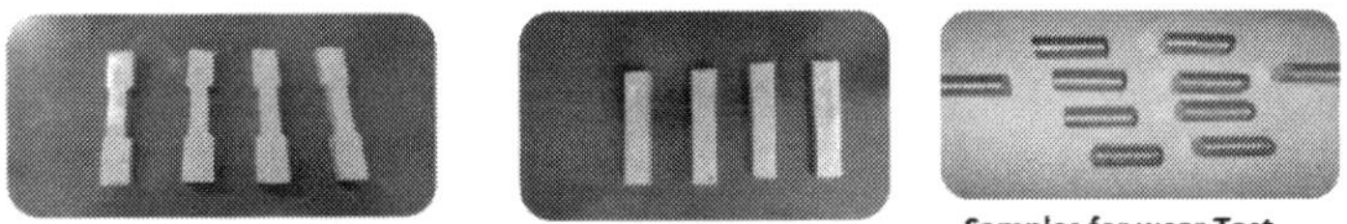

Samples for Tensile Test **Flexural Test Samples** **Samples for wear Test**

Figure 2 :Samples for Mechanical Tests

S1 S2 S3 S4

Figure 3: Microhardness Images of Composite Samples

4. Characterization techniques

4.1 Atomic Force Microscopy

Atomic force microscopy was used to study the composites' surface topography. AFM images of artificially synthesised al-nano graphene composites are shown in Figure4. The root-mean-square (RMS) roughness parameter, which was used to measure surface roughness, determined how smooth the surface was. The observed RMS value of the pure Al matrix is 763.1 nm. The surface roughness of the composites is reduced when reinforcement made of nanographene (160 nm) is added. The RMS value increases considerably further when GNPs (227.7 nm) are taken into account. As the amount of GNPs in the Al-nano graphene composites rises, the RMS value peaks at 313.3 nm at the highest compositions (Al-nano graphene, S4). FE-SEM pictures of the development of agglomerated nanographene and copper within the Al matrix., and the Al-nano graphene composites' behaviour of gradually growing surface roughness may be caused by the insertion of nanoscale carbonaceous reinforcement into the Al matrix. An increase in roughness was found to be a common trend in earlier studies [7,8].

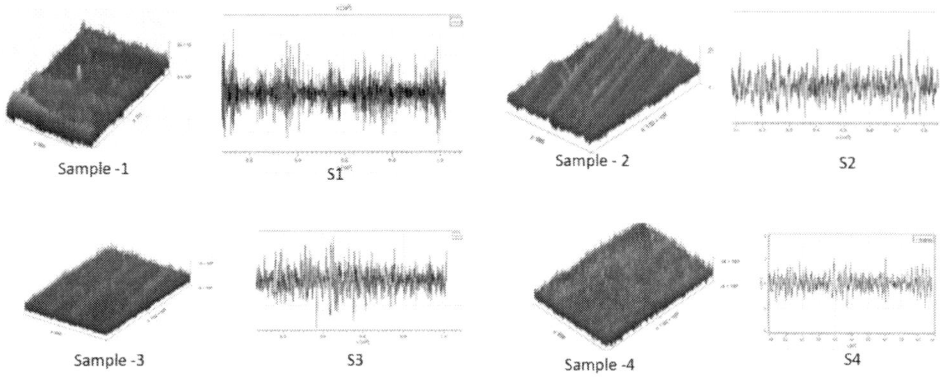

Figure 4: 2D and 3D atomic force microscope images

4.2 Brauner-Emett-Teller (BET)

It was found that nanographene has a surface area of 4.4086E+01 [m^2 g^{-1}] and a total pore volume of 3.3550E-02 [cm3 g-1]. Both of these figures are quite large. The BJH method was applied in order to carry out the computations necessary to determine the related pore size distribution data for nanographene. These results led to the conclusion that the pores are uniform and have a restricted pore size distribution that is centred at 3 nm. The findings of the BET isotherm plot indicate that the particles have a type II isotherm out of a possible 6, which indicates that they are capable of unrestricted mono-multilayer adsorption. This conclusion was reached as a consequence of the fact that the particles have a type II isotherm. The vast majority of the time, this phenomena takes place when adsorption is taking place on powders that are either nonporous or have powders whose diameters are larger than micropores. The inflection point takes place at a moment that is quite near to the end of the process of forming the first monolayer of adsorbed particles. Strong van der Waals interactions between the graphene sheets cause graphene particles to have a tendency to stack on top of one another. Because functionalization increases the likelihood that graphene particles would stack on top of one another, the BET surface area of nanographene decreases as a direct consequence of this behaviour. The setup required for performing BET analysis and analysis graph are shown in the figures 5 and 6 respectively.

Figure 5: BET Analysis Set-up

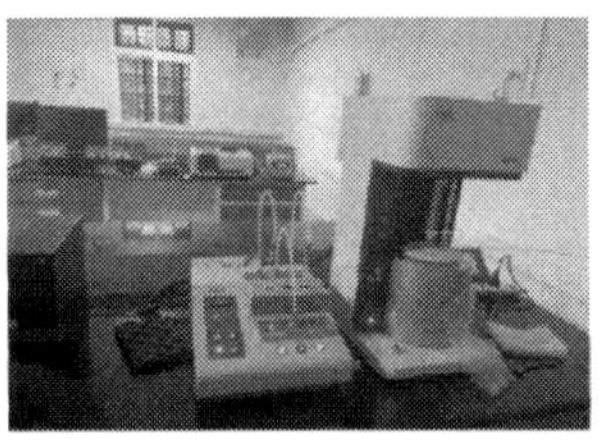

Figure 6: Graph showing BET Analysis

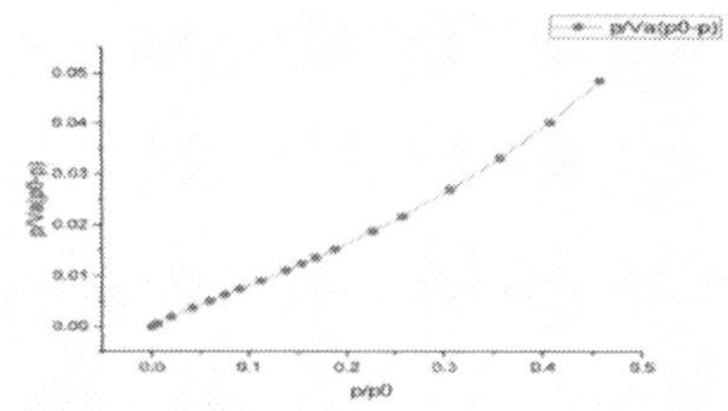

5. Tribological study

5.1 Pitting corrosion

According to ASTM guidelines, the studies on pitting corrosion were completed. 120 hours were spent submerging the Al/Nano GO composite samples in a 3.5% NaCl solution. Figure 18 displays the polarisation curves for various Al-nanographene composite compositions. For studying the corrosion behaviours of Al/nano GO composites under various circumstances, polarisation curves are absolutely necessary. Plotting the current density (I) vs. electrode potential (E) for a particular

IOP Conf. Series: Materials Science and Engineering 1300 (2024) 012028 doi:10.1088/1757-899X/1300/1/012028

electrode-electrolyte combination resulted in the polarisation curve. Table 2 gives estimates of the corrosion potential and density based on the polarisation curve. The polarisation curves show that when the amount of nanographene in the Al alloy grows, the current density rises and the corrosion potential falls. Corrosion resistance is found to diminish with rising corrosion current density. The experiment's findings showed that how effectively composites resist corrosion depends on the weight proportion of the reinforcements in the matrix. Figure 7 depicts the pitting corrosion setup.

The mean Ecorr and Epit values of the Al/nano GO composites are calculated using potential

SAMPLE	Ecorr (mV)	Epit(mV)	Passive Range(Epit – Ecorr)
S1	-1420	-890.03	529.97
S2	-1425	-720.56	704.44
S3	-1400	-690.7	709.3
S4	-1090	-778.02	311.98

dynamic techniques (Table 3).

Table 2: Pitting Corrosion Values of Samples

The resistance of a material can be calculated using the values of Epit, Ecorr, and the passive range. With this figure, the material's corrosion resistance rises.Despite graphene being a nobler metal than aluminium, which would be predicted to boost corrosion resistance and shift Ecorr values to higher positive values, no appreciable variations were detected in the solution potentials of the graphene-containing and graphene-free materials. However, there are noticeable improvements in passive range values between S2 and S3, as well as between alloys with and without reinforcement. Due to its substantially higher Epit value compared to the other materials examined, the S3's passive range value was the highest. According to these results, composite materials have better corrosion resistance than unreinforced alloys electrochemically and composites are electrochemically superior to unreinforced alloys. Copper-

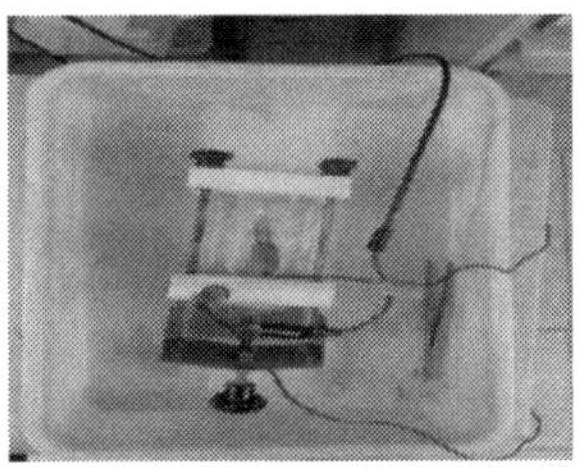

Figure 7: Pitting Corrosion Setup

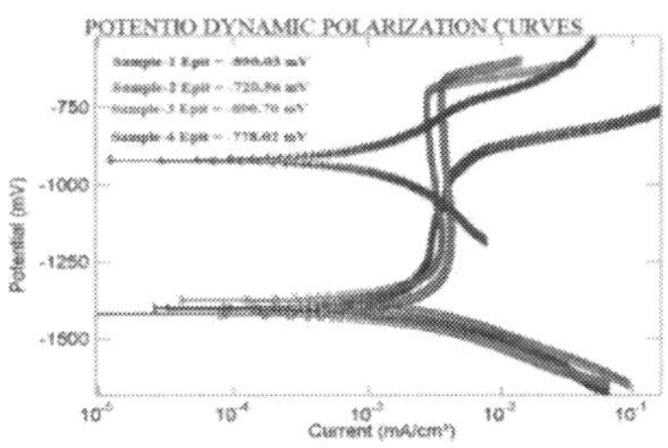

Figure 8: Polarization Curves of Samples

containing materials exhibit higher corrosion resistance than unreinforced materials, outperforming electrochemically. The addition of alumina fibres affects corrosion behavior by physically restricting the course of corrosion and resulting in a finer microstructure. This results in less micro segregation and inhomogeneities, preventing localised corrosion in the specimen[9,10,11]

Conclusion

- By changing the weight proportion of reinforcements, stir casting composites of Al, Copper, and Nano Graphene may be made successfully.
- Higher tensile strength is observed in composite S4 (AL 73% + 21%Cu + 6% Nano Graphene) due to the formation strong bond among Copper, nano Graphene and Alcompound.
- The addition of varied amounts of reinforcement to the base material is found to boost hardness; the composite S4 (AL 73% + 21% Cu + 6% Nano Graphene) achieves the highest value. This is because Cu/Nano Graphene is present in the Al compound.

- It has been discovered that, in comparison to other combinations, adding copper to a nanographene compound increases the attribute of flexural strength because a strong link is formed.
- It has been shown that the addition of various reinforcements to the base material increases wear resistance. The composite S3 (AL 82% + 14% Cu + 4% Nano Graphene) exhibits the lowest wear loss; this is because Cu/Nano Graphene is present in the Al compound.
- The composite S3 (AL 82% + 14% Cu + 4% Nano Graphene) has much higher corrosion resistance compared to other composites because graphene increases corrosion resistance, which is a crucial quality for maritime applications and automotive radiators, among other places.
- Characterization of samples of Al/Copper/Nano Graphene composites made by varying wt% of reinforcements are done by AFM, BET, SEM and EDAX.

References

[1] Anthony Xavior M, Prashantha Kumar HG. Processing and Characterization Techniques of Graphene Reinforced Metal Matrix Composites (GRMMC); A Review. Materials Today: Proceedings, Elsevier Ltd 2017, 3334–3341.

[2] Kumar V, Nagegowda KU, Boppana SB, Sengottuvelu R, Kayaroganam P. Wear behavior of Aluminium 6061 alloy reinforced with coated/uncoated multiwalled carbon nanotube and graphene. Journal of Metals, Materials and Minerals 2021; 31: 17–24.

[3] Wang X, Zhang L, Zhou X, Wu W, Jie X. Corrosion behavior of Al2O3-reinforced graphene encapsulated Al composite coating fabricated by low pressure cold spraying. Surf Coat Technol 2020; 386.

[4] Venkatesan S, Anthony Xavior M. Tensile behavior of aluminum alloy (AA7050) metal matrix composite reinforced with graphene fabricated by stir and squeeze cast processes. Science and Technology of Materials 2018;

[5] Radha A, Suresh S, Ramanan G, Mohanavel V, Emmy Prema C. Processing and characterization of mechanical and wear behavior of Al7075 reinforced with B4C and nano graphene hybrid composite. Mater Res Express 2019; 6.

[6] Bajargan G, Singh G, Sivakumar D, Ramamurty U. Effect of temperature and strain rate on the deformation behavior and microstructure of a homogenized AZ31 magnesium alloy. Materials Science and Engineering A 2013; 579: 26–34.

[7] Khan A, Matli PR, Nawaz M *et al.* Microstructure and mechanical behavior of hot extruded aluminum/tin-bismuth composites produced by powder metallurgy. Applied Sciences (Switzerland) 2020; 10.

[8] Ashraf PM, Thomas SN, Edwin L. Development of graphene–nanometre-sized cerium oxide-incorporated aluminium and its electrochemical evaluation. Applied Nanoscience (Switzerland) 2016; 6: 149–158.

[9] Mortensen A, Gungor MN, Cornie JA, Flemings MC. Alloy Microstructures in Cast Metal Matrix Composites. .

[10] Mortensen A, Michaud V. Infiltration of Fiber Preforms by a Binary Alloy: Part I. Theory. .

[11] Mortensen A, Cornie JA, Flemings MC. Columnar Dendritic Solidification in a Metal-Matrix Composite.

Switching Behavior of the Composite Low Dimensional Structural Hybrids of Carbon After UV Exposure

Sonia Saini[1,2,3], Kuntala Bhattacharjee[1,3,*] and Girish M Gouda[2]
[1]Indian Institute of Space Science and Technology (IIST), Thiruvanthapuram, 695 547, India
[2]Laboratory for Electro-Optics Systems (LEOS), ISRO, Bengaluru, 560 058, India
[3]Institute of Physics, Sachivalaya Marg, Bhubaneswar 751 005, Odisha, India
Corresponding author: kbhattacharjee@iopb.res.in , kuntala.iopb@gmail.com

Abstract. Low dimensional multi structural components of carbon are of great interest currently owing to their applications in developing flexible plastic electronics. In this work, we discuss emergence of different structural hybrids of carbon from as-prepared HiPCO SWCNTs subjecting them to oxidative acid purification and covalent functionalization process. Transmission electron microscopy (TEM) investigations reveal single wall carbon nanotubes (SWCNTs) serve as a building block to obtain multi wall carbon nanotubes (MWCNTs), graphene sheets (GS), carbon nano scrolls (CNS) structures which coexist in the sample. These structures when grafted with polymer binder and spray coated on Si substrate provide highly sustainable thin film coatings that are stable even in adverse space conditions [1]. These CNT and CNS based composite coatings are promising candidates for stray light control space applications exhibiting a low reflectance of the order of 2-3% in the visible spectral range [1]. Electrical switching behaviour of these films were investigated by capturing current(time) (I(t)) response, displaying promising switching character with distinct ON-OFF cycle. These coatings offer opportunities for the development of facile, cost-effective carbon-based devices without going into the nontrivial task of separating different the structures.

1. Introduction

The discovery of carbon nanotubes (CNTs) by S. Iijima stands as a pivotal event that revolutionized nanomaterials research on both fundamental and practical fronts [2]. In 1991, S. Iijima made a groundbreaking observation of multi-walled carbon nanotubes (MWCNTs) while conducting transmission electron microscopy (TEM) studies [3]. He characterized these structures as carbon allotropes with a distinctive hollow cylindrical-tube shape. In 1993, Iijima and Bethune independently identified single-walled carbon nanotubes (SWCNTs) [2]. Subsequently, the development of high-pressure carbon monoxide (HiPCO) SWCNTs introduced a novel synthesis method, spearheaded by Richard Smalley and his team at Rice University in 2000 [4]. This innovative approach represented a major advancement in the production of high-quality SWCNTs, known for their exceptional quality and structural consistency with fewer defects and a narrower diameter distribution compared to alternative synthesis techniques [5]. These remarkable properties, encompassing high electrical and thermal conductivity, coupled with impressive mechanical strength, have positioned HiPCO SWCNTs as highly coveted materials for applications in materials science, electronics, and nanotechnology [6], [7], [8]. Although production costs may be relatively higher, persistent research and development initiatives determinedly explore HiPCO SWCNTs' potential for next-generation materials and technologies, reinforcing their substantial impact in the realm of advanced materials. Recently, carbon nanomaterials have shown promising photo responsive behaviour using white light (WL) as a stimuli which is readily available and is a spontaneous choice for many applications owing to its wide spectral range. Photoactive structural forms of carbon, for example functionalized SWCNTs and graphene with azobenzene can induce switching of electrical conductivity when irradiated by UV and visible light, showcasing potential utility in sensors, memory and energy storage devices [9], [10], [11], [12], [13], [14], [15].

In this work, we explore the current switching characteristics of modified low-dimensional derivatives of carbon - polymer composite thin films under WL before and after treating with UV-Ozone exposure. We demonstrate light induced electrical switching behaviour of these films, an effect that is more prominent in the UV-Ozone exposed samples compared to the untreated ones. This research holds significant importance as it extends the application of these high-absorber thin film coatings as electrical switch with a well demonstrated light ON-OFF cycle, documenting their capability beyond stray light control applications in the realm of space domain reported by us earlier [1]. Further, this hybrid composite films were fabricated using facile, cost-effective methods of post-synthesis purification, functionalization and coating process along with the flexibility of choosing substrate as per requirement which make them robust for any practical implementations.

2. Experimental section

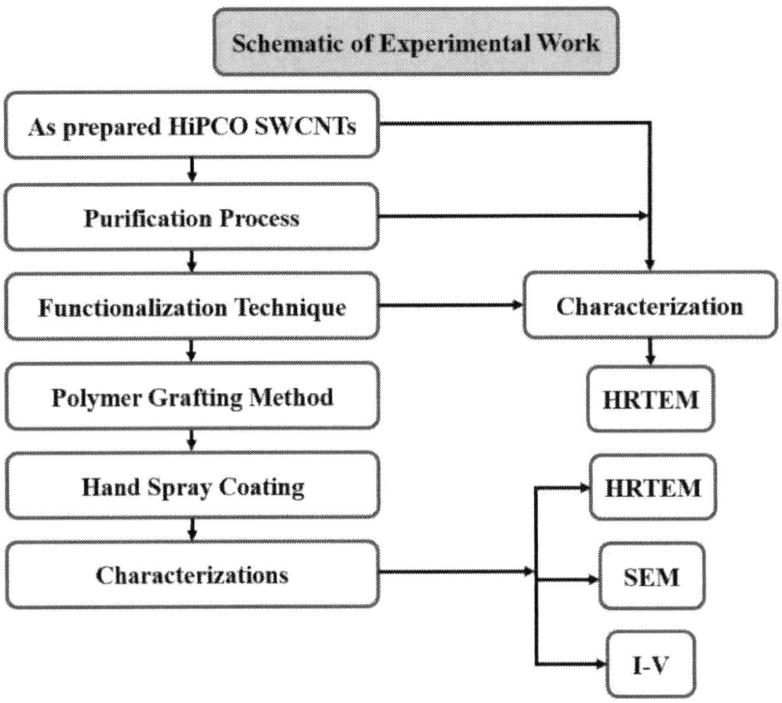

Fig. 1. Flow chart of the experimental process adapted in this work.

The as-prepared HiPCO SWCNTs required purification and functionalization before subjecting them for any applications. We employed an oxidative acid purification process to remove the embedded Fe catalyst particles and amorphous carbon contaminations from the initial HiPCO SWCNT samples. The purification procedure consisted of three primary stages, which included wet oxidation at 300°C, followed by a concentrated hydrochloric acid (HCl) wash, and concluded with annealing at 900°C in an inert atmosphere [16]. Functionalization process is conducted by combining concentrated H_2SO_4 and HNO_3 (3:1 volume ratio) and weightage amount 100 mg of purified sample in a beaker and stirring the mixture with a magnetic stirrer for 8 hours at 70 °C. Subsequently, the mixture was allowed to cool down to room temperature before rinsing with deionized water until a neutral pH is reached. The final functionalized product was obtained by filtering the solution with Whatman filter paper and drying at room temperature under vacuum [1]. To make the coated film, functionalized powder was blended with organic polyurethane binder and solvent, and subjected to ultrasonication for 60 minutes using a Q500 ultrasonic probe sonicator from Q-Sonica. This ensures creation of a well-dispersed solution, free from agglomerations. The resulting solution was employed for hand spray coating onto pre-cleaned Si substrate with a 2 μm SiO_2 layer. This whole process of purification, functionalization and coating is discussed elsewhere [1]. A visual representation of the experimental process is shown in Figure 1. The samples were investigated using transmission electron microscopy (TEM) by drop-casting the samples on Cu grids in an FEI-Titan Themis instrument operated at 300 kV, and a two-probe probe station was used to acquire the current-time (I-t) response under white light (WL) with an Agilent B2912A source/measure unit. Two-terminal electrodes of gold (Au) were deposited on the sample surface in a thermal evaporation chamber using shadow mask technique.

3. Results and discussion

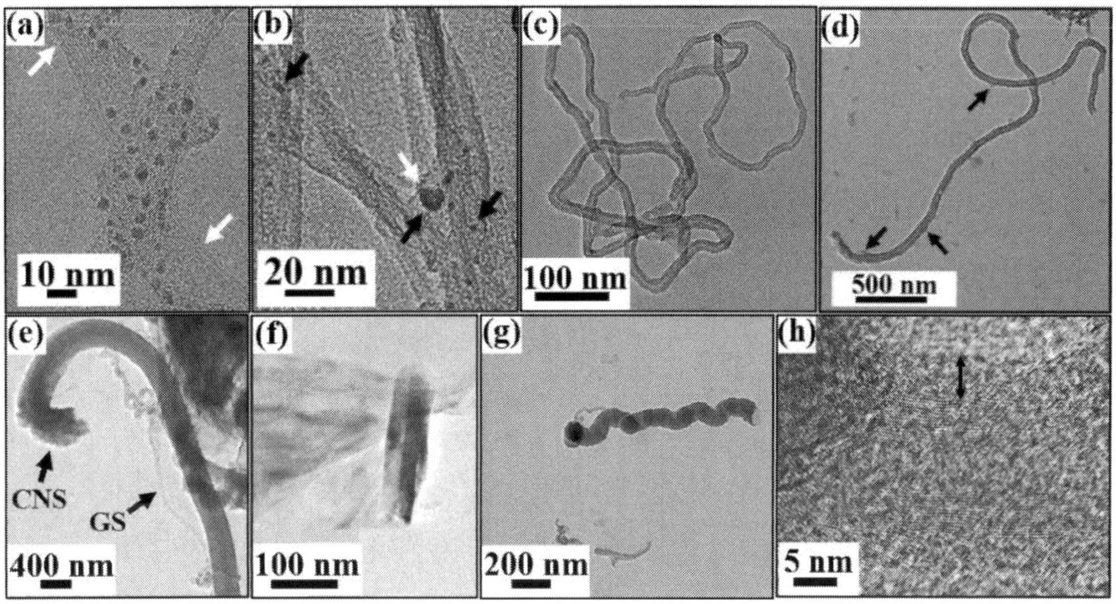

Fig. 2. (a) High-resolution TEM image of SWCNT bundles, revealing the presence of FeNP contaminants. (b) Purified SWCNTs with huge reduction in Fe catalyst particles. (c, d) MWCNTs in the specimen after purification and functionalization process. Attachment of functional groups can be seen in (d) indicated by arrows. (e, f) Presence of GS and CNS in the functionalized and coated sample. Rolling of GS to CNS is seen in (f). (g) Evidence of twisted structure. (h) TEM images of polymer-coated CNTs with a polymer thickness of around 5 nm.

Figure 2 shows high resolution TEM (HRTEM) micrographs obtained from the as-prepared, purified, functionalized and polymer grafted samples [12]. The embedded FeNPs in the initial HiPCO SWCNTs can be seen as dark spots in Figure 2(a). Figure 2(b) reveals huge reduction of the FeNPs after the purification process, though, there are some traces of the Fe particles still evident in the purified data. We observe emergence of various low dimensional structural forms of carbon like MWCNTs [Figure 2(c, d)], attachment of functional groups [Figure 2(d)], graphene sheet (GS) and carbon nanoscrolls (CNS) [Figure 2(e, f)] [15], [17] in the TEM investigations as a result of purification and functionalization process. Rolling of suspended graphene layers to form CNS can be clearly seen in Figure 2(f). During purification, FeNPs undergo oxidation and are converted into soluble ferric salts that can be washed away [18]. However, removal of these impurities is associated by generating localized stress, leading to the rupture of the domains around the FeNPs. Also, the FeNPs that are present in the purified sample, albeit in a small amount can act as a catalytic scissor and help in tearing or cutting the SWCNT bundles [Figure 2(b), shown as white arrow].

This unzipping or tearing process [17], [18] in turn, is further intensified when functional groups are introduced, causing separation of SWCNT bundles via ion intercalation [7], [19]. Consequently, various functionalized low-dimensional structures of carbon emerge due to exfoliation and merging of the initial CNT networks. Functionalization adds on various functional groups such as carboxyl (-COOH), hydroxyl (-OH), carbonyl (-C-O-C-) etc. which we have reported elsewhere [1], [16]. Figure 2(g) shows a twisted structure in the sample, while Figure 2(h) is an HRTEM image of a polymer grafted MWCNT with a polymer layer of thickness around ~5 nm.

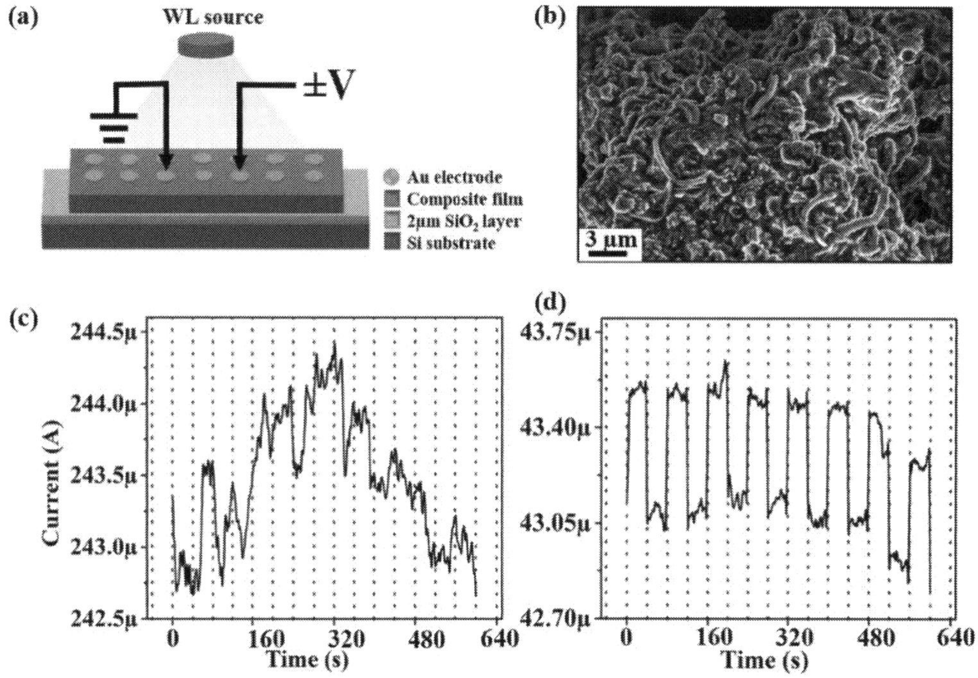

Fig. 3. (a) A schematic of the electrical characterizations. (b) SEM micrograph of the coated film. (c) I(t) photoresponse of the composite film using WL with a ON-OFF cycle of 40s before, and (d) after UV-Ozone treatment. The composite film shows rectified ON-OFF switching behaviour of the UV-Ozone irradiated sample.

We studied the electrical switching behaviour of these functionalized multi-structural all-carbon coated films under WL exposure by performing I(t) measurements and looking into its On-OFF states. These results are shown in Figure 3. Figure 3(a) is the schematic of the two probe measurement, while Figure 3(b) is the SEM micrograph obtained from the spin coated film on the Si substrate. I(t) behaviour under WL before and after irradiation with UV-Ozone exposure are shown in Figure 3 (c) and (d) respectively. These I(t) characteristics were acquired for a Voltage of 12 V. We observe a drastic improvement in the switching characteristic of the current with proper On-OFF cycle when the sample is treated with UV-Ozone radiation. While, CNTs are metallic or semiconducting in nature, the electrical photo-response behaviour of this conjugated multi-structural derivatives of carbon and polymer composite remains unreported. The study becomes particularly intriguing when after UV treatment, light induced electrical switching of the films show promising outcome. This experiment comprehends us about two key aspects that could be responsible for the observed results. Firstly, in the case of CNTs, there is a transition from π to π^* under UV exposure [20], [21]. When WL illuminates the composite film, it assists these excited electrons in transitioning to the conduction band more swiftly compared to the untreated sample without UV-Ozone treatment [20].

Secondly, during HRTEM analysis, we observe a coating of polymer layer on the CNTs, which thins down when exposed to UV radiation. Given that polymers are generally insulator [22], this thinning of the polymer shield over the CNTs facilitates the movement of charge carriers in the localized domains of the polyurethane network within the film [22], manifesting a promising light induced electrical photo-response and switching characteristics. Jintoku et al. [9] reported on SWCNT, graphene hybrid films of having better absorbance by controlling the intensity and irradiation time of UV light and are switchable after visible light irradiation and thermal treatment, making them potential candidates for optoelectronic devices and sensors.

4. Conclusion

This study aims to unravel the photosensing and electrical switching properties of composite films containing low dimensional structures of carbon that are derived from HiPCO SWCNTs. Previously, we have conducted a comprehensive analysis of these coated samples, highlighting their low reflectance and remarkable resilience in harsh space conditions [1]. Here, we show that subsequent to UV exposure, the coated films display rectified photosensing and switching behavior, introducing an intriguing additional feature. The enhanced switching characteristic is attributed to the electronic transitions and a thinner polymer coating of the UV-Ozone irradiated sample. It marks a notable step in customizing the optical and electronic properties of the carbon composite films without going into non-trivial process of separating individual structures, and enabling us to selectively modulate the properties by choosing and controlling the purification, functionalization, polymer grafting, and coating techniques. These developments hold significant importance in manufacturing photodetectors, black absorber coatings, and optoelectronic devices [5]. This may herald a fresh era in the realm of nanotechnology research associated with carbon nanomaterials.

5. Acknowledgement

K. B. sincerely thanks DST Mobility Scheme for offering essential professional and personal stability, as well as funding for comprehensive support.

6. References

[1] S. Saini, S. Reshmi, G.M. Gouda, A. Kumar S., S. K. V., K. Bhattacharjee, Low reflectance of carbon nanotube and nanoscroll-based thin film coatings: a case study, Nanoscale Advances. 3 (2021) 3184–3198. https://doi.org/10.1039/d0na01058h.

[2] M. Monthioux, V.L. Kuznetsov, Who should be given the credit for the discovery of carbon nanotubes?, Carbon. 44 (2006) 1621–1623. https://doi.org/10.1016/j.carbon.2006.03.019.

[3] S. Iijima, C. Brabec, A. Maiti, J. Bernholc, Structural flexibility of carbon nanotubes, Journal of Chemical Physics. 104 (1996) 2089–2092. https://doi.org/10.1063/1.470966.

[4] M.J. Bronikowski, P.A. Willis, D.T. Colbert, K.A. Smith, R.E. Smalley, Gas-phase production of carbon single-walled nanotubes from carbon monoxide via the HiPco process: A parametric study, Journal of Vacuum Science & Technology A: Vacuum, Surfaces, and Films. 19 (2001) 1800–1805. https://doi.org/10.1116/1.1380721.

[5] L. Golubewa, I. Timoshchenko, O. Romanov, R. Karpicz, T. Kulahava, D. Rutkauskas, M. Shuba, A. Dementjev, Y. Svirko, P. Kuzhir, Single-walled carbon nanotubes as a photo-thermo-acoustic cancer theranostic agent: theory and proof of the concept experiment, Scientific Reports. 10 (2020) 1–9. https://doi.org/10.1038/s41598-020-79238-6.

[6] M.M. Stylianakis, J.A. Mikroyannidis, E. Kymakis, A facile, covalent modification of single-wall carbon nanotubes by thiophene for use in organic photovoltaic cells, Solar Energy Materials and Solar Cells. 94 (2010) 267–274. https://doi.org/10.1016/j.solmat.2009.09.013.

[7] M.T. Martínez, M.A. Callejas, A.M. Benito, M. Cochet, T. Seeger, A. Ansón, J. Schreiber, C. Gordon, C. Marhic, O. Chauvet, J.L.G. Fierro, W.K. Maser, Sensitivity of single wall carbon

nanotubes to oxidative processing: Structural modification, intercalation and functionalisation, Carbon. 41 (2003) 2247–2256. https://doi.org/10.1016/S0008-6223(03)00250-1.
[8] S. Kruss, M.P. Landry, E. Vander Ende, B.M.A. Lima, N.F. Reuel, J. Zhang, J. Nelson, B. Mu, A. Hilmer, M. Strano, Neurotransmitter detection using corona phase molecular recognition on fluorescent single-walled carbon nanotube sensors, Journal of the American Chemical Society. 136 (2014) 713–724. https://doi.org/10.1021/ja410433b.
[9] H. Jintoku, Y. Matsuzawa, M. Yoshida, Switching the optical and electrical properties of carbon nanotube hybrid films using a photoresponsive dispersant as a dopant, RSC Advances. 8 (2018) 11186–11190. https://doi.org/10.1039/c8ra01447g.
[10] J. Zhao, J. Ma, Theoretical Designs of Photoresponsive Energy-Storage Materials Based on Attachment of π-Conjugated Molecules onto Sulfur-Doped Graphene, Journal of Physical Chemistry C. 120 (2016) 25131–25141. https://doi.org/10.1021/acs.jpcc.6b09247.
[11] S. Li, Y. Feng, P. Long, C. Qin, W. Feng, The light-switching conductance of an anisotropic azobenzene-based polymer close-packed on horizontally aligned carbon nanotubes, Journal of Materials Chemistry C. 5 (2017) 5068–5075. https://doi.org/10.1039/c7tc00142h.
[12] A.M. Kolpak, J.C. Grossman, Azobenzene-functionalized carbon nanotubes as high-energy density solar thermal fuels, Nano Letters. 11 (2011) 3156–3162. https://doi.org/10.1021/nl201357n.
[13] A.M. Kolpak, J.C. Grossman, Hybrid chromophore/template nanostructures: A customizable platform material for solar energy storage and conversion, Journal of Chemical Physics. 138 (2013). https://doi.org/10.1063/1.4773306.
[14] M. Kim, N.S. Safron, C. Huang, M.S. Arnold, P. Gopalan, Light-driven reversible modulation of doping in graphene, Nano Letters. 12 (2012) 182–187. https://doi.org/10.1021/nl2032734.
[15] L. Zhou, Q. Liu, X. Lv, L. Gao, S. Fang, H. Yu, Photoinduced triple shape memory polyurethane enabled by doping with azobenzene and GO, Journal of Materials Chemistry C. 4 (2016) 9993–9997. https://doi.org/10.1039/c6tc03556f.
[16] S. Saini, S. Reshmi, G.M. Gouda, K. Bhattacharjee, Emergence of carbon nanoscrolls from single walled carbon nanotubes: an oxidative route, Physical Chemistry Chemical Physics. 23 (2021) 27437–27448. https://doi.org/10.1039/d1cp03945h.
[17] J.L. Li, K.N. Kudin, M.J. McAllister, R.K. Prud'homme, I.A. Aksay, R. Car, Oxygen-driven unzipping of graphitic materials, Physical Review Letters. 96 (2006) 5–8. https://doi.org/10.1103/PhysRevLett.96.176101.
[18] X. Fan, Z. Peng, Y. Yang, H. Zhou, X. Guo, Atomic H-induced cutting and unzipping of single-walled carbon nanotube carpets with a teepee structure and their enhanced supercapacitor performance, Journal of Materials Chemistry A. 3 (2015) 10077–10084. https://doi.org/10.1039/c5ta01426c.
[19] C. Bower, A. Kleinhammes, Y. Wu, O. Zhou, Intercalation and partial exfoliation of single-walled carbon nanotubes by nitric acid, Chemical Physics Letters. 288 (1998) 481–486. https://doi.org/10.1016/S0009-2614(98)00278-4.
[20] K. Drozdowska, A. Rehman, A. Krajewska, D. V. Lioubtchenko, K. Pavłov, S. Rumyantsev, J. Smulko, G. Cywiński, Effects of UV light irradiation on fluctuation enhanced gas sensing by carbon nanotube networks, Sensors and Actuators B: Chemical. 352 (2022) 131069. https://doi.org/10.1016/j.snb.2021.131069.
[21] G.A. Rance, D.H. Marsh, R.J. Nicholas, A.N. Khlobystov, UV-vis absorption spectroscopy of carbon nanotubes: Relationship between the π-electron plasmon and nanotube diameter, Chemical Physics Letters. 493 (2010) 19–23. https://doi.org/10.1016/j.cplett.2010.05.012.
[22] T. Gurunathan, C.R.K. Rao, R. Narayan, K.V.S.N. Raju, Polyurethane conductive blends and composites: Synthesis and applications perspective, Journal of Materials Science. 48 (2013) 67–80. https://doi.org/10.1007/s10853-012-6658-x.

UV-assisted green synthesis of silver nanoparticles and their characterizations

K Kavyasree, H Basappa, J M Lavita, S B Suma and Y Sangappa*
Department of Studies in Physics, Mangalore University-574199, Karnataka, India
*Corresponding author: *sangappay@gmail.com*

Abstract. Silver nanoparticles are receiving discernible attention because of their potential applications in different fields of science and technology. However, chemical synthesis methods leave behind harmful byproducts. Here, we report an alternative method for the rapid synthesis of silver nanoparticles through radiation-assisted green synthesis. Aqueous leaf extract of *Ocimum sanctum* is used as a reducing and stabilizing agent with silver nitrate solution under UV-B irradiation. UV-visible spectroscopy (UV-Vis), transmission electron microscopy (TEM), energy dispersive x-ray spectroscopy (EDX), photoluminescence spectroscopy (PL) and fourier transform infrared spectroscopy (FTIR) collectively revealed that bio reduced silver nanoparticles were polydisperse, polycrystalline, spherical in shape, had an average diameter of 40.1 nm, and were stable more than six months. These observations suggest that radiation-assisted green synthesis is a quick and eco-friendly method for the large-scale production of stable silver nanoparticles without any harmful byproducts.

1. Introduction

Silver nanoparticles (AgNPs) are one of the noble metal nanoparticles attaining tremendous applications in many branches of science and technology, including photovoltaics, bioengineering and biomedical science, electronics, textile industrial science, and environmental science [1–2]. These are due to the unique physical, chemical, mechanical, optical, and electrical properties of AgNPs [3-4]. The AgNPs can be synthesized through different approaches, such as chemical reduction, sonochemical, microwave-assisted, and physical [5]. However, these approaches include several disadvantages that may have harmful effects on the environment [6]. Thus, to avoid harmful reagents, the safest alternative is biological synthesis, as this method leads to less toxicity, is low-cost, and is more eco-friendly [7-8]. Biological synthesis of AgNPs can be achieved using various precursors like algae, microorganisms, lichens, and plant-based products [9]. These biological entities are suitable for the synthesis of nanoparticles at ambient temperature and a regulated pH. The biochemicals present in them serve a dual role as reducing and stabilizing agents [10]. Among these entities, plant-based synthesis has more benefits, as the production cost is lower and the purification processes are also simple [11].

Electromagnetic radiation is another key parameter that affects synthesis rates. Different types of radiation, namely microwave, visible, UV, combinations of UV and visible, and gamma rays, are used for the synthesis of AgNPs [12-16]. The use of radiation for the production of semiconductor and nano-metal particles has become very popular in chemistry, but the role of irradiation in nanoparticle biosynthesis has not yet been systematically studied [17]. Studies have reported the synthesis of AgNPs under microwave irradiation using *Musa paradisiaca, Ficus racemosa, Elephantopus scaber,* and *Biophytum sensitivum* leaf extracts [18-21]. AgNPs were also successfully synthesized through sunlight irradiation utilizing the leaf extracts of *Nagia nagi, Sida retusa, Jasmine subtriplinerve blume,* and *Polygonatum graminifolium* [22–25]. Bulent et al. employed combinations of UV and visible radiation for the synthesis of AgNPs using *Ficus carica* leaf extract [26]. Pradeep et al. used *Lantana camara* leaf extract in the UV-assisted production of AgNPs [27].

Ocimum sanctum (OS) is a diploid plant belonging to the Lamiaceae family, commonly known as Tulasi. It has been regarded as a divine herb in the Indian Ayurvedic system of traditional medicine. Tulasi extract is widely used in tonics as immune stimulators, anti-oxidants, neuromodulators, and for the treatment of various types of heart diseases, poisoning, inflammation, malaria, the common cold, and cough. Recently, studies reported the synthesis of AgNPs using Tulasi leaf extract, and eugenol, quercetin, luteolin, apigenin, naringin, and kaempferol are a group of flavonoids found in the extract that function as reducing and capping agents [28–29].

Considering all these things, the present study was carried out to synthesize simple, low-cost, eco-friendly, and easily renewable AgNPs using Tulasi leaf extract under UV-B irradiation.

2.Experimental

2.1 Materials: The chemicals silver nitrate (>99%) and sodium hydroxide pellets, that were employed in this study were acquired from Sigma Aldrich and used directly without additional purification. Fresh and healthy OS leaves were gathered from Bela village in Kasaragod, Kerala, India. To remove particulate debris, leaves were washed numerous times with tap water, finally being washed with double-distilled water. To remove excess water from their surface, leaves were shade dried for thirty minutes.

2.2 Preparation of OS leaf extract: 10g of weighted leaves were pasted with 5mL of double-distilled water using a mortar and pestle. The solution was diluted up to 300mL, boiled for thirty minutes, and then filtered through Whatman No.1 filter paper. For the synthesis of AgNPs, clear OS leaf extract is utilized.

2.3 Synthesis of OS-AgNPs: 0.0016g of silver nitrate was dissolved in 10mL of double-distilled water to prepare a solution of 1mM silver nitrate [40]. 2mL of leaf extract was combined with 4mL of freshly made 1mM silver nitrate solution. By adding 0.1N NaOH solution, pH of the mixture was maintained at 9. The reaction mixture was exposed to UV-B (wavelength=280-315nm) irradiation for the quick and complete conversion of Ag^{+} to Ag^{0}.

2.4 Characterizations of synthesized AgNPs: The formation of AgNPs was confirmed by the Shimadzu UV-1800 (Japan) UV-visible spectrophotometer. Further, they were characterized by the Titan Themis 300kV from FEI transmission electron microscope and the IR prestige-21 FTIR instrument. The PL study was carried using HITACHI Japan model F7000 spectrophotometer. The Oxford instrument JSM6390LV scanning electron microscope with energy dispersive spectrum detector is used for EDX analysis.

3. Results and Discussions

3.1 UV-Vis. absorbance analysis: With the addition of silver nitrate solution and UV-B irradiation, leaf extract underwent a color change from clear brown to black. The reason for this is surface plasmon resonance (SPR) phenomenon [30]. It generated a peak having center close to 419nm (figure 1). The widening of the SPR curve denotes the creation of polydisperse nanoparticles [31].

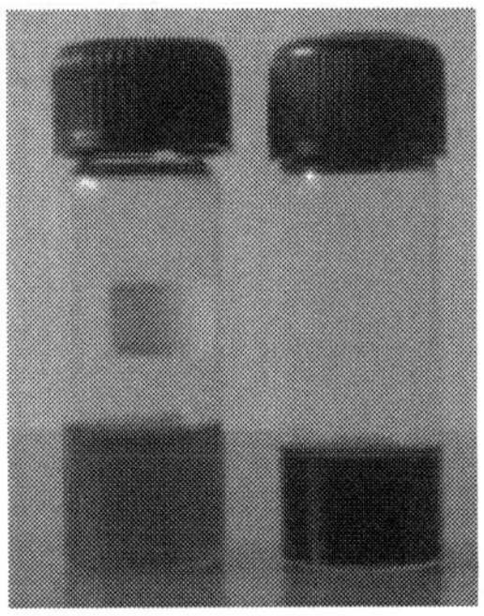

Figure 1. Leaf extract and synthesized OS-AgNPs solution.

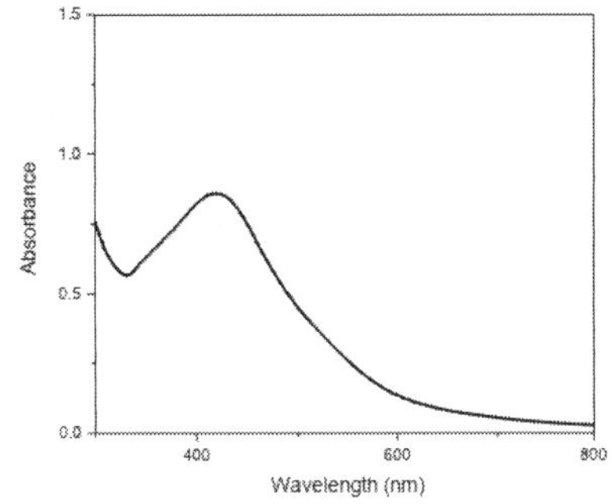

Figure 2. UV-Vis. absorbance spectrum of AgNPs.

3.2 TEM analysis: TEM analysis revealed that biosynthesized AgNPs were spherical in shape, almost non-agglomerated, and polydisperse, having an average diameter of 40.1nm (figures 3 and 4). According to the selected area electron diffraction (SAED) pattern, synthesized AgNPs were polycrystalline, and the calculated d-spacing for five concentric circles matches JCPDS file number 00-004-0783 (figure 4).

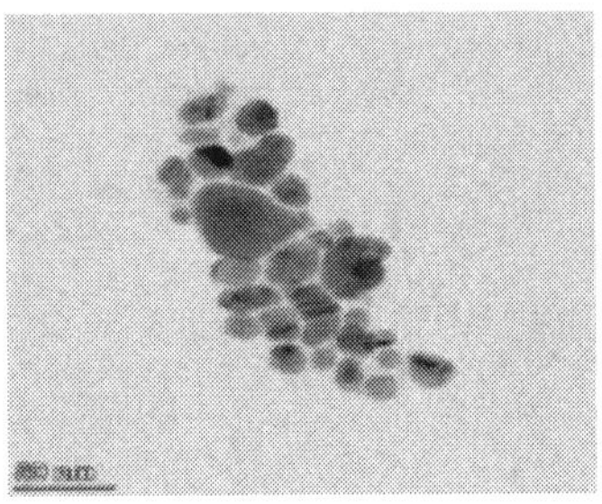

Figure 3. TEM image of OS-AgNPs.

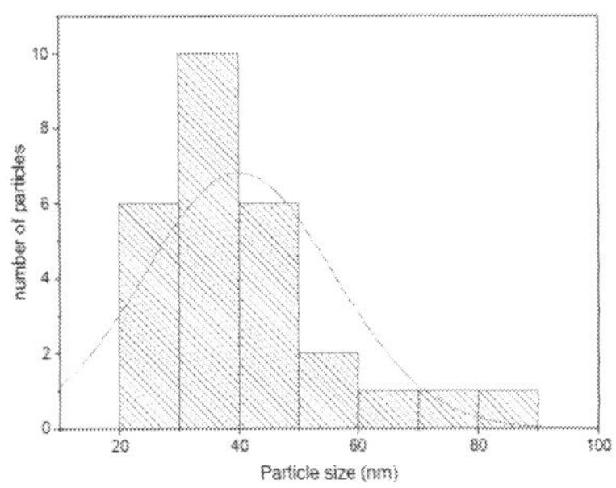

Figure 4. Particle size histogram.

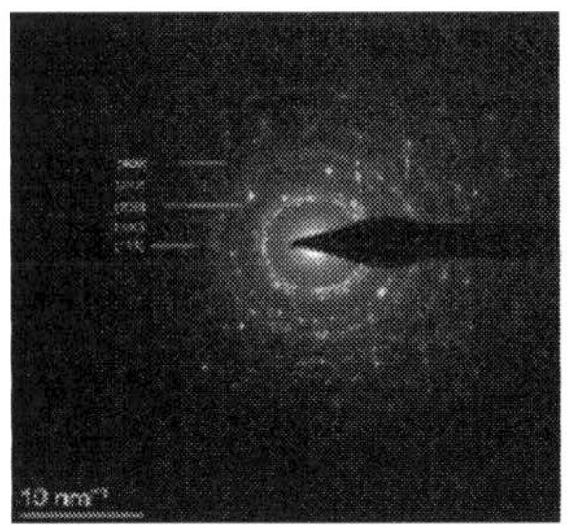

Figure5. SAED pattern.

3.3 EDX analysis: A spectral signal in the silver region was found by EDX analysis (figure 5). Silver made up 0.40 of the total atomic percentage. Additionally, a spectral signal in the oxygen region is noticed. This is because extracellular molecules that were adsorbate on the surface of AgNPs and were derived from the leaf extract [32]. Also, a peak for carbon is detected because of the grid utilized for the analysis.

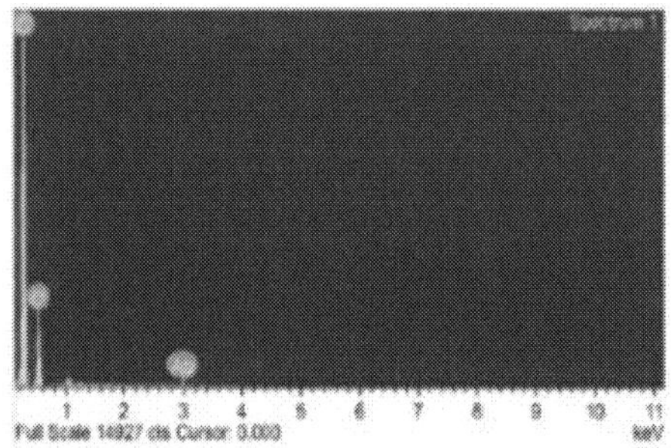

Figure 6. EDX spectrum of OS-AgNPs

3.4 FTIR and stability analysis: Figure 7. denotes FTIR spectrum of OS-AgNPS. Due to alcohol and phenol O-H stretching vibrations, the FTIR spectrum displayed a peak at 3424.3cm^{-1}. Peaks detected at 1616.3cm^{-1} and 1390.3cm^{-1} are because of the bending of -C-N- and stretching of -C-H- bonds respectively. The spectrum also revealed two additional peaks at 906.28cm^{-1} and 756.62cm^{-1}. These results are from out-plane bending of aromatic -CH bonds [33].

It has been investigated the stability of the OS-AgNPs colloidal solution. Even after six months of storage at room temperature, no color or visual aggregate has been seen. Additionally, there was no change in the SPR peak position as indicated by the absorbance spectrum (figure 8). Thus, AgNPs formed using this approach are extremely stable.

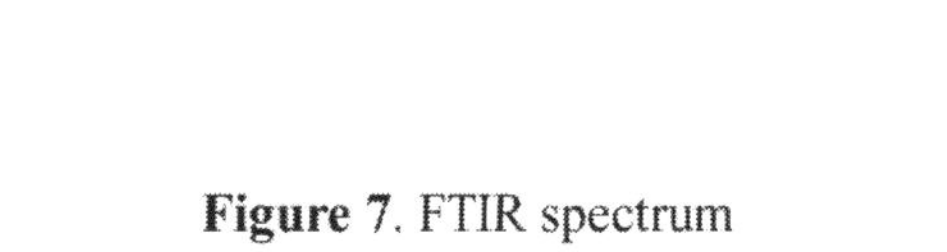

Figure 7. FTIR spectrum

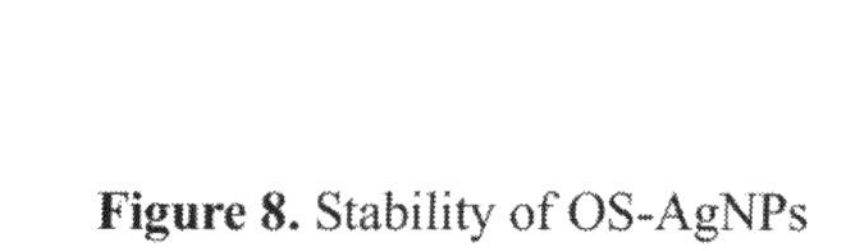

Figure 8. Stability of OS-AgNPs

4. Conclusion

In summary, we have demonstrated a simple method for the quick synthesis of AgNPs employing an aqueous leaf extract of *Ocimum sanctum* as the reducing and stabilizing agent. When compared to chemical synthesis, the remarkable repeatability of these nanoparticles without the need for extra chemicals will be advantageous. The findings showed that radiation-assisted green synthesis provides an eco-friendly, efficient, and alternative method for producing AgNPs on a large scale without creating hazardous by-products

References

[1] Kalfagiannis N, Karagiannidis P, Pitsalidis C, Panagiotopoulos N, Gravalidis C, Kassavetis S, Patsalas P and Logothetidis S 2012 *Solar energy materials & solar cells* **104** 165-174

[2] Chanhua L, Xuipei Y, Hongyan Y, Zaide Z and Dan X 2007 *Sensors* **7(5)** 708-718

[3] Toudert J, Fernandez H, Babonneaue D, Camelio S, Girardeau T and Solis J 2009 *Nanotechnology* **20** 475-705.

[4] Zhang X, Sun H, Tan S, Gao J, Fu Y and Liu Z 2009 *Inorganic Chemistry Communications* **100** 44-50.

[5] N. Esmaili, P. Mohammadi, M. Abbaszadeh, H. Sheibani 2020 *Applied organometallic chemistry* **34** 5547

[6] Ragaa A Hamouda, Mervat H Hussein, Rasha A Abo-elmagd and Salwa S Bawazir 2019 *Scientific Reports* **9** 1-17 13071

[7] Vignesh V, Kulandaiyesu F Anbarasi, Karthikeyeni S, Sathiyanarayanan G, Subramanian P andThirumuruga *Colloids and Surfaces A: Physicochemical and Engineering Aspects* **439** 184-192.

[8] Kalimuthu K, Ramkumarpandian S Babu, Venkataraman D, Bilal M and Gurunathan S 2008 *Colloids and Surfaces B: Biointerfaces* **65** 150-153.

[9] Sai A Dahoumane, Mourad M, Kushlani W, Carlos D Filipe, Sicard C, Bazylinski D and Jeffryes C 2017 *Green Chemistry* **19(3)** 552-587.

[10] Debashish B *et al.* 2020 *Applied organometallic chemistry* **34(5)** 5597.

[11] Kannan B Narayanan and Natarajan S 2010 *Advances in colloid and interface science* **156(1-12)** 1-13

[12] Fu-Ken L, Fu-Hsiang K, Pei-Wen H, Chien-Hou Wu and Tieh-Chi C 2005 *J. Chromat. A* **1062 (1)** 139–145.

[13] Dan R Clary, Mills G 2011 *J. Phys. Chem. C* **115 (30)**14656–14663

[14] Khan E, Langford S, Dickinson J, Boatner L and Hess W 2009 *Langmuir* **25 (4)** 1930–1933

[15] Lilia C. Courrol, Flavia R deOliveirsilva and Laercio G 2007 *Coll. Surf. A: Physicochem. Eng. Aspects* **305 (1)** 54–57

[16] Temgire M and Joshi S *Rad. Phys. Chem.* **71 (5)** 1039–1044

[17] Nim T, Azam I, Mahdavi S and Reza-Esmaili M 2005 *Phys. E: Low-Dimen. Syst. Nanostruct.* **30 (1–2)** 114–119.

[18] Raghvendra N, Leena V Hublikar, Patil S M and Preetam B 2021 *Journal of water and environmental nanotechnology* **6** 49-61

[19] Maurya A K, Upadhyaya S, Gupta A, Jeva K Ebnezer and Rubina L 2021 *Res. J. Chem. Environ.* **25** 111-119

[20] Sijo F, Siby J, Koshy E P and Beena M 2017 *Nanomedicine and Biotechnology* **4** 1-10.

[21] Siby J and Beena M 2015 *Journal of Molecular Liquids* **204** 184-194.

[22] Liu Y, Hussain M, Memon H and Yasin S 2015 *Dig. J. Nanomater. Biostruct.* **10** 1019-24.

[23] Sooraj M P, Archana S Nair and Vineetha D 2021 *Chemical Papers* **75** 351-363

[24] Paung A N, Hong P P, Trung D B, Nguyen V M and Nhat L D 2021 *Journal of Plant Biochemistry and Biotechnology* **31** 1-4

[25] Vanshika R, Anjna S, Vijay P Bhatt, Raghavendra P Singh and Indresh K 2020 *Materials Today: Proceedings* **29** 911-916

[26] Bulent U, Haluk M Turkdemir, Ahmet C and Ahmet M 2015 *Spectrochemica Acta Part A: Molecular and Biomolecular Spectroscopy* **135** 153-161.

[27] Pradeep K Singh, Kirti B, Parul D and Asmita P 2015 *RSC advances* **5(31)** 24513-20

[28] Shrinath B Manjeshwar, Rosmy J, Ramdas T Karadka, Sunitha V and Neeta R Bhat 2013 *Nutr. and Cance*r. **65** 26–35

[29] Giriraj T, Yadav B L, Jyoti C, Manoj J and Chetana S 2020 *Biochemistry and Biophysics Reports* **24** 100848

[30] Mulvaney P 1996 *Langmuir* **12** 788–800.

[31] Philip D and Unni C 2011 *Phys. E* **43** 1318-22.

[32] Malapermal V, Joyce N Mbatha, Robert M Gengan and Anand K 2015 *Adv. Mater. Lett.* **6(12)** 1050-5

[33] Jaya M Jacob, Merin S John, Aleena J, Abitha P, Sruthy S Kumar, Reju R, Suganthy N and Arivalagan P 2019 *Mater. Res. Express* **6** 045401

IOP Conf. Series: Materials Science and Engineering 1300 (2024) 012031 doi:10.1088/1757-899X/1300/1/012031

Performance of Hybrid Nanofluids of Mineral Oil and Vegetable oil based Ester

Prabhat Kumar Maiti*, Manish Kumar and Moumita Naskar
Central Power Research Institute, Regional Testing Laboratory, Kolkata 700091, India
*Email:maiti@cpri.in

Abstract. The prosperity and economic development of a nation depend on its capability to maintain reliable and high-quality electric power management. It is achieved through the availability of extremely reliable ancillary electrical equipment. A power transformer is one such piece of crucial equipment that transfers electricity from one circuit to another without altering the frequency. Mineral oil has been conventionally used as a liquid dielectric in transformers since the last century. However, in recent days, vegetable oil-based esters have emerged as a favourable alternative to mineral oils due to their biodegradability and sustainability. The introduction of nanotechnology led to massive studies to enhance the quality of liquid dielectrics. Further, the addition of more than one nanoparticle to the base fluid has broadened the scope of nanofluids' property enhancement. This study focused on the assessment of the dielectric properties of hybrid nanofluids consisting of ZnO and reduced graphene. Mineral oil and vegetable-based transformer oil were chosen as base fluids to prepare the nanofluids. The mass of the added nanoparticles was 0.01%. The ratio between the nanoparticles was selected as 1:1. The hybrid nanofluids were characterized for breakdown voltage viscosity at temperatures ranging from 40 °C to 100 °C, interfacial tension, tan delta, and resistivity. The overall properties of hybrid nanofluids were found to be improved due to the inclusion of ZnO and reduced graphene.

Key words- mineral oil, vegetable ester oil, blended oil, nanofluids

1. Introduction

It is essential to improve the performance of power transformers for the upgrade and advancement of electrical power systems. One of the most common varieties of power transformers is the oil-filled transformer. Transformer oil serves as both an electrical insulator and a thermal coolant. Therefore, enhancing the power transformer's thermal and dielectric qualities has a positive impact on its overall performance. Mineral oils (MO), being inexpensive, having a high thermal capacity, and having a good pour point at low temperatures, have been used in power transformers for a long time. MOs are less sustainable for extended periods due to a number of drawbacks. For instance, these materials are derived from a non-renewable source, are low-eco-friendly and have a significant fire hazard [1]. It is reported that the biodegradability of MO is less than 30 % [2]. A significant increase in the demand for petroleum products is also a result of society's increased industry and motorization. The aforementioned issues have drawn more attention to the need to find alternatives that can be prepared from resources that are abundant in nature and may provide the best long-term chances.

Vegetable oil esters (VO) are regarded as a good replacement for mineral oil. These insulating liquids are derived from both seeds and flowers. VOs are the subject of several studies in an effort to use them as insulating oils in transformers and create eco-friendly environments [3]. VOs are thought to be more nature-friendly liquids due to their high biodegradability (>95 %), high flash and fire points (both >300 °C), and lesser flammability [4]. In addition, compared to MOs, these VOs absorb more moisture [5]. The viscosity of MO is lower than that of VO. Moreover, the VOs show lower oxidation stability. Therefore, blending VO and MO is another area of research interest [6].

Numerous researchers have focused on the enhancement of the properties of liquid dielectrics in light of the technological qualities outlined above and ecological protection considerations [7]. Nanofluid (NF) is a colloidal liquid containing suspended nano-scale particles in a base liquid such as water, ethylene glycol, or oil [8]. Recent literature makes it clear that many scientists throughout the world have been interested in the addition of nanoscale particles to VOs. Some of the nanoparticles, according to reports and consensus, are able to change the characteristics of VOs to improve their performance in transformers. Therefore, the significance of nanoparticles in modifying the characteristics of VOs has received particular attention. It should be noted that numerous assessment reports are available in the literature focused on the use of nanoparticles to enhance the characteristics of insulating oil.

The nanoparticles are classified depending on their chemical, physical, and electrical properties, such as metal oxide, nitrides,carbides, non-conducting, semiconducting, conducting, spherical, tube, layer, etc. [9]. Even though there has been extensive study on single-particle-based NFs, enhancing their thermophysical characteristics and stability remains a significant problem. Utilising the hybridization process to create hybrid nanofluids is one way to get around this constraint.

Hybrid NFs are prepared by suspending two or more different types of nanoparticles in a base fluid. Hybrid NFs based on water or ethylene glycol are the most commonly reported [10]. However, much attention has not been given to studies of VO-based hybrid NFs. A large number of metal oxides, such as Al_2O_3, SiO_2, TiO_2, Fe_2O_3, and CuO are investigated for their performance in VOs as compared to ZnO. Reduced graphene-based bio-nanofluids appear to be one of the potential liquid dielectrics. Therefore, adequate research is needed for the development of this material [11–12]. A lower viscosity is preferred for a good liquid insulator. It is reported that the addition of NPs resulted in escalating viscosity as compared to the base fluid [1]. So, the effect of the mixing of nanoparticles with an ester on its viscosity property is required for detailed studies.

Therefore, it is proposed to include blended VO and MO in this study. This study focused on the physicochemical and electrical characterization of ZnO-reduced graphene-based blended NFs of VO and MO. The mass of the added nanoparticles was 0.01%. The ratio between the nanoparticles was selected as 1:1. The performance of hybrid NFs was assessed by estimating their breakdown voltage, viscosity, interfacial tension, tan delta, and volume resistivity.

2. Results Materials and methods

The base fluids selected for the study were mineral-oil based transformer oil (MO) and vegetable oil-based ester oil (VO). The VO used in this work was obtained from M/s. Savita Oil Technology, India. This VO is commercially known as Biotransol-HF, and it conforms to IEC 62770. Table 1 describes the characteristics of the base fluids VO and MO. The nanoparticles picked for the work were ZnO of size less than 50 nm and reduced graphene (RG). The mass of the added nanoparticles was 0.01%. The ratio between the nanoparticles was selected as 1:1. The details of the composition of the samples prepared for the work are shown in table 2. Nanofluids for this study were prepared following a two-step method. The required amount of nanoparticles was added to the base fluids. The first step for the proper dispersion of the mano particles was performed through magnetic stirring of the composite fluid for 2 hours. The second step of the process of further homogenization was conducted using an ultrasonic bath for 1 hour. The prepared samples were kept in an air oven for 48 hours for drying. The hybrid NFs were characterized for breakdown voltage viscosity at temperatures ranging from 40°C to 100°C, interfacial tension, tan delta, and resistivity.

Table 1. Oil specifications.

Parameter	VO	MO
Appearance	Light green	Transparent
Density @ 20°C (kg/m^3)	918	837
Kinematic Viscosity@40°C (m^2/s)	3.4385×10^{-5}	8.124×10^{-6}
Interfacial tension, @ 27°C (mN/m)	24.79	40.62
Flash Point (°C)	>280	170.5
Pour Point (°C)	- 27	- 36
Acidity (mg KOH/g oil)	0.027	0.0092
Electric Strength (kV)	62.2	56.4
DDF (Tan delta) at 90°C	0.0117	0.0010
Specific Resistance (Resistivity) (GΩm)	15.55	3290

Table 2. Sample details.

Samples	Compositions
S1	100 % VO
S2	100 % MO
S3	100 % VO + 0.01 % ZnO
S4	100 % MO + 0.01 % ZnO
S5	100 % VO + 0.01 % Reduced graphene (RG)
S6	100 % MO + 0.01 % Reduced graphene (RG)
S7	100 % VO + 0.005 % ZnO + 0.005 % Reduced graphene (RG)
S8	100 % MO + 0.005 % ZnO + 0.005 % Reduced graphene (RG)
S9	50 % VO + 50 % MO + 0.005 % ZnO + 0.005 % Reduced graphene (RG)

3. Results and discussions

The inner machinery of a transformer becomes hot during its operation as a result of energy losses in the form of hysteresis loss, eddy current loss, and copper loss. The insulating oil acts as a coolant to dissipate the excess heat. The viscosity of oil indicates its resistance to flow. A dielectric fluid with low viscosity is a preferred material as it provides higher heat dissipation through an enhanced rate of circulation. Therefore, assessing liquid insulators for their viscosity is essential. Figure 1 shows the viscosity of the samples measured at various temperatures, such as 40 °C, 60 °C, 80 °C, and 100 °C. It is observed from figure 1 that at 40 °C, the reduction in viscosity was significant in VO-ZnO nanofluids, with a decrease of about 4*E-6 m^2/s. The decrease in viscosity for VO-RG nanofluid and VO-ZnO-RG hybrid nanofluid was about 2*E-6 m^2/s. However, there was an insignificant effect on MO-based nanofluids and hybrid nanofluids. The viscosity of VO and its nanofluids appeared to have higher values as compared to MO and its nanofluids. Among the VO-based nanofluids, the ZnO-RG hybrid nanofluid showed a lower viscosity up to about 75 °C. The viscosity of blended and hybrid nanofluids (S9) was in between the values of VO-based and MO-based nanofluids, indicating the possibility of the usefulness of this nanofluid.

The specific resistance is the difference in DC resistance between two sides of an oil block that is one cm^3 in size. A substance with higher resistance has better insulation properties. The nanoparticles that are capable of tapping larger numbers of electrons are better for the improvement of the nanofluid. Figure 2(a) shows the specific resistance (SR) of VO and its nanofluids, whereas figure 2(b) shows the SR for MO and its nanofluids at 90°C. The resistivity of MO and its nanofluids was higher as compared to VO and its nanofluids. It has been observed that the resistivity of VO-RG was much better than that of the other VO-based nanofluids, though it was still higher in the base fluid. On the other hand, a significant improvement in SR was seen in MO-based nanofluids. The SR value was almost double in MO-RG (S8) nanofluid as compared to MO.

IOP Conf. Series: Materials Science and Engineering 1300 (2024) 012031 doi:10.1088/1757-899X/1300/1/012031

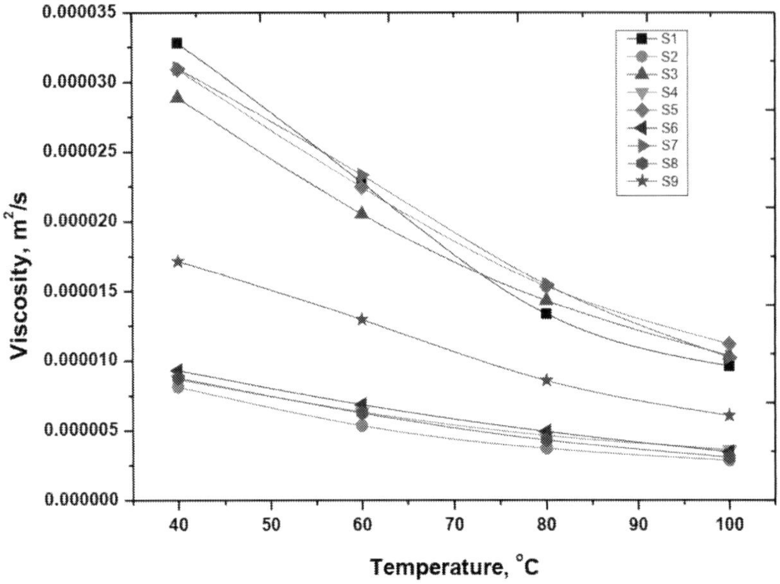

Figure 1. Viscosity of the samples at the temperature range from 40°C to 100°C.

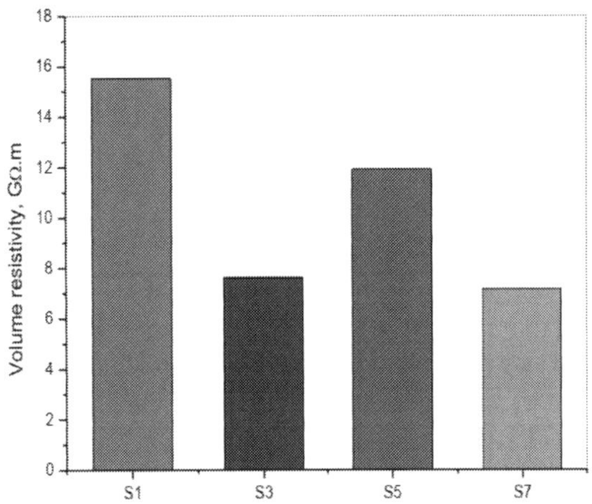

Figure 2(a). Specific resistance (SR) of VO and its nanofluids at 90°C.

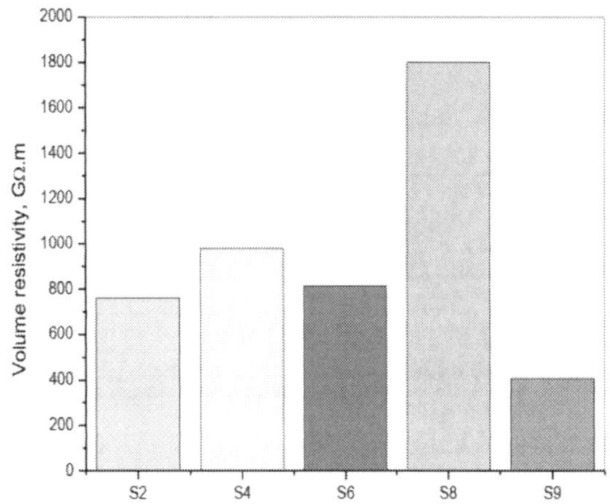

Figure 2(b). Specific resistance (SR) of MO and its nanofluids at 90°C.

The dielectric dissipation factor (DDF) and dielectric constant of the specimen are shown in table 3. DDF measures the dielectric losses in insulating materials. A good liquid insulator should have low dielectric losses. The DDF of a sample increases due to the presence of polar materials. An insulating liquid with a DDF of <0.1 is preferred for use in power transformers. The DDF values of VO-based nanofluids did not exhibit any significant improvement. However, a noticeable reduction in DDF was seen in MO-ZnO nanofluid. This observation indicates that MO-ZnO nanofluid is capable of controlling dielectric losses. The capacity of a material to transport electrical potential energy is determined by its dielectric constant. A dielectric substance is one that, although having poor conductivity, can store a charge when an electric field is applied. The dielectric constant is defined by the ratio of the capacitance formed by two plates with a material kept between them to the capacitance of the same plates placed in a vacuum. As the additives themselves may have different dielectric constants than oil molecules, adding additives should influence the resulting dielectric constant of the oil. The dielectric constant value of VO was reduced from 3.039 to 2.895 and 2.971 by the addition of 0.01 % ZnO and reduced graphene, respectively, indicating no significant effect of nanoparticle addition in VO. A similar inference is drawn for MO-based nanofluids.

Table 3. Dielectric dissipation factor and dielectric constant of oil samples at 90°C.

Samples	DDF (Tan Delta)	Dielectric Constant (Ɛ)
S1: (VO)	0.0122	3.039
S2: (MO)	0.000505	2.071
S3: (VO+ZnO)	0.0155	2.8945
S4: (MO+ZnO)	0.00035	2.098
S5: (VO+GP)	0.02328	2.955
S6: (MO+GP)	0.0105	2.0995
S7: (VO+ZnO+GP)	0.0441	2.971
S8: (MO+ZnO+GP)	0.00253	2.0865
S9: (VO+MO+ZnO+GP)	0.00733	2.494

Figures 3(a), 3(b), and 3(c) depict the graphical presentation of the Weibull probability distribution for the dielectric breakdown strength the oil samples. The shape and scale values of the Weibull probability distribution for BDV of oil samples are shown in table 4. The shape values of the plot show the linear fit features' slope. The scale values demonstrate the degree of breakdown. Figure 3(a) describes the Weibull probability distribution curve of the base fluids VO and MO. The shape and scale values of VO were 4.71578 and 62.13511, whereas for MO, they were 6.89011 and 56.3964, respectively. The Weibull probability distribution curves of the ZnO and reduced graphene-based nanofluids VO and MO are shown in figure 3(b). Figure 3(c) illustrates the Weibull probability distribution curve of the hybrid nanofluids VO and MO. There was a significant improvement in shape values for hybrid nanofluids.

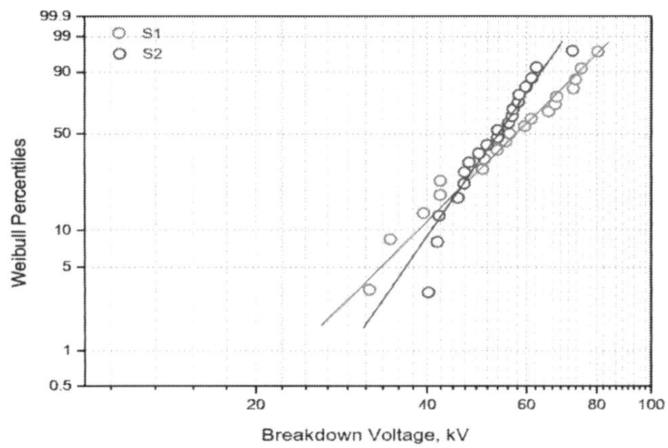

Figure 3(a). Weibull probability distribution curve of the base fluids VO and MO.

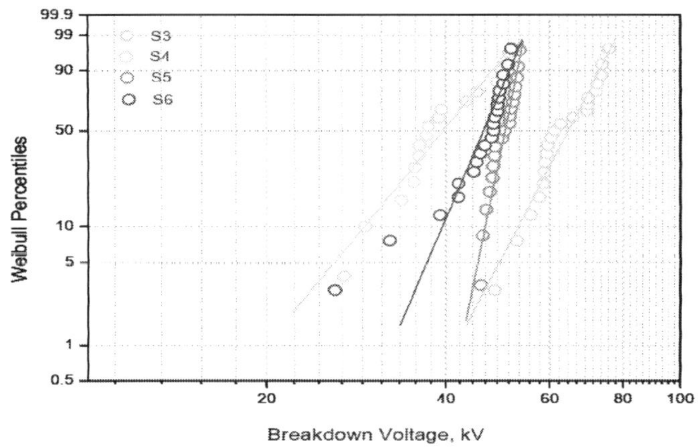

Figure 3(b). Weibull probability distribution curve of the ZnO and reduced graphene based nanofluids VO and MO.

Table 4. Dielectric Shape and scale values of Weibull probability distribution for ACBDV of oils.

Samples	Shape	Dielectric Scale
S1: (VO)	0.0122	3.039
S2: (MO)	0.000505	2.071
S3: (VO+ZnO)	0.0155	2.8945
S4: (MO+ZnO)	0.00035	2.098
S5: (VO+GP)	0.02328	2.955
S6: (MO+GP)	0.0105	2.0995
S7: (VO+ZnO+GP)	0.0441	2.971
S8: (MO+ZnO+GP)	0.00253	2.0865
S9: (VO+MO+ZnO+GP)	0.00733	2.494

4. Conclusions

This study focused on the characterization of ZnO and RG-based blended NFs of MO and VO. The mass of the added nanoparticles was 0.01%. The ratio between the nanoparticles was selected as 1:1. The hybrid nanofluids were characterized for breakdown voltage viscosity at temperatures ranging from 40 °C to 100 °C, tan delta, and resistivity. The findings of the investigation are:

A reduction in viscosity was significant in VO-ZnO nanofluids, with a decrease of about 4 cSt at 40°C. The decrease in viscosity for VO-RG nanofluid and VO-ZnO-RG hybrid nanofluid was about 2 cSt. Among the VO-based nanofluids, the ZnO-RG hybrid nanofluid showed a lower viscosity up to about 75°C. The viscosity of blended and hybrid nanofluis (S9) was in between the values of VO-based and MO-based nanofluids, indicating the possibility of the usefulness of this nanofluid.

The resistivity of VO-RG was much better than that of the other VO-based nanofluids, though it was still higher in the base fluid. A substantial improvement in SR was seen in MO-based nanofluids. The SR value was almost double in MO-RG (S8) nanofluid as compared to MO. MO-ZnO nanofluid exhibited a noticeable reduction in DDF. This observation indicates that MO-ZnO nanofluid is capable of controlling dielectric losses. The Weibull probability distribution for the dielectric breakdown strength showed a major improvement in shape values for hybrid nanofluids.

Acknowledgment

The author expresses his gratitude to the CPRI management for granting him permission to carry out the work and share the data. The author is thankful to M/s. Savita Oil Technology, India, for providing vegetable ester for this study.

References

[1] Oparanti S, Rao U and Fofana I 2023 *Energies* **16** 61 1-23
[2] Kumar S, Iruthayarajan M and Bakrutheen M 2014 *IEEE International Conference on Science Engineering and Management Research (ICSEMR),* at Chennai, India, 27-29 Nov.
[3] Wanatasanappan V, Rezman M and Abdullah M 2022 *Nanomaterials* **12** 20, 3621
[4] Lv M, Zhou Ma K, Wang W, Li C and Wang O 2015 *Renew. Sust. Energ. Rev.* **52** 308– 324
[5] Li J, Zhang Z, Grzybowski S, Liu Y 2012 *IEEE Trans. Dielectr. Electr. Insul.* **19** 1650–6
[6] Maiti P 2022 *97th IEEE CEIDP 2022 (97th IEEE Conference on Electrical Insulation and Dielectric Phenomena),* Denver, USA Oct. 30-Nov. 02
[7] Shen Z, Wang F, Wang Z and Li J 2021 *Renew. Sustain. Energy Rev.* **141** 110783
[8] Choi S and Eastman J 1995 *ASME-Pub ications-Fed, United States* **231** 99–106
[9] Koutras K, Peppas G, Fetsis T, Tegopoulos S, Charalampakos V, Kyritsis A, Yiotis A, Gonos I and Pyrgioti E 2022 *IEEE Access* **10** 79222 - 79236
[10] Babu J, Kumar K and Rao S 2017 *Renew. Sust. Energ. Rev.* **77** 551–565
[11] *Duzkaya H and Beroual A 2021* Energies ***14*** *99*
[12] Suhaimi N, Ishak M, Rahman A, Din M, Abidin M, and Kh A 2022 *IEEE Access* **10** 1033563-103586

IOP Conf. Series: Materials Science and Engineering 1300 (2024) 012032 doi:10.1088/1757-899X/1300/1/012032

Influence of Zirconium substitution on electrical properties of Nickel-Zinc Ferrites

Soumya Sajjan[1], Ananda Kulkarni[2], Shivraj G. Gounhalli*[1]

[1]Department of Physics, Smt. Veeramma Gangasiri College for Women, Kalaburagi - 585102, Karnataka, India.
[2]Department of Physics, Government First Grade College, Vijayapur, Karnataka, India.
Email address:* sgg19777@gmail.com

Abstract. The Zirconium substituted Nickel Zinc Ferrite nano particles with chemical composition of $Ni_{(0.5-x)}Zn_{(0.5-x)}Zr_xFe_2O_4$, where the value of x varies from 0 to 0.2, have been synthesized by the sol-gel auto combustion method, using high purity nitrates and fuelling agent citric acid. The structural property was studies from X-ray diffraction spectra. The crystallite size decreases with increase in Zirconium substitution. The electrical property of the same was investigated by studying the dielectric constant, and AC conductivity. There are fluctuations in these properties with variation in Zirconium substitution in Ni-Zn Ferrite.
Keywords: Sensors, Spinel Ferrites, sensing response, calcination, fabrication.

1. Introduction

In recent years nanoparticles have proved to be better from their bulk counter parts with respect to their extensive applications in the field of electronics, medical, agriculture, pharmacy and engineering fields [1]. These nanoparticles have helped in the rapid progress in technology of 21st Century because of their flexibility in tailoring required size, structure, and composition. The materials with diverse applications with specific electrical and magnetic properties have received considerable attention [1-4]. Among them Ferrite is one which can be engineered so that they prove promising materials to be used in fields with varying electric as well as magnetic behaviour.

Among the Spinel ferrite group Nickel ferrite, Zinc ferrite and Nickel Zinc ferrite are one with wide range of applications. These are the ferrites that can be explored with different chemical compositions and substitutions, and can get varied properties. Nickel Zinc ferrite are used in applications like electronic devices[5], high frequency devices [5-6]., microwave devices [7-8], as sensors [9-10], for photo catalysis in waste water treatment [11-12], in biomedical applications such as magnetic carriers [13], drug delivery, anti microbial activity and many more. The method of synthesis also has an essential role in the determination of property of ferrites. The methods like Sol-gel method, Ceramic method, Hydrothermal method, Co-precipitation method, and many other are known for their specializations in deciding the nature of nanoparticles formed.

2. Experimental Procedure

2.1. Chemicals used
The analytical grade Nickel (II) Nitrate Hexahydrate [$Ni(NO_3)_2.6H_2O$], Zinc (II) Nitrate Hexahydrate [$Zn(NO_3)_2.6H_2O$], Iron (III) Nitrate Nanohydrate [$Fe(NO_3)_3.9H_2O$], Zirconly (IV) Oxynitrate [$ZrO(NO_3)_2.xH_2O$] and Citric Acid [$C_6H_8O_7$] were used for synthesis without further purification.

2.2. Method of synthesise
The nanoparticles of Zirconium substituted Nickel Zinc Ferrite were prepared by sol-gel auto combustion method. In this method the fuel/reducing agent used was Citric acid and metal nitrate precursor acts as oxidizer are dissolved in deionized water. The stoichiometric ratio of fuel to oxidizers was kept constant at 3:5. The precursor of metal nitrates in deionized water along with citric acid was continuously stirred for 30 minutes. Then the solution was then heated at 75^{0}C so that the water in it evaporates forming a gelly solution. The gel is further heated to 150^{0}C. As the temperature rises, foam starts to form from the reaction mixture, which latter forms a powder of desired ferrite on complete combustion. The powder is then grounded and calcinated at 500^{0}C for 2 hours to remove the unreacted chemicals if any. These calcinated powders are the made into pellets and further sintered at 700^{0}C for 2 hours.

2.3. Characterization techniques
The crystal structure of synthesized nano particles of Zirconium substituted Nickel-Zinc Ferrite is analyzed from X-ray diffraction measurements (XRD Model: SmartLab SE) recorded with X-rays of wavelength 1.5406Å. The AC Conductivity and impedance analysis was done using Wayne Kerr Precision Impedance analyzer 6500B.

3. Results and discussion
3.1. Structural analysis
The Crystal structure of Zirconium substituted Nickel Zinc Ferrite was examined by X-ray diffraction studies. The XRD spectra of ferrite nanoparticles under investigation are as shown in Figure 1. These well defined sharp peaks in the spectra show the high crystallinity of synthesized ferrite.

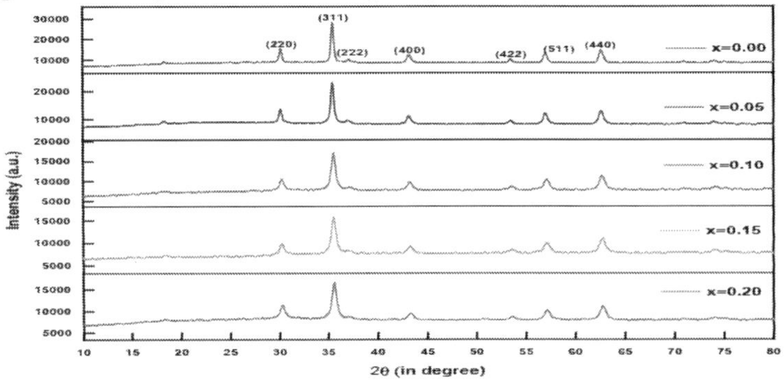

Figure 1: X-Ray diffraction graph of Zirconium substituted Ni-Zn Ferrite $Ni_{(0.5-x)}Zn_{(0.5-x)}Zr_xFe_2O_4$

The X-ray diffraction spectra also validates the formation of Spinel Phase of Zirconium substituted Nickel-Zinc Ferrites as we can observe the intense peak at (311) plane and other peaks corresponding to (220), (222), (400), (422), (511), and (440) planes. The formation of crystalline phase of the ferrite under investigation was verified by referring to the particulars from the JCPDS card no. 00-062-0437.

The size of crystallite of the nanoparticles was determined by employing Scherrer equation [14]:

$$t = \frac{k\lambda}{\beta \cos\theta} \quad (1)$$

in Eq. (1) 't' is crystallite size, 'k' is Scherrer constant, 'λ' is wavelength of X-ray, 'β' is the full width at half maximum (FWHM) of the peak for (311) plane and 'θ' is Bragg angle.

The lattice parameter of crystal system is obtained from the relation [14]:

$$a = d_{hkl}\sqrt{h^2 + k^2 + l^2} \qquad (2)$$

in Eq. (2) d_{hkl} is the interplanar spacing of corresponding plane of Miller Indices (hkl).

We can notice that the average size of crystallite of ferrite under study decreases from 29.98 nm to 14.73 nm as the Zirconium substitution increases. The size of crystallite and Lattice constant of nanoparticles under discussion are presented in Table 1. It has to be noted that both size of crystallite and Lattice constant decrease with increase in Zirconium substitution.

Table 1: Particle size and Lattice parameter of Zirconium substituted Ni-Zn Ferrite $Ni_{(0.5-x)}Zn_{(0.5-x)}Zr_xFe_2O_4$

x	Particle Size t (in nm)	Lattice Parameter a (in Å)
0.00	29.9852	8.3915
0.05	26.4466	8.3950
0.10	16.0387	8.3838
0.15	14.9892	8.3754
0.20	14.7301	8.3640

3.2. Dielectric Studies

The Dielectric studies are helpful in the investigation of electrical properties of materials. These dielectric properties are dependent on composition of the material, method involved in synthesis, temperature under which the study is carried out and also electric field frequency. These investigations help in knowing the nature and transport of electric charge carriers. Under this discussion of dielectric properties we study the frequency dependency of dielectric constant, dielectric loss and AC conductivity of the materials.

The relative permittivity of material is studied as Dielectric constant (ε_r). Dielectric constant of a material gives us the measure of its ability to store electrical energy. The relative permittivity or dielectric constant is a complex parameter which consists of two parts; viz. real part of ε_r represented by ε\` and imaginary part of $\varepsilon_r$ represented by ε\`\`.

The calculation of real part (ε\`) of dielectric constant is done by utilizing the relation that follows

$$\varepsilon` = \frac{C_p t}{\varepsilon_0 A} \qquad (3)$$

In Eq. (3) 'C_p' represents parallel capacitance (in farad),'t' represents pellet's thickness, 'A' denotes pellet's cross section area and 'ε_0' denotes absolute permittivity. The variance of dielectric constant's real part with frequency is represented in figure 2. An increase in frequency brings down the value of ε\`. This value of 'ε\`' is high at lower frequency, however we can observe that these values decrease as frequency goes up and almost saturates at higher frequency.

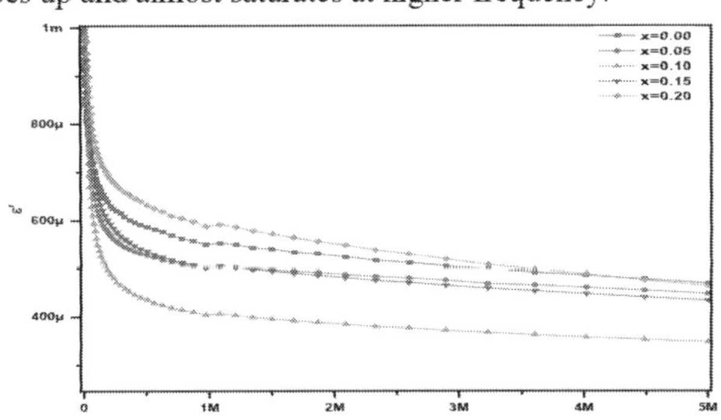

Figure 2: Real part of dielectric constant (ε') versus Frequency of Zirconium substituted Ni-Zn Ferrite $Ni_{(0.5-x)} Zn_{(0.5-x)} Zr_x Fe_2 O_4$

IOP Conf. Series: Materials Science and Engineering 1300 (2024) 012032 doi:10.1088/1757-899X/1300/1/012032

We can obtain the dielectric constant's imaginary part (ε``) by employing the relation given below

$$\varepsilon`` = \varepsilon` \tan \delta \quad (4)$$

According to Eq. (4), 'tan δ' is dielectric loss tangent. The frequency dependent change of dielectric constant's imaginary part is shown in figure 3. As observed for ε`, the ε`` also exhibits same behavior. The values of ε`` also goes down as the frequency rises up. 'ε`' is high for lower frequency and the values decrease with increase in frequency. This values of ε`` almost saturates at higher frequency.

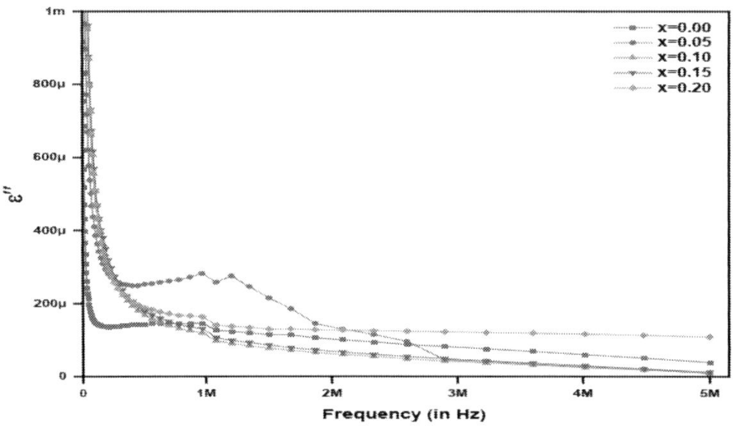

Figure 3: Dielectric constant's Imaginary part (ε``) versus Frequency of Zirconium substituted Nickel Zinc Ferrite $Ni_{(0.5-x)}$ $Zn_{(0.5-x)}$ Zr_x Fe_2 O_4

The dielectric loss (tan δ) is the energy lost due to varying electric field. The figure 4 shows the frequency dependency of tan δ of Zirconium substituted Nickel-Zinc Ferrite at room temperature. The plots show that lower frequencies result in a high dielectric loss, which decreases as frequency increases.

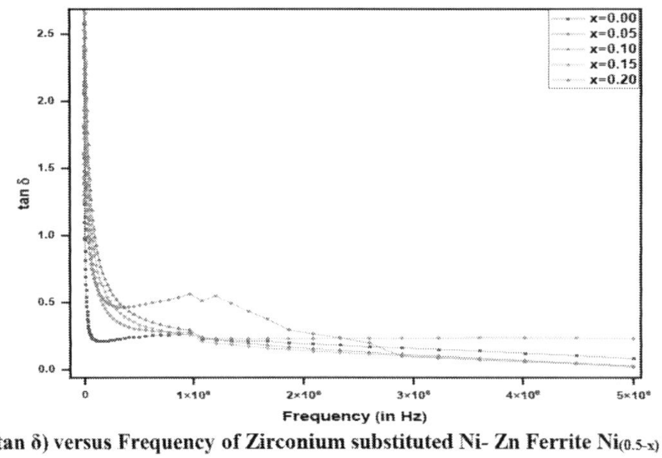

Figure 4: Dielectric loss (tan δ) versus Frequency of Zirconium substituted Ni- Zn Ferrite $Ni_{(0.5-x)}$ $Zn_{(0.5-x)}$ Zr_x Fe_2 O_4

3.3. AC conductivity

The AC conductivity (σ_{ac}) of Zirconium substituted Nickel-Zinc Ferrites are measured at different frequencies at room temperature. The relation between the dielectric constant and dielectric losss of the nanoparticles is utilized to study AC conductivity (σ_{ac}),

$$\sigma_{ac} = \omega \varepsilon_0 \varepsilon` \tan\delta \quad (5)$$

In Eq. (5) 'ω' represents the angular frequency (ω=2πf), 'ε_0' represents permittivity of free space (8.869 x 10^{-12} F/m), 'ε`' represents real part of dielectric constant and 'tan δ' represents dielectric loss. This relation (5) was helpful in representing the deviations in AC conductivity as frequency changes and is depicted in Figure (5). We can observe that at lower frequency the dependency of conductivity of frequency is less significant, but as frequency increases the conductivity also increases at higher frequencies. Maxwell-Wagner Theory and Koop Theory explain this change in electrical conductivity, according to which we can observe the increase in AC conductivity as frequency increases. The theories say that the grain boundaries are more active at lower frequencies which results in lowering of hopping frequency of the electrons between Fe^{3+} and Fe^{2+}. Whereas conductive grains at higher frequency are more active, resulting in increased hopping of Fe^{3+} and Fe^{2+} with frequency [15-16]. As we can also observe that at lower frequency the Zirconium substitution has increased conductivity, which shows that Zirconium substituted Nickel-Zinc Ferrite, can be useful in applications that need high AC conductivity at low frequency.

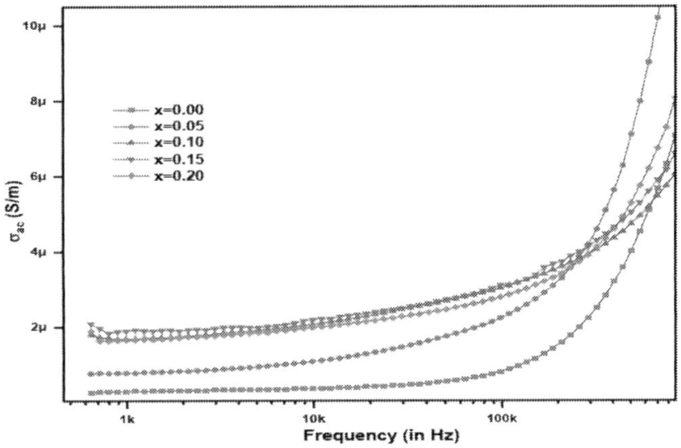

Figure 5: AC Conductivity versus Frequency of Zirconium substituted Ni-Zn Ferrite $Ni_{(0.5-x)}$ $Zn_{(0.5-x)}$ Zr_x Fe_2 O_4

4. Conclusion

Our focus in this work has been on the synthesis of Zirconium substituted Nickel-Zinc Ferrite employed by Sol gel Auto combustion method for the investigation of its structural and electrical properties. The structure of synthesized ferrite nanoparticles was analyzed from XRD studies, which confirms spinel structure of ferrite, where the increase in Zirconium concentration raises the crystallite size and lattice parameter. The dielectric studies show that the dielectric constant decreases and AC conductivity increases with increase in frequency. We can also observe a remarkable impact of Zirconium substitution on the dielectric properties of Nickel-Zinc ferrite. These properties are some indications that could the material is suitable for the applications in sensors and in microwave devices.

Acknowledgement

The authors are thankful to SAIF Dharwad, GFGC Vijayapur for availing analysis facilities.

References

[1] Sulabha K. Kulkarni, 2016, Nanotechnology: Principles and Practices, 3rd edition, (*Capital Publishing Company*, New Delhi, India).
[2] Lawrence H. Van Vlack, 2014, Elements of Material Science and Engineering, 6th edition, (*Pearson India Education Services Pvt. Ltd.*).
[3] Najahi-Missaoui, W.; Arnold, R.D.; Cummings, B.S. Safe Nanoparticles: Are We There Yet? *Int. J. Mol. Sci.* 2021, 22, 385.
[4] Sharma, S.; Jaiswal, S.; Duffy, B.; Jaiswal, A.K. Nanostructured Materials for Food Applications: Spectroscopy, Microscopy and Physical Properties. *Bioengineering (Basel)* 2019, 6.
[5] Kebede K. Kefeni et al, 2017, Review Ferrite Nanoparticles: Synthesis, Characterization and Applications in Electronic Devices, *Material Science and Engineering*, B 215, 37-55.

[6] Preeti Thakur, Shilpa Taneja, Deepika Chahar, Blaise Ravelo, Atul Thakur, 2021, Recent advances on synthesis, characterization and high frequency applications of Ni-Zn ferrite nanoparticles, *J. Magn. Magn. Mater.* 530 (2021)167925.

[7] Neha Aggarwal, Sukhleen B N, 2021, Microwave absorption analysis of Mg-Zr substituted Ni-Zn spinel ferrites in the X-band, *j. Phy. Chem. Sol.* 159 (2021) 110289.

[8] Anas H, Zharmenov A A, Yomen A, Z Bagasharova T, Mirzalieva S, Kadyrakunov K, 2021, Microwave absorbing properties and their composites: A review, *J. Magn. Magn. Mater.* 529 (2021)167839.

[9] J. Y. Patil, D. Y. Nadargi, J. L. Gurav, I. S. Mulla, S. S. Suryavanshi, 2014, Synthesis of glycine combusted $NiFe_2O_4$ Spinel ferrite: A highly versatile gas sensor, *Mat. Let.*, 124 (2014) 144-147.

[10] Jayvant Y. P., Digambar Y. N., Jyoti L. G., I. S. Mulla, Sharad S. S., 2014, Glycine combusted $ZnFe_2O_4$ gas sensor: Evaluation of structural, morphological and gas response properties, *Cer. Intern.*

[11] K. K. Kefeni, B. B. Mamba, T. A. M. Msagati, 2017, Application of Spinel Ferrite Nanoparticles in Water and Wastewater Treatment: A Review, *Separation and Purification Technology*, (17) 30806-7.

[12] D. Harikishore Kumar Reddy, Yeoung-Sang Yun, 2016, Spinel ferrite magnetic absorbents: Alternative future materials for water purification?, *Coord Chem Rev.*, 315 (2016) 90-111.

[13] Mahnaz A., Masoud S-N, Ahmad A. 2019, Magnetic nanocarriers: Evolution od spinel ferrites for medical applications, *Adv. Coll. Int. Sci.*

[14] Cullity, B.D. (1978) Elements of X-Ray Diffraction. 2nd Edition, (*Addison-Wesley Publishing Company Inc., Phillippines*).

[15] J. C. Maxwell, 1929, Electricity and Magnetism, vol.1, (*Oxford University Press, Oxford*).

[16] Amrutesh K., Avinash P., S. R. Manohara, Mohsina T., Kotresh M. G., 2023, In-depth study of Zinc nanoferrite particles at different calcination temperatures and their behavior in the presence of electric and magnetic fields, *J. Magn. Magn. Mater.* 584 (2023) 171079.

IOP Conf. Series: Materials Science and Engineering 1300 (2024) 012033 doi:10.1088/1757-899X/1300/1/012033

TiO_2/ Graphitic carbon nitride Z- scheme catalyst for binary dye degradation by persulphate activation

Shalini Viswanathan[1, a]), Anooj Anil Pallikkara[2, a)], Fermil Muhammed[2, a)], Majji Kavya Sri[2, a)], Aparna K[*a)]

Material Science and Environmental Sustainability Group, Department of chemical engineering, National Institute of Technology, Calicut. India.
Corresponding author: aparnak@nitc.ac.in

Abstract
In this work, we describe the preparation of the TiO_2 (P25) /Graphitic Carbon Nitride composite and the degradation of pollutants using this composite with persulphate activation. The materials are characterized by XRD, FTIR analysis, etc. XRD analysis confirms the presence of TiO_2 anatase and GCN and the characteristic functional groups of both TiO_2 and GCN are determined with the help of FTIR. The degradation of crystal violet and Rhodamine B in binary mixtures over this synthesized catalyst is easier than individual catalysts separately. The degradation efficiency of crystal violet over TiO_2/GCN composite is 1.5 times higher than GCN and 3 times higher than TiO_2 alone and the degradation efficiency of Rhodamine B over TiO_2/GCN composite is 1.1 times higher than GCN and 2.4 times higher than TiO_2 alone. In flat band analysis, a Z-Scheme catalyst is formed between TiO_2 and GCN. The electron transfer between the bands of TiO_2 and GCN occurs because of proper band alignment between these materials.

1. Introduction

As a consequence of economic development, the manufacture and use of organic compounds in sectors such as printing, dyeing, medicinal products, clothing, and skincare are expanding. Research on environmental and energy issues has moved clean energy generation and wastewater treatment to the forefront in recent years. Inappropriate use or storage of water resources can nevertheless lead to the development of organic pollutants in water ecosystems which is liable to have a detrimental effect on their efficiency and conservation. Due to their delayed biodegradation and deleterious consequences, hazardous pollutants such as organic dyes, heavy metals, and pesticides cause major health hazards for people as well as aquatic ecosystems. Modern methods of pollution removal from industrial waste, like photocatalysis and adsorption, have shown promise in these difficulties. Advanced oxidation based on radical reactions was demonstrated as an effective method to degrade organic refractory compounds in the field of environmental remediation.[1] Transition metal ions, thermal, photolytic, oncolytic, and radiolytic methods are conventional methods of persulfate (PS) activation to generate SO4•− radicals. The fuel activation mechanism consists of the fission of OO bonds, which results in high energy consumption as they are fed from outside sources. Titanium dioxide (TiO_2) is a very affordable, stable, and non-toxic semiconductor photocatalyst. TiO_2's drawbacks, such as its wide band gap and the need for UV light to activate and aggregate nanoparticles, have prompted efforts to alter it to enhance performance.[2] Graphitic carbon nitride g-C_3N_4 has a narrow band gap, is easy to manufacture, contains an extensive raw material source, and can be produced cheaply due to these it was regarded as the best carbon-based material.[3] The use of carbonaceous materials is especially favored, as they have a negligible amount of toxicity, cost-effectiveness, wide areas surface, and excellent chemical stability. Carbonaceous materials are examined for the activation of persulfates when applied in contaminated groundwater remediation as well as wastewater treatment facilities, and economic concepts and perspectives on this issue are presented. Based on the composition and microstructure of these materials, they are divided into three categories, including pristine carbon materials and their derivatives, heteroatom-doped carbon materials, and carbon materials with metal particles.[4] For persulphate (PS) activation, g-C_3N_4 can be excited by visible light-producing electrons in the conduction band. Persulphate activation processes involve the formation of the highly reactive sulfate radical ($SO_4^{\bullet-}$). Sulfate radicals ($SO_4^{\bullet-}$) have been utilized for the degradation of organic pollutants due to the advantages of $SO_4^{\bullet-}$ over •OH. $SO_4^{\bullet-}$ has a longer half-time (3–4 × 10^{-5}s) and has selectivity to the electron-rich compounds. $SO_4^{\bullet-}$ could take effect in a wide range of pH (2.0–8.0). In addition, $SO_4^{\bullet-}$ has a redox potential (2.5 V–3.1 V). Persulphates have a peroxide bonding environment that is vulnerable to one-electron reduction, allowing for $SO_4^{\bullet-}$ formation through reductive transformation.[5] An effective method of enhancing photogenerated electron–hole pair separation is using g-C_3N_4, in combination with TiO_2, which forms a heterojunction. It was proven that the energy level positions of g-C_3N_4 and TiO_2 correspond to each other quite well for TiO_2 to be able to accept the electrons from g-C_3N_4. The efficiency of pollutant degradation in the catalytic system can be significantly enhanced by the presence of persulphate.[6] A simple method for the preparation of P25/ g-C_3N_4 catalysts which have a uniform distribution of TiO_2 on a layer of g-C_3N_4 has been reported in this study. More than one technology has been applied to evaluate the physical and chemical properties of a Synthetic Photocatalyst.

2. Materials and methods

TiO2 (P25, Degussa), Melamine (Merck), and Ammonium Chloride(Merck) were purchased and used.

Graphitic Carbon Nitride was synthesized by calcining melamine at 550°C for four hours. P25/GCN composite was synthesized by mixing a pre-determined amount of TiO_2 with 100mg of prepared GCN and dispersing it in ethanol. After evaporating ethanol, the obtained slurry was calcined at 250°C to get P25/GCN composite.

3. Results and Discussion

Below is a plot of the XRD of GCN (graphitic carbon nitride), with the distinctive peak of GCN being at 12.89°, 27.4°, and corresponding to planes (100) and (002) of GCN with ICDD number 01-087-1526. The peaks in the XRD of TiO_2 located at Bragg's angle 25.5°, 38.04°, 48.16°, 54.05°, 62.96°, 69.01°, and 75.29° corresponding to the planes (100), (004), (200), (105), (204), (116), and (215) of anatase TiO_2 with ICDD number 01-073-1764.[7]The appearance of characteristic peaks of both TiO_2 and GCN samples confirms the presence of both samples in the P25/GCN composite. The observed XRD results are well matched with already reported literature

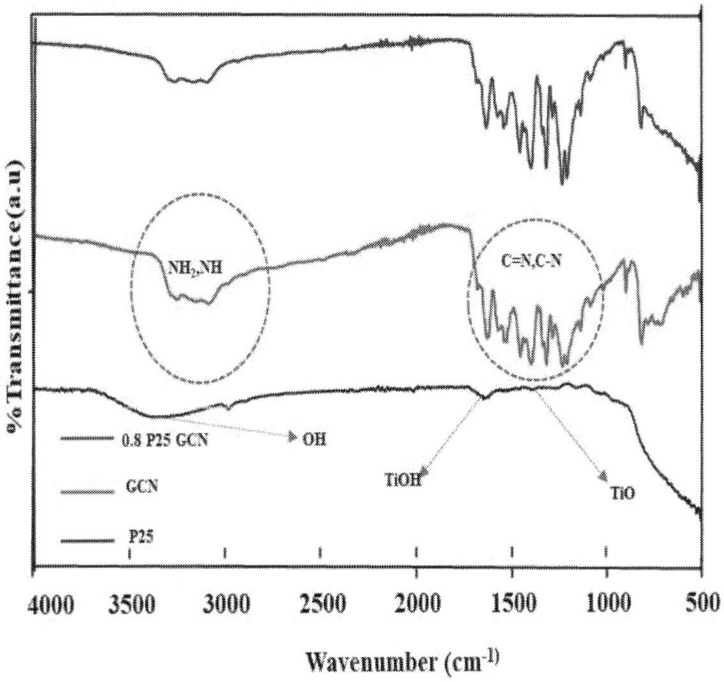

FIGURE 1a XRD diffraction pattern of TiO_2, GCN, and P25/GCN composite

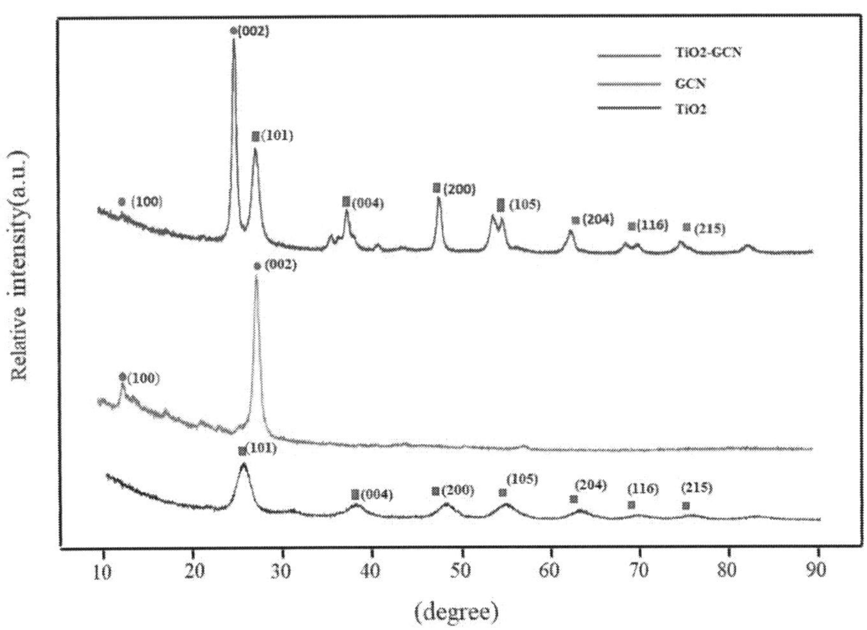

Figure 1b shows the FTIR data of P25 and GCN composites.

The characteristic peak of the GCN sample got was 816.7cm-1 which indicates the presence of the tri-s-triazine functional group. The second bandwidth range is the prominent one, the wavenumber observed came in the range 2981.899-3399.27 cm-1 which corresponds to the bending modes vibration of NH_2 and NH functional groups resulting from hydrogenation on nitrogen atoms during thermal polymerization or the presence of terminal NH and NH_2 at the defect regions of the GCN. The last group of peaks observed was in the range 1202.83-1632.87 cm-1 which corresponds to the stretching vibration of C=N and C-N functional groups.[8]

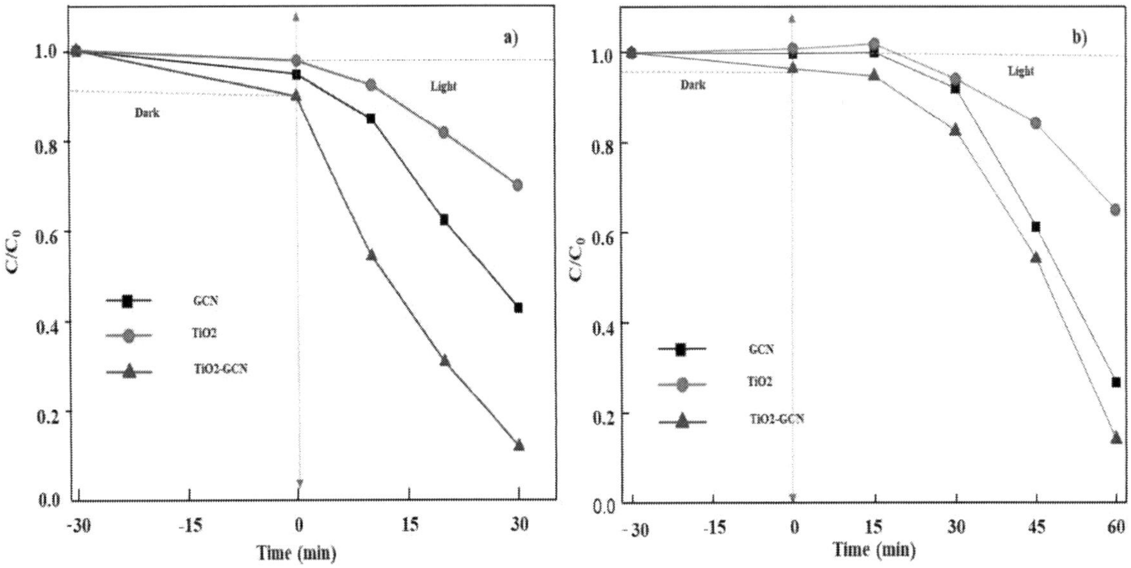

Figure 2:Degradation efficiencies of Crystal violet and Rhodamine B in binary mixtures over the synthesized catalysts. a) crystal violet, **b):**Rhodamine B

Figure 2a shows the degradation efficiency of 5ppm Crystal violet in a binary mixture of RhB and CV over synthesized catalysts over some time. From the graph, it is evident that the P25/GCN composite has a degradation efficiency of 88%. The remaining individual catalysts which are GCN and TiO_2 show degradation efficiency of 57% and 30% respectively. So from the plot, it is clear that the P25/GCN composite has the maximum degradation efficiency compared to the other two individual catalysts.

Figure 2b shows the degradation efficiency of crystal violet in the binary mixture of CV and RhB over the synthesized catalysts. From the graph, it is evident that the P25/GCN composite has a degradation efficiency of 85%. The remaining individual catalysts which are GCN and TiO_2 show degradation efficiency of 73% and 35% respectively. So from the plot, it is clear that the P25/GCN composite has the maximum degradation efficiency compared to the other two individual catalysts.[9]

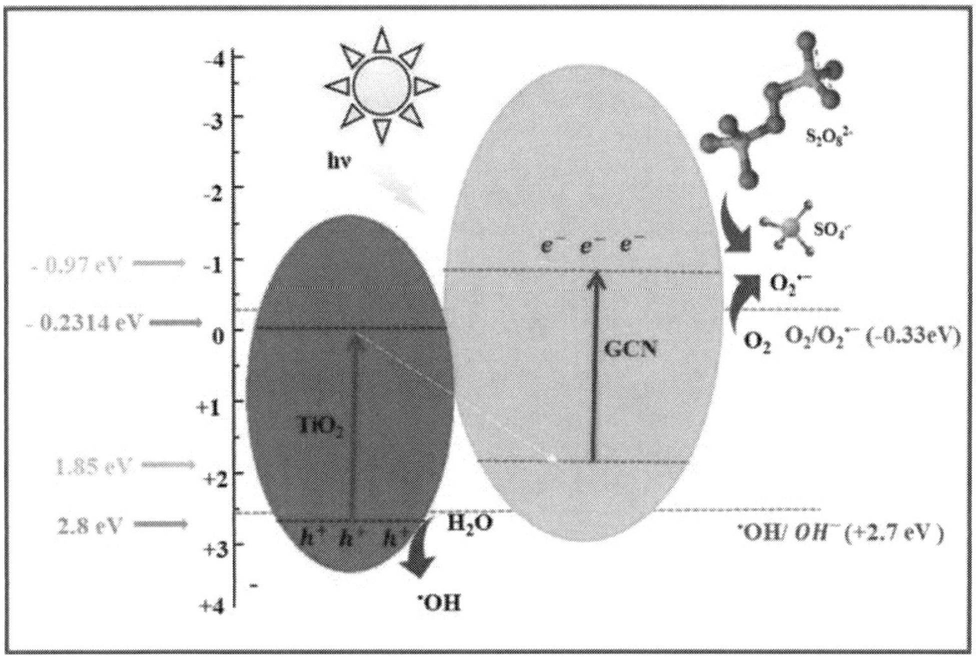

Figure 3: A plausible mechanism for photocatalytic degradation over P25/GCN composite

The valence band and conduction band potentials of P25 and GCN have been calculated and from the figure, it is evident that the transfer of electrons from the conduction band of TiO_2 to the valence band of GCN is possible due to appropriate band alignment between two materials.[10] Thus the effective charge separation and charge migration in the composite results in the degradation of organic pollutants, The active oxygen species involved in photocatalytic degradation include superoxide radical, hydroxyl radical, and sulfate radical. 11 The active oxygen species can be successfully generated over the composite since the redox potentials are sufficiently high to produce superoxide and hydroxyl radicals from the solution.[12]

4. Conclusion

In this study, we synthesized P25/GCN composite by successfully incorporating TiO_2 into GCN and finally synthesized it into a Z-scheme catalyst. We observed XRD diffraction patterns of TiO_2, GCN, and P25/GCN composite individually and the results are matched with already available literature. We also studied the degradation of the pollutants by testing the degradations of crystal violet and Rhodamine B in binary mixtures over synthesized catalysts (P25/GCN composite) and respective degradation efficiencies were calculated. The degradation efficiencies of crystal violet and Rhodamine B were calculated separately for individual species and composite and compared. For crystal violet, the P25/GCN composite has a degradation efficiency of 88% while the individual catalysts GCN and TiO_2 show degradation efficiencies of 57% and 30% respectively. Similarly for Rhodamine B, the P25/GCN composite has a degradation efficiency of 85% while the individual catalysts GCN and TiO_2 have degradation efficiencies of 73% and 35% respectively. The photocatalytic degradation of pollutants using P25/GCN composite with persulphate activation has been studied.

References

(1) Xu, X. R.; Li, S.; Hao, Q.; Liu, J. L.; Yu, Y. Y.; Li, H.-B. Activation of Persulfate and Its Environmental Application. *Int J Environ Bioenergy* **2012**, *1*, 60–81.

(2) Sofyan, N.; Ridhova, A.; Yuwono, A. H.; Udhiarto, A. Preparation of Anatase TiO2 Nanoparticles Using Low Hydrothermal Temperature for Dye-Sensitized Solar Cell. *IOP Conf. Ser.: Mater. Sci. Eng.* **2018**, *316*, 012055. https://doi.org/10.1088/1757-899X/316/1/012055.
(3) Khan, M.; Farah, H.; Iqbal, N.; Noor, T.; Amjad, M. Z. B.; Ejaz Bukhari, S. S. A TiO_2 Composite with Graphitic Carbon Nitride as a Photocatalyst for Biodiesel Production from Waste Cooking Oil. *RSC Adv.* **2021**, *11* (59), 37575–37583. https://doi.org/10.1039/D1RA07796A.
(4) Huang, W.; Xiao, S.; Zhong, H.; Yan, M.; Yang, X. Activation of Persulfates by Carbonaceous Materials: A Review. *Chemical Engineering Journal* **2021**, *418*, 129297. https://doi.org/10.1016/j.cej.2021.129297.
(5) Ike, I. A.; Orbell, J. D.; Duke, M. Activation of Persulfate at Waste Heat Temperatures for Humic Acid Degradation. *ACS Sustainable Chem. Eng.* **2018**, *6* (3), 4345–4353. https://doi.org/10.1021/acssuschemeng.7b04840.
(6) Du, X.; Bai, X.; Xu, L.; Yang, L.; Jin, P. Visible-Light Activation of Persulfate by TiO2/g-C3N4 Photocatalyst toward Efficient Degradation of Micropollutants. *Chemical Engineering Journal* **2020**, *384*, 123245. https://doi.org/10.1016/j.cej.2019.123245.
(7) Yu, H.; Ye, L.; Zhang, T.; Zhou, H.; Zhao, T. Synthesis, Characterization and Immobilization of N-Doped TiO_2 Catalysts by a Reformed Polymeric Precursor Method. *RSC Adv.* **2017**, *7* (25), 15265–15271. https://doi.org/10.1039/C6RA27187A.
(8) Matouke, M. M. FTIR Study of the Binary Effect of Titanium Dioxide Nanoparticles (NTiO2) and Copper (Cu2+) on the Biochemical Constituents of Liver Tissues of Catfish (Clarias gariepinus). *Toxicology Reports* **2019**, *6*, 1061–1070. https://doi.org/10.1016/j.toxrep.2019.10.002.
(9) Raja, V.; Jaffar Ali, B. M. Synergy of Photon Up-Conversion and Z-Scheme Mechanism in Graphitic Carbon Nitride Nanoparticles Decorated g-C3N4-TiO2. *Colloids and Surfaces A: Physicochemical and Engineering Aspects* **2021**, *611*, 125862. https://doi.org/10.1016/j.colsurfa.2020.125862.
(10) Qi, H.; Ji, X.; Shi, C.; Ma, R.; Huang, Z.; Guo, M.; Li, J.; Guo, Z. Bio-Templated 3D Porous Graphitic Carbon Nitride Hybrid Aerogel with Enhanced Charge Carrier Separation for Efficient Removal of Hazardous Organic Pollutants. *Journal of Colloid and Interface Science* **2019**, *556*, 366–375. https://doi.org/10.1016/j.jcis.2019.08.072.
(11) Baranowska, D.; Kędzierski, T.; Dworczak, M.; Baca, M.; Mijowska, E.; Zielińska, B. Bifunctional Catalyst Based on Molecular Structure: Spherical Mesoporous TiO2 and GCN for Photocatalysis. *Materials Research Bulletin* **2022**, *146*, 111608. https://doi.org/10.1016/j.materresbull.2021.111608.
(12) Yousaf, M. U.; Pervaiz, E.; Minallah, S.; Afzal, M. J.; Honghong, L.; Yang, M. Tin Oxide Quantum Dots Decorated Graphitic Carbon Nitride for Enhanced Removal of Organic Components from Water: Green Process. *Results in Physics* **2019**, *14*, 102455. https://doi.org/10.1016/j.rinp.2019.102455.

IOP Conf. Series: Materials Science and Engineering 1300 (2024) 012034 doi:10.1088/1757-899X/1300/1/012034

Predicting shock hugoniot and equation of state of single-crystal molybdenum by molecular dynamics simulation

Anuradha Singla[1,2,*] **and Aditi Ray**[1,2]

[1] Theoretical Physics Section, Bhabha Atomic Research centre, Mumbai 400085, India

[2] Homi Bhabha National Institute, Mumbai 400094, India

E-mail: anuradhas@barc.gov.in

Abstract.

In the present paper, we employ non-equilibrium molecular dynamics approach to simulate impact-induced shock propagation in single-crystal Molybdenum. Shock hugoniot, generated by simulation of impact with varied strength shows excellent agreement with experimental data in the strong shock regime. The resulting hugoniot parameters obtained by linear fitting of shock velocity *vs.* particle velocity data are then used to estimate ambient pressure Grüneisen coefficient. Finally Mie-Grüneisen equation of state (EOS) with hugoniot as reference state, is employed to express pressure-volume-energy relationship for [001] single-crystal Mo. The influence of different analytical forms for volume dependence of Grüneisen parameter on EOS is investigated.

1. Introduction

Investigation of material properties at extreme thermodynamic conditions of pressure and temperature is of interest to many fields dealing with high energy density, such as astrophysics, geophysics, inertial confinement fusion, explosive and impact driven phenomena, etc. Theoretical study of these dynamic systems require detail knowledge of Equation of State (EOS) at wide range of pressures and energies.

Molybdenum is a refractory metal which is known to be stable in a body-centred cubic structure up to high pressures. Mo, its alloys and other body-centered cubic (BCC) refractory metals find wide range of applications in high pressure physics, e.g. planetary systems, and hypervelocity impacts of rocky planets, aerospace and armor/anti-armor applications, because of their excellent strength, mechanical stability, creep resistance and ductile nature. Most of the existing theoretical and experimental study on Mo are aligned towards examining structural stability, phase transition and melting under static pressures [1]. Current work is motivated by recent interest in understanding response of single-crystal Mo under shock compression - a relatively unexplored domain.

While most of the shock compression experiments carried out in laboratory are limited to strain rates $\sim 10^6/s$, ultrashort pulsed laser extends this limit to a great extent. However, the same for still higher strain rates, as encountered in many high energy density systems remains unexplored largely due to lack of experimental facilities. For this purpose large-scale nonequilibrium molecular dynamics (NEMD) study is preferred which facilitates extracting information about shock evolution and associated microstructural changes at such exotic environment.

Due to recent advances in computational resources classical MD simulations are being increasingly used to study the shock compression behaviour of crystalline solids. In this regard, recently we have examined the elastic-plastic behavior of Ta, another BCC metal, at ultra-high strain rates using NEMD simulation [2]. Added advantage is that MD simulation provides atomistic insight of underlying physical phenomena.

Aim of the present work is to gain microscopic insights of shock propagation, construction of shock hugoniot of single-crystal Mo and utilize it to generate off-hugoniot states in the compressed region through Mie-Grüneisen EOS (MGEOS) ($P - V - E$ relation).

2. Simulation details

All the results presented in this work are generated using open source classical molecular dynamics simulation package, LAMMPS [3]. Atomistic visualization of shock propagation is facilitated by OVITO software [4]. The simulation box contains $32 \times 32 \times 320$ unit cells which corresponds to a dimension of 100 nm along z-direction and about 10 nm along x and y directions. As required for NEMD simulation, free boundary condition is applied along shock propagation direction (z), while periodic boundary conditions are used for the transverse directions. The interaction between atoms is treated by N-body semi-empirical potential parameterized by Ackland et al [5]. Before shock simulations, the sample is initially thermalized at 300 K and zero pressure for 30 ps under NPT ensemble. The time step of integration is 1 fs.

Shock is imparted to the sample by setting a pre-decided constant velocity to atoms in first four unit cells of the simulation box on one end. This atomic layer behaves like a rigid-piston that compresses the sample. Thus generated shock is allowed to propagate along positive z-direction which coincides with the (001) crystallographic orientation of Mo crystal. Determination of resultant thermodynamic and macroscopic flow variables are described in Ref. [2] .

3. Results and Discussion

3.1. Shock Hugoniot

Shock is a propagating discontinuity traveling at certain high velocity (more than the sound velocity of the medium), resulting in an abrupt and irreversible change in state variables, such as density ($\rho = 1/V$), pressure (P_H) and temperature

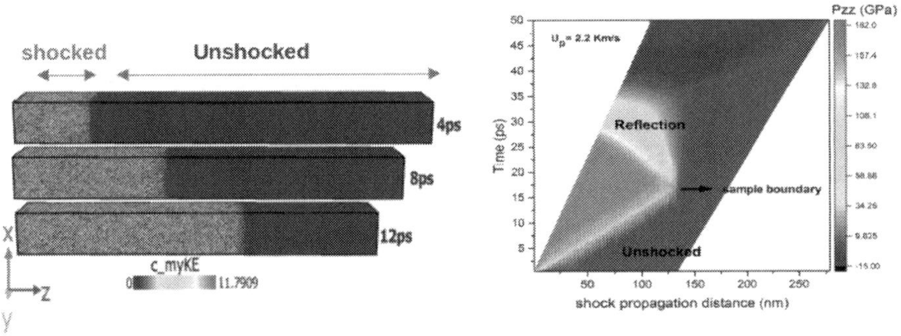

Figure 1: (a) Atomistic visualization of shock propagation due to impact at $U_p = 2.2$ km/s for three different times post-impact. Atoms are colored according to their kinetic energies with red being maximum and blue minimum (unshocked). (b)Spatio-temporal evolution of pressure inside Mo target for same impact velocity.

(T_H). The thermodynamic variables on either side of the wave are related through conservation relations with the parlance Rankine-Hugoniot jump conditions, mathematically expressed as:

$$\rho(U_s - U_p) = \rho_0 U_s; \quad P_H = \rho_0 U_s U_p; \quad E_H = \frac{P_H}{2}(V_0 - V) \tag{1}$$

Hugoniot is the locus of all the points attained in single shock experiments performed at various impact velocities. In the present work, piston velocity is tuned to 1.8 - 3.0 km/s so as to capture only pure plastic wave for the chosen crystal orientation. For illustration, we present visualization plot in Fig. 1(a) demonstrating the generation of shock and its propagation at different instances within the simulation box due to impact at 2.2 km/s. Shock front travels from left to right which can be identified from the abrupt change in kinetic energy of the atoms. Corresponding spatio-temporal evolution of impact generated pressure in the target (bin-averaged values) is displayed in 2D contour plot of Fig. 1(b).

Shock velocity (U_s) vs. particle velocity (U_p) hugoniot is displayed in Fig. 2(a). In case of uniaxial compression it is appropriate to use the component of pressure along shock direction, i.e., P_{zz} as hugoniot pressure. Pressure (P_H)-particle velocity hugoniot is shown in Fig.2(b). Both the hugoniot obtained in NEMD simulation show good agreement with experimental data [6, 7] pertaining to polycrystalline sample. Available single-crystal data for [001] direction [8] is also shown. This establishes the validity of the chosen interatomic potential for the present requirement in high-pressure applications. The linear-fitting of U_s U_p curve, $U_s = C + S\,U_p$ gives the hugoniot parameters C, i.e. bulk sound speed, and slope, S. Using our simulations carried out on Mo crystal, we get $C = 5.24\,km/s$ and $S = 1.27$.

Using linear ($U_s - U_p$) relation in Eq. 1 we can write hugoniot pressure as

$$P_H(V) = \frac{C^2(V_0 - V)}{[V_0 - S(V_0 - V)]^2}. \tag{2}$$

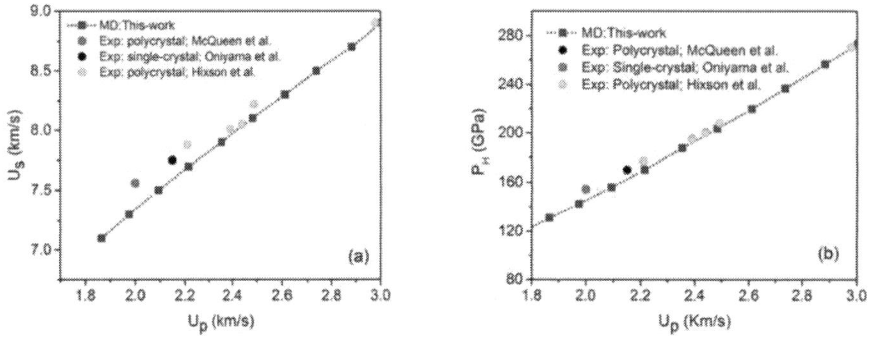

Figure 2: Hugoniot curves of single-crystal Mo obtained by NEMD simulation: (a) $U_s - U_p$; (b) $P_H - U_p$. Experimental data for single and polycrystalline Mo are also shown.

3.2. Mie-Grüneisen EOS

Previously, we had constructed thermodynamically complete global EOS model [9] that utilizes additive property of helmholtz free energy. EOS data in tabular form was generated for range of materials including Mo. Thermal EOS which relates thermodynamic variables for materials can offer insightful details about their phase diagrams, melting, etc. Mie-Grüneisen EOS is one such analytical form of global EOS that utilizes shock hugoniot or cold isotherm as reference state. This specially finds its practicality in hydrodynamic simulations of shock-induced phenomena [10].

With hugoniot as reference, MGEOS can be written as

$$P - P_H = \frac{\gamma(V)}{V}\left(E - E_H\right). \tag{3}$$

In Eq. 3, $\gamma(V)$ is the Grüneisen parameter and is a function of specific volume $V = 1/\rho$.

The analytical form of Grüneisen parameter is the well known expression,

$$\gamma(V) = \frac{t-2}{3} - \frac{V}{2}\frac{d^2\left[P_c(V)V^{2t/3}\right]/dV^2}{d\left[P_c(V)V^{2t/3}\right]/dV}. \tag{4}$$

The constant 't' takes the value of 0, 1 and 2 for Slater, Dugdale-MacDonald and Vashchenko-Zubarev models respectively [11]. $P_c(V)$ is the pressure along zero temperature isotherm. We employ D-M model ($t = 1$) to determine ambient pressure Grüneisen coefficient (γ_0) given by

$$\gamma(V_0) = \gamma_0 = -1 - \frac{V_0}{2}\frac{d^2\left[P_H(V_0)\right]/dV^2}{dP_H(V_0)/dV}, \tag{5}$$

wherein we have assumed hugoniot curve approaches cold isotherm at $V \to V_0$.
Taking first and second derivative of Eq. 2 and putting $V = V_0$ we get,

$$\gamma_0 = 2S - 1 \tag{6}$$

IOP Conf. Series: Materials Science and Engineering 1300 (2024) 012034 doi:10.1088/1757-899X/1300/1/012034

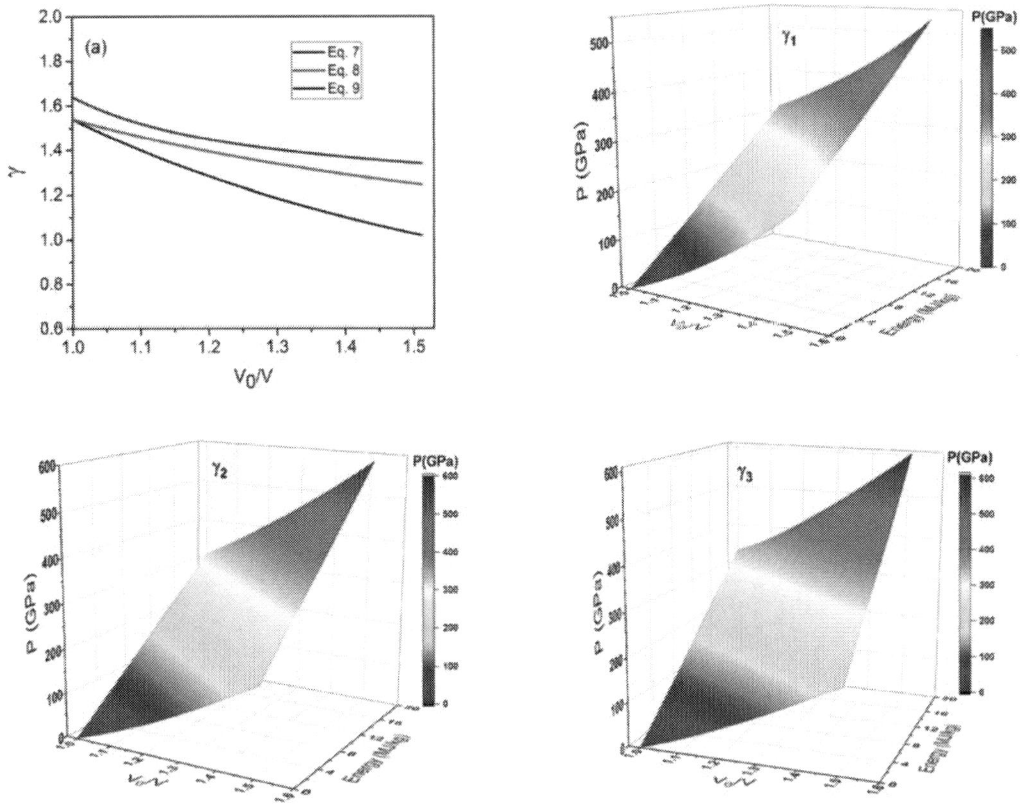

Figure 3: (a) Variation of Grüneisen parameter with compression (V_0/V) for three different models. (b, c, d) MGEOS data for three different models of $\gamma(V)$ as given in Eqs. 7, 8 and 9.

Using the above calculated value of S, we obtain $\gamma_0 = 1.54$. Incidentally, this value is in close agreement with experimental value (1.58) of Hixson [7].

Numerous empirical relations have been proposed to describe the volume dependence of $\gamma(V)$. In the present work, we have used three models of Ref. [11, 12, 13] to obtain MGEOS data given as:

$$\gamma_1(V) = \gamma_0 \frac{V}{V_0}, \tag{7}$$

$$\gamma_2(V) = \frac{2}{3} + \left(\gamma_0 - \frac{2}{3}\right) \frac{V}{V_0}, \tag{8}$$

$$\gamma_3(V) = \frac{1}{2} + g_1 \left(\frac{V}{V_0}\right)^{\frac{1}{3}} + g_2 \left(\frac{V}{V_0}\right)^{\alpha}. \tag{9}$$

Following Ref. [11], we take $g_1 = 0.95$, $g_2 = 0.19$ and $\alpha = 6.8$ for Mo. The volume dependence of γ as predicted by Eqs. 7, 8 and 9 is shown in Fig. 3(a). It can be noted that γ_0 estimated using Ref. [11] is higher than other two models, whereas γ when

Table 1: Comparison of $P-V-E$ data for three different models of Grüneisen parameter

E (MJ/kg)	V_0/V	$\gamma_1(V)$	$\gamma_2(V)$	$\gamma_3(V)$	P_1 (GPa)	P_2 (GPa)	P_3 (GPa)
5.0	1.2	1.28	1.3	1.42	144.49	150.48	153.41
	1.5	1.02	1.18	1.36	289.57	290.70	291.17
15.0	1.2	1.28	1.3	1.42	301.57	321.25	330.87
	1.5	1.02	1.18	1.36	446.65	482.05	496.89

calculated using Ref. [12] predicts faster decay with volume. Above three models of $\gamma(V)$ are then used to generate the P-V-E data (Eq. 3) in the compressed regime as displayed in Fig. 3(b), (c) and (d).

Table 1 shows quantitative comparison of pressure values estimated by three models of γ for representative energy values of 5.0 and 15.0 MJ/kg. There is no appreciable difference among the pressure values estimated by three models.

4. Conclusion

The shock hugoniot of Mo single-crystal along [001] orientation has been obtained by NEMD simulation which shows good agreement with experimental results. The hugoniot parameters are utilized to estimate the Grüneisen coefficient at ambient pressure. Finally, MGEOS has been constructed using different models for volume dependent Grüneisen parameter.

References

[1] Zeng Z Y *et al.* 2010 *J. Phys. Chem. B* **114** 298
[2] Singla A and Ray A 2022 *Phys. Rev. B* **105**(6) 064102
[3] Thompson A P *et al.* 2022 *Comp. Phys. Comm.* **271** 108171
[4] Stukowski A 2010 *Modelling Simul. Mater. Sci. Eng* **18** 015012
[5] Ackland G J and Thetford R 1987 *Philos. Magazine A* **56** 15–30
[6] McQueen R G *et al.* 1970 *High velocity Impact Phenomena ed. by R. Kinslow* (Academic Press.)
[7] Hixson R S and Fritz J N 1992 *J. Appl. Phys.* **71** 1721–1728
[8] Oniyama T, Gupta Y M and Ravichandran G 2021 *J. Appl. Phys.* **129** 245906
[9] Ray A, Srivastava M, Kondayya G and Menon S 2006 *Laser and Particle Beams* **24** 437–445
[10] Ray A and Singla A 2021 *Phys. Rev. B* **103**(5) 054301
[11] Burakovsky L and Preston D L 2004 *J. Phys. and Chem. Solids* **65** 1581–1587
[12] Meyer M A 1994 *Dynamic behaviour of materials* (John Wiley & Sons, Inc.)
[13] Young D A and Corey E M 1995 *J. Appl. Phys.* **78** 3748–3755

Structural and Optical Properties of AgO Nanoparticles Synthesized using Solid State Combustion

Thejas Ramakrishnaiah [a], Vinay D [a, b], Ganesh S Hegde [a, b], Siddivinayaka T S [a], K T Vasudevan [a], Abhiram Jagannathan [a, c*]

[a)] Department of Post Graduate Studies and Research in Physics, The National College, Jayanagar, Bangalore- 560070, Karnataka, India
[b)] Department of Physics, KLE's S Nijalingappa College, Rajajinagar, Bangalore-560010
[c)] Department of Physics, Indian Academy Degree College- Autonomous, Hennur Cross, Hennur Main Road, Kalyan Nagar, Bangalore-560043 Karnataka, India

*Corresponding Author: abhi.ram.jagan@gmail.com

Abstract: Silver oxide nanoparticles (AgO NPs) were synthesized by solid state combustion technique with $AgNO_3$ as a precursor and cow dung cake and cow urine as fuel. XRD results exhibit prominent peaks at 28°, 32°, 46° respectively indicating the AgO nanoparticles in cubic phase. Furthermore, structural elucidation through Rietveld refinement has confirmed the existence Ag_2O_3 nanoparticles, thereby indicating the formation of silver oxide nanoparticles. UV-Vis-NIR results are indicative that that the silver nanoparticles have exhibited a strong SPR peak at 435 nm, indicating the existence of AgNPs and it is interesting to note that solid state combustion of AgO NPs with cow dung cake has yielded to size of Ago Nps around 4.44 nm and AgO Nps with cow dung cake and cow urine has yielded to size of AgO Nps about 6.31 nm.

Keywords: Solid state combustion method, silver nanoparticles, surface plasmon resonance.

1. Introduction:

Thermal decomposition had been an effective method to synthesis metallic nanoparticles (MNPs) and it has gained much attention because of their ease of accessibility to tailor the shape and size of the particles through temperature or choice of an appropriate fuel cells [1]. These nanoparticles exhibit unique physical, chemical and biological properties, as their relatively larger surface area to the volume ratio is tuned, there is a possibility of either increased reactivity or stability in a chemical process, enhanced mechanical strength, etc., which enables them to venture into various applications. The electronic structure of nanoparticles reveals either *semiconducting or metallic, or insulating character* [2]. MNPs play a vital role in research domains in the field of *materials, chemistry, medicine, agriculture, information technology, biomedical, optical, electronics, catalysis, environment etc.,*

MNPs in particular have potential application in the biomedical field due to its anticancer properties, enhancement in radiotherapy, drug delivery, thermal ablation, antibacterial properties and so on. Nanoparticles provide effective carriers for biomolecules such as *DNA, RNA, or proteins*, protecting these materials from degradation and transporting them across the *cell-membrane barrier*. The size of nanoparticles used for the drug delivery was greater than 100 nm. As the *size of nanoparticles* is less they can penetrate through the cell membrane and deliver the drug to the target [3].

Collective oscillation of conduction electrons at a resonant frequency lead to a phenomenon called as surface plasmon resonance (SPR). The material absorbs light at the SPR resonant wavelength, the free electrons in the conduction band start to vibrate and dissipate energy. This process usually occurs at the surface of the material. The size of the nanoparticle increases when the plasmon resonance wavelength moves towards longer wavelength (red shift). These properties find potential applications in biosensors as they tend to cover water window regions [4]. Silica-Au Nano shells are tuned to perform SPR at near-infrared regions [5].

There is also a trend in enhancement of luminescence behaviour in the AgO nanoparticles leads towards development of *surface enhanced Raman scattering (SERS) technologies*. These nanoparticles have been widely used for plasmonic surface enhancement and also amplify the optical response of an electrochemical sensors. The electrochemical mechanism is based on the SPR which is influences the plasmonic metal nanoparticles with desired size and shape [6].

In general, cancer therapy is based on the *destruction of cancer cells* by lasers. When the high-intensity laser is incident on cancer cells the neighbouring cells might get affected. To control this, *nanoparticles are used such that damage to the healthy cell is controlled* [3].

This work discusses the synthesis of AgO nanoparticles from cow urine and cow dung cakes as fuels in the thermal decomposition process. As part of the Indian knowledge, traditionally agnihotri's use cowdung cakes for the sacrifices (havan's). This cow dung cake as fuel has been an inspiration to perform this research work in understanding the structure-property relationship of AgO nanoparticles. The prepared samples were subjected towards XRD and UV-Vis-NIR studies.

2. Experimental Methods:

2.1. Materials used:

Precursor salt silver nitrate {$Ag(NO_3)_2\ 6H_2O$- Loba Chemie, purity 99.9 % }, cow urine, cow dung cake, were used as fuels.

2.2. Synthesis of Ag_2O_3 nano-composites:

Silver chloride of 0.01mol% was transferred into a agate mortar pestle and grinded to a fine homogenous powder mixed with the appropriate amount of cow dung cake and cow urine (Bos Indicus) and the later mixture was transferred into a porcelain crucible. These samples were subjected to heat treatment in the muffle furnace at 600 °C and held at the operating temperature for four hours. During this process, as part of the thermal decomposition process, as the operating temperature is higher than the boiling point of the precursor, they tend to fragment from bulk particles and thereby result in formation of nanoparticles and thus it is indicative to see, that these particles have relatively lesser crystallite size. Thus conventional top-down method may cause significant *crystallographic damage* in the system [7]. The obtained sample was crushed again using agate mortar pestle. These powdered samples were stored in an airtight vial and sent for further characterizations. The sample prepared from cow dung cake is coded as A1, silver chloride with cow dung cake is coded as A2, silver chloride, cow dung cake and cow urine is coded as A3. The schematic process for synthesising Ag Nps with Bos Indicus and cow dung cake as fuel is shown in Figure 1.

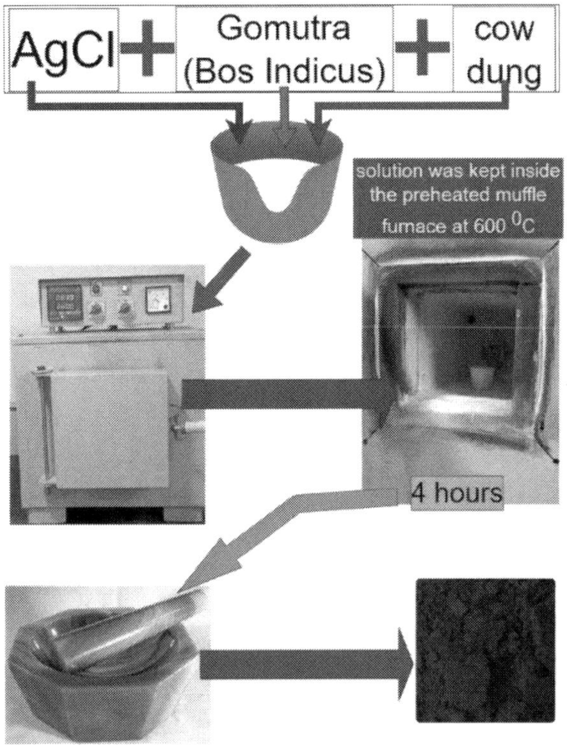

Figure 1. Schematic diagram of thermal decomposition method

3. Results and Discussion

3.1. XRD

The peaks of AgO particles indicates the Miller planes viz., (102), (024), (134), (22-6), (230), (26-1) (1-8-1) of the crystal lattice, which indicates that they have a simple cubic structure [8]. Existence of sharp peak around 22°, in sample A1 which is heat treatment of the fuel i.e., cow dung cake indicates that there could be an existence of carbon nanotubes [Ref] with its crystallite size around 28 nm, which is calculated using Debye-Scherrer equation [9]. It is also clear that, the Ag-Nps play a dominant role in the synthesis of nanoparticles as in case of samples A2 and A3, the featured peak at 22° is missing. To elucidate its crystal structure, these samples were subjected to Rietveld refinement using expo2014 software on the raw XRD data [10]. It was found that the raw XRD data fitted the standard model with the least error. Williamson-Hall analysis was performed on the XRD data of the samples and their crystallite size and the strain were calculated using the formula $C = \frac{0.9\lambda}{D}$ $m = \varepsilon$. Where 'C' is the y-intercept obtained from the graph plotted of 4sin θ versus βcos θ and m is the slope of the graph which were depicted in figure 2 and values were recorded in table 1. The slope gives the strain of the nanoparticle. We can notice that the size of nanoparticles was found to be 77.9 nm and 75.5 nm for A2 and A3 respectively. The strain was found to be 0.0017 and 0.00183 for A2 and A3 respectively.

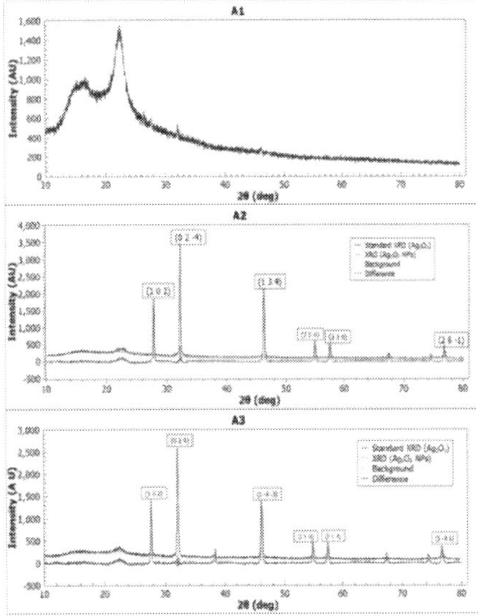

Figure 2. XRD patterns of the CNTs and nanosized Ag_2O_3 samples

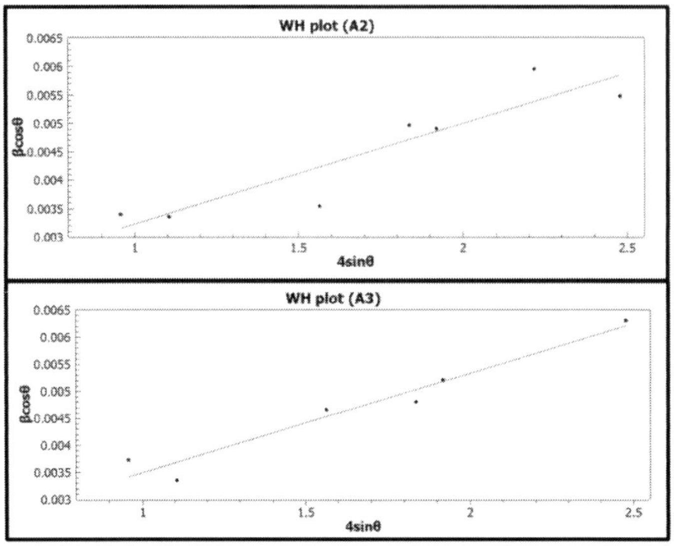

Figure 3. W-H Plot of A2 and A3 respectively.

Table 1. Results of W-H Plot.

Sample Code	A2	A3
Slope (Strain)	0.0017771	0.0018357
Y-intercept	0.001445374	0.0016586
D(m)	77.9922×10^{-9}	75.5025×10^{-9}
χ^2	1.08×10^{-6}	2.86×10^{-7}
R^2	0.844972267	0.948868412

3.2. UV-Visible Spectroscopy

The UV-Vis absorption spectrum of the sample A1 shows absorption peak at 220 nm is assigned to the π-plasmon of the graphitic structure, and 259 nm originates from the absorption features of CNT. The absorption spectrum obtained in our experiment provides evidence that the nanocarbons in the collision products comprise many defective carbon nanostructures instead of fine carbon nanostructures [11]. The synthesised AgO nanoparticles (A2 and A3) has shown higher absorption peak at 251 to 271 nm respectively which indicates the role of organic bond from the fuels and the existence of silver oxide is reflected through the characteristic SPR peak around 431-440 nm [1]. The position and shape of the surface plasmon absorption of noble metal nanoclusters are strongly dependent on the particle size, dielectric medium and surface adsorbed species, and may shift due to various factors during the production process. AgO nanoparticles have free electrons, which give SPR absorption band due to the combined vibration of electrons of AgO nanoparticles as well as from the result it can be tuned with the synthesis process and also with change in composition of the reducing agents [12], [13]. Additionally the energy bandgap values of the prepared CNT is 2.73eV and AgO nanoparticles (A2 and A3) is 3.08 and 2.91eV [14].

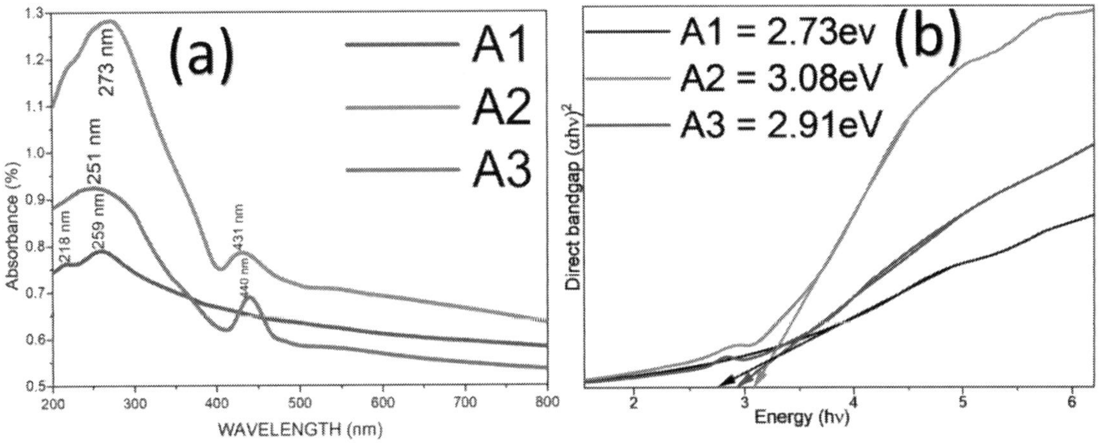

Figure 4 (a). UV -Vis Spectra **(b).** bandgap energy of Ag_2O_3 samples.

The size of NPs estimated theoretically by applying Mie theory of scattering as mentioned in the reference [15];

$$r = \frac{A\,V_f}{2\pi c\left\{\frac{\Delta\lambda}{\lambda_{SPR}^2}\right\}}$$

Where, V_f is the Fermi velocity, c is speed of light in vacuum, $\Delta\lambda$ is FWHM, λ^2_{SPR} is the surface plasmon resonance wavelength.

The size of the nanoparticle was found to be 4.41 nm and 6.31 nm for A2 and A3 samples respectively. The reduction in bandgap also indicates the red shift in its energy which correlates with higher wavelength shift in its SPR positions and the results correlate with the increase in the size of nanoparticle for sample A3.

4. Conclusion

The Silver nanoparticles were synthesised through thermal decomposition method using gomutra and cow dung cake as fuel agents. XRD analysis of the sample A1 indicates the formation CNT, A2 and A3 dominates with the formation of Ag_2O_3 nanoparticles. The size of the nanoparticles was found to vary due to the fuel agents. The UV-Vis characteristic peaks of CNT were found at 220 and 259 nm, the bandgap is at 2.73eV and AgO nanoparticles were found at 251 and 271 nm and 431 and440 nm, and the bandgap of the nanoparticles were found to be 2.91 and 3.08eV and the red shift in the SPR positions are clearly indicative that that the size of nanoparticle is slightly larger with combination of cow dung cake and gomutra together than cow dung cake alone.

References

[1] H. S. Sumantha, S. Rajagopal, G. Nagaraju, M. Shashank, and B. L. Suresha, 'Facile and eco-friendly combustion synthesis of NiO particles for photodegradation studies', *Chemical Physics Letters*, vol. 779, p. 138837, Sep. 2021, doi: 10.1016/j.cplett.2021.138837.

[2] M. S. Chavali and M. P. Nikolova, 'Metal oxide nanoparticles and their applications in nanotechnology', *SN Applied Sciences*, vol. 1, no. 6, p. 607, May 2019, doi: 10.1007/s42452-019-0592-3.

[3] O. Salata, 'Applications of nanoparticles in biology and medicine', *Journal of Nanobiotechnology*, vol. 2, no. 1, p. 3, Apr. 2004, doi: 10.1186/1477-3155-2-3.

[4] X. Luo, A. Morrin, A. J. Killard, and M. R. Smyth, 'Application of Nanoparticles in Electrochemical Sensors and Biosensors', *Electroanalysis*, vol. 18, no. 4, pp. 319–326, Feb. 2006, doi: 10.1002/elan.200503415.

[5] P. K. Jain, I. H. El-Sayed, and M. A. El-Sayed, 'Au nanoparticles target cancer', *Nano Today*, vol. 2, no. 1, pp. 18–29, Feb. 2007, doi: 10.1016/S1748-0132(07)70016-6.

[6] M. Švecová, O. Volochanskyi, M. Dendisová, D. Palounek, and P. Matějka, 'Immobilization of green-synthesized silver nanoparticles for micro- and nano-spectroscopic applications: What is the role of used short amino- and thio-linkers and immobilization procedure on the SERS spectra?', *Spectrochimica Acta Part A: Molecular and Biomolecular Spectroscopy*, vol. 247, p. 119142, Feb. 2021, doi: 10.1016/j.saa.2020.119142.

[7] S. Anu Mary Ealia and M. P. Saravanakumar, 'A review on the classification, characterisation, synthesis of nanoparticles and their application', *IOP Conference Series: Materials Science and Engineering*, vol. 263, no. 3, p. 032019, Nov. 2017, doi: 10.1088/1757-899X/263/3/032019.

[8] M. Thenmozhi, K. Kannabiran, R. Kumar, and V. Gopiesh Khanna, 'Antifungal activity of Streptomyces sp. VITSTK7 and its synthesized Ag2O/Ag nanoparticles against medically important Aspergillus pathogens', *Journal de Mycologie Médicale*, vol. 23, no. 2, pp. 97–103, Jun. 2013, doi: 10.1016/j.mycmed.2013.04.005.

[9] F. Xin and L. Li, 'Decoration of carbon nanotubes with silver nanoparticles for advanced CNT/polymer nanocomposites', *Composites Part A: Applied Science and Manufacturing*, vol. 42, no. 8, pp. 961–967, Aug. 2011, doi: 10.1016/j.compositesa.2011.03.024.

[10] G. S. Hegde, A. N. Prabhu, R. Y. Huang, and Y. K. Kuo, 'Reduction in thermal conductivity and electrical resistivity of indium and tellurium co-doped bismuth selenide thermoelectric system', *J Mater Sci: Mater Electron*, vol. 31, no. 22, pp. 19511–19525, Nov. 2020, doi: 10.1007/s10854-020-04383-7.

[11] Z. Tan *et al.*, 'Interstellar Analogs from Defective Carbon Nanostructures Account for Interstelar Extinction', *AJ*, vol. 140, no. 5, p. 1456, Oct. 2010, doi: 10.1088/0004-6256/140/5/1456.

[12] O. Pawar, N. Deshpande, S. Dagade, S. Waghmode, and P. Nigam Joshi, 'Green synthesis of silver nanoparticles from purple acid phosphatase apoenzyme isolated from a new source Limonia acidissima', *Journal of Experimental Nanoscience*, vol. 11, no. 1, pp. 28–37, Jan. 2016, doi: 10.1080/17458080.2015.1025300.

[13] Z. H. Dhoondia and H. Chakraborty, 'Lactobacillus Mediated Synthesis of Silver Oxide Nanoparticles', *Nanomaterials and Nanotechnology*, vol. 2, p. 15, Dec. 2012, doi: 10.5772/55741.

[14] J. Banua and J. I. Han, 'Biogenesis of Prism-Like Silver Oxide Nanoparticles Using Nappa Cabbage Extract and Their p-Nitrophenol Sensing Activity', *Molecules*, vol. 25, no. 10, Art. no. 10, Jan. 2020, doi: 10.3390/molecules25102298.

[15] A. Jagannathan *et al.*, 'Investigations on nonlinear optical properties of gold nanoparticles doped fluoroborate glasses for optical limiting applications', *Journal of Non-Crystalline Solids*, vol. 538, p. 120010, Jun. 2020, doi: 10.1016/j.jnoncrysol.2020.120010.

Green synthesis and characterization of Zirconium Oxide with antimicrobial activities

B N Veerabhadraswamy[1]*, H K Pradeep[2], K Swaroop[3], K M Manoj[4], M V Dhanush Nadigar[4], M P Akash Patel [5] and N P Bhagya[6]

[1]*Department of Chemistry, GM Institute of Technology, P B Road, Davanagere, India – 577 006*
[2]*Department of Pharmaceutics, GM Institute of Pharmaceutical Sciences and Research, P B Road, Davanagere, India – 577 006*
[3]*Department of Physics, GM Institute of Technology, P B Road, Davanagere, India – 577 006*
[4]*Department of Chemistry, Pooja Bhagavat Memorial Mahajana. P. G Centre Metagalli, K. R. S. Road, Mysuru, India - 570016*
[5]*Department of Biotechnology, GM Institute of Technology, P B Road, Davanagere, India – 577 006*
[6]*Department of Chemistry, Sai Vidya Institute of Technology, Bangalore, India - 560064*

***Correspondence Email**: vswamy136@gmai.com

Abstract: Green synthesis methods have garnered considerable attention due to their eco-friendly and sustainable nature. In this study, we report a green synthesis approach to fabricate Zirconium Oxide nanoparticles (ZrO_2 NPs) using cinnamon extract and citric acid as reducing and stabilizing agents. The green synthesis process was carried out through a sol-gel reaction method. The obtained ZrO_2 NPs were characterized using various analytical techniques, including UV-Vis, XRD and FTIR. The characterization results confirmed the successful synthesis of crystalline ZrO_2 NPs with a narrow size distribution and the presence of functional groups from cinnamon and citric acid on the NP surface. The synthesized ZrO_2 NPs were further investigated for their antimicrobial properties against bacteria and fungi. The results demonstrated potent antibacterial and antifungal activities of the green-synthesized ZrO_2 NPs. Overall; this study showcases a green and sustainable approach to synthesize ZrO_2 NPs using cinnamon extract and citric acid as reducing and stabilizing agents. The resulting ZrO_2 NPs exhibited promising antibacterial and antifungal properties, making them potential candidates for various biomedical and environmental applications, particularly in the development of novel antimicrobial agents and coatings.

Keywords: Green synthesis, Zirconium oxide nanoparticles, Cinnamon, Citric acid, Antibacterial activity, Antifungal activity, Nanomaterials, Eco-friendly synthesis.

1. Introduction

In recent years, the field of nanotechnology has witnessed a remarkable shift towards the development of eco-friendly and sustainable synthesis methods for nanoparticles, driven by the growing concerns over environmental impact and the need for greener technologies [1–4]. Nanoparticles have garnered significant attention due to their unique properties and wide-ranging applications, including in biomedicine, electronics, catalysis, and environmental remediation [4-7]. Among the various nanoparticles, Zirconium Oxide nanoparticles (ZrO_2 NPs) have emerged as a promising material with diverse potential applications, owing to their exceptional properties, such as high thermal stability, excellent mechanical strength, and remarkable chemical inertness [5-7]. Traditionally, the synthesis of ZrO_2 NPs has relied on conventional chemical methods, often involving the use of hazardous chemicals, high temperatures, and energy-intensive processes. These conventional approaches not only pose environmental risks but also raise concerns about the safety of the synthesized nanoparticles for various applications, particularly in biomedicine and environmental contexts [5-7]. Therefore, there is an urgent need to develop greener and more sustainable methods for the synthesis of ZrO_2 NPs. In response to this demand, the present study introduces a novel and environmentally friendly approach for the fabrication of ZrO_2 NPs using cinnamon extract and citric acid as reducing and stabilizing agents, respectively. The utilization of natural extracts and bio-derived compounds as reducing and stabilizing agents has gained prominence as an eco-friendly alternative to traditional chemical reagents. Cinnamon, a commonly available spice, has demonstrated remarkable potential in this regard due to its rich content of bioactive compounds, including polyphenols and flavonoids, which possess reducing and capping properties [8]. Citric acid, on the other hand, is a widely recognized green and biocompatible compound that can serve as an effective stabilizing agent for nanoparticles. These two environmentally friendly components in the synthesis of ZrO_2 NPs highlights the sustainable nature of the proposed method.

The synthesis of ZrO_2 NPs in this study was achieved through a sol-gel reaction method, a versatile and widely used approach for the production of nanoparticles. Sol-gel processes offer several advantages, including precise control over particle size and morphology, ease of scalability, and the ability to synthesize nanoparticles at relatively low temperatures [9-13]. These advantages align well with the principles of green chemistry, which emphasize resource efficiency, reduced energy consumption, and minimized waste generation. The sol-gel approach, combined with the use of cinnamon extract and citric acid, sets the stage for an eco-friendly and sustainable nanomaterial synthesis route. Characterization of the synthesized ZrO_2 NPs is of paramount importance to assess their quality, structure, and chemical composition. In this study, a comprehensive set of analytical techniques, including UV-Vis spectroscopy, X-ray diffraction (XRD) and Fourier-transform infrared spectroscopy (FTIR), was employed to characterize the ZrO_2 NPs [14-16]. These techniques provide valuable insights into the physical and chemical properties of the nanoparticles, confirming the successful synthesis of crystalline ZrO_2 NPs with a narrow size distribution. Additionally, the presence of functional groups derived from cinnamon extract and citric acid on the surface of the nanoparticles was elucidated through FTIR analyses, further supporting the green synthesis approach. Beyond the synthesis and characterization, this study also explores the potential applications of the green-synthesized ZrO_2 NPs, focusing on their antimicrobial properties against bacteria and fungi [15-16]. The emergence of antimicrobial resistance is a global health concern, necessitating the development of new antimicrobial agents and strategies. Nanoparticles have shown promise as effective antimicrobial agents due to their unique size-dependent properties and high surface area. Therefore, the investigation of the antimicrobial properties of ZrO_2 NPs synthesized using eco-friendly methods is of particular significance [14-16]. The results of our study reveal potent antibacterial and antifungal activities of the green-synthesized ZrO_2 NPs. These findings suggest that the nanoparticles exhibit promise as novel antimicrobial agents, which can find applications in various biomedical and environmental contexts. In particular, the development of antimicrobial coatings and materials using ZrO_2 NPs could have a profound impact on industries such as healthcare, food packaging, and water treatment, where the prevention of microbial contamination is of paramount importance [17-18].

In summary, this study represents a significant contribution to the field of green nanotechnology by introducing a sustainable approach for the synthesis of ZrO_2 NPs using cinnamon extract and citric acid. The environmentally friendly sol-gel method employed in this research yields crystalline ZrO_2 NPs with a narrow size distribution and surface functionalization with bio-derived compounds [19-20]. Moreover, the potent antibacterial and antifungal properties of the synthesized ZrO_2 NPs underscore their potential as innovative antimicrobial agents and coatings. As such, this study not only showcases a greener and more sustainable route to nanomaterial synthesis but also opens up exciting possibilities for the development of eco-friendly solutions to pressing challenges in biomedicine and environmental protection.

2. Results and Discussion

2.1. Synthesis

The target ZrO_2 nanoparticles were prepared according to Scheme-I (Fig.1). As can be seen, ZrO_2 synthesis required the key precursors Zirconyl nitrate and cinnamon extract / citric acid in 1:5 mole ratio, dissolved in 100 mL of distilled water in the beaker. The resulting solution is subjected to stirring at a speed of 450-500 rpm, pH of 14. The solution is then heated to a temperature range

IOP Conf. Series: Materials Science and Engineering 1300 (2024) 012036 doi:10.1088/1757-899X/1300/1/012036

between 75 °C and 80 °C for duration of 6 hours. The dull white products obtained (**MD-1** and **MD-2**) by the process been calcinated in muffle furnace under the temperature of 450 °C to get the target nanoparticles.

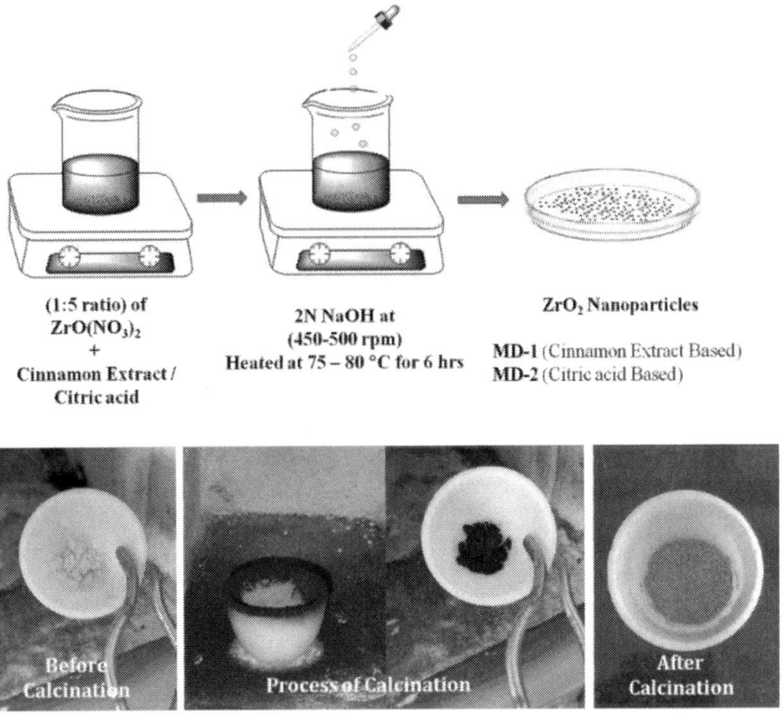

Figure 1. Schematic representation of Synthesis of ZrO_2 (**MD-1** and **MD-2**) and photographs of Calcination process

2.2. Evaluation of ZrO_2 nanoparticles

2.2.1. Fourier Transform Infrared (FTIR) Spectroscopy

The chemical structures of Zirconium oxide nanoparticles MD-1 and MD-2 are characterized by FTIR spectroscopy. The FTIR study confirms the functional groups present in the prepared Zirconium Oxide nanoparticles. Fig.2 shows the FTIR spectrum of a prepared sample. The weak peaks from ~ 3000 to ~ 3200 cm^{-1} correspond to stretching vibrations of O-H on the surface of Zirconium Oxide nanoparticles this may be due to the presence of moisture, while the peaks at ~700 cm^{-1}. Zr-O-Zr assymmetric and Zr-O-Zr stretching modes respectively

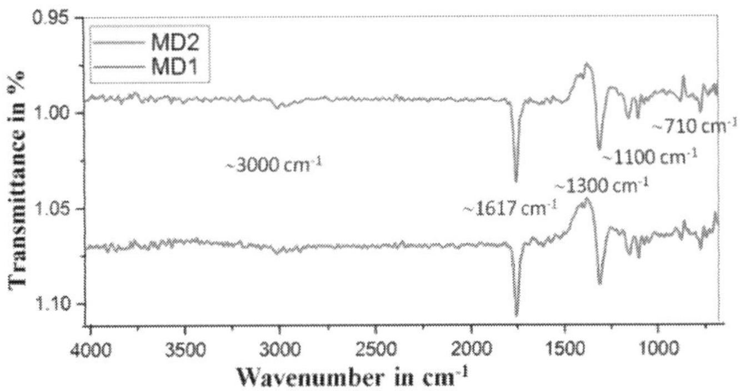

Figure 2. FTIR Spectrum of MD-1 (Gray line) and MD-2 (Red line) of ZrO_2 NPs.

confirm the formation of Zr-O The weak bands observed at ~1617 cm^{-1} and ~ 3000 cm^{-1} are explained by the stretching and bending vibrations of O-H bands. The bands absorbed by water molecules at ~1300 cm^{-1} confirm the absorption of non-bridging O-H groups. The peak observed at ~700 cm^{-1} occurs at a characteristic of *m*-ZrO.

2.2.2. X-ray diffraction

In order to elucidate the structure of the ZrO_2 synthesized by green method, powder X-ray measurements were performed in room temperature shown in Fig.3. The XRD experiments were carried out with CuK_α (λ = 0.15418 nm) radiation. The nanoparticles exhibit the tetragonal phase from diffraction peaks at 29.8 (111), 35 (200), 50.7 (220), and 60 (311) for MD-1 nanoparticles which synthesized using Cinnamon extract and 30.4 (111), 38.3 (200), 50.7 (220) and 59.9 (311) for MD-2. Also, the diffraction bands were confirmed their crystalline nature.

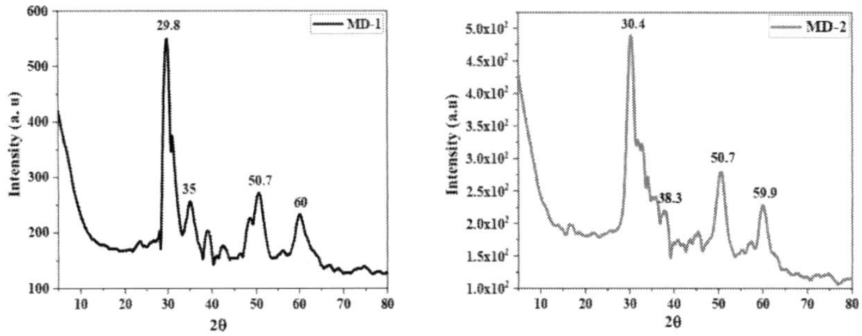

Figure 3. The intensity *vs* 2θ profile obtained for MD-1 and MD-2 nanoparticles.

2.2.3. UV-Vis Spectroscopy

Considering the fact that the synthesized nanoparticles are insoluble in any solvents, they are uniformly dispersed in water by sonification to get absorption peak using UV-Vis radiation. The solution UV-Vis spectra (~2 gm nps in 10 ml water) of MD-1 and MD-2 nanoparticles were found to be virtually identical; as representative cases MD-1 is shown in the Fig.4, spectrum comprises peak at ~ 263 which can be assigned to π-π* transitions, display distinct 650 nm up peak signifying the transformation of $ZrO(NO_3)_2$ into ZrO_2 NPs.

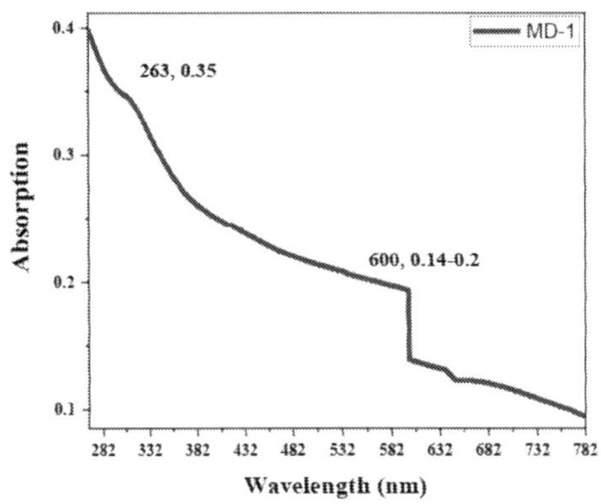

Figure 4. UV-Vis spectrum recorded for MD-1 and MD-2, dispersed in water

3. Microbial activity of ZrO_2 nanoparticles

The ZrO_2 (MD-2) nanoparticles were tested for antibacterial efficacy against gram negative bacteria (Pseudomonas aeruginosa and E. coli) and gram positive bacteria (Staphylococcus aureus and

Bacillus subtilis) using Well diffusion method. Cell suspension were prepared and grown on media and cultures were incubated for 24 hrs at 37°C. The cell suspensions of all the cultures were adjusted

Test Organisms	Test Compounds	Conc. per well	Zone of inhibition (mm)	Image of ZOI
Staphylococcus aureus (Gram +Ve)	Tetracycline (Standard)	25 µg/ml	18	
	MD 2	250 µg/ml	09	
Pseudomonas aeruginosa (Gram -Ve)	Tetracycline (Standard)	25 µg/ml	18	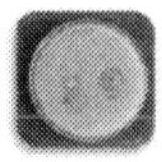
	MD 2	250 µg/ml	11	

Note: Results are average of three trials

to 1-2 × 10^6 cells/ml. The prepared MD-2 nanoparticles showed a good inhibitory action against Pseudomonas aeruginosa (inhibition zone size of 12 mm) at 250 g/mL compared to other bacteria. Furthermore, the nanoparticles were directly proportional to their inhibitory actions beside the tested microorganism. The biological applications of MD-2 nanoparticles have been confirmed more significantly in Pseudomonas aeruginosa (Gram –ve).

4. Conclusion

In the presented study, demonstrated that a green and sustainable method for the synthesis of Zirconium Oxide nanoparticles, MD-1 and MD-2 using Cinnamon extract and Citric acid as both reducing and stabilizing agents respectively. The utilization of a sol-gel reaction method allowed for the preparation of ZrO_2 NPs. The synthesized ZrO_2 NPs were subjected to thorough characterization using UV-Vis, XRD, FTIR spectroscopic analytical techniques, confirming structural properties, support the quality and suitability of the green-synthesized ZrO_2 NPs for further applications. Most notably, this study explored the antimicrobial properties of the green-synthesized ZrO_2 (MD-1 and MD-2) NPs and revealed their significant efficacy against bacteria.

Acknowledgement

The authors are grateful to the Principal and Management, GMIT for financial support. Also, authors are thankful to Dr. Girish Bolakatti, Principal, GMIPSR for providing laboratory facilities.

5. Reference

1. Ahmed S F, Mofijur M, Rafa N, Chowdhury A T, Chowdhury S, Nahrin M, Islam A B M S, Ong H C, 2022, *Environmental Research*, **204**, 111967.
2. Grace A V M- G, Alessandra M de O, Jean P S N, Rosângela B, Raquel G G, 2021, *Environmental Nanotechnology*, **16**, 100526.
3. Rajender S V, 2016, *ACS Sustainable Chem. Eng.* **4**, 11.
4. Ruma A S, Rizwan M A, Singh S, 2022, *Nanotechnology for Environmental Engineering*, **7**, 661.
5. Vishal D, Jyothis D, Ankush C, Jayalakshmi M, Vasantha V L, Ayush Jha, Aatika N, Kun-Yi A Lin, Suresh G, 2022, *Chemical Engineering Journal Advances*, **11**, 100353.
6. Kim D- Y, Avinash K, Surendra S, Rijuta G S, Jayanta P, Gajanan G, 2017, *Journal of the Science of Food and Agriculture*, **98**, 849.

7. Zainab A, Rooh U, Mustafa T, Saif U, Abdur R, Tawfik A S, 2023, *Trends in Environmental Analytical Chemistry*, **37**, e00187
8. Alessandra De G, Stefania P, Jessica Z, Matilde F, Paola F, Ilaria B, Luca C and Patrizia Di G, 2022, *Foods*, **11**, 3202.
9. Ajay V R, Krishnan K, Abitha V K, Sabu T, 2018, *Synthesis of Inorganic Nanomaterials*, **Chapter 5**, pp 121-139, Elsevier.
10. Singh L P, Bhattacharyya S K, Rahul K, Geetika M, Usha S, Garima S, Saurabh A, 2014, *Advanced in Colloid and Interface Science*, **214**, 17.
11. M. Yazdanbakhsh, I. Khosravi, E.K. Goharshadi, A. Youssefi, 2010, *J. Hazard. Mater.* **184**, 684.
12. I. Ferino, M. Casula, A. Corrias, M. Cutrufello, R. Monaci, G. Paschina, 2000, *Phys. Chem. Chem. Phys*. **2**, 1847.
13. C. Suciu, L. Gagea, A. Hoffmann, M. Mocean, 2006, *Chem. Eng. Sci.* **61**, 7831.
14. Mohammed A M, Ali M M, 2021, *Research Journal pf Pharmacy and Technology*, **14**, 938.
15. Anandhi S, Edward Leo M, Jaisanka V, 2021, *Materialstoday:proceedings*, **40**, S93.
16. Nabil Al- Z, Muthuvel A, Jothibas M, Ali A, Fahad A A, Mohana V, 2021, *Inorganic Chemistry Communications*, **127**, 108507.
17. Duygu K, Gökhan G G, Gulsun A E, Tamer A, Tomáš V, Piotr K, Anet R J, Fatih O, 2023, *Trends in Food Science & Technology*, **135**, 144.
18. Michał M, Małgorzata G- M, 2013, *Chemical Engineering Journal*, **228**, 596.
19. Fatimah I, Ika Y, Totok E S, Suresh S, 2022, *Inorganic Chemistry Communications*, **143**, 109808.
20. Shiyun M, Kai L, Hong L, Zhenyu Z, Xiaoqi L, Yilai J, Xingang L, Xin G, Xiaolei F, 2022, *Fuel Processing Technology*, **233**, 107292.

IOP Conf. Series: Materials Science and Engineering 1300 (2024) 012037 doi:10.1088/1757-899X/1300/1/012037

Virtual substrate method: synthesis and growth kinetics of 2D metal oxide nanosheets

S M Aiswarya[1,2], Somesh Chandra[1,2], Ashish Jain[3], Shamima Hussain[1], S Ganesamoorthy[4], G M Bhalerao[1*] and R Govindaraj[4]

[1]UGC-DAE CSR Kalpakkam Node, Kokilamedu, Tamil Nadu – 603104, India
[2] University of Madras, Chennai, Tamil Nadu – 600005, India
[3] MC & MFCG, IGCAR, HBNI, Kalpakkam, Tamil Nadu – 603102, India
[4] Material Science Group, IGCAR, HBNI, Kalpakkam, Tamil Nadu – 603102, India
* *Corresponding author: gmbhalerao@csr.res.in*

Abstract. We present the experimental decomposition kinetics for the synthesis of metal oxide 2D nanosheets by the virtual substrate method. In this method, acetates of Mg and Cu were used as precursors for the growth of prototype MgO and CuO nanosheets, while the filter paper was utilized as a virtual substrate. The synthesized samples were characterized by X-ray diffraction for the phase analysis of the nanostructures, which confirms the cubic phase of MgO and monoclinic phase of CuO; with a minor Cu_2O phase. Field emission scanning electron microscopy was used to determine the surface morphology of the nanosheets. Fourier transform infrared spectroscopy was utilized to identify the vibrational modes of the metal oxides and the functional groups present in the sample. Thermogravimetric analysis revealed the evolution of combustive oxidation of filter paper with the thermal decomposition of acetates situated on its surface.

1. Introduction

Two-dimensional (2D) nanosheets are freestanding nanomaterial with a high lateral size-to-thickness ratio. Its remarkable electrical, mechanical and catalytic characteristic distinguishes it from bulk counterparts or nanoparticles. This is due to their distinct structure, peculiar surface chemistry, and quantum size effect [1–3]. Following the discovery of graphene, 2D nanosheets have been produced and used extensively as fillers in polymer composite [4]. In many applications, including energy storage, catalysis, and sensing, metal oxide nanosheets have drawn a lot of interest; for example, the use of nanosheets in Li-ion battery anodes [5], in sensor materials with high selectivity toward formaldehyde and acetone against interfering gases [6], and in development of synthetic photosynthetic systems [7]. Even though the potential and significance of the applications of metal oxide nanosheets are escalating, there are only limited synthesis techniques. These techniques generally require a lot of efforts, time, expenditure and a need of pre- and post- processing [8–13]. Recently an effortless, low cost and sustainable method for the synthesis of copper oxide (CuO) and magnesium oxide (MgO) nanosheets was reported, namely virtual substrate method, in which ashless filter paper was utilised as the virtual substrate [14].

Both CuO and MgO have their own remarkable application level importance in the field of nanomaterials. For instance, CuO nanosheets serve as a promising material to make high sensitive gas sensors [15], exhibits an incredible electrocatalytic response to H_2O_2 [16], photocatalytic activity [17]. The MgO is used in catalysis, as an additive in refractory, paint and superconducting products [18]. Besides, the mixed CuO-MgO shows highly efficient and excellent photocatalytic properties [19].

Our present work elucidates the experimental decomposition kinetics for synthesis of copper oxide and magnesium oxide 2D nanosheets by virtual substrate method. Here, ashless filter paper acts as a supporting material for the growth of metal oxide nanosheets.

2. Experimental

2.1 Materials

For the formation of metal oxide nanosheets, 99% pure copper acetate monohydrate ($Cu(CH_3COO)_2.H_2O$ or $Cu(AcO)_2$) and magnesium acetate tetrahydrate ($Mg(CH_3COO)_2.4H_2O$ or $Mg(AcO)_2$) were used as precursors. A virtual substrate, specifically Whatman Grade 41 ashless filter paper with an ash residue rating of 0.007%, was utilized to facilitate the growth of the nanosheets. De-ionized water was used as a solvent.

2.2 Synthesis

Three samples using precursors and one control sample were prepared. The first two samples of CuO (S1) and MgO (S2) were prepared using 0.2 M aqueous solutions of pure copper acetate and magnesium acetate respectively. A mixture of 0.2:0.1 M concentration of copper acetate and magnesium acetate was used in the synthesis of third sample (S3). All the three solutions were independently loaded to the ashless filter papers and allowed to dry in normal atmospheric conditions, according to the literature [14]. Dried precursor loaded filter papers were combusted in air, leaving metal oxide ash residues. These residues were collected and used for further characterization. Bare filter paper was used as control sample.

2.3 Characterization

X-ray diffraction (XRD) measurement of thoroughly grinded samples were performed to examine phase purity and crystallinity of the synthesized samples. XRD was done using GNR APD 2000 PRO powder X-ray diffractometer with Cu-kα radiation of wavelength 1.5406 Å in step of 0.1°from 30° - 85° of 2θ-range. Investigation of surface morphology of the synthesized samples were done using a CARL ZEISS SUPRA 55 field-emission scanning electron microscope (FE-SEM) with an operating voltage of 5 kV. The FE-SEM samples were ultrasonically processed in ethanol, then placed on Si-substrate and coated by gold. Fourier transform infrared (FTIR) measurement was performed on Bruker Vertex 80v spectrometer in the range of 300 to 4000 cm^{-1} to determine the purity and functional group present in the sample. Potassium bromide (KBr) pellet technique was used to measure the infrared spectra. The thermogravimetry and differential thermal analyses was carried out by using a TGA / DSC 3+ combined differential thermal analysis and thermogravimetric analyzer supplied by M/S. Mettler Toledo, Switzerland. The temperature calibration of thermogravimetric analyzer was carried out by the method of fixed melting points employing International Confederation for Thermal Analysis and Calorimetry (ICTAC) recommended standards such as indium, aluminum and gold. This equipment has a mass sensitivity of 0.1 μg and a temperature sensitivity of 0.01 K. A platinum crucible of 70 μl capacity was used as the sample container. For each experiment 5 mg of the sample was used. The temperature was measured using a Pt-10 % Rh thermocouple which was in firm contact with the sample holder on which the platinum sample container was placed. The inner muffle of the furnace was constantly purged with air, the flow of which was controlled by a mass flow controller. A constant gas flow of 20 sccm, a pre-determined rate of heating of 5 K/min and a temperature range of 303 -1073 K were used to carry out dynamic experiments.

3. Results and Discussion

The XRD patterns of the produced powdered samples S1, S2, and S3 are illustrated in Figure 1. It shows monoclinic phase of CuO and a small Cu_2O phase in Cu containing samples S1 and S3 (Figure 1(a, c)). As demonstrated in Figure1 (b, c), the XRD pattern of S2 and S3 reveals the cubic phase of MgO. Figure 1(c) confirms the existence of both phases of CuO and MgO in the sample S3. Peaks profile indicates towards good crystallinity as well as purity in the synthesized samples. The XRD pattern of all samples indicates that composites have no detectable secondary phase or impurity peaks, except for a minor Cu_2O phase in Cu containing samples.

FE-SEM micrographs of the samples are shown in Figure 2. It displays the surface morphology of each samples as layered structures for CuO (S1), thin sheets for MgO (S2), and folded sheet like structures for CuO-MgO (S3) samples. Notably, in contrast to the completely sheet-like structure found in the S1 (Figure 2(a)) and S2 (Figure 2(b)) samples, S3 contains granular, fibrous as well as highly porous features. The filter paper is made of fibres of cellulose. Therefore, when it is drenched into a solution containing the precursors, it gets on its surface fully coated by these precursors. In case of S1 and S2, precursors were coated on surface of cellulose and result into formation of metal oxide nanosheets followed by combustive heating. The mixed sample S3 has two different kind of nanostructure formation; that of MgO nanosheets and CuO nanoparticles. In a previous report, it is established that it happens because the magnesium precursor has higher solubility than copper acetate [14]. Thus, it confirms the role of filter paper's (virtual substrate) surface characteristics in determining the product morphology.

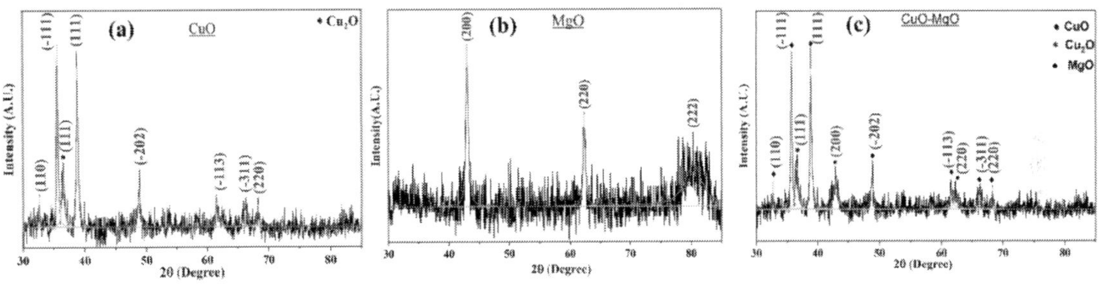

Figure 1. XRD of synthesized samples **(a)** CuO(0.2M - $Cu(AcO)_2$), **(b)** MgO(0.2M - $Mg(AcO)_2$) and **(c)**mixture of CuO and MgO (0.2M:0.1M -$Cu(AcO)_2$: $Mg(AcO)_2$).

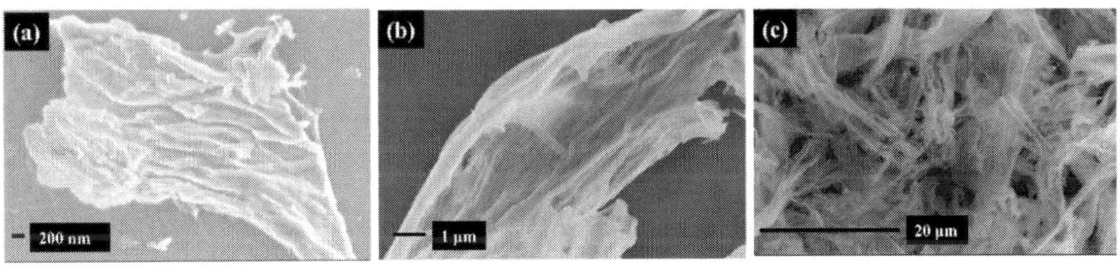

Figure 2. FESEM micrographs of **(a)** S1 (CuO) **(b)** S2 (MgO) and **(c)** S3 (CuO-MgO).

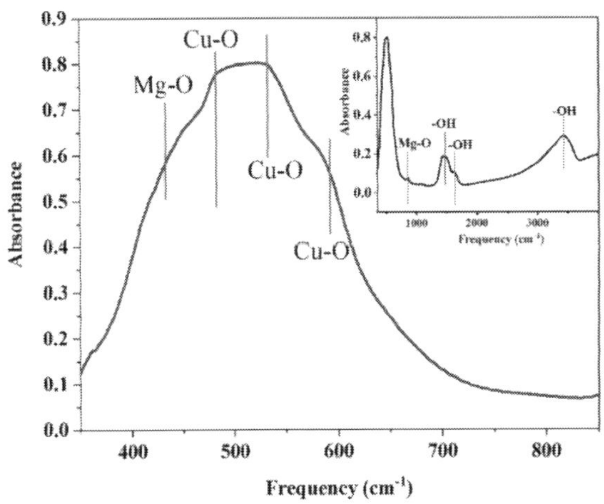

Figure 3. FTIR spectrum of S3 (CuO-MgO). Inset graph showing the –OH functional group present at the surface of nanosheets.

The functional groups present on the surface and chemical structure of S3 were further investigated using the FTIR spectrum, as shown in Figure 3. Bands at 432 and 862 cm^{-1}indicates the stretching vibrations of Mg-O bond [20] and peaks at 482, 532, 592 cm^{-1} confirms the stretching and bending vibrations of the Cu-O bonds [21]. Presence of the bending (1455, 1633 cm^{-1}) and the stretching modes (3434 cm^{-1}) of hydroxyl group indicate the absorption of water molecules from the air by the nanosheets [22,23].

TG and DTG thermograms of four samples, copper acetate loaded filter paper (Cu/FP), magnesium acetate loaded filter paper (Mg/FP), mixed loaded filter paper (Cu-Mg/FP) and bare filter paper (FP), are shown in Figure 4. While considering TG curve, around 373- 450 K, all the thermograms show a mass loss of around 9-10%, which is related to the loss of moisture and the production of anhydrous acetates [24,25]. As we heat further, there is a huge and an evident decrease in sample's weight between 495 K and 750 K. It is due to the decomposition of precursors [24] as well as depolymerisation and volatilization of organic components in filter paper [25,26]. Table 1 shows the values of the initial temperature (T_i), the temperature at which the reaction rate is maximum (T_m), and the final temperature (T_f) of the major decomposition process. It is observed that the Cu/FP has the lowest thermal stability, as it starts to decompose at the lowest temperature when compared with other samples, including the bare filter paper. The decomposition temperature range 453-573 K of $Cu(AcO)_2$ [27] is lower than that 548-613 K of $Mg(AcO)_2$ [28] and 573-873 K of filter paper [29]. This may be the reason behind the accelerating decomposition pattern of Cu/FP sample.

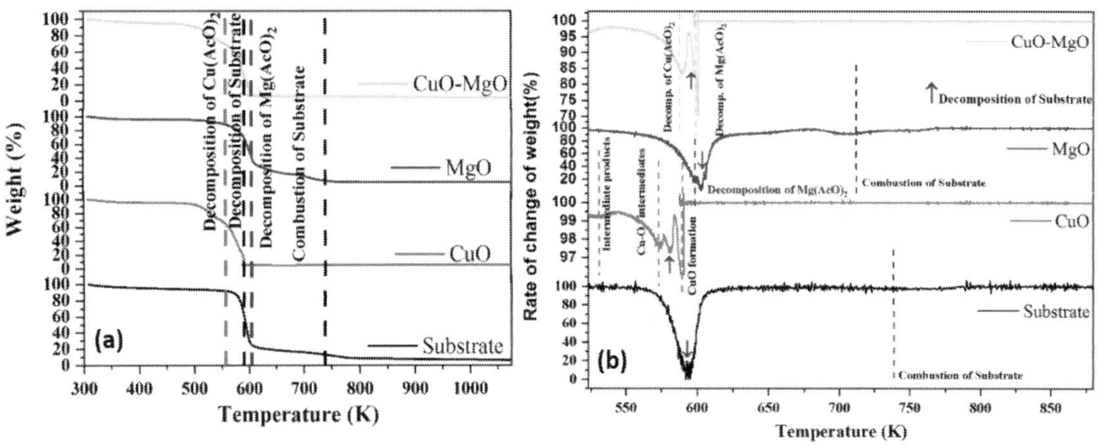

Figure 4. (a) TG and (b) DTG curves of precursor loaded filter paper and bare filter paper. DTG curve used to examine the major decomposition in the temperature range 520 – 880 K.

Table 1. Characteristic temperatures from TG curves

Samples	T_i(K)	T_m(K)	T_f(K)
FP	573	594	603
Cu/FP	502	585	590
Mg/FP	565	603	610
Cu-Mg/FP	509	595	600

The exact decomposition pattern can be obtained from the DTG curve, as shown in Fig 4 (b). For filter paper there are two decomposition phases. The one around 593 K corresponds to the loss of cellulose, and the other one in the range 700-750 K corresponds to the gradual decomposition (combustion) [29]. For Cu/FP, there is a series of many steps, probably corresponding to intermediate decomposition product formation processes from 534-588 K [27], as can be clearly observed in the DTG curve (Figure 4(b)). This indicates that initially the copper acetate starts to decompose forming intermediates around 530-574 K. This is then closely followed by the loss of cellulose which happens at around 581 K. We propose that while the cellulose is still decomposing into carbon, this newly formed carbon will take part in the formation of Cu_2O. In this process, the carbon will be catalytically consumed by the copper intermediates. Because of the carbon influence in the formation of Cu_2O, gradual combustion phase of filter paper is absent in Cu/FP. On the other hand, this phase of gradual combustion is present in both the FP and Mg/FP specimens, which do not have any copper content, corroborating the proposed mechanism. Further increase in the temperature around 590 K, oxidization of Cu_2O into CuO occurs, during which mass increases considerably. This is observed in the DTG curve (Figure 4(b)), in agreement with the literature value of 573-673 K [27].

In the case of Mg/FP, the decomposition of magnesium acetate around 599 K followed by the loss of cellulose occurring around 603 K. However, these temperature values are very close and it can be stated that these two events occur almost simultaneously, with a clear precedence for the $Mg(AcO)_2$ decomposition. Here, the gradual decomposition of filter paper is visible around 712 K, as observed in FP (Figure 4(b)). This is because carbon has no interference in the formation of magnesium oxides, as previously discussed in the case of copper containing precursors.

From the above discussion, it is clear that Cu and Mg cations can alter the thermal stability of filter paper. Also, the temperature of FP to initiate decomposition i.e. 573 K (from Table 1), is higher than the other samples. This means that before the decomposition of FP, the process of decomposition of acetates to form oxides gets started. This prevents the agglomeration of the newly formed nanostructures [14].

4. Conclusion

We have synthesised metal oxide nanosheets by virtual substrate method and examined the detailed thermal decomposition kinetics of metal oxide and bare filter paper by TG/DTG analysis. The study indicates that the decomposition temperatures of acetates are less than that of filter paper, which supports the formation of nanosheets and prevents their agglomeration. The DTG analysis shows the involvement of carbon in the formation of Cu_2O in copper containing samples while a carbon independent decomposition process is noticed in case of magnesium acetate. It indicates the effects of the substrate on the growth process, which provides a significant insight into the fundamental mechanisms governing the growth of nanosheets on filter paper.

Acknowledgements

Help of Dr. Bijay Kumar Das, IGCAR in FTIR characterizations is gratefully acknowledged.

References

[1] Ten Elshof J E 2017 Electronic band structure and electron transfer properties of two-dimensional metal oxide nanosheets and nanosheet films *Curr. Opin. Solid State Mater. Sci.* **21** 312–22

[2] Moghadassi A R, Koranian P, Hosseini S M, Askari M and Madaeni S S 2014 Surface modification of heterogeneous cation exchange membrane through simultaneous using polymerization of PAA and multi walled carbon nano tubes *J. Ind. Eng. Chem.* **20** 2710–8

[3] Safarpour M, Arefi-Oskoui S and Khataee A 2020 A review on two-dimensional metal oxide and metal hydroxide nanosheets for modification of polymeric membranes *J. Ind. Eng. Chem.* **82** 31–41

[4] Li J, Liu X, Feng Y and Yin J 2022 Recent progress in polymer/two-dimensional nanosheets composites with novel performances *Prog. Polym. Sci.* **126** 101505

[5] AbdelHamid A A, Yu Y, Yang J and Ying J Y 2017 Generalized Synthesis of Metal Oxide Nanosheets and Their Application as Li-Ion Battery Anodes *Adv. Mater.* **29** 1701427

[6] Kim R, Jang J S, Kim D H, Kang J Y, Cho H J, Jeong Y J and Kim I D 2019 A General Synthesis of Crumpled Metal Oxide Nanosheets as Superior Chemiresistive Sensing Layers *Adv. Funct. Mater.* **29** 1903128

[7] Maeda K and Mallouk T E 2018 Two-Dimensional Metal Oxide Nanosheets as Building Blocks for Artificial

Photosynthetic Assemblies *https://doi.org/10.1246/bcsj.20180258* **92** 38–54

[8] Sun Z, Liao T, Dou Y, Hwang S M, Park M S, Jiang L, Kim J H and Dou S X 2014 Generalized self-assembly of scalable two-dimensional transition metal oxide nanosheets *Nat. Commun.* **5**

[9] Ding Y L, Wen Y, Van Aken P A, Maier J and Yu Y 2015 Rapid and Up-Scalable Fabrication of Free-Standing Metal Oxide Nanosheets for High-Performance Lithium Storage *Small* **11** 2011–8

[10] Yang J, Zeng Z, Kang J, Betzler S, Czarnik C, Zhang X, Ophus C, Yu C, Bustillo K, Pan M, Qiu J, Wang L W and Zheng H 2019 Formation of two-dimensional transition metal oxide nanosheets with nanoparticles as intermediates *Nat. Mater.* **18** 970–6

[11] Zhao H, Zhu Y, Li F, Hao R, Wang S and Guo L 2017 A Generalized Strategy for the Synthesis of Large-Size Ultrathin Two-Dimensional Metal Oxide Nanosheets *Angew. Chemie Int. Ed.* **56** 8766–70

[12] Hou C, Zhang M, Zhang L, Tang Y, Wang H and Chi Q 2017 Reagent-Free Electrophoretic Synthesis of Few-Atom-Thick Metal Oxide Nanosheets *Chem. Mater.* **29** 1439–46

[13] Peng L, Xiong P, Ma L, Yuan Y, Zhu Y, Chen D, Luo X, Lu J, Amine K and Yu G 2017 Holey two-dimensional transition metal oxide nanosheets for efficient energy storage *Nat. Commun.* **8**

[14] Chandra S, Ganesamoorthy S, Dasgupta A, Bhalerao G M and Shekar N V C 2023 Synthesis of large 2D oxide nanosheets by virtual substrate method and growth-kinetics studies using prototype CuO/MgO *Thin Solid Films* **768** 139726

[15] Umar A, Alshahrani A A, Algarni H and Kumar R 2017 CuO nanosheets as potential scaffolds for gas sensing applications *Sensors Actuators B Chem.* **250** 24–31

[16] Hsu Y K, Chen Y C and Lin Y G 2015 Spontaneous formation of CuO nanosheets on Cu foil for H2O2 detection *Appl. Surf. Sci.* **354** 85–9

[17] Li F, Liu X, Zhang Q, Kong T and Jin H 2012 Fabrication and photocatalytic property of CuO nanosheets via a facile solution route *Cryst. Res. Technol.* **47** 1140–7

[18] Shah M A 2013 Large Scale Production of MgO Nanostructures and their Possible Applications *Mater. Sci. Forum* **760** 69–71

[19] Priya R, Stanly S, Dhanalekshmi S B and Sagadevan S 2019 Fabrication of photocatalyst MgO: CuO composite and enhancement of photocatalytic activity under UV light *Mater. Res. Express* **6** 125023

[20] Sutapa I W, Wahab A W, Taba P and Nafie N La 2018 Synthesis and structural profile analysis of the MgO nanoparticles produced through the sol-gel method followed by annealing process *Orient. J. Chem.* **34** 1016–25

[21] Askari P, Faraji A, Khayatian G and Mohebbi S 2017 Effective ultrasound-assisted removal of heavy metal ions As(III), Hg(II), and Pb(II) from aqueous solution by new MgO/CuO and MgO/MnO2 nanocomposites *J. Iran. Chem. Soc.* **14** 613–21

[22] Reddy K R 2017 Green synthesis, morphological and optical studies of CuO nanoparticles *J. Mol. Struct.* **1150** 553–7

[23] Madejová J, Janek M, Komadel P, Herbert H-J and Moog H C 2002 FTIR analyses of water in MX-80 bentonite compacted from high.pdf *Appl. Clay Sci.* **20** 255–71

[24] Zhang K, Hong J, Cao G, Zhan D, Tao Y and Cong C 2005 The kinetics of thermal dehydration of copper(II) acetate monohydrate in air *Thermochim. Acta* **437** 145–9

[25] Ahmad I, Kamal T, Khan S B and Asiri A M 2016 An efficient and easily retrievable dip catalyst based on silver nanoparticles/chitosan-coated cellulose filter paper *Cellulose* **23** 3577–88

[26] Arseneau D F 1971 Competitive Reactions in the Thermal Decomposition of Cellulose *Can. J. Chem.* **49** 632–8

[27] Naktiyok, Jale and Özer A K 2019 SYNTHESIS OF COPPER OXIDE (CuO) FROM THERMAL DECOMPOSITION OF COPPER ACETATE MONOHYDRATE (Cu(CH3COO)2.H2O) *Niğde Ömer Halisdemir Üniversitesi Mühendislik Bilim. Derg.* **8** 1292–8

[28] Afzal M, Butt P K and Ahmad H 1991 Kinetics of thermal decomposition of metal acetates *J. Therm. Anal.* **37** 1015–23

[29] Wang B, Han N, Meng D, Yue R, Yan J and Chen Y 2011 Filter paper-templated preparation of ZnO thin films and examination of their gas-sensing properties *Particuology* **9** 253–9

IOP Conf. Series: Materials Science and Engineering 1300 (2024) 012038 doi:10.1088/1757-899X/1300/1/012038

Physicochemical Investigation of Synthesized Bismuth and Silver-Doped Bismuth Nanoferrites, and their Dielectric Properties

Amrutesh Kannolli[1], Avinash P[1*]

[1]*Department of Studies in Physics, Vijayanagara Sri Krishnadevaraya University, Ballari, 583 105, Karnataka, INDIA*
Email: avinashp@vskub.ac.in

Abstract
Bismuth nano ferrite and silver-doped bismuth nano ferrite are synthesized by the sol-gel method. The synthesized compound is characterized through several characterization methods such as the X-ray diffraction method, Fourier transform infrared spectroscopy (FTIR), Scanning electron microscopy (SEM), and UV-Visible spectral studies. The X-Ray diffraction confirms bismuth nano ferrite has a Rhombohedral structure and belongs to the R-3m space group, and the silver-doped bismuth nano ferrite has an orthorhombic structure and belongs to the Cmcm space group. The crystallite size is calculated through Scherrer's formula, the bismuth nano ferrite has a crystallite size of 20 nm, and the silver-doped bismuth nano ferrite has a crystallite size of 36.33 nm. FTIR studies confirm the formation of ferrite and indicate the metal ion group observed the peak shift after doping in the fingerprint region. The metal ion peaks are presented at 446.7 cm^{-1} and 527.8 cm^{-1} but after doping they shifted to 453 cm^{-1}, 534 cm^{-1}. The UV-Visible spectral analysis identifies the change in the energy band gap through a tauc plot, initially the energy band gap of bismuth nano ferrite is about 2.79 eV but after doping it reduces to 1.61 eV. The surface morphology is determined by SEM analysis and it provides 30 nm and for the doped compound 37 nm average particle size. Dielectric study confirms doping of silver to the bismuth gives more improved results which will be helpful at the application level such as in memory devices, photocatalysis, water splitting, and gas detection.

1. Introduction

Ferrites are a special type of magnetic material that are extensively employed in several technological applications. In addition to iron oxide, they may also contain other metal oxides [1]. High magnetic permeability, electrical conductivity, and exceptional chemical stability are all characteristics of ferrites Based on their crystal structure, ferrites may be categorized into several groups, such as spinel ferrites (AB_2O_4), hexagonal ferrites ($BaFe_{12}O_{19}$), and garnet (perovskite) ferrites ($Bi_3Fe_5O_{12}$) [2]. Bismuth nano ferrite has a perovskite structure with a cubic unit cell and a corner-centered lattice. Bismuth ions (Bi^{3+}) and iron ions (Fe^{3+}) are arranged in precise patterns in the unit cell. The magnetic and electrical characteristics of bismuth nano ferrite are determined by the arrangement of these ions. Because of these arrangements, the ions show both ferroelectric and magnetic characteristics. Bismuth nano ferrite has an array of applications in the files of spintronics, data storage, energy harvesting, and sensing [3]. It is often employed in high-temperature sensors because of its high Curie temperature. Bismuth nano ferrite can also be utilized in solar cells and as a catalyst in chemical processes, as well as in biological applications like drug delivery systems and imaging agents [4, 5].

The fundamental disadvantage of bismuth nano ferrite is its high leakage current caused by oxygen vacancies and mixed iron valence. These flaws compromise the efficiency and stability of bismuth nano ferrite devices. To compensate for the shortcomings of bismuth nano ferrite, silver can be doped into it to produce silver-doped bismuth nano ferrite. Silver doping decreases the band gap of bismuth nano ferrite and enhances its electrical conductivity. Silver-doped bismuth nano ferrite has potential uses in photocatalysis, water splitting, and gas detection [6].

There are numerous ways for synthesizing bismuth and silver-doped bismuth nano ferrite, including the sol-gel approach, co-precipitation method, hydrothermal method, sonochemical method, microwave-assisted method, and combustion method. However, the sol-gel technique is an essential, cost-effective, and simple approach to adopt since it provides good control over the size and form of the particles so this method is very convenient, and in the present study, the nanoferrites are synthesized utilizing the sol-gel method. The motivation of this study is to reduce the band gap of bismuth nano ferrite by doping and to study the physicochemical properties of both synthesized compounds [7, 8].

2. Materials and Method

2.1 Materials

Ferric nitrate nano hydrate ($Fe(NO_3)_3.9H_2O$), Bismuth (III) nitrate pentahydrate ($Bi(NO_3)_3.5H_2O$), Silver nitrate ($AgNO_3$), Citric acid ($C_6H_8O_7$), Ammonia (NH_3). All are procured by SD-fine chemicals.

2.2*Procedure for Synthesis*

In the present study, the synthesis of bismuth nano ferrite and the silver-doped bismuth nano ferrite is done through the Sol-Gel method. Bismuth nitrate, Ferric nitrate, and Citric acid are initially taken in the ratio of 1:2:3. While doping, silver nitrate is added to the mentioned compounds and the new ratio becomes 0.06:0.94:2:3. For a homogeneous mixture of the solutions, all the purchased nitrates are first dispersed separately in deionized water for three hours using a magnetic stirrer. Next, the solutions are combined into a large glass beaker and stirred for 3 hours. To maintain the pH value, Ammonia is added dropwise until the pH value reaches nine. The solution mixture is then heated for 2 hours at 70°C, and then raised to 90°C for an hour, where gel formation occurs, and then rise the temperature to 110°C the gel is converted into fluffy ash which will be collected and ground with the help of mortar and pestle to achieve a homogeneous compound. The compound is calcined at 200°C for 2 hours to remove impurities after which the compound is collected and characterized [9].

3. Various methods of characterization

3.1. The X-ray diffraction method

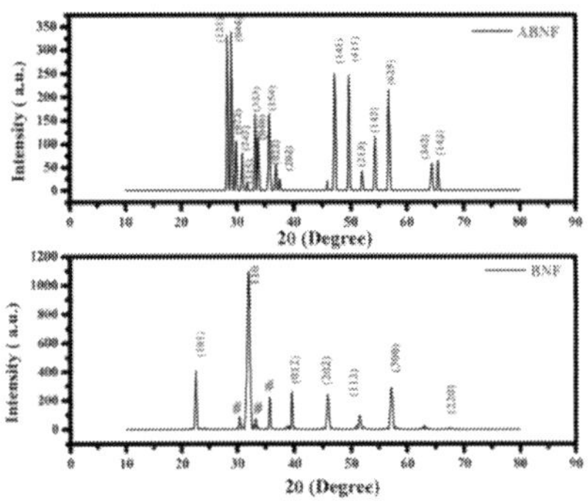

Figure 1. XRD patterns of BNF and ABNF

X-ray diffractometry (XRD) is an important tool for analyzing the crystallographic structure of materials by observing how X-rays interact with the sample. XRD study of bismuth nano ferrite revealed the existence of 10 distinctive intensity peaks at 2θ angles of 22.52°, 30.34°, 32.06°, 33.38°, 35.84°, 39.64°, 45.92°, 51.34°, 57.28°, and 67.24° correspond to the miller indices (101), (110), (012), (202), (113), (300), and (220), and impurities are denoted by the mark '#' for some 2θ angles; where α-Fe_2O_3 are the impurities which are found in the synthesized compound. By comparing the synthesized Bismuth nanoferrite (BNF) to the reference data from the JCPDS card no. 00-014-0181, which has a Rhombohedral structure belongs to the R-3m space group which validates the crystal system, and the crystalline phase information was discovered.

IOP Conf. Series: Materials Science and Engineering 1300 (2024) 012038 doi:10.1088/1757-899X/1300/1/012038

The silver-doped bismuth nano ferrite (ABNF) compound exhibits 17 peaks at 2θ angles of 28.28°, 29.08°, 29.94°, 31.06°, 31.92°, 33.32°, 33.84°, 35.8°, 36.98°, 37.64°, 47.2°, 49.7°, 52.04°, 54.26°, 56.84°, 64.48°, and 65.48° matches the miller indices of (121), (044), (022), (242), (315), (333), (046), (154), (022), (202), (141), (411), (213), (142), (625), (342), and (143). The crystalline phase information of the synthesized ABNF was determined by comparing it to reference data from JCPDS card nos. 01-078-1518 and 00-025-0090, which confirm the crystal system as an orthorhombic structure and belongs to the Cmcm space group. In the presence of silver ions (Ag^{+}), new crystal planes and interatomic distances are formed as a result of changing the crystal structure. In contrast to pure bismuth nano ferrite, these alterations in the lattice structure result in additional peaks in the XRD pattern. Scherrer's formula is used to determine crystallite size; In this formula, the XRD peak broadening is inversely proportional to the crystallite size of the material [10].

$$D = \frac{k\lambda}{\beta \cos\theta} \quad (1)$$

Where D is the average crystallite size, K is the Scherrer constant (0.89), λ is the wavelength of the X-rays employed, β is the diffraction peak's full width at half maximum (FWHM) value, and θ is the diffraction angle. Utilizing the Debye-Scherrer formula, we can calculate crystallite size for BNF and ABNF. It was found that BNF is 20 nm in crystallite size, while ABNF is 36.33 nm in crystallite size [11].

3.2. FTIR analysis

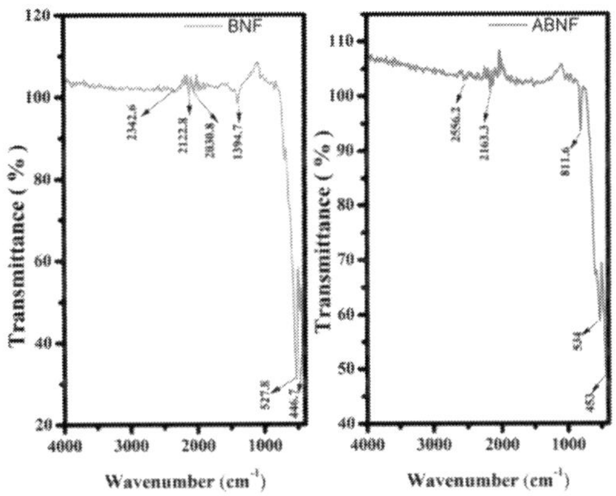

Figure 2. FTIR graph of BNF and ABNF

From Figure 2, the FTIR spectra of BNF and ABNF can reveal many different functional groups, such as Bi-O, Fe-O, and Ag-O. It is confirmed that ferrites form at about 446.7 cm^{-1}, 453 cm^{-1}, 527.8 cm^{-1}, and 534 cm^{-1}, respectively, corresponding to the stretching and bending modes of Bi-O, Fe-O, and Ag-O. A peak near 811.6 cm^{-1} and a peak near 1394.7 cm^{-1} correspond to nitrate ions and a peak near 2342.6 cm^{-1} corresponds to nitrile ions. The O-H groups were found as a result of water molecules stretching and bending at about 2122.8 cm^{-1}, and 2030.8 cm^{-1} as a result of insufficient water removal during synthesis or adsorption of water molecules from the air. Accordingly, FTIR confirms doping by revealing the shifting of the peaks from initially 446.7 cm^{-1} and 527.8 cm^{-1} to 453 cm^{-1}, and 534 cm^{-1} after doping [12].

3.3. UV- Visible spectral analysis

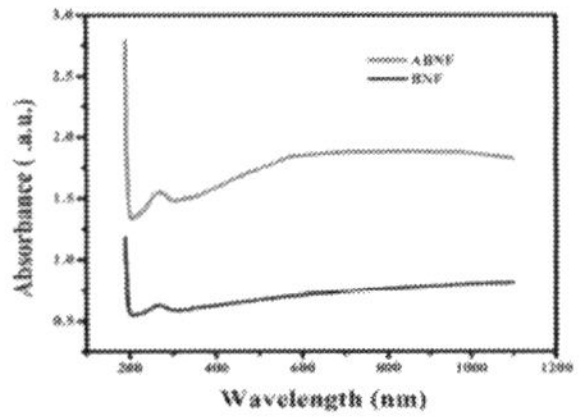

Figure 3. UV-visible absorption spectra graph of BNF and ABNF

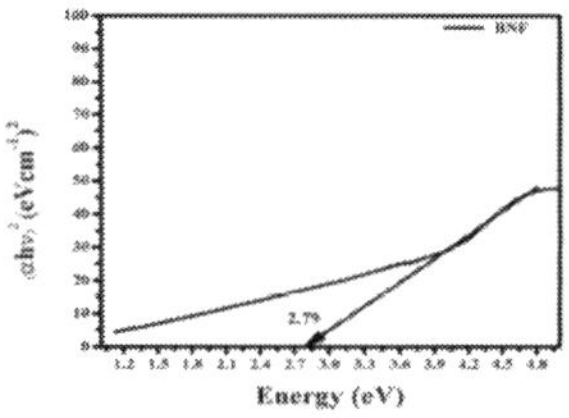

Figure 3.1. The direct band gap of BNF

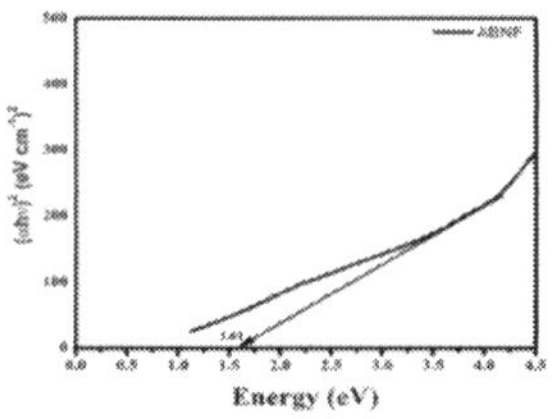

Figure 3.2. The direct band gap of ABNF

UV-visible absorption spectroscopy was employed to investigate BNF and ABNF nanoferrites (NFS) in the range of 200-1100 nm. Figure 3 demonstrates that the effective optical absorbance seen in the UV area may be brought on by a synthetic interband transition between BNF and ABNF nano ferrites. The energy bandgap of the synthesized NFS was estimated using the tauc plot and the subsequent equation.

$$(\alpha h\nu)^n = K(h\nu - Eg) \qquad (2)$$

In equation (2) "α" denotes the absorption coefficient, "h" represents the Planck's constant, "ν" is the frequency, "Eg" denotes the energy bandgap, and the "n" value is taken as 2 for the direct band gap and 1/2 for indirect band gap. The tauc plot of prepared NPs for direct permitted transition is shown in Figures 3.1 and 3.2. The energy band gap is obtained by projecting the tangent of the linear area towards the X-axis. The band gap for BNF samples is 2.79 eV, but the band gap for ABNF sample is 1.61 eV, owing to a change in electronic structure caused by the oxygen vacancy near the band edge. It has been discovered that when crystalline size increases, the energy band gap decreases. Furthermore, the shift in the energy band gap might be attributed to the phase transformation of BNF to ABNF. Because of this, ABNF is well suited for sensors, spintronics, data storage, and so forth [13].

3.4. SEM Analysis

Figure 4.1. SEM image of BNF

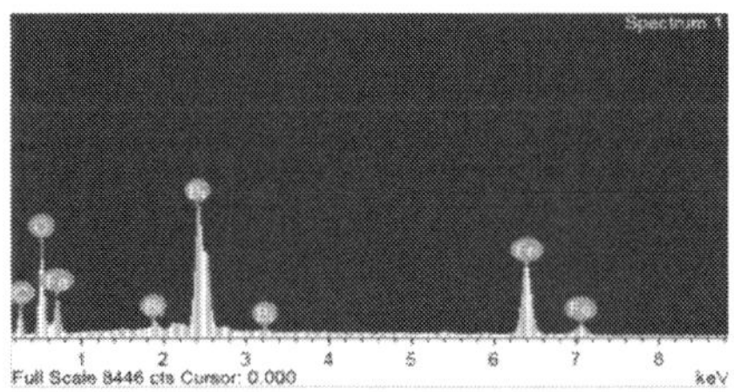

Figure 4.2. EDX image of BNF

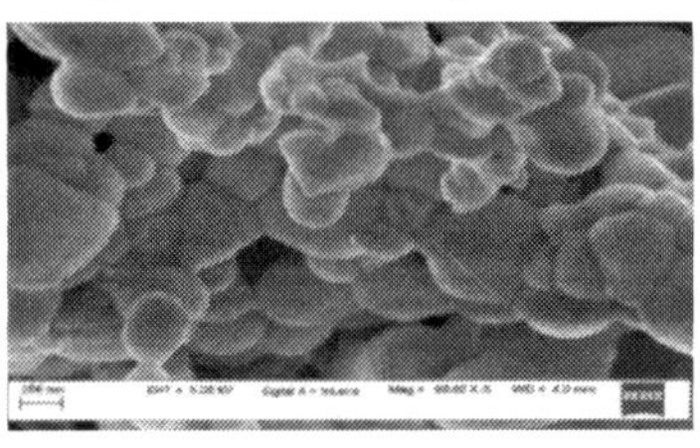

Figure 4.3. SEM image of ABNF

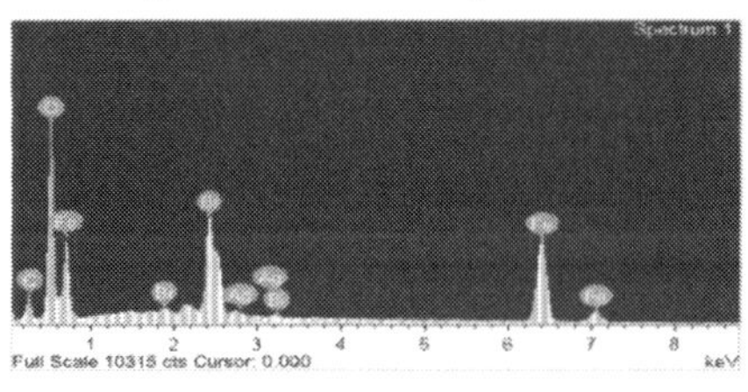

Figure 4.4. EDX image of ABNF

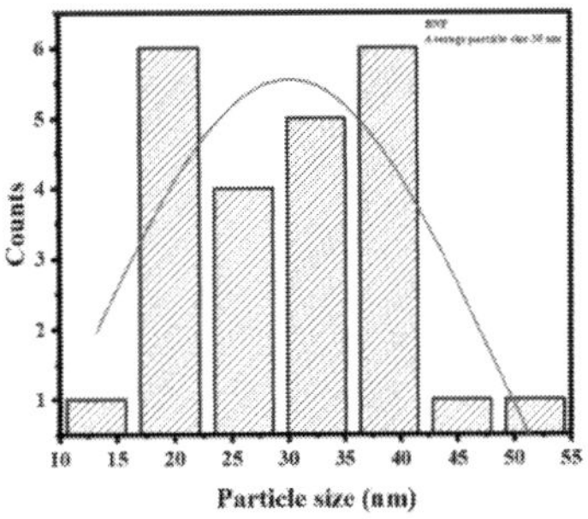

Figure 4.5. Histogram plot for BNF

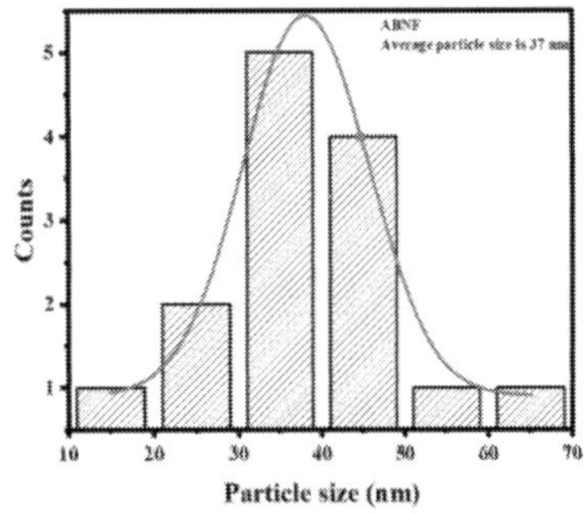

Figure 4.6. Histogram plot for ABNF

The surface morphology of the BNF and ABNF is determined by using the scanning electron microscope. The SEM images of BNF, i.e. Figure 4.1 shows the clear grain-like patterns in the image and confirm the formation of BNF through EDX which is shown in Figure 4.2. This indicates the presence of bismuth, iron, and oxygen compounds in the composite, and a little bit of carbon content is found in the composite it may be due to calcination. The SEM images of ABNF show a little bit of agglomerated grain-like patterns as shown in Figure 4.3 and the doping of Ag to the compound BNF is confirmed by the EDX through Figure 4.4. Here silver (Ag) peak appears along with the compound's bismuth, iron, and oxygen hence it confirms the doping. The average size of the nanoparticles is calculated through the histogram graph, which is shown in Figures 4.5 and 4.6. The average particle size of BNF is 30 nm and for ABNF is 37 nm [14].

3.5. Dielectric properties

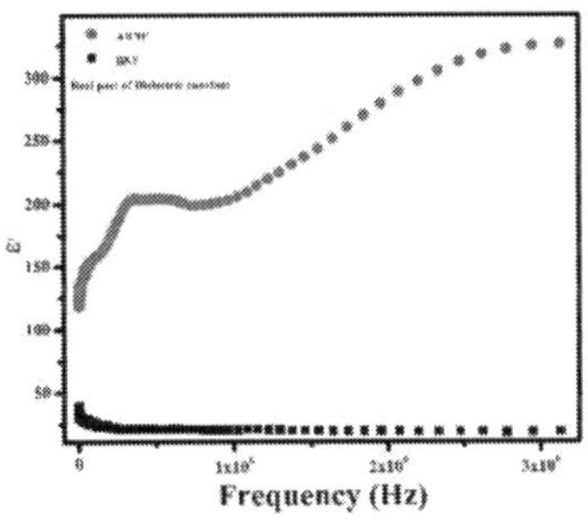

Figure 5.1. The real part of the dielectric constant for BNF& ABNF compounds

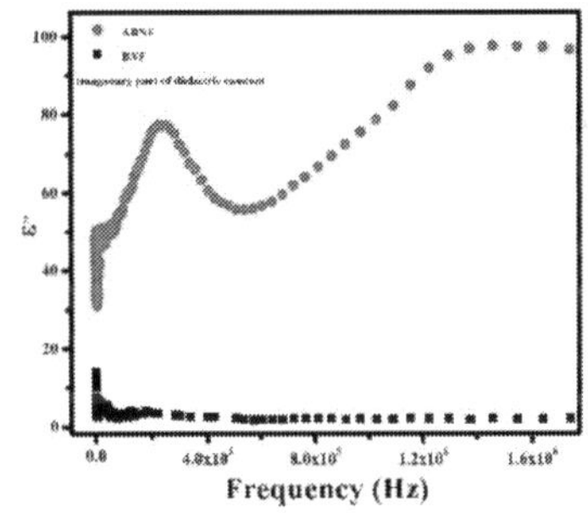

Figure 5.2. The imaginary part of dielectric constant for BNF & ABNF compounds

Materials' behavior in an electric field can be understood through the study of dielectric properties. The dielectric constant, often denoted as ε, is a measure of a material's ability to store electrical energy in an electric field. A capacitor's capacitance with material as the dielectric is compared to the capacitance with vacuum as the dielectric of the same capacitor. The real part (ε') represents the ability to store energy, whereas the imaginary part (ε") accounts for the dissipation of energy as heat. The real and the imaginary part of the dielectric constant can be explained by the following equations.

$$\varepsilon' = \frac{Ct}{\varepsilon_0 A} \quad (3)$$

IOP Conf. Series: Materials Science and Engineering 1300 (2024) 012038 doi:10.1088/1757-899X/1300/1/012038

Where the ε' represents the real component of the dielectric constant. *C* is identified as the capacitance in the above equation, and the thickness of the pellet is represented by *t*, whereas *A* represents the cross-section area of the pellet. Figure 5.1 shows the real part of the dielectric constant for BNF & ABNF compounds.

$$\varepsilon'' = \varepsilon' \tan\delta \qquad (4)$$

Equation (4) represents the imaginary component of the dielectric constant (ε") and the dielectric loss tangent is represented by tan*δ*. Figure 5.2 shows the imaginary part of the dielectric constant for BNF & ABNF compounds.

$$\sigma_{ac} = 2\pi f \varepsilon' \varepsilon_0 \tan\delta \qquad (5)$$

Equation (5) represents the AC electric conductivity of the synthesized compounds, whereas *f* represents the frequency in Hertz. The real component ε' exhibits frequency-dependent behavior in the dielectric analysis of BNF. Due to the presence of multiple polarization mechanisms at low frequencies, such as domain wall motion, space charges, and impurities, ε' exhibits a high value. The value of ε' decreases with increasing frequency, demonstrating the opposition to polarization. The relaxation mechanisms within the material are thought to be the cause of this frequency-dependent activity. ABNF has a significantly lower real part ε' than BNF, suggesting a lower electrical energy storage capacity suggests a lower capacity for electrical energy storage. This decline can be attributable to the increased conductivity brought on by the doping of silver. The flow of charge carriers is made easier by the presence of Ag ions, which lowers ε' by reducing polarization [15].

Additionally, frequency-dependent is the imaginary portion ε" of the dielectric constant for BNF. When exposed to an electric field, it exhibits heat as a representation of energy loss. Low energy dissipation is suggested by the fact that ε" is very modest at low frequencies. Nevertheless, when the frequency rises, ε" grows as a result of the greater contribution of relaxation processes, such as dipolar relaxation and ionic conduction [16]. The imaginary portion ε" of the dielectric constant in ABNF will change in a similar manner as the real part. Higher values of ε" are obtained as a result of the conductivity improvement brought on by the doping of silver. As shown in BNF, this higher energy dissipation is accompanied by an enhanced frequency dependency of ε".

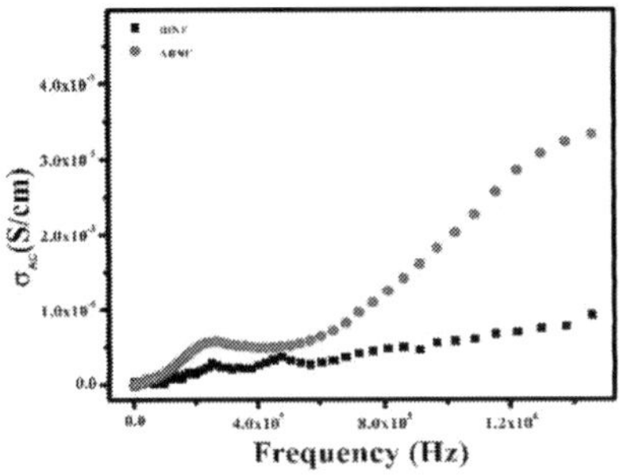

Figure 5.3. AC electrical conductivity (σ_{AC}) for BNF & ABNF compounds

Figure 5.3 displays the compounds BNF and ABNF's AC electrical conductivity. Thus, these graphs show that when the frequency rises, conductivity likewise rises, confirming to the universal power law. In this occurrence, the growth in electrical conductivity is caused by electron migration. The primary cause of the rise in conductivity is a result of the charge carrier's hopping frequency continuously rising. The Maxwell-Wagner and Koop phenomenological theories describe the idea of AC electric conductivity [17]. While it was claimed that the frequency of electron hopping between Fe^{3+} and Fe^{2+} is lower and that the grain boundaries are more active at low frequencies. Due to the increased activity of conductive grains at higher frequencies, the hopping of electrons between Fe^{3+} and Fe^{2+} ions rise with frequency. From the Figure 5.3, we can see that the ABNF exhibits more electric conductivity than the BNF combination. This indicates that adding silver to the bismuth nano ferrite boosts its AC electric conductivity, making it a better material for application in electronic devices [18].

4. Conclusion

In the present study, the bismuth and silver doped bismuth nano ferrite is synthesized by sol-gel method and to know its physio-chemical properties several characterizations have been done such as XRD, FTIR, UV-visible study, SEM, and the dielectric properties has been observed and studied. Through the XRD study, the crystallite size found, for BNF is 20 nm, and for the ABNF is 36.66 nm crystallite size. The FTIR plot confirms the metal ion group presence in the fingerprint region. The SEM gives the enhanced image of the synthesized particles where the BNF has the agglomerated images and the ABNF has the grain-like structures. The energy band gap for BNF is 2.7 eV and for the ABNF the band gap is about 1.61 eV. From this, we can observe that doping enhances the electric properties of the synthesized nanoferrites. In the dielectric properties, the real part and the imaginary part of the dielectric constant were determined and the AC electric conductivity study has also been done. Hence from these dielectric studies it is found that as the frequency increases the real and the imaginary part of the dielectric constant also increases and the same thing happens in the AC-electric study. Hence these things lead to several applications like photocatalysis, water splitting, gas detection, and many more.

Acknowledgment:

The author Amrutesh Kannolli is grateful to Vijayanagara Sri Krishnadevaraya University, Ballari, for providing financial assistance through the fellowship under the SCP Scheme.

References

1. K Sarkar, S Mukherjee, and M K Mitra 2014 Synthesis, Characterization and Studies on Optical, Dielectric and Magnetic Properties of undoped and Cobalt doped Nanocrystalline Bismuth Ferrite, Journal of the Institution of Engineers (India) Series D, vol. 95, no. 2, pp. 135–143
2. P. Dhiman, G C Rana, D Goyal, and A Goyal 2021 Basics of Ferrites: Types and Structures, in Engineering materials, pp. 1–25
3. T Vigneswari, P Raji, and P Thiruramanathan 2022 Correction to: Magnetic targeting carrier applications of Bismuth-Doped nickel ferrites nanoparticles by co-precipitation method,Transactions of the Indian Institute of Metals, vol 74,2255-2265
4. R Dilip and R Jayaprakash 2018 Synthesis and characterization of $BaFe_2O_4$ nano-ferrites for gas sensor applications, Energy, Ecology and Environment, vol. 3, no. 4, pp. 237–241
5. LV Hublikar, S V Ganachari, and V B Patil 2023 Zn and Co ferrite nanoparticles: towards the applications of sensing and adsorption studies, Environmental Science and Pollution Research, vol. 30, no. 25, pp. 66994–67007
6. T K Rao, I V K Viswanath, and Y L N Murthy 2019 Study of Bismuth Ferrite-Silver Ferrite Nanocomposite about structure, characterization, magnetic properties and band Gap Evaluation, in Lecture notes on multidisciplinary industrial engineering
7. A Hajalilou and S A Mazlan 2016 A review on preparation techniques for synthesis of nanocrystalline soft magnetic ferrites and investigation on the effects of microstructure features on magnetic properties, Applied Physics A, vol. 122, no. 7
8. M. Hussain et al. 2020 Ferrite nanoparticles for biomedical applications, in Nanomedicine and nanotoxicology, pp. 243–265
9. S E Shirsath, D Wang, S S Jadhav, M L Mane, and S Li 2017Ferrites obtained by Sol–Gel method, in Springer eBooks, pp. 1–41
10. Amrutesh Kannolli, Avinash P, SR Manohara, Mohsina Taj, Kotresh MG 2023 In-depth study of zinc nanoferrite particles at different calcination temperatures and their behavior in the presence of electric and magnetic fields, Journal of Magnetism and Magnetic Materials, vol.584, 171079
11. Honarmand M, Golmohammadi M, & Hafezi-Bakhtiari J. 2020 Synthesis and characterization of SnO2 NPs for photodegradation of eriochrome black-T using response surface methodology, Environmental Science and Pollution Research, 28(6), 7123–7133
12. Kumar NH, Ravinder D, & Edukondalu A 2023 Effect of Ce^{3+} ion doped Ni-Zn ferrites: Structural, optical and low temperature magnetic properties, Chinese Journal of Physics, 81, 171–180
13. Basavanagoudra H, Jangannanavar V D, Patil M K, Shettar A K, Hanamanta Hegde, B G, InamdarS R, & Goudar K M 2023 Multifunctional role of engineered tin oxide nanoparticles with the variation of calcination temperatures, Materials Today Communications, 35, 105546
14. Rahman M M, Hasan N, Hoque M A, Hossen M B, &Arifuzzaman M 2022 Structural, dielectric, and electrical transport properties of Al^{3+} substituted nanocrystalline Ni-Cu spinel ferrites prepared through the sol–gel route, Results in Physics, 38, 105610
15. M Sayer, A Mansingh, A Arora, and AY Lo 1992 Dielectric response of ferroelectric thin films on non-metallic electrodes, Integrated Ferroelectrics, vol. 1, no. 1, pp. 129–146
16. G Sathishkumar, C Venkataraju, and K Sivakumar 2012 Magnetic and dielectric properties of cadmium substituted nickel cobalt nanoferrites, Journal of Materials Science: Materials in Electronics, vol. 24, no.3, pp. 1057–1062
17. C G Koops 1951 On the Dispersion of Resistivity and Dielectric Constant of Some Semiconductors atAudiofrequencies, Physical Review, vol. 83, no. 1, pp. 121–124
18. S K Abbas, Aslam M A, Amir M, Atiq S, Ahmed Z, Siddiqi S A, Naseem S 2017 Electrical impedance functionality and spin orientation transformation of nanostructured Sr-substituted BaMnO3 hexagonal perovskites, Journal of Alloys and Compounds, vol. 712, pp. 720–731

Synthesis and characterization of $NiFe_2O_4$ nanoparticles by tartaric acid assisted sol-gel auto-combustion method

Prabhakar Ningayya Patil[1], Sarvesh Kumar[2], V M Jali[1*] and B Sahoo[2]

[1] Department of Physics, Gulbarga University, Kalaburagi, 585106, India
[2] Materials Research Center, Indian Institute of Science, Bengaluru, 560012, India
* E-mail:vmjali@gmail.com

Abstract : Nickel ferrite nanoparticles were synthesized by modified sol-gel auto-combustion method. The final product obtained was annealed at 600 ℃. The thermal, structural, morphological, vibrational, optical and magnetic properties were studied using TGA, XRD, SEM, FTIR, UV-DRS and VSM techniques, respectively. The thermo gravimetric analysis shows three step decomposition of residual nitrate and tartrate complexes with a total weight loss of 6.69% below 600 ℃. Further, there was no significant weight loss above 600 ℃. XRD confirmed the spinel cubic phase with *Fd3m* space group. The average crystallite size was 14 nm. An agglomerated random morphology with porous flake like microstructure was observed. The vibrational analysis confirmed the presence of Ni-O, Fe-O stretching vibrational modes in the wave number range 400-600 cm^{-1}. Tetrahedral site (k_{Td}) and octahedral site (k_{Oh}) force constant values were 2.20×10^5 and 1.31×10^5 dynes/cm, respectively. From UV-DRS spectra, the Kubelka-Munk function was used to calculate the optical band gap and it was 1.71 eV. The magnetic saturation, retentivity and coercive field values: M_s = 22.35 emu/gm, M_r = 5.20 emu/gm, H_c = 165 Gauss were obtained from VSM analysis. The calculated magnetic moment value was 1.115 μ_B. The law of approach to saturation was employed to analyze the VSM data in saturation region. The calculated magneto-crystalline anisotropy constant, K_1 value was 0.857×10^5 erg/cc.

1. Introduction

Ferrites have received considerable attention among the Scientists and Researchers. They continue to be the subject of technical research due to their remarkable chemical and thermal stability, as well as their structural, magnetic, electrical, optical, and catalytic properties. They find several applications in technological fields: magnetic hyperthermia, catalysis, permanent magnets, EMI shielding, sensors, waste water treatment, energy storage and microwave devices. The type of the ions, their charges, and the distribution of cations among tetrahedral (T_d) and octahedral (O_h) sites significantly impact the characteristics of ferrites nanoparticles.

Nickel ferrite is technologically valuable soft ferrite material with an inverse spinel structure, and it has n-type semiconducting nature. In $NiFe_2O_4$, Ni^{2+} ions will be distributed among available O_h-sites and Fe^{3+} ions will be evenly distributed among both T_d- and O_h-sites. In recent years, preparation of ferrite nanoparticles with different preparation conditions and their effects on structure and morphology has been explored [1–4]. In order to produce stoichiometric and chemically pure spinel ferrite nanoparticles, widely used chemical methods, such as hydrothermal [5], microwave combustion [2], combustion [3], co-precipitation [1] and sol-gel technique [6] have been developed. Because of its simple preparation procedure, easy to manage stoichiometry, good chemical homogeneity, high purity and crystallinity with minimal processing time, the sol-gel combustion method is one of the most often utilized synthesis approaches [7]. The process of synthesis is easy, involving an exothermic reaction between metal salts and fuels. Appropriate fuels like citric acid [8], ethylene glycol, urea, glycine [6], aloe-vera [9] and honey [10] have been used.

Herein, we report a modified sol-gel auto-combustion method that uses tartaric acid (TA) and ethylene glycol (EG) as fuel/chelating agent to prepare Nickel ferrite sample. Additionally, thermal, structural, morphological, vibrational, optical, and magnetic properties have been invetigated.

2. Experimental method

2.1 Synthesis

A modified sol-gel auto combustion method was used for synthesizing $NiFe_2O_4$ nanoparticles. The procured chemicals are of analytical grade. The stoichiometric proportion of Nickel nitrate hexahydrate, Iron nitrate nonahydrate and L-tartaric acid (TA) in the molar ratio 1:2:4 were taken. In 100 ml of de-ionized water, $Ni(NO_3)_2.6H_2O$ and $Fe(NO_3)_3.9H_2O$ were dissolved. Then, aqueous solution of tartaric acid was added to nitrate solution. EG was added in 70:30 (TA:EG) ratio. The solution was stirred for 1 hr at 70 °C. The mixture was next neutralized with aqueous Ammonia and heated to 100 °C until the solution turned into a gel. The gel underwent an auto-combustion reaction after which fluffy powder was produced. It was then dried for 1 hr at 200 °C in a Muffle furnace. The obtained yield was pulverized in an Agate Mortar and Pestle. Then, the powder was annealed for 2 hr at 600 °C to obtain Nickel ferrite nanoparticles. It was designated as NF-600.

2.2 Characterization

Thermal analysis was carried out to understand the thermal stability of the sample using Thermo Gravimetric Analyser (*TGA* Q50 Series). *PANalytical X'Pert MPD X-ray diffractometer* with CuK_α radiation (λ=1.5406Å) was used to record XRD (X-day diffraction) data for identifying the structure and phase. To investigate the surface morphology, a *ZIESS Scanning Electron Microscope* (SEM) was employed. Fourier Transform Infrared (FTIR - *Agilent Technologies Cary 600 Series*) spectra was recorded to analyze the vibrational bands. UV-DRS data was recorded in reflectance mode to study the optical properties using *Perkin Elmer UV-Vis Spectrometer-Lambda 750.* To examine magnetic characteristics, A *Lakeshore 665* Vibrating Sample Magnetometer (VSM) was used to record the M-H curve.

3. Results and Discussion

3.1 Thermo gravimetric analysis

As-prepared Nickel ferrite powder was subjected to thermo gravimetric analysis between 27 °C and 850°C with a ramping rate of 10 °C/minute. Figure 1(a) shows the TGA curve of non-calcined Nickel ferrite. The weight loss was noticed up to 600 °C in three steps. In the first step, the decomposition continued upto 120 °C due to the adsorbed water content amounting to 3.04% weight loss. In the second step, the decomposition continued upto 300 °C corresponding to residual nitrates and ethylene glycol precursors with 1.84% weight loss. In the third stage, 1.81% weight loss, observed at higher temperature from 300°C to 600 °C was due to the decomposition of residual tartrate precursors. Therefore, the total weight loss observed was ~ 6.69%. Further, the as-prepared Nickel ferrite sample was calcined at 600 °C for 2 hr. Thus, the weight loss up to 600 °C indicates oxidation of metal ions and formation of ferrite phase [1,3], which was confirmed by XRD, FTIR analysis and are shown in figures 1(b) and (d), respectively.

3.2 XRD analysis

Figure 1(b) depicts an XRD pattern of Nickel ferrite sample annealed at 600 °C. The observed Bragg peaks match well with the standard crystal structure ICSD data (00-010-0325). A cubic spinel phase (*Fd3m* space group) was observed. The high crystallinity of the sample is shown by the strong and distinct reflection peaks. The absence of secondary peaks represent the sample's phase purity. The average crystallite size (D) was calculated using the Debye-Scherrer formula and the structural parameters such as lattice constant (a), interplanar spacing (d_{hkl}), cell volume (V), dislocation density (δ), and x-ray density (ρ_x) were determined [11]. Table 1 summarizes the computed structural parameters. The reported results agree well with the literature values [1,12].

Table 1. Structural parameters of NF-600.

Parameter	Value
d_{hkl} (Å)	2.506
a (Å)	8.312
V (Å^3)	574.27
D (nm)	14.00
δ (nm^{-2})	0.0051
ρ_x (gm/cc)	5.420

3.3 Morphological analysis

The scanning electron micrograph of NF-600 is shown in figure 1 (c). An agglomerated random morphology with porous flake like microstructure was observed. The synthesis conditions may have an impact on the microstructure of the nanoparticles. In addition, pores and voids that have been formed because of the emission of gases that were released during combustion reaction may be noticed [13].

3.4 FTIR analysis

The FTIR spectra of spinel ferrite show two primary absorption bands in the 400–600 cm^{-1} region. The lower wavenumber band (ν_2 ~400 cm^{-1}) is assigned to intrinsic stretching vibrations of metal ions present at the O_h-site. The higher wavenumber band (ν_1 ~600 cm^{-1}) is related to the intrinsic stretching vibrations of the metal ions at the T_d-site [2]. The absorption occurs at different frequencies depending on the mass of metal cations and strength of the bond between metal cations and oxygen [14]. The recorded infrared spectra of NF-600 sample in the 400-4000 cm^{-1} wave number range is shown in figure 1(d). Inset in figure 1(d) displays magnified IR spectra in the 400-700 cm^{-1} range. The absorption bands (table 2(a)) at ν_1- 550 cm^{-1}, ν_2(1)- 408 cm^{-1}, ν_2(2)- 423 cm^{-1}, and ν_2(3)- 470 cm^{-1} are assigned to the intrinsic stretching vibrational modes of metal cations with oxygen ions ((Fe^{3+}-O^{2-})$_A$, (Fe^{3+}-O^{2-})$_B$ and (Ni^{2+}-O^{2-})$_B$) present at T_d (A) and O_h (B) –sites, respectively. The related bands at 1363 cm^{-1} and 3349 cm^{-1} were also seen due to the stretching vibrations of the C-O bond and O-H group due to the residual traces and water molecules adsorbed on the surface of the sample. The vibrational bands that have been detected agree well with the values that have been reported [9]. The values of the force constants were calculated using the formula: $k=4\pi^2 c\nu^2\mu$, where, c is the speed of light, ν is the wavenumber (frequency of ions at the T_d and O_h-sites) and μ is the reduced mass for (M-O) Metal-Oxygen bonds [15]. Tetrahedral site (k_{Td}) and octahedral site (k_{Oh}) force constants (table 2 (b)) were found to be 2.20 $\times 10^5$ and 1.31 $\times 10^5$ dynes/cm, respectively.

Table 2 (a). FTIR vibrational band assignment.

Wave number (cm^{-1})	Assignment	Ref.
408	Octahedral metal stretching vibration (Fe^{3+}-O^{2-})$_B$	[16]
423	Octahedral stretching vibration of Ni^{2+}-O^{2-}	[17]
470	Octahedral asymmetric vibration of Ni^{2+}-O^{2-}	[14]
550	Tetrahedral metal stretching vibration (Fe^{3+}-O^{2-})$_A$	[3]
1363	Stretching vibrations of C-O bond	[2]
3349	Stretching vibration of O-H group	[2]

Table 2 (b). FTIR vibrational modes and force constant values.

Wave number (cm^{-1})		k_{Td} $\times 10^5$ dynes/cm	k_{Oh} $\times 10^5$ dynes/cm
ν_1	ν_2		
550	423	2.20	1.31

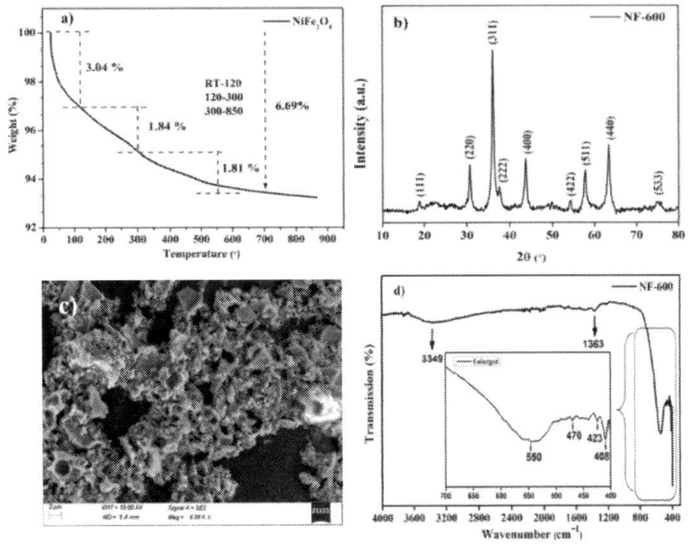

Figure 1. (a) TGA curve of non-calcined $NiFe_2O_4$, (b) XRD pattern (c) SEM image and (d) FTIR spectra of NF-600.

3.5 Optical analysis

The optical characteristics of the NF-600 sample was studied using UV-Visible diffused reflectance spectra. Figure 2 (a) depicts the reflectance spectra of NF-600 from 200 to 850 nm. Maximum reflectance was detected in the 200-300 nm wavelength region, with a significant rise in 700-850 nm wavelength region. As shown in figure 2 (b), the optical energy band gap (E_g) was determined by plotting the Kubelka Munk function $(F(r)\ h\upsilon)^2$ v/s Energy and extrapolating the linear portion of the curve to $F(r)^2=0$. Typically, ferrite structures have two bands: an O-2p orbital at the valance band and a Fe-3d orbital at the conduction band [18]. The obtained direct energy band gap value of 1.71 eV is substantially within the range of previously reported values [8,19,20]. As a result, it can be concluded that $NiFe_2O_4$ nanoparticles are optically active in the visible region.

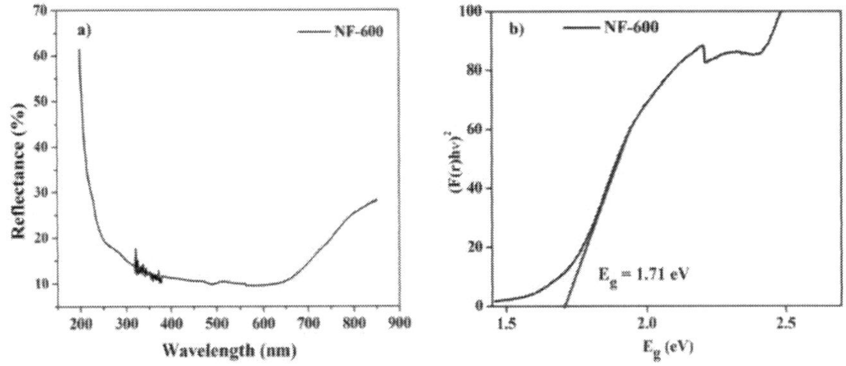

Figure 2. (a) Reflectance spectra of NF-600, (b) Extrapolated curve: $(F(r)\ h\upsilon)^2$ v/s E_g.

3.6 VSM analysis

The magnetic properties of NF-600 sample was studied by VSM at RT at 1.5 Tesla applied magnetic field. Based on the hysteresis curve shown in figure 3 (a), the magnetic saturation, retentivity and coercive field values were obtained from VSM analysis: M_s= 22.350 emu/gm, M_r= 5.200 emu/gm, H_c= 165 Gauss and M_r/M_s=0.195. The obtained magntic parameters are given in table 3. The results show that an S-shaped hysteresis curve exhibits ferrimagnetic behavior at room temperature. Magnetic moment per formula unit (f.u) i.e, μ_B in Bohr magneton was calculated using the relation: $\mu_B=M_{wt}\times M_s/N_A\times\beta$, where, M_{wt}- molecular weight in mole/gm, M_s- saturation magnetization, $N_A= 6.022\times10^{23}$ mol^{-1} (Avagadro's number), $\beta= 9.27\times10^{-21}$ (conversion factor) and the magnetic moment was estimated to be $1.115\mu_B$ [8,21].

The magnetic measurements data at higher magnetic field (~above 1 Tesla) are fitted to the law of approach in order to comprehend the magnetic anisotropy of NF-600 sample. The law of approach method (LA) was introduced to analyze the data in the saturation part. LA denotes the dependency of magnetization M on the applied field for $H>>H_c$. The higher magnetic field parts (above 1 Tesla), the M-H curve is fitted to the following equation : $M = M_s [1 - b/H^2] + kH$, where, $b=(8/105)\times(K_1^2/M_s^2)$ (cgs system), M–magnetization, H–applied field, M_s–magnetic saturation, K_1–magneto-crystalline cubic anisotropy constant and the term kH–forced magnetization. 8/105–the numerical co-efficient is used to describe the cubic anisotropy of random polycrystalline samples. Typical curve fit to LA is shown in figure 3 (b). Using the M_s and b values obtained from LA fitting, the K_1 value was 0.857×10^5 erg/cm^3, is in accordance with the inverse spinel ferrites [15,22–24].

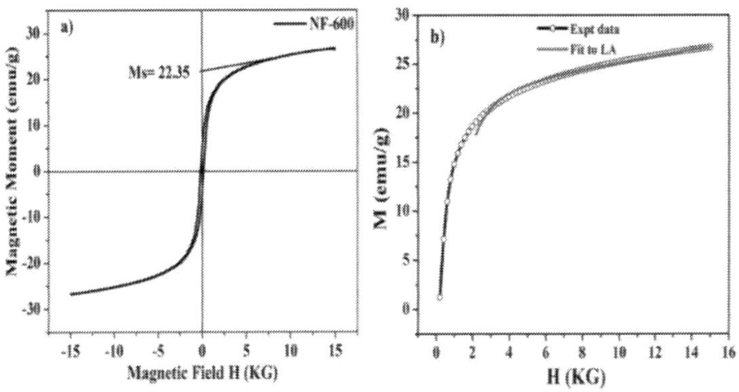

Figure 3. (a) M-H curve, (b) Law of Approach (LA) fit for NF-600.

Table 3. Magnetic parameters of NF-600.

Parameter	Value
M_s (emu/g)	22.350
M_r (emu/g)	5.200
H_c (G)	165
M_r/M_s	0.195
μ_B (Bohr Mag.) f.u.	1.115
b (Oe^2)	1.122×10^6
K_1(erg/cm^3)	0.857×10^5

4. Conclusion

$NiFe_2O_4$ nanoparticles were successfully prepared using a modified sol-gel autocombustion method with tartaric acid and ethylene glycol. TGA investigation revealed that the remaining precursors decomposed up to 600 °C, implying that the sol-gel auto-combustion process occur at lower temperatures to produce spinel ferrite phase. XRD investigation revealed the formation of a single phase cubic spinel structure with a lattice constant of 8.312 Å and average crystallilte size, 14 nm. An agglomerated random morphology with a porous flake-like microstructure was found. The observed FTIR characteristic absorption bands of the Nickel ferrite phase were in the 400-600cm^{-1} range. Kubelka-Munk function was used to estimate the optical energy band gap and was found to be 1.71 eV. The VSM study of produced Nickel ferrite reveals ferrimagnetic behavior with a saturation magnetization of 22.35 emu/gm. The law of approach method was used for analyzing VSM data in the saturation region. The calculated magneto-crystalline anisotropy constant, K_1 value was 0.857×10^5 erg/cc.

References

[1] Hariharasuthan R, Chitradevi S, Radha K S and Chithambaram V 2022 *Appl. Phys. A* **128** 1045

[2] Suresh Kumar G, Akbar J, Govindan R, Girija E K and Kanagaraj M 2016 *Journal of Science: Advanced Materials and Devices* **1** 282–5

[3] Prabhakaran T and Hemalatha J 2011 *Journal of Alloys and Compounds* **509** 7071–7

[4] Kumari S, Manglam M K, Shukla A, Kumar L, Seal P, Borah J P and Kar M 2021 *Physica B: Condensed Matter* **621** 413280

[5] Nejati K and Zabihi R 2012 *Chemistry Central Journal* **6** 23

[6] Bhagwat V R, Humbe A V, More S D and Jadhav K M 2019 *Materials Science and Engineering: B* **248** 114388

[7] Sutka A and Mezinskis G 2012 *Front. Mater. Sci.* **6** 128–41

[8] Baykal A, Esir S, Demir A and Güner S 2015 *Ceramics International* **41** 231–9

[9] Laokul P, Amornkitbamrung V, Seraphin S and Maensiri S 2011 *Current Applied Physics* **11** 101–8

[10] Yadav R S, Kuřitka I, Vilcakova J, Havlica J, Masilko J, Kalina L, Tkacz J, Enev V and Hajdúchová M 2017 *Journal of Physics and Chemistry of Solids* **107** 150–61

[11] Cullity B D 1978 *Elements of x-ray diffraction* (Reading, Mass: Addison-Wesley Pub. Co)

[12] Kamble R B, Varade V, Ramesh K P and Prasad V 2015 *AIP Advances* **5** 017119

[13] Nadumane A, Shetty K, Anantharaju K S, Nagaswarupa H P, Rangappa D, Vidya Y S, Nagabhushana H and Prashantha S C 2019 *Journal of Science: Advanced Materials and Devices* **4** 89–100

[14] De M, Mukherjee A and Tewari H 2015 *PAC* **9** 193–7

[15] Channagoudra G, Saw A K and Dayal V 2021 *Journal of Physics and Chemistry of Solids* **154** 110086

[16] Manju B G and Raji P 2018 *Ceramics International* **44** 7329–33

[17] Aliahmad M and Noori M 2013 *Indian J Phys* **87** 431–4

[18] Kombaiah K, Vijaya J J, Kennedy L J, Bououdina M and Al-Najar B 2018 *Journal of Physics and Chemistry of Solids* **115** 162–71

[19] Ortiz-Quiñonez J-L, Pal U and Villanueva M S 2018 *ACS Omega* **3** 14986–5001

[20] Madhukara Naik M, Bhojya Naik H S, Nagaraju G, Vinuth M, Vinu K and Viswanath R 2019 *Nano-Structures & Nano-Objects* **19** 100322

[21] Güner S, Amir Md, Geleri M, Sertkol M and Baykal A 2015 *Ceramics International* **41** 10915–22

[22] Kumar L, Kumar P and Kar M 2013 *Journal of Alloys and Compounds* **551** 72–81

[23] Amir Md, Gungunes H, Slimani Y, Tashkandi N, El Sayed H S, Aldakheel F, Sertkol M, Sozeri H, Manikandan A, Ercan I and Baykal A 2019 *J Supercond Nov Magn* **32** 557–64

[24] Choudhary H K, Kumar R, Anupama A V and Sahoo B 2018 *Ceramics International* **44** 8877–89

IOP Conf. Series: Materials Science and Engineering 1300 (2024) 012040 doi:10.1088/1757-899X/1300/1/012040

Investigation on the crystal structure, magnetic, and electrical transport properties of magnetron sputtered Co_2FeGe Heusler alloy films

P.V. Midhunlal[1], Joshy Joseph[2], J. Arout Chelvane[3], A. Rajani Kanth[4], N. Harish Kumar[5]

1. Department of Sciences and Humanities, Christ (Deemed-to-be University), Bangalore- 560074, India.
2. Department of Physics, St. Joseph's college (Autonomous), Devagiri,Kozhikode, India
3. Defence Metallurgical Research Laboratory, Kanchanbagh (PO), Hyderabad, India.
4. School of Physics, University of Hyderabad, Hyderabad-500046, India
5. Department of Physics, Indian Institute of Technology-Madras, Chennai-600036, India.

Abstract. Crystal structure, magnetic, and electrical resistivity behaviour of Co_2FeGe Heusler alloy films deposited at different sputtering parameters have been studied using X-ray diffraction, VSM, and standard four-probe techniques. Though the expected structure was $L2_1$, X-ray diffraction studies indicate the A_2-type disordered structure. All films exhibited soft ferromagnetic characteristics having a coercive field of 5-65 Oe and a high ferromagnetic ordering temperature (More than 700 K). The electrical resistivity of the films deposited on the Si substrates was influenced by the substrate temperatures. Out of the different scattering mechanisms present in the low and high-temperature regimes, the two-magnon scattering effect is dominant in all the films. The scattering mechanisms are the same in all films irrespective of the substrate temperature. The optimum sputter deposition parameters that yields good quality Co_2FeGe Heusler alloy thin films were found to be 50 W power, 5 mTorr pressure, and 400 ^{0}C substrate temperature.

1. Introduction

The attraction towards Heusler alloys started after the discovery of the ferromagnetic intermetallic compound Cu_2MnAl by F. Heusler in 1903 in which no constituent elements are ferromagnetic [1]. Heusler alloys are ternary intermetallic compounds having the chemical formula X_2YZ for the full Heusler alloy and XYZ for the half Heusler alloy. Here X and Y are mostly transition elements and Z is a sp valent element. These alloys are widely used in spintronic devices due to their high Curie temperature, and high saturation magnetic moment and some of them exhibit a half-metallic nature in which it shows a metallic behavior for either up or down spin carriers and insulating behavior for the other type [2]. Half metallic property gives complete spin polarization at the Fermi level which is important for magnetic multilayer structures such as spin valves and magnetic tunnel junctions (MTJ). For these devices, half-metallic Heusler alloy films are used as spin injection electrodes to inject spin-polarized charge carriers into a nonmagnetic layer [3]. The efficiency of spin injection will greatly depend on the quality of individual film layers as well as the interface. Hence the optimization of the film deposition parameters such as power, pressure, substrate temperature, and post-annealing conditions are crucial for the device fabrication. Due to high T_c and magnetic moment, Heusler alloys having Co in the X positions are well studied theoretically as well as experimentally [4–7]. T. Daibou *et al* have shown the significant influence of annealing conditions on the tunnel magnetoresistance (TMR) value of Co_2MnSi-based MTJ [8]. Enhanced TMR value was due to the better crystalline ordering of Co_2MnSi at higher annealing temperatures. Zhu Qin *et al* [9] have reported the influence of substrate temperature on the crystallinity of Co_2FeSi thin films and K. Srinivas *et al* [10] have reported a detailed study on the influence of sputtering conditions on the physical properties of the same alloy.

Researchers have also reported the correlation between sputtering parameters and physical properties of Co_2FeAl [11] Heusler alloy films.

Even though the experimental and theoretical study carried out by K. Ramesh Kumar *et al.* [12] indicated the absence of half metallicity in Co_2FeGe Heusler alloy, some of the recent experimental studies and first principle calculations reported in the literature have predicted the possibility of having half metallic characteristics in this system [13,14]. As per the report, it has a Curie temperature of 981 K and a saturation magnetic moment of 5.71 μ_B. Moreover, point contact Andreev reflection measurement shows a spin polarization value of 0.58 which is comparabale to the values reported for the well known Heusler alloys Co_2FeAl (0.59) and Co_2MnAl (0.6) [15,16]. These factors make this alloy one of the attractive candidates for spintronic applications. Even though reports on structural, magnetic [17,18], optical [19], and tunnel magnetoresistance [20] characteristics have been reported on Co_2FeGe films, a systematic study on the influence of sputtering parameters such as sputtering pressure and substrate temperature on the physical properties of the films has not been reported so far. Since the alloy possesses an $L2_1$ crystal structure, high Curie temperature, and high saturation magnetic moment, we are motivated to do an investigation on the dependence of structural, magnetic, and electrical transport properties on the sputtering parameters. Optimized parameters reported in this paper may be useful for the fabrication of spintronic devices based on this material.

2.Experimental

Thin films of the Heusler alloy Co_2FeGe were deposited on 1cm×1cm Si (100) substrate by using a DC magnetron sputtering system. Before the deposition, substrates were cleaned through ultrasonication by using acetone followed by isopropanol for five minutes. A two-inch Co_2FeGe target was used for the deposition. For all depositions, the base pressure achieved was 8×10^{-8} Torr. Argon gas was used as the sputtering gas. The sputtering power was maintained at 50 Watt for all deposition. Films were deposited with sputtering pressures 5, 10, and 15 mTorr. For each pressure, the substrate temperature was varied as room temperature (RT), 300°C, 400°C, and 500°C. For all films substrate rotation was set at 10 rpm during the deposition. Films were allowed to cool after the deposition and were cut into pieces having dimensions 0.5cm×1cm for the characterization. For estimating the thickness, a step has been made on the film deposited at 5, 10, and 15 mTorr pressure by keeping the substrate at room temperature. The film thickness was measured using Bruker's contourGT 3D optical microscope. The structural analysis was carried out using a Rigaku smartLab high-resolution X-ray diffractometer with Cu- K_α radiation (λ=1.5418 Å) in grazing incident angle mode (Incident angle ω=1 for all measurements). The film stochiometry was identified by using an EDS attachment by Horiba Ltd. with Hitachi S-4800 Type II SEM. Magnetic measurements were carried out by using Lake Shore VSM. Resistivity measurement has been done using a standard four-probe measurement technique using a Cryogenic Probe Station (CRX-4K, Lake Shore Cryotronics).

3.Results and Discussion

3.1 Crystal Structure

The measured film thicknesses through the optical method are 78 nm, 64 nm, and 44 nm for the films deposited at 5, 10, and 15 mTorr respectively. Figure 1 shows the grazing incident angle (with ω=1) XRD patterns of Co_2FeGe films synthesized at various sputtering parameters. Diffraction patterns did not show (111) and (200) superlattice peaks which are the characteristic peaks for the $L2_1$ crystal structure. The absence of both peaks indicates the swapping of all the atoms in the unit cell which can be taken as evidence for *A*2 disorder [21]. But the similar atomic scattering factors of the elements Co, Fe, and Ge can also lead to the reduction in super lattice peak intensity. This makes it difficult to figure out the crystal structure of the alloy through X-ray analysis alone. Through EXAFS studies, Benjamin Balke *et al* have reported that Co_2FeGe alloy has the ordered $L2_1$ structure even though the alloy appeared to be in *A*2 disordered structure at first glance [22].

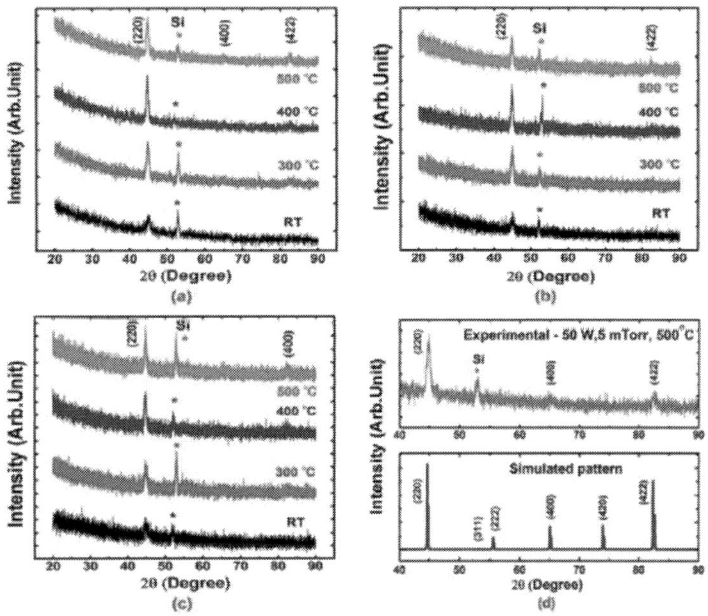

Figure 1: Grazing incident angle x-ray spectrum recorded using Cu-Ka radiation for Co_2FeGe films synthesized at (a) 5mTorr (b) 10 mTorr (c) 15 mTorr pressures and different substrate temperatures. Figure 1d shows the comparison between simulated X-ray spectrum (Bottom) and experimental X-ray spectrum (Top).

Films deposited at 5 mTorr pressure and substrate temperatures up to 400°C have (220), (311) and (422) peaks and a low intense (400) peak was observed at 500°C (Figure 1a). Also, the intensity of the main peak which is (200) has shown an enhancement at 400°C for all sputtering pressures which indicates a better crystallization at this temperature). X-ray spectrums showed a better crystalline order for thin films synthesized at 5 mTorr pressure and higher substrate temperatures (400°C and 500°C). As per the reported values, the lattice parameter for the Co_2FeGe alloy in the bulk form is in the range of 5.70-5.73 Å. The lattice parameters for the films investigated here are in the range 5.69-5.73 Å which is close to the values reported by other researchers.

3.2 Magnetic Properties

Figure 2 a, b, and c shows the magnetic hysteresis curves recorded at room temperature for the samples deposited at various parameters. Except for the film deposited at room temperature, all films exhibited soft ferromagnetic characteristics. Magnetization was saturated within 500 Oe. However, the samples synthesized with RT substrate temperature showed saturation only at higher fields which is around 2000 Oe as shown in figure 2d. This is possibly due to the presence of mixed magnetic interactions such as ferromagnetic and spin/cluster glass type. When the major ferromagnetic phase allows the magnetic domain to align parallel with the field direction the minor spin/cluster glass phase tries to oppose the parallel alignment. The ZFC and FC magnetization measurements showed a maxima around 80 K which is possibly due to the presence of a glassy phase in addition to the dominant ferromagnetic phase This competing interaction results in a lack of magnetic saturation up to 2000 Oe for all the room-temperature deposited films. All films exhibited soft ferromagnetic characteristics having a coercive field of 5-65 Oe. As per the previous reports, Co_2FeGe Heusler alloy should possess a magnetic moment value of around 6 μB. Our investigation showed a systematic variation in saturation magnetic moment value from 2.98 μB to 4.95 μB with respect to sputtering parameters. The difference in the magnetic moment values might be because of the atomic disorder and off-stoichiometry in the films. The estimated moment value based on the EDS results and Slater-Pauling rule are 5.08, 5.05, 5.42, and 5.46 μB for the films deposited at 5 mTorr pressure and (RT), 300°C, 400°C, and 500°C substrate temperatures respectively.

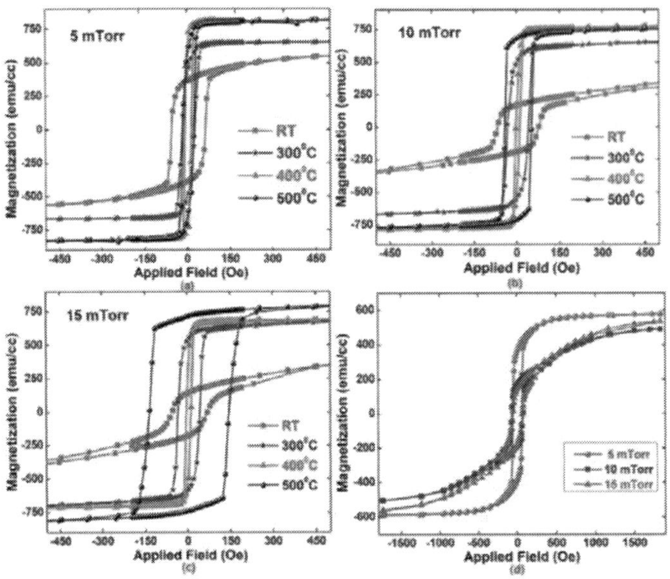

Figure 2: Isothermal magnetization curves for Co_2FeGe films deposited at 50 W power and (a) 5 (b) 10 and (c) 15 mTorr pressures. Figure 5d shows the saturation at higher fields for the samples synthesized at room temperature and at different sputtering pressures.

3.3 Electrical Transport Properties

The contribution to the electrical resistivity of a material mainly arises due to electron scattering by defects and impurities, electron-electron interaction (T^2 dependence), electron-phonon interaction (T), and electron-magnon interaction (In the case of magnetic material only). The magnon contribution could be either due to one magnon scattering (T^2) or due to two magnon scattering ($T^{9/2}$ at lower temperature and $T^{7/2}$ at higher temperature). In addition to the above-mentioned contributions, spin wave excitations (T^2 at lower temperature and T at higher temperature) and Stoner fluctuations (T^2 and $T^{5/3}$) can also contribute to electronic resistivity [23]. Figure 3(a)-(d) shows the experimental (in black) and fitted (in red and green) ρ-T graphs for the films deposited on Si substrate at 5 mTorr pressure. The surface roughness of the films obtained through AFM measurements were in the range of 0.4 to 4.1 nm.

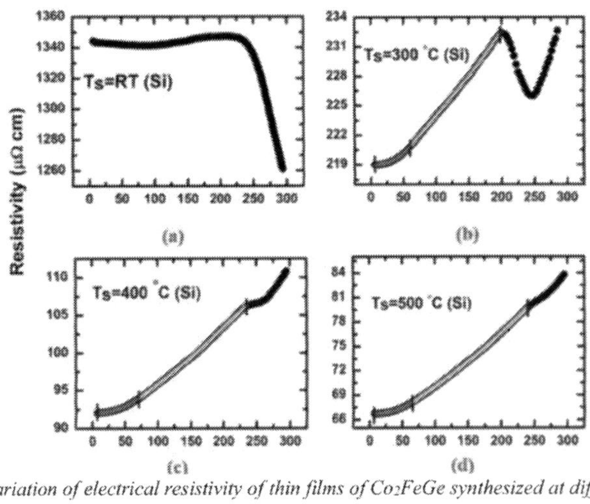

Figure 3: Temperature variation of electrical resistivity of thin films of Co_2FeGe synthesized at different substrate temperatures

The RT films deposited film has shown a semiconducting behavior as the temperature decreases and is followed by an almost temperature-independent resistivity. This unusual behavior has restricted us from fitting the curves. Another interesting observation is the presence of a minimum around 240 K for the film deposited at 300°C. This minimum in the resistivity curve has become a point of inflection for the samples synthesized at 400°C and 500°C. For fitting, curves were divided into two regions where the scattering contributions are different. Different combinations of scattering terms were considered for fitting and the least $\chi 2$ fit was chosen for the discussion. The respective equations and temperature ranges used for the fitting are

$$\rho=\rho o+AT^2+BT^{9/2} \quad \text{for } 7\text{ K} \leq T \leq 60\text{ K}$$

and

$$\rho=\rho o+CT+DT^2+ET^{7/2} \quad \text{for } T > 60\text{ K}$$

Here A, B, C, D, and E are the respective coefficients for various scattering mechanisms which are obtained as fit parameters. It was evident that the prominent contribution to the resistivity at low temperatures arises from two magnon scattering ($T^{9/5}$). Since electron-electron, one magnon scattering (absence of half metallicity), and magnetic excitations such as spin wave and Stoner fluctuations have the same temperature dependence (T^2) at low temperatures, it is difficult to separate the individual scattering mechanisms. The second region of the resistivity curve is contributed by three terms having T, $T^{2,}$ and $T^{7/5}$ dependence. The first term could be due to the electron scattering with phonon and spin wave excitations and the second term could be due to one magnon scattering. Since the phonon energy is directly dependent on the temperature, the emergence of T dependent term at high temperature regime is understandable. The third term shows that there is a presence of strong two magnon scattering in the film samples.

4. Conclusion

In this article, the effect of magnetron sputtering parameters on the physical properties of Co_2FeGe Heusler alloy has been investigated. X-ray analysis revealed a disordered *A2* crystal structure for all the deposited films. Better crystalline order and desired stoichiometry were observed at 400°C for all sputtering pressures. DC magnetization measurements revealed a soft ferromagnetic nature for all films. The optimum sputtering parameters for for obtaining films with low coercivity and high saturation magnetization values are 50 W power, 5 mTorr pressure, and 400 °C substrate temperature. Anomalous electrical transport behavior was observed for the films deposited at lower substrate temperatures.

5. Acknowledgement

One of the authors, P.V.Midhunlal thanks IIT-Madras and DMRL-Hyderabad, for their help in the fabrication and characterization of the films. Midhunlal P V also thank Dr R.B. Gangineni for the AFM characterization.

6. References

[1] Graf T, Casper F, Winterlik J, Balke B, Fecher G H and Felser C 2009 Crystal structure of new heusler compounds *Z Anorg Allg Chem* **635** 976–81

[2] Kawasaki J K, Chatterjee S and Canfield P C 2022 Full and half-Heusler compounds *MRS Bull* **47** 555–8

[3] Akmal A, Arshad F, Shakeel R, Shabir T, Riaz S and Naseem S 2023 Study of electronic, structural and magnetic properties of electrodeposited Co_2MnSn Heusler alloy thin films *Journal of Materials Research and Technology* **22** 1–16

[4] Elphick K, Frost W, Samiepour M, Kubota T, Takanashi K, Sukegawa H, Mitani S and Hirohata A 2021 Heusler alloys for spintronic devices: review on recent development and future perspectives *Sci Technol Adv Mater* **22** 235–71

[5] Hu J, Granville S and Yu H 2020 Spin-Dependent Thermoelectric Transport in Cobalt-Based Heusler Alloys *Ann Phys* **532**

[6] Balke B, Wurmehl S, Fecher G H, Felser C and Kübler J 2008 Rational design of new materials for spintronics: Co_2FeZ (Z=Al, Ga, Si, Ge) *Science and Technology of Advanced Materials* vol 9

[7] Kandpal H C, Fecher G H and Felser C 2007 Calculated electronic and magnetic properties of the half-metallic, transition metal based Heusler compounds *J Phys D Appl Phys* **40** 1507–23

[8] Daibou T, Shinano M, Hattori M, Sakuraba Y, Oogane M, Ando Y and Miyazaki T 2006 Tunnel magnetoresistance effect in CoFeB/MgO/Co_2FeSi and Co2MnSi tunnel junctions *IEEE Trans Magn* **42** 2655–7

[9] Qin Z, Liu X D and Li Z Q 2012 Anomalous Hall effects in Co_2FeSi Heusler compound films and Co_2FeSi-Al_2O_3 granular films *J Appl Phys* **111**

[10] Srinivas K, Manivel Raja M, Sridhara Rao D V. and Kamat S V. 2014 Effect of sputtering pressure and power on composition, surface roughness, microstructure and magnetic properties of as-deposited Co2FeSi thin films *Thin Solid Films* **558** 349–55

[11] Yadav A and Chaudhary S 2014 Structural and dynamical magnetic response of co-sputtered Co $_2$FeAl heusler alloy thin films grown at different substrate temperatures *J Appl Phys* **115**

[12] Kumar R K, Bharathi K K, Chelvane A J, Venkatesh S, Markandeyulu G and Harishkumar N 2009 First-principles calculation and experimental investigations on full-heusler alloy Co_2FeGe *IEEE Trans Magn* **45** 3997–9

[13] Mahat R, KC S, Karki U, Regmi S, Law J Y, Franco V, Galanakis I, Gupta A and LeClair P 2021 Structural, electronic, magnetic, transport and mechanical properties of the half-metal-type quaternary Heusler alloy $Co_2Fe_{1-x}V_xGe$ *J Magn Magn Mater* **539**

[14] Ahmad A, Srivastava S K and Das A K 2021 First-principles calculations and experimental studies on Co_2FeGe Heusler alloy nanoparticles for spintronics applications *J Alloys Compd* **878**

[15] Varaprasad B S D C S, Srinivasan A, Takahashi Y K, Hayashi M, Rajanikanth A and Hono K 2012 Spin polarization and Gilbert damping of $Co_2Fe(Ga_xGe_{1-x})$ Heusler alloys *Acta Mater* **60** 6257–65

[16] Felser C and Hirohata A *Springer Series in Materials Science 222 Heusler Alloys Properties, Growth, Applications*

[17] Vovk A, Bunyaev S A, Štrichovanec P, Vovk N R, Postolnyi B, Apolinario A, Pardo J Á, Algarabel P A, Kakazei G N and Araujo J P 2021 Control of structural and magnetic properties of polycrystalline Co_2FeGe films via deposition and annealing temperatures *Nanomaterials* **11**

[18] Nakatani N, Imai S, Tanaka M A, Kubota T, Takanashi K and Mibu K 2018 Deposition temperature dependence of interface magnetism of Co_2FeGe-Heusler-alloy/Ag films studied with 57Fe Mössbauer spectroscopy *J Magn Magn Mater* **464** 71–5

[19] Tanaka M A, Maezaki D, Ishii T, Okubo A, Hiramatsu R, Ono T and Mibu K 2014 Interface magnetism of Co_2FeGe Heusler alloy layers and magnetoresistance of Co_2FeGe/MgO/Fe magnetic tunnel junctions *J Appl Phys* **116**

[20] Uvarov N V., Kudryavtsev Y V., Kravets A F, Vovk A Y, Borges R P, Godinho M and Korenivski V 2012 Electronic structure, optical and magnetic properties of Co_2FeGe Heusler alloy films *J Appl Phys* **112**

[21] Ahmad A, Mitra S, Srivastava S K and Das A K 2021 Structural, magnetic, and magnetocaloric properties of Fe_2CoAl Heusler nanoalloy *J Magn Magn Mater* **540**

[22] Balke B, Wurmehl S, Fecher G H, Felser C, Alves M C M, Bernardi F and Morais J 2007 Structural characterization of the Co_2FeZ (Z=Al, Si, Ga, and Ge) Heusler compounds by x-ray diffraction and extended x-ray absorption fine structure spectroscopy *Appl Phys Lett* **90**

[23] Yu L, Li Z, Zhu J, Liu H, Zhang Y, Cao Y, Xu K and Liu Y 2022 Electrical and Magnetic Transport Properties of Co_2VGa Half-Metallic Heusler Alloy *Materials* **15**

IOP Conf. Series: Materials Science and Engineering 1300 (2024) 012041 doi:10.1088/1757-899X/1300/1/012041

Effect of Sn on Physical and Optical Properties of Lithium Zinc Borate Glasses

Mohamed Haneen K[a], Akshay K Thammaiah[a] Abhishek R[b], Shivamurthy C B[b], Vivek Thirumalaimurugan[a], Jagadeesha B Gangadaraiah[b,c], Abhiram Jagannathan[a,b*]

a) *Department of Physical Sciences, Indian Academy Degree College - Autonomous, Hennur Cross, Kalyan Nagar, Bangalore-560 043, Karnataka, India*

b) *Department of Post Graduate Studies and Research in Physics, The National College, Jayanagar, Bangalore- 560 070, Karnataka, India*

c) *Department of Physics, Sri Kuvempu Mahavidyalaya First Grade College, Kengal Channapatna 562 161, Ramanagara(Dist.), Karnataka, India*

*Corresponding author: abhi.ram.jagan@gmail.com

Abstract: 22.5 Li_2O_3-(20-x) ZnO-57.5B_2O_3 – x$SnCl_2$ (x = 0, 0.5. 1) glasses were synthesized using the conventional melt quench technique. Glass composition with larger concentration of lithium oxide exhibits excellent transport properties. Introducing intermediate element like zinc into the matrix will lead towards slight modification in structure by not varying much in its intrinsic properties. Thus, incorporation of Sn as dopant into the matrix has tailored the physical and optical properties of glasses. In this study, it is interesting to note that, the density of the glass has reduced for 0.5 mol % $SnCl_2$ in the matrix and on the contrary, the density of glass was found to be increased for the $SnCl_2$ with 1 mol % concentration. It indicates that, there could be a possibility of slight modification in the structure of glasses which eventually leads towards tuning of the optical bandgap due to existence of Sn.

1. Introduction

Glass is an inorganic non-crystalline solid that is often transparent, hard, brittle, and chemically inert. Glass find its applications in variety of fields like optics, architecture, medical, house wares etc[1–3]. Borate glass network consists of BO_3 and BO_4 structural units and the combination of these units gives rise to di, tri, tetra, and penta borate groups in glass network[4,5]. Borate glasses have many advantages over silica glasses and soda-lime glasses like lower melting and softening temperature[1,6], better thermal shock resistance, chemical durability and higher electric resistivity[1]. Li_2O in B_2O_3 network modify the host structure through the transformation of the structural units of the borate network from [BO_4] to [BO_3] with the creation of non bridging oxygen(NBO) which forms more ionic bonds which results in higher ionic conductivity and refractive index values[5,7] and also results in the decrease of optical energy band gap for both direct and indirect band gap[8]. Addition of transition metals in borate glasses has shown great potential in technological applications like electrooptic, electronic, electrochemical devices and radiation dosimetry[9–12]. Addition of zinc to lithium borate glass will increase the strength and enhance the electron emission[11]. Presence of zinc in lithium borate glasses can alter or increase the glass transition temperature[13], thermal expansion coefficient and density. SnO_2 is a semiconductor material, and its incorporation into borate glasses improves their electrical conductivity which has very important applications in sensors, optoelectronic devices, and solid state batteries[14].SnO_2 acts as a network modifier and also as network former in borate glasses[15]. This work discusses about the physical and

IOP Conf. Series: Materials Science and Engineering 1300 (2024) 012041 doi:10.1088/1757-899X/1300/1/012041

optical properties of Sn doped in lithium zinc borate glass system and the work is further intended to examine towards transport behavior of such glass systems.

2.**Experimental Method:** 57.5B_2O_3 –(20-x)ZnO-22.5Na_2O- xSnO (where, x = 0, 0.5,1 mol %) glass composition were synthesized through conventional melt quench technique using precursors of analytical reagent grade chemicals from with a purity of 99.9% was procured from Loba Chemie. The batch composition was measured with an appropriate stoichiometric calculation and the powders were mixed thoroughly and transferred to the porcelain crucibles. These crucibles were placed inside the muffle furnace for obtaining the molten glass at 1050 °C and were constantly stirred for 15 minutes in its melting temperature in order to achieve homogeneous bubble-free glasses; the melt was then poured onto the brass plate through stainless steel washers in order to obtain the button shaped glasses which are coded as LZB, 0.5SnLZB, 1SnLZB. These samples were polished using sand paper of its grade ranging from 220, 400, 600, 800 and 1500 and then was subjected to analyze physical properties and optical properties.

3. **Results and Discussions**

3.1 **Physical Properties:** Density measurements were calculated using Archimedes principle with toluene as immersing liquid with its density around 0.866 g/cm^3. Density related parameters such as Molar volume, ion concentration, interionic distance, polaron radius, field strength and average boron- boron distance were evaluated using the reference[16,17]. Figure 1 indicates the anomalous relationship has been observed with the incorporation of $SnCl_2$ in the matrix, as it is observed that initially at 0.5 mol % of Sn in LZB matrix, the reduction in density suggests that there could be a possibility of creation of more NBO's in the network. Further addition of Sn of 1 mol% in the matrix, the density of glass has increased which suggests that there could be possibility of formation of bridging oxygens in the network. Further, elucidation on the BO's and NBO's in the network can be probed by evaluating the optical properties of glasses.

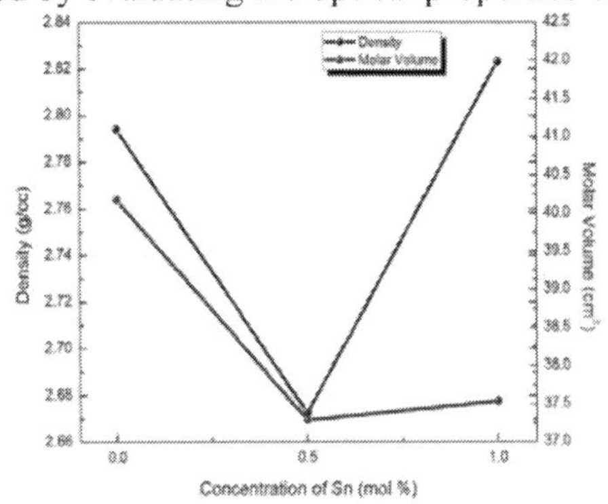

Figure 1: Concentration of Sn versus Density and Molar Volume

Table 1: Refractive Index related parameters for Lithium Zinc Borate glasses doped with Sn

Density Related Parameters	**0 SnLSB**	**0.5 SnLSB**	**1.0 SnLSB**
Density (g/cm^3)	2.794	2.672549	2.822791
Molar Volume (cm^3/ mol)	40.177	37.289	37.536

Sn Ion Concentration ($\times 10^{20}$ ions/cm^3)		8.05	17.05
Interionic Distance r_i ($\times 10^{-9}$m)		1.08	8.38
Polaron Radius r_p($\times 10^{-10}$m)		4.33	3.38
Field Strength cm^{-2} ($\times 10^{20}$)		1.6	2.63
Average Boron-Boron Distance ($\times 10^{-10}$ m)	3.93	4.26	4.27

3.2 Optical Properties

Well-polished lithium zinc borate glasses doped with $SnCl_2$ were subjected to UV-Vis-NIR spectrometer. Figure 3 (a) indicates the characteristic absorption profile of the synthesized glasses. Figure 3 (b) and 3 (c) depicts the direct and indirect bandgap of samples which were estimated with a plot of $(\alpha h\nu)^n$ against energy $(h\nu)$. In case of direct band gap, the exponent' n' accounts for 2, while in case of indirect band gap exponent value are ½. Figure 3 (d) indicates the Urbach energy of glasses. The anomalous behavior of glasses with incorporation of Sn at 0.5 and 1 mol% are indicative with the trend in bandgap as, 0.5SnLZB shows least optical bandgap and higher Urbach energy, thereby it is clear that the non-bridging oxygens in the glasses are found to be quite high. On the contrary, it is found that at 1 mol % of Sn in lithium zinc borate glass matrix the bandgap value has increased and there is drastic reduction in Urbach energy. This clearly suggests that, $SnCl_2$ plays a crucial role in establishing network bridging[18].

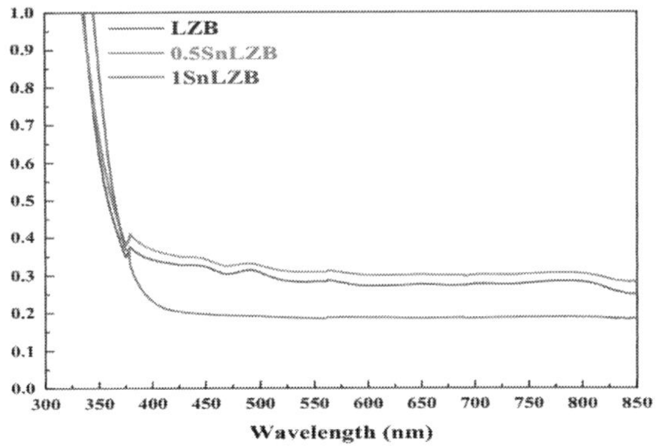

Figure 2 (a) : Absorption spectra of Lithium zinc borate glasses doped with Sn

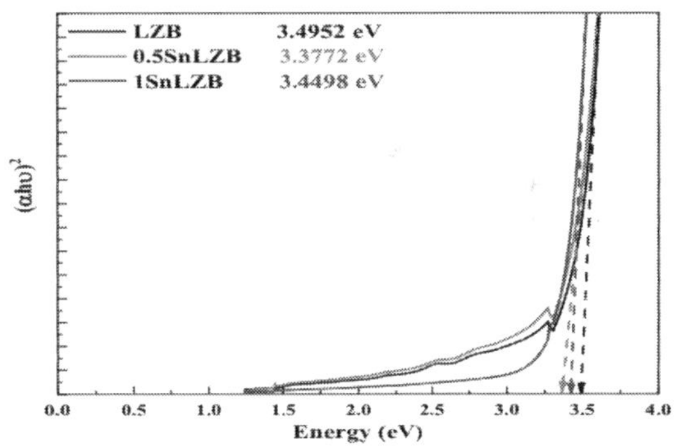

Figure 2 (b) : Direct bandgap of lithium zinc borate glasses doped with Sn

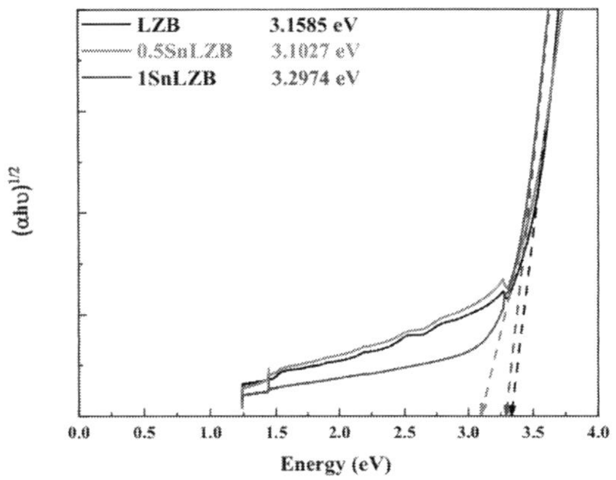

Figure 2 (c) : Indirect bandgap of lithium zinc borate glasses doped with Sn

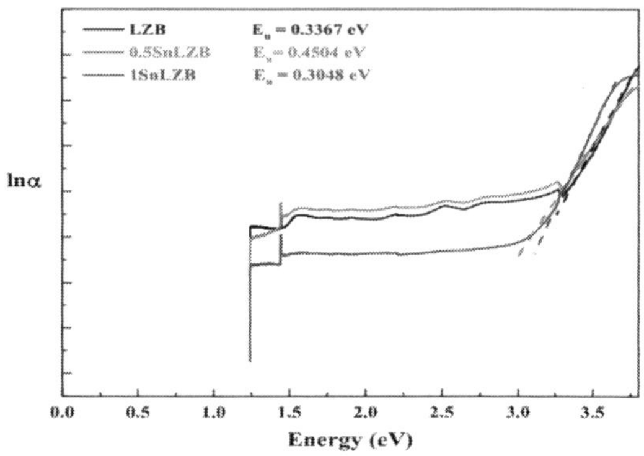

Figure 2 (d) : Urbach energy of lithium zinc borate glasses doped with Sn

Table 2: Refractive Index related parameters for Lithium Zinc Borate glasses doped with Sn

Refractive Index Related Parameters	0SnLZB	0.5SnLZB	1.0SnLZB
Refractive Index (n)	2.336	2.414	2.383
Molar Refraction R_m (cm^3 /mol)	24.016	23	22.875
Molar Polarizability $(Å)^3$	9.530	9.127	9.077
Metallization Criteria (M)	0.402	0.383	0.390
Cation Polarizability α_{cat} $(Å)^3$	6.22	6.08	5.94
Oxide Ion Polarizability $(Å)^3$	0.0451	0.0425	0.0424
Interaction Parameter $(Å)^3$	6.228	6.556	6.901
ε_0 (F/m)	5.458	5.829	5.680
ε_{stat} (F/m)	44.394	28.160	33.748

Based on the bandgap with the reference [19,20] , refractive index related physical properties of glass were evaluated. These values very well corellate with the fact that 0.5SnLZB glass has prononunced non-bridging oxygens in the network. The dielectric constant of the glass is found to be high and it clearly indicative that this glass could exhibit excellent charge transport properties as well as nonlinear optical properties [21,22].

4. Conclusion:

Lithium Zinc Borate Glasses doped with $SnCl_2$ and coded as LZB, 0.5SnLZB, 1SnLZB were synthesized by conventional melt quench technique. These glasses were examined for physical and optical properties. It is interesting to note that the tin plays a role of intermediate in the lithium zinc borate glass matrix for which, at 0.5 mol % of $SnCl_2$, the glass tends to create more NBOs in the network and on the contrary at 1 mol % of $SnCl_2$, there is an increase in density and optical bandgap which enbales glass to bridge the network. Thus, 0.5SnLZB glass has a great potential in charge-transport or nonlinear optical properties.

5. References

[1] Shelby J E 2020 *Introduction to Glass Science and Technology* (The Royal Society of Chemistry)

[2] Hench L L 2006 The story of Bioglass® *Journal of Materials Science: Materials in Medicine* **17** 967–78

[3] Anon Fernandes, Hugo R et al. "Bioactive Glasses and Glass-Ceramics for Healthcare Applications in Bone Regeneration and Tissue Engineering." Materials (Basel, Switzerland) vol. 11,12 2530. 12 Dec. 2018, doi:10.3390/ma11122530

[4] Meera B N and Ramakrishna J 1993 Raman spectral studies of borate glasses *Journal of Non-Crystalline Solids* **159** 1–21

[5] Naresh P 2020 Influence of TeO2 on the UV, electrical and structural studies of Li2O–ZnO– B2O3 glasses *Journal of Molecular Structure* **1213** 128184

[6] Bobkova N 2003 Thermal Expansion of Binary Borate Glasses and Their Structure *Glass Physics and Chemistry* **29** 501–7

[7] Krogh-Moe J 1959 *On the structure of boron oxide and alkali borate glasses* (Chalmers Tekniska Hogskola (Sweden))

[8] Mohamed Kamari H, Chiew W H, Aziz S, Daud W M, Wahab zaidan, Matori K and Shahrim I 2014 Optical Properties of lithium borate glass (Li2O)x (B2O3)1-x *Sains Malaysiana* **43** 899–902

[9] Kafadar V E, Yazici A N and Yildirim R G 2009 Determination of trapping parameters of dosimetric thermoluminescent glow peak of lithium triborate (LiB3O5) activated by aluminum *Journal of Luminescence* **129** 710–4

[10] G Naga Raju, N Ch Ramesh, P Naresh, T L Krishna, K Srinivasulu, K S V Sudhkar, and P Venkateswara Rao 2009 Spectroscopic and dielectric properties of titanium doped MgO-Al2O3-B2O3 glass system *IOP Conference Series: Materials Science and Engineering* **2** 012031

[11] Saidu A, Wagiran H, Saeed M A and Alajerami Y S M 2014 Structural properties of Zinc Lithium borate glass *Optics and Spectroscopy* **117** 396–400

[12] Gautam C, Yadav A K and Singh A K 2012 A Review on Infrared Spectroscopy of Borate Glasses with Effects of Different Additives ed M Kawashita, C-F Yang, Y S Cho and D Jia *ISRN Ceramics* **2012** 428497

[13] Kim N J, La Y H, Im S H, Han W-T and Ryu B K 2009 Effect of ZnO on physical and optical properties of bismuth borate glasses *Electronic Materials Letters* **5** 209–12

[14] Abouhaswa A S, Rammah Y S, Ibrahim S E and El-Hamalawy A A 2018 Structural, optical, and electrical characterization of borate glasses doped with SnO2 *Journal of Non-Crystalline Solids* **494** 59–65

[15] Hayashi A, Nakai M, Tatsumisago M, Minami T, Himei Y, Miura Y and Katada M 2002 Structural investigation of SnO–B2O3 glasses by solid-state NMR and X-ray photoelectron spectroscopy *Journal of Non-Crystalline Solids* **306** 227–37

[16] Venugopal A R, Rajaramakrishna R, Abhiram J, Pattar V, Rajashekara K M and Kaewkhao J 2019 Sm3+ Doped Lithium Strontium Borate Glasses for Solid State Lighting Applications *Glass Physics and Chemistry* **45** 472–84

[17] Kaewjaeng S, Kothan S, Chaiphaksa W, Chanthima N, Rajaramakrishna R, Kim H J and Kaewkhao J 2019 High transparency La2O3-CaO-B2O3-SiO2 glass for diagnosis x-rays shielding material application *Radiation Physics and Chemistry* **160** 41–7

[18] Jagannathan A, Rajaramakrishna R, Rajashekara K M, Gangareddy J, K V P, S V R, B E, V J A, Kaewkhao J and Kothan S 2020 Investigations on nonlinear optical properties of gold nanoparticles doped fluoroborate glasses for optical limiting applications *Journal of Non-Crystalline Solids* **538** 120010

[19] Roopa and Eraiah B 2021 Experimental and theoretical approach on the physical, structural and optical properties of ZrO2 - Na2O - B2O3 glasses doped with Dy2O3 *Journal of Non-Crystalline Solids* **551** 120394

[20] Mallur S B, Czarnecki T, Adhikari A and Babu P K 2015 Compositional dependence of optical band gap and refractive index in lead and bismuth borate glasses *Materials Research Bulletin* **68** 27–34

[21] Shanmugavelu B, Ravi Kanth Kumar V V, Kuladeep R and Narayana Rao D 2013 Third order nonlinear optical properties of bismuth zinc borate glasses *Journal of Applied Physics* **114** 243103

[22] Chandrashekaraiah G, Sivasankara Reddy N, Sujatha B, Viswanatha R and Narayana Reddy C 2018 Role of Er3+ and Bi3+ ions on thermal and optical properties of Li2B4O7 glasses: Structural correlation *Journal of Non-Crystalline Solids* **498** 252–61

IOP Conf. Series: Materials Science and Engineering 1300 (2024) 012042 doi:10.1088/1757-899X/1300/1/012042

Unveiling the Structural and Optical Properties of Samarium Titanates Pyrochlore, $Sm_2Ti_2O_7$

Rachana Sain* **and Chandan Upadhyay**

School of Materials Science and Technology, Indian Institute of Technology (Banaras Hindu University), Varanasi 221005, India

*Email: rachanasain.rs.mst20@itbhu.ac.in

Abstract. A magnetic system whose geometry prevents all pairwise spin interactions from being fulfilled simultaneously to minimize the energy is said to be frustrated. These systems are prevalent throughout all disciplines of physical and chemical study. This article reports a detailed structural and optical analysis of such frustrated $Sm_2Ti_2O_7$ (STO), a pyrochlore structure. The ordered pyrochlore, STO of cubic symmetry, space group $Fd\bar{3}m$ [no. 227], has been synthesized using the standard solid-state reaction method. The Tauc's plot reveals a direct band gap value of 3.8 eV from ultraviolet-visible reflectance data. Photoluminescence studies for STO using 280 nm excitation wavelength show the most intense emission peak at 395 nm, 468 nm, and other weak emissions at 368 nm, 412 nm, 430 nm, 449 nm, and 514 nm corresponding to the respective electronic transitions, indicates that the materials show photoluminescence properties. However, the unique characteristics of $Sm_2Ti_2O_7$, including its wide band gap, photoluminescence, and high refractive index, make it a promising candidate for various magneto-optical, solid oxide fuel cell device fabrication applications.

1. Introduction

The cubic-structured titanate oxides, $RE_2Ti_2O_7$ (RE: Rare-Earth) pyrochlore are noteworthy materials due to their unique ground states, including spin ice, spin glass, cluster glass, spin liquid, and Griffiths phase-like behaviour resulting from magnetic frustration and coupling among various ordered parameters. The interplay of various ordered parameters makes them promising for diverse applications such as optical device manufacturing, solid oxide fuel cells, water purification, nuclear waste containment, and spintronics [1-5]. Geometrically frustrated rare-earth pyrochlore magnets have been intensely studied in recent years, rich in various unconventional ground states and quantum phenomena due to their special structural character at low temperatures [6]. The frustrated pyrochlore structure is extremely susceptible to various external impacts, such as crystal field, pressure, magnetic field, frequency, temperature, composition, and chemical pressure [7-9]. Only a few research works are concerned with the structural, optical, electrical, and magnetic properties of samarium-based pyrochlores. Most of these related to pure and doped oxides as $Sm_2X_2O_7$, X=Ti, Zr [10-12], ising antiferromagnetic in $Sm_2Ti_2O_7$ pyrochlore [5], ternary oxides Ln_2CrTaO_7, Ln = Y, Sm, Gd [13], $Sm_2Ti_2O_7/SmCrO_3$ heterojunction photocatalyst [14,15].

The idealized chemical formula $A_2B_2X_6X'$ describes the pyrochlore structure, where the A-site element belongs to the rare earth metal family, and the B-site element is typically from the transition metal class. And pyrochlore family can be more accurately represented as $A_2B_2X_6X'$ [6].

These rare earth titanate pyrochlores, $RE_2Ti_2O_7$ create an endless 3-dimensional corner-sharing tetrahedral network with a very high potential for frustration. The pyrochlore compound $Sm_2Ti_2O_7$ exhibits ordered atomic arrangements and possesses a face-centred cubic (FCC) crystal structure. It can be described by the stoichiometry $A_2B_2X_6X'$ classified under the space group $Fd\bar{3}m$ [no. 227] shown in Figure 1, in which Sm^{3+} 3p and Ti^{4+} 4p ions reside at 16d and 16c sites, respectively. In the structure, 16d crystallographic site is occupied by larger Sm 3p cations, coordinated with eight oxygen anions (6X and 2X') positioned inside a distorted cubic polyhedron formed by eight oxygen atoms. While 16c crystallographic site is occupied by smaller Ti 4p cations coordinated with six oxygen anions (6X) and placed inside distorted octahedron formed by six oxygen atoms, BX_6. The 48f site, with a 2mm symmetry, is occupied by oxygen anions, and each is coordinated with two Ti and two Sm cations. As for the X' anions, they are positioned at the 8a site, exhibiting 43m symmetry, and are tetrahedrally coordinated with four A cations [5,11].

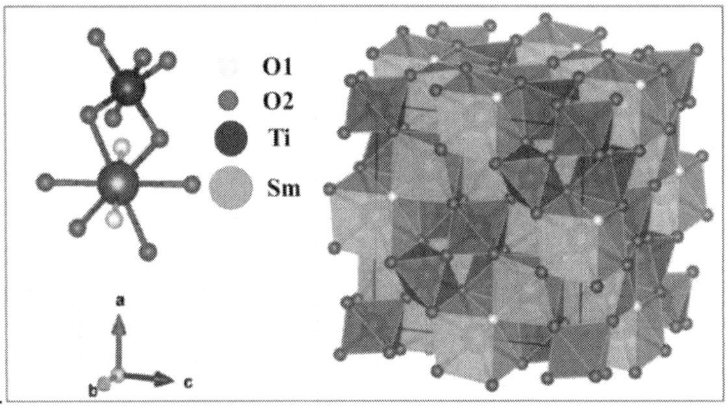

Figure 1. Schematic $Sm_2Ti_2O_7$ pyrochlore crystal structure with oxygen crystal field of rare earth atom, Sm at A site and transition metal atom, Ti at B site.

The present article aims to detail the synthesis and characterizations (structural and optical) of $Sm_2Ti_2O_7$ pyrochlore material with Sm^{+3} $4f^5$ and Ti^{+4} $3d^0$ ions on an fcc lattice of edge-shared network of tetrahedra. The ordered pyrochlore exhibits cubic symmetry, used to perform optical properties and band gap value using Wood and Tauc's equation of reflectance data. It is one of the most concise structural and optical studies of an ordered frustrated samarium titanate pyrochlore system.

2. Experimental Details

2.1. Synthesis method

The ordered polycrystalline samples of $Sm_2Ti_2O_7$ pyrochlore oxide material were prepared by firing and sintering stoichiometric quantities of Sm_2O_3 (~99.999%, Sigma Aldrich), and TiO_2 (~98%, Sigma Aldrich) under open air for 20 hours at 1200, 1400 and 1475°C in alumina, Al_2O_3 crucibles. The step-wise synthesis of samarium titanate pyrochlore oxide by standard solid-state reaction synthesis process is detailed in the flow chart, as shown in Figure 2. The resultant product was found to be impure, so subsequent regrinding, pelletization, and sintering stages were needed to get a pure single-phase product. The expected synthesis reaction of rare earth titanate oxide is given as,

$$Sm_2O_3 + 2\,TiO_2 \rightarrow Sm_2Ti_2O_7$$

IOP Conf. Series: Materials Science and Engineering 1300 (2024) 012042 doi:10.1088/1757-899X/1300/1/012042

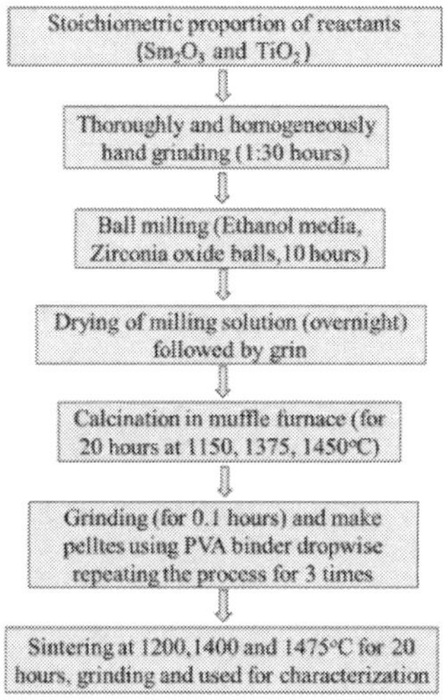

Figure 2. Flow chart showing the detailed synthesis of $Sm_2Ti_2O_7$ pyrochlore by solid-state synthesis method.

2.2. Characterizations

The crystallographic phase analysis (crystal structure) of the resultant product was confirmed using a powder x-ray diffraction and high-resolution x-ray diffraction, HRXRD (Smart Lab 3 kW and 9 kW, Rigaku, Japan) with Cu K_α radiation, which was initially standardized by a high-purity Si powder sample. Photoluminescence (PL) emission spectra for the STO powder sample were recorded at CIF, IIT(BHU) on the F-4600 spectrometer, Hitachi, Japan, equipped with a Xenon lamp with a scan rate of 30 nm s^{-1} at room temperature. The PL emission spectra were measured in the wave number range from 300 to 3700 cm^{-1}. Absorption spectra (at room temperature) were recorded using a UV-visible (double-beam) spectrophotometer of SHIMADZU-2450, and reflectance data were collected in the 200–1200 nm range. A standard $BaSO_4$ (Barium sulphate) standard sample was initially used as a reference sample.

3. Results and discussions

3.1. Powder XRD and HRXRD

Powder XRD and high-resolution x-ray diffraction (HRXRD) pattern of STO polycrystalline powder samples is illustrated in Figure 3 (a) and (b). To eliminate unwanted phase impurities and ensure pure cubic phase evolution, the firing temperature was optimized at different calcination temperatures. After the initial firing, the product sample consisted of a few unwanted reflection indices of very small intensity up to 1400°C temperature. The calcination process optimization was done by recording successive XRD patterns. Regrinding and sintering pellets under open air yielded only the cubic phase as the resultant product. The determined space group of the final product is $Fd\bar{3}m$ [227] consistent with the $Sm_2Ti_2O_7$ phase reported by Subramanian e. al. and Kun Yang et al. [16,17] where the most prominent reflection peak indexed in diffraction pattern as [222] with the International Centre for Diffraction Data (ICDD) file No. 73-1699. Moreover, all reflection indices for STO obtained from their high-resolution powder x-ray diffractions agree with the previously reported rare earth titanate pyrochlores. Also, similar cubic structures have been presented before by Geioushy et al. and Surbhi et al. [18,19] for STO prepared by the sol-gel method of synthesis and the related compounds $RE_2Ti_2O_7$ analogues to samarium titanate pyrochlore extensively reported in hundreds of research articles [3,6,20,21].

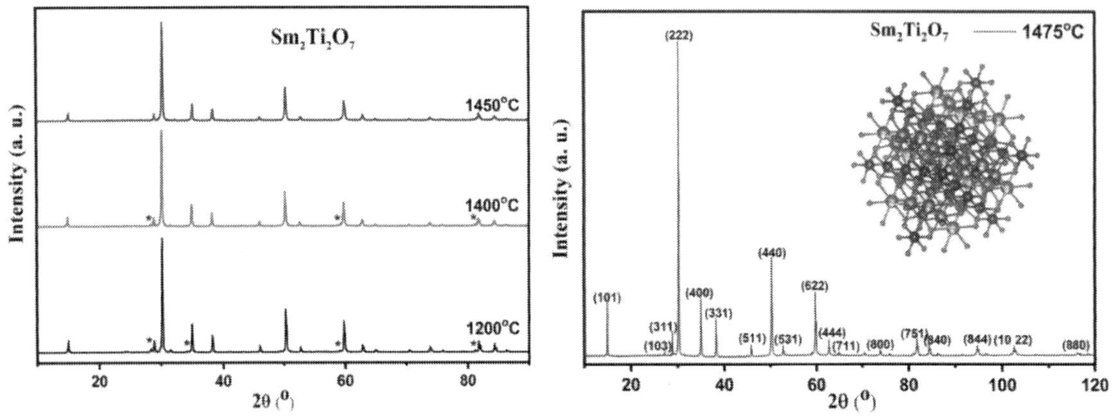

Figure 3. (a) XRD patterns for the illustration of optimization of sintering temperature of $Sm_2Ti_2O_7$ pyrochlore at different temperatures for 20 hours. (b) HRXRD pattern of STO pyrochlore sintered at 1475°C for 20 hours showing pure cubic pyrochlore phase with reflection indices.

3.2. *Photoluminescence Spectroscopy*

The absorption and emission properties of $RE_2Ti_2O_7$ have been widely investigated in previous experiments and theoretical studies [12,22]. Figure 4 depicts the emission spectrum at room temperature of a samarium titanate sample that had undergone sintering at 1475°C for 20 hours, and recorded in the range 350-580 nm when excited with 280 nm wavelength. This spectrum was recorded within the wavelength range of 350 to 580 nm when excited with light at 280 nm. The STO pyrochlore sample's PL (emission) spectra consist of numerous emission peaks corresponding to several transitions from the CB (conduction band) edge to the ground state. The higher wavelength emission edges from 449 nm correspond to the charge transfer transition (CTT) within Sm states [23], as depicted in Figure 4 (a).

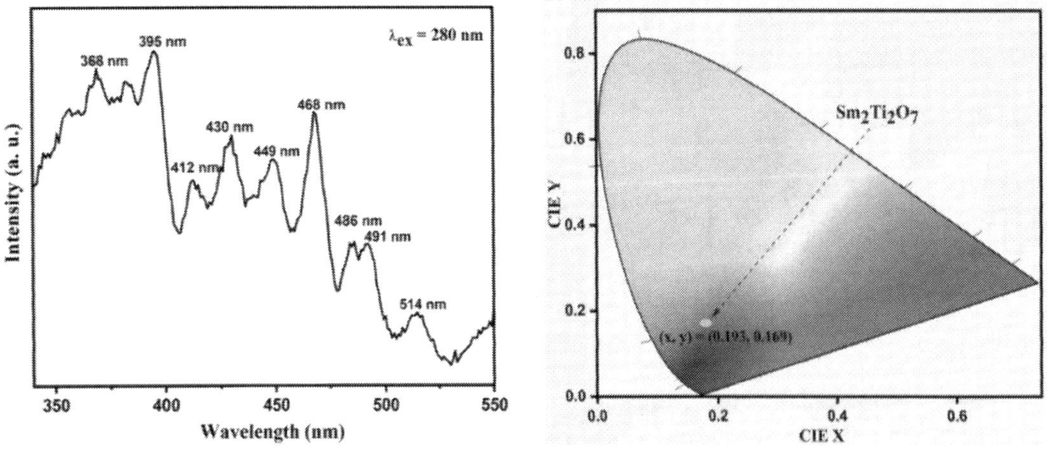

Figure 4. (a) Room temperature PL emission spectrum of polycrystalline $Sm_2Ti_2O_7$ pyrochlore oxide in 340-550 nm wavelength range. (b) CIE (colour space) diagram of $Sm_2Ti_2O_7$ under excitation of 280 nm.

PL emission phenomenon was further assured by the chromatic response (CIE) chromaticity colour model, where the schematic depicts a 2-dimensional (x, y) CIE chromaticity colour model of STO pyrochlore under an excitation wavelength of 280 nm depicted in Figure 4 (b). The CIE chromaticity (x, y) coordinate comes out to be (0.193, 0.169), which is close to light blue colour emission. This outcome provides additional evidence supporting the idea that the emission spectra are being suppressed in the visible and near-infrared (IR) range due to the arrangement of oxygen vacancies surrounding the B site ion.

3.3. Ultraviolet-Visible Spectroscopy

Further, Figure 5 presents the room temperature ultraviolet (UV) -visible absorption spectra recorded for the polycrystalline samarium titanate oxide sample sintered at 1475°C for 20 hours in the 200-700 nm wavelength range. The STO sample's absorption spectrum contains high and low-intensity absorption edges, which agrees with the previous work for rare earth titanates pyrochlore [18,23]. In addition, the absorption edges centred at 302 nm in the STO powder sample is associated with the ligand-to-metal charge transfer (LMCT) process. This involves the transition of electrons from the O-2p orbital in the valence band to the Ti-3d orbital in the conduction band. The optical band gap has been determined from UV-visible absorption data employing Wood and Tauc's equation and by extrapolating a graph of $(\alpha h\nu)^2$ vs. hν considering the direct transition [19], as depicted in the inset of Figure 4. The Wood and Tauc's equation is written as

$$(\alpha h\nu)^n = A\,(h\nu - Eg)$$

where it consists of α, which is an absorption coefficient, A is a material-dependent constant parameter, hν is an incident photon energy, and here, the value of n decides the nature of the transition (2-corresponds to direct). For determining the direct optical band gap, the linear portion of the curve was extrapolated to the x-axis yielding a value of 3.8 eV for the STO sample. It corresponds to the electronic transition of the O - 2p to the Ti - 3d orbital and it is equated to the absorption edge value at 302 nm, i.e. 1240/300 = 4.1 eV.

We can estimate the conduction band (CB) and valence band (VB) edge energies using the following equations,

$$E_{CB} = \chi - E^C - \frac{1}{2}E_g \quad \text{and} \quad E_{VB} = E_{CB} + E_g$$

Where χ is the absolute electronegativity of STO, band gap energy E_g, and E^C is the energy of the free electrons on the hydrogen scale about (~4.5 eV). The computed position of the conduction band edge (E_{CB}) for STO sample is -1.14 eV, with the valence band edge (E_{VB}) situated at 2.66 eV.

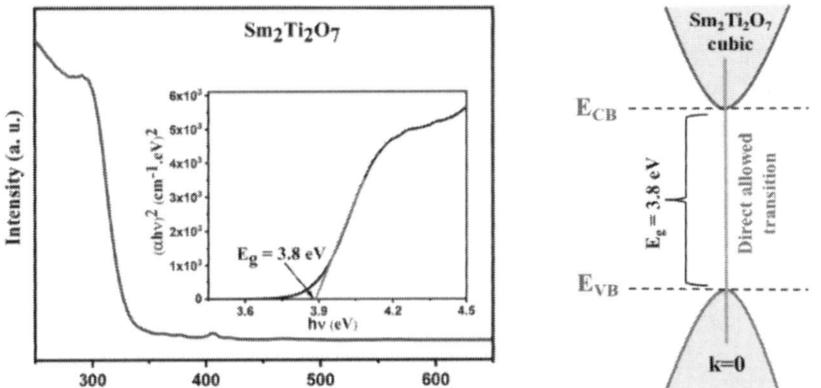

Figure 5. (a) Room temperature UV–vis spectrum of $Sm_2Ti_2O_7$ pyrochlore oxide in the wavelength range 200-700 nm. (Inset shows Tauc's plot [$(\alpha h\nu)^2$ vs hν] for band gap). (b) An energy band model shows the maxima of VB and the minima of CB in E-k space.

4. Conclusion

In summary, using the well-established method of synthesis and characterization, we discovered the evolution of crystalline $Sm_2Ti_2O_7$ pyrochlore and structure upon calcination of initially mixed oxides by a complex of powder XRD and HRXRD. Further, the direct band gap of STO powder sample is estimated using Tauc's plot that is comparable to the theoretically and experimentally reported value. Moreover, the chromaticity colour model CIE diagram reveals emission edge colour coordinates towards the blue part of the spectrum, implying that the ordering of oxygen vacancies has a significant role in tuning the optical characteristics of rare earth titanate oxides. These findings introduce a promising approach to manipulating the absorption and emission properties, particularly in the visible to near-infrared (IR) spectrum, which is important for magneto-optical device fabrication.

Acknowledgement

The author gratefully acknowledges the Central Instrumental Facility (CIF), Indian Institute of Technology (BHU), for any support. Also, the author wishes to thank the School of Materials Science and Technology, IIT(BHU).

References

1 R. C. Ewing, W. J. Weber, and J. Lian, Journal of Applied Physics **95,** 5949 (2004).

2 A. P. Ramirez, A. Hayashi, R. J. Cava, R. Siddharthan, and B. S. Shastry, Nature **399,** 333 (1999).

3 R. Upadhyay, M. Shukla, R. Sain, M. Tolkiehn, and C. Upadhyay, Journal of Magnetism and Magnetic Materials **566,** 170308 (2023).

4 S. T. Bramwell and M. J. P. Gingras, Science **294,** 1495 (2001).

5 C. Mauws, A. M. Hallas, G. Sala, A. A. Aczel, P. M. Sarte, J. Gaudet, D. Ziat, J. A. Quilliam, J. A. Lussier, M. Bieringer, H. D. Zhou, A. Wildes, M. B. Stone, D. Abernathy, G. M. Luke, B. D. Gaulin, and C. R. Wiebe, Physical Review B **98,** 100401 (2018).

6 J. S. Gardner, M. J. P. Gingras, and J. E. Greedan, Reviews of Modern Physics **82,** 53 (2010).

7 F. X. Zhang and S. K. Saxena, Chemical Physics Letters **413,** 248 (2005).

8 F. X. Zhang, B. Manoun, S. K. Saxena, and C. S. Zha, Applied Physics Letters **86** (2005).

9 M. Ruminy, E. Pomjakushina, K. Iida, K. Kamazawa, D. T. Adroja, U. Stuhr, and T. Fennell, Physical Review B **94,** 024430 (2016).

10 K. Sandeep, J. K. Thomas, and S. Solomon, Materials Science and Engineering: B **254,** 114512 (2020).

11 S. Singh, S. Saha, S. K. Dhar, R. Suryanarayanan, A. K. Sood, and A. Revcolevschi, Physical Review B **77,** 054408 (2008).

12 M. Irfan, S. Azam, S. Hussain, S. A. Khan, M. Makhdoom, B. Gul, S. Khan, I. V. Kityk, S. Muhammad, and S. M. Siddeeg, Computational Condensed Matter **21,** e00414 (2019).

13 A. V. Egorysheva, O. G. Ellert, D. I. Kirdyankin, E. F. Popova, and R. D. Svetogorov, Journal of Magnetism and Magnetic Materials **513,** 167226 (2020).

14 C. Magdalane, D. Jayakumar, Y. Samson, A. K. H. Bashir, M. Maaza, D. Letsholathebe, A. Mahmoud, and J. Kennedy, Journal of King Saud University - Science **32** (2019).

15 A. Nashim and K. M. Parida, Chemical Engineering Journal **215-216,** 608 (2013).

16 K. Yang, P. Lei, T. Yao, B. Gong, Y. Wang, M. Li, J. Wang, and J. Lian, Corrosion Science **185,** 109394 (2021).

17 M. A. Subramanian, G. Aravamudan, and G. V. Subba Rao, Progress in Solid State Chemistry **15,** 55 (1983).

18 R. A. Geioushy, O. A. Fouad, M. M. Rashad, D. A. Rayan, and A. T. Kandil, New Journal of Chemistry **45,** 7799 (2021).

19 S. Verma, S. Rani, and S. Kumar, Journal of Alloys and Compounds **750,** 902 (2018).

IOP Conf. Series: Materials Science and Engineering 1300 (2024) 012042 doi:10.1088/1757-899X/1300/1/012042

20 P. K. Yadav, M. Tolkiehn, and C. Upadhyay, Journal of Physics and Chemistry of Solids **134,** 201 (2019).

21 H. Liu, Y. Zou, L. Ling, L. Zhang, W. Tong, C. Zhang, and Y. Zhang, Journal of Magnetism and Magnetic Materials **369,** 107 (2014).

22 B. Winkler, A. Friedrich, W. Morgenroth, E. Haussühl, V. Milman, C. R. Stanek, and K. J. McClellan, Chinese Science Bulletin **59,** 5278 (2014).

23 W. T. Carnall, P. R. Fields, and K. Rajnak, The Journal of Chemical Physics **49,** 4424 (2003).

IOP Conf. Series: Materials Science and Engineering 1300 (2024) 012043 doi:10.1088/1757-899X/1300/1/012043

Effect of immersion liquids on physical properties of Pr^{3+}-ions doped lithium zinc borate glasses

Uthayakumar T[1], Namrata Yaduvanshi[1], Madhu A[*,1], Srinatha N[2]

[*,1] Department of Physics, Dayananda Sagar college of engineering, Kumaraswamy Layout, Bangalore- 560 078
[2]Department of Physics, RV Institute of Technology and Management, Bangalore-560076
*Email: mmathi.33@gmail.com

Abstract. Our research focused on the creation of lithium zinc borate glasses that were doped with Pr^{3+} ions. Specifically, our study focused on the effects of substituting lithium with Pr_2O_3. Our x-ray diffractometer analysis indicated that the glasses were amorphous, which was as expected. We have also evaluated the impact of adding Pr_2O_3 using toluene and xylene immersion liquids, and we found that these liquids had unique effects on the glass properties. Furthermore, we examined density changes using the Archimedes principle and estimated physical parameters. We carefully analysed the variations in NBO that were caused by Pr_2O_3, and our results provide valuable insights into the structural behaviour of these glasses.

1. Introduction

Recent research [1][2][3]has shown that borate-based glasses have various practical applications. These include creating thin amorphous films for batteries, developing bioactive glasses for tissue engineering, nuclear waste disposal[4], photonic applications, tuneable or short pulse lasers, optic fibre amplifiers and fibre lasers. The thermal activation of alkali ions within the glass enables them to move swiftly from one side to another. This facilitates the replacement of alkali ions near the glass surface by other ions of the same valence[5]. By incorporating alkali ions into the borate network, the Tg is increased while the thermal expansion coefficient is lowered[6]. The addition of transition metal ions, like ZnO, to glassy systems creates various dopant sites through strong interactions. This leads to optical and spectral properties with high intensity. ZnO-doped glasses are a popular choice for developing optoelectronic devices, solar converters, ultraviolet-emitting lasers, and gas sensors due to their favourable properties[7][8]. These glasses possess visual, electrical, and magnetic properties, while also being non-toxic and non-hygroscopic. Pr^{3+} ions have great potential in photonic devices, particularly in white light emitting diodes (WLEDs). As far as we know, besides Eu^{3+} and Sm^{3+} rare earth (RE) ions, Pr^{3+} is another well-known RE ion with an ideal luminescent center at around 600 nm for emitting red light. These emission centres are highly beneficial for producing high-color rendering index glasses for WLED applications[9]. This application requires high emission line intensities and is becoming increasingly popular in industries that use X-rays[10] and gamma radiation for technical and medical purpose. The use of gamma rays in medicine, agriculture, and industries such as nuclear reactors and nuclear waste storage [4]requires the development of new and improved shielding materials[3][10]. Laser technology, created using Pr^{3+} ions doped/codoped glass systems, has great

IOP Conf. Series: Materials Science and Engineering 1300 (2024) 012043 doi:10.1088/1757-899X/1300/1/012043

potential for eye-safe remote sensing and medical surgery applications[11]. These reports have inspired us to select the composition and rare-earth for preliminary studies aimed at understanding the physical and optical parameters of lithium-zinc-borate glass.

2. Synthesis and characterization

By utilizing the high-quality precursors of 99.9% purity, lithium zinc borate glasses doped with different concentrations of Pr^{3+} were prepared using the melt-quenching process. The initial precursors of the glass composition include lithium carbonate (Li_2CO_3), zinc oxide (ZnO), boric acid crystals (H_3BO_3), and praseodymium oxide (Pr_2O_3). The stoichiometric mixture of glass LZBPr series prepared includes (20-x) Li_2CO_3 –15 ZnO – 65 H_3BO_3– x Pr_2O_3 (where, x = 0.25, 0.5, 0.75, and 1.0 mol %). These starting materials measuring 10gram stoichiometric batches, consisting of different concentrations of Pr^{3+} were mixed thoroughly and ground using an agate motor with the pestle, to obtain smooth and fine powders. These fine powders were then transferred to the unglazed porcelain crucibles and melted in a furnace. During this melting process, the crucibles were taken out at the temperature of around 900°C and swirled for few times to ensure the homogeneity of the mixtures. Next, the obtained molten mixtures at 980°C were moulded into small circular discs by following rapid quench through the preheated brass plates. The flow chart for the preparation of glass samples and prepared glass samples is illustrated in Figure 1. The uniform and smooth surface required for the characterization of the samples was obtained by fine polishing with suitable grades of emery paper. Lastly, the synthesized glass samples are annealed at 250°C for 4 hours for the removal of the thermal stress and strain.

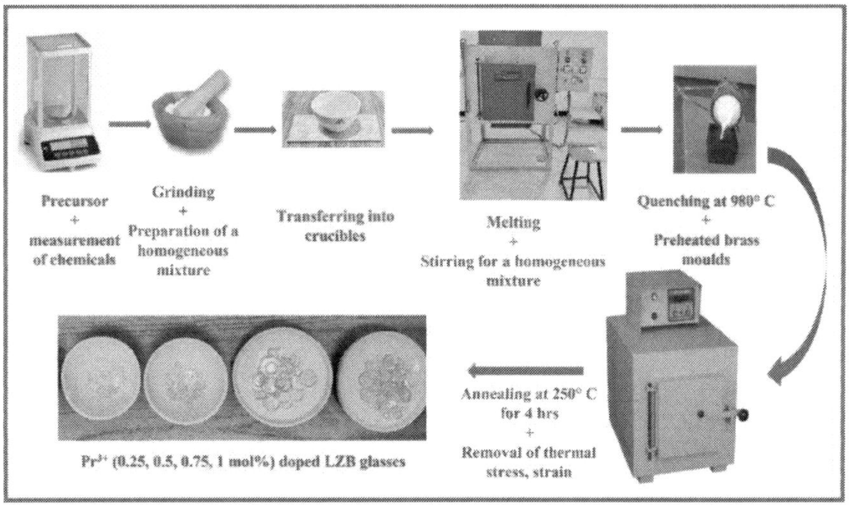

Figure 1. Flow chart for glass preparation and the synthesised Pr^{3+} doped LZB glasses.

2.1. *Density measurement and thickness measurement*

Densities of the LZBPr series samples synthesized were determined through Archimedes principle using xylene and toluene as immersion liquids, separately. The density is evaluated using the relation[12–15],

$$\rho_{\mathrm{LZBPr}} = \frac{W_{air}}{W_{xylene\ or\ toluene}} \times \rho_{xylene\ or\ toulene}\ \mathrm{g/cm^3} \quad (1)$$

Where, W_{air}, W_{xylene}, and $W_{toluene}$ are the weights in air, weight of xylene, and weight of toluene for LZBPr glass series. ρ_{xylene} and $\rho_{toulene}$ are 0.876 and 0.866 g/cm^3. The apparatus used for the determination of density is displayed in Figure 2 (a). The thickness of the individual samples is measured using the digital vernier caliper shown in Figure 2 (b).

2.2. *Molar volume (V_m)*

Molar volume (V_m) of the prepared LZBPr glass series is calculated through the following expression,

$$V_m = \frac{M.W}{\rho_{LZBPr}} \text{ cm}^3/\text{mol} \quad (2)$$

Where, $M.W$ (g/mol) will be molar weight of the LZBPr series.

(a)

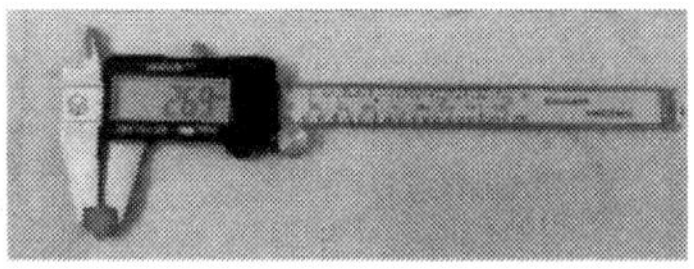
(b)

Figure 2. Apparatus used for the determination of (a) density and (b) thickness.

2.3. *Reflection loss ($R_L\%$), Dielectric constant (ε), and Optical dielectric constant ε_{op}*

The refractive index of the LZBPr glass series is determined via the Abbe refractometer. The Reflection loss ($R_L\%$) of synthesized glass samples can be calculated from the refractive index given by

$$R_L = \left(\frac{n-1}{n+1}\right)^2 \quad (3)$$

Dielectric constant (ε) can be expressed in terms of refractive index for individual samples using

$$\varepsilon = n^2 \quad (4)$$

Optical dielectric constant can be obtained using the expression,

$$\varepsilon_{op} = n^2 - 1 \quad (5)$$

IOP Conf. Series: Materials Science and Engineering 1300 (2024) 012043 doi:10.1088/1757-899X/1300/1/012043

2.4. *Molar refraction (R_m) and molar polarizability (α_m)*

Molar refraction (R_m) can be calculated in terms of refractive index and molar volume V_m using the relation

$$R_m = \frac{n^2-1}{n^2+2}(V_m)\mathrm{cm^3/mol} \tag{6}$$

The polarizability (α_m) of the prepared glasses can be calculated using the relation,

$$\alpha_m = \frac{3}{4\pi N_A} R_m \text{Å} \tag{7}$$

Where, Z = 3 is the charge of rare earth ion.

2.5. *Concentration of Praseodymium ions ($C_{Pr^{3+}}$), inter-ionic distance (r_i), Polaron Radius (r_p), and Field strength (F_s)*

The concentration of praseodymium ions ($C_{Pr^{3+}}$), inter-ionic distance (r_i), polaron radius (r_p), Field strength (F_s) is determined using the following equations

$$C_{Pr^{3+}} = \frac{M_{Pr^{3+}} \times \rho \times N_A}{W \times TM} \times 10^{20}\,\mathrm{ions/cm^3} \tag{8}$$

$$R_i = \left[\left(\frac{1}{C_{Nd^{3+}}}\right)^{1/3}\right] \text{Å} \tag{9}$$

$$r_p = \left[\frac{1}{2}\left(\frac{\pi}{6C_{Nd^{3+}}}\right)^{1/3}\right] \text{Å} \tag{10}$$

$$F_S = \left(\frac{Z}{r_p^2}\right) \times 10^{14}\,\mathrm{cm^{-2}} \tag{11}$$

Where, $M_{Pr^{3+}}$ is mass of Pr^{3+}-ions, *W* is molecular weight of the Pr^{3+} powder, *TM* is the total mass of chemical composition, and Z is the charge of Pr^{3+}- ion.

Table 1. Density of two different immersion liquids

Glass	Density	
	Toluene	Xylene
LZBPr025	2.6207	3.1692
LZBPr05	2.7378	3.1069
LZBPr075	2.7270	3.3608
LZBPr1	2.7772	3.0699

Table 2. Physical and optical parameters unaffected by density

Glass	Thickness	Refractive Index	Molecular Weight	Reflection loss	Dielectric constant	Numerical Aperture	Electric susceptibility	Optical dielectric constant
LZBPr025	0.210	1.6115	69.5441	5.4829	2.5969	0.2279	0.1270	5.7440
LZBPr05	0.294	1.6120	69.5615	5.4898	2.5985	0.2279	0.1271	5.7524

LZBPr075	**0.242**	**1.6120**	**74.2813**	**5.4898**	**2.5985**	**0.2279**	**0.1271**	**5.7524**
LZBPr1	**0.269**	**1.6125**	**76.6510**	**5.4966**	**2.6001**	**0.2280**	**0.1273**	**5.7608**

3. Results and discussions

All the physical and optical parameters which where estimated[15] are tabulated in Table 1, 2 & 3. The immersion liquid effect is evidenced in Table 1. for LZBPr series. Density and molar volume serve as significant physical parameters that have been utilized to investigate the level of compactness or openness within the glass network's structure. The LZBPr series glass variations demonstrate densities ranging from 2.62 to 2.77 g/cm³ for Toluene and 3.06 to 3.36 g/cm³ for Xylene. This disparity in densities is attributed to the fact that the density of Xylene is higher than that of Toluene. The molar volume of LZBPr series glass variants falls within the range of 25.40 to 27.59 cm³ for Toluene and 21.94 to 24.96 cm³ for Xylene. This variation is attributed to the significantly lower molecular weight of lithium compared to praseodymium. This expansion in molar volume occurs due to the incorporation of Pr^{3+} ions into the network's interstitial spaces, causing structural transformations. This expansion in volume is primarily attributed to the formation of non-bridging oxygens (NBOs) that emerge within the glass structure during the modification process. From Table 2. you can see that the physical and optical parameters are unaffected by the density. The refractive loss (%) is minimal, and the numerical aperture value falls within the required range for fiber fabrication [16]. The variation in density among immersion liquids leads to differences in all recorded physical parameters (molar refraction, molar polarizability, concentration, polaron radius, ionic radius, field strength), as detailed in Table 3.

Table 3. Physical parameters which are affected by two different immersion liquid

	Glass →	LZBPr025	LZBPr05	LZBPr075	LZBPr1
Xylene	Molar volume	21.9431	22.3958	22.1023	24.9679
	Molar refraction	7.6228	7.7852	7.6831	8.6850
	Molar polarizability	3.0215	3.0859	3.0454	3.4426
	Concentration	1.6465	3.2265	4.9039	5.7885
	Polaron radius	7.4696	5.9704	5.1935	4.9145
	Interionic radius	1.8530	1.4811	1.2883	1.2191
	Field strength	5.3767	8.4160	11.1223	12.4211
Toluene	Molar volume	26.5355	25.4076	27.2382	27.5992
	Molar refraction	9.2182	8.8321	9.4685	9.6003
	Molar polarizability	3.6539	3.5009	3.7531	3.8054
	Concentration	1.3615	2.8440	3.9792	5.2366
	Polaron radius	7.9576	6.2266	5.5678	5.0812
	Interionic radius	1.9740	1.5446	1.3812	1.2605
	Field strength	4.7375	7.7377	9.6774	11.6193

4. Conclusions

Our research involved the preparation of lithium-zinc-borate glasses doped with Pr^{3+} ions, through the use of the melt quenching method. Our main focus was on the effects of substituting lithium with Pr_2O_3. After conducting x-ray diffractometer analysis, it was clear that the glasses were indeed amorphous. We also investigated the influence of adding Pr_2O_3 with different immersion liquids, toluene and xylene, and found that each had unique effects on the glass properties. Additionally, we used the Archimedes principle to measure density changes and estimate physical parameters. Through careful analysis, we were able to identify the variations in NBO caused by Pr_2O_3, providing valuable insights into the structural behaviour of these glasses.

References

[1] de Souza N C A, Santos C C, Guedes I, Dantas N O and Vermelho M V D 2013 Structural and optical properties of $60B_2O_3$–(20−x)Na_2O–10PbO–$10Al_2O_3$:xTiO_2:yNd_2O_3 glasses *Opt. Mater. (Amst).* **35** 2544–50

[2] Chen Y J, Gong X H, Lin Y F, Luo Z D and Huang Y D 2010 Quasi-continuous-wave 1065nm laser oscillation in Nd^{3+}-doped bismuth borate glasses *Opt. Mater. (Amst).* **33** 71–4

[3] Shinohara K, Empizo M J F, Cadatal-Raduban M, Yamanoi K, Shimizu T, Yoshimura M, Sarukura N, Murata T, Uy M A, Abe H, Yoshikawa A, Boulon G and Dujardin C 2023 Radiation resistance of praseodymium-doped aluminum lithium fluorophosphate scintillator glasses for laser fusion experiments *Jpn. J. Appl. Phys.* **62** 010613

[4] Baccaro S, Catallo N, Cemmi A and Sharma G 2011 Radiation damage of alkali borate glasses for application in safe nuclear waste disposal *Nucl. Instruments Methods Phys. Res. Sect. B Beam Interact. with Mater. Atoms* **269** 167–73

[5] Annapoorani K, Basavapoornima C, Suriya Murthy N and Marimuthu K 2016 Investigations on structural and luminescence behavior of Er^{3+} doped Lithium Zinc borate glasses for lasers and optical amplifier applications *J. Non. Cryst. Solids* **447** 273–82

[6] Nanda K, Berwal N, Kundu R S, Punia R and Kishore N 2015 Effect of doping of Nd^{3+} ions in BaO–TeO_2–B_2O_3 glasses: A vibrational and optical study *J. Mol. Struct.* **1088** 147–54

[7] Ahmad A U, Hashim S and Ghoshal S K 2020 Physical, thermal and absorption traits of lithium strontium zinc borate glasses: Sensitiveness on Dy^{3+} doping *J. Alloys Compd.* **844** 156176

[8] Moustafa M G, Ammar M H, Saad M, Qasem A and Abdallah E M 2023 Distinctive spectroscopic features of samarium ions in the lithium zinc-borate glass *Opt. Mater. (Amst).* **141** 113904

[9] Luewarasirikul N and Kaewkhao J 2022 Pr^{3+}-Doped Ba-Na-B Glasses: Luminescence and Judd–Ofelt Analysis for Photonic Applications *Integr. Ferroelectr.* **225** 34–41

[10] Lacomme E, Sayyed M I, Sidek H A A, Matori K A and Zaid M H M 2021 Effect of bismuth and lithium substitution on radiation shielding properties of zinc borate glass system using Phy-X/PSD simulation *Results Phys.* **20** 103768

[11] Vighnesh K R, Ramya B, Nimitha S, Wagh A, Sayyed M I, Sakar E, Yakout H A, Dahshan A and Kamath S D 2020 Structural, optical, thermal, mechanical, morphological & radiation shielding parameters of Pr^{3+} doped ZAlFB glass systems *Opt. Mater. (Amst).* **99** 109512

[12] Madhu A, Eraiah B, Manasa P and Srinatha N 2018 Nd^{3+}-doped lanthanum lead boro-tellurite glass for lasing and amplification applications *Opt. Mater. (Amst).* **75** 357–66

[13] Hemalatha S, Nagaraja M, Madhu A, Suresh K and N S 2021 The role of Sm_2O_3 on the structural, optical and spectroscopic properties of multi-component ternary borate glasses for orange-red emission applications *J. Non. Cryst. Solids* **554** 120602

[14] A M, Al-Dossari M, EL-Gawaad N S A, Gelija D, Ganesha K N and Srinatha N 2023 Structural, optical and luminescence properties of Nd^{3+} ions in B_2O_3+SiO_2+TeO_2+Na_2O glasses *Opt. Mater. (Amst).* **136** 113436

[15] DIMITROV V and KOMATSU T 1999 Effect of Interionic Interaction on the Electronic Polarizability, Optical Basicity and Binding Energy of Simple Oxides. *J. Ceram. Soc. Japan* **107** 1012–8

[16] Madhu A, Eraiah B, Manasa P and Basavapoornima C 2018 Er^{3+}-ions doped lithium-bismuth-boro-phosphate glass for 1532 nm emission and efficient red emission up conversion for telecommunication and lasing applications *J. Non. Cryst. Solids* **495** 35–46

IOP Conf. Series: Materials Science and Engineering 1300 (2024) 012044 doi:10.1088/1757-899X/1300/1/012044

Dielectric relaxation studies in Sm_2O_3 doped boro-zinc-vanadate glasses

D Ashwini[1], T Sankarappa[1,*], M Amarkumar[1], H Mohansingh[1], J Pallavi[1], D Arvind and B Raghavedra[1]

[1]Physics Department, Gulbarga University, Kalaburagi

*Author for correspondence, e-mail: sankarappa@rediffmail.com

Abstract. In this work, the conventional melting procedure was adopted to synthesize glasses with a composition of $(ZnO)_{0.3} - (V_2O_5)_{0.3} - (B_2O_3)_{0.4\text{-}x} - (Sm_2O_3)_x$, x = 0.002, 0.005, 0.007, 0.01 and 0.02 mol%. The dielectric properties of these glasses were measured. The dielectric constant (ε′) and dielectric loss (ε″) of glasses are found to vary in the range from 1.9177×10^2 to 3.9456×10^3 and from 1.5751 to 6.9095×10^5 respectively for the temperature range 343 K–573 K and frequency range 50 Hz to 10 MHz. With increase of frequency, both dielectric constant and loss decreased and increased with increase of temperature. The analysis of dielectric constant and dielectric loss confirmed that the phenomenon of dielectric relaxation is mainly due to the frequency-dependent polarization mechanism. Electric modulus and impedence spectroscopy revealed non-Debye type and a single phase relaxation process. Activation energy for dc conductivity and dielectric relaxation are found to be of same size indicating that the potential barrier encountered by the charge carriers is same in both the processes. The master curves suggested that temperature-independent relaxation is occurring in the present glasses. The high values of measured dielectric parameters suggest that the present glasses are suitable in semiconducting and energy storage applications.

Key words. Dielectric relaxation, Electric modulus, Impedence Spectroscopy, Master curves, Zinc-vanadate glasses.

1. Introduction

Zincvanadate glasses are very significant as they are widely used in many applications such as electrochromic devices, electrical memory switching devices and solid state electrolytes etc., [1-2]. They posses low melting point and good thermal stability etc. Although these glasses are dielectric materials, the presence of transition metal ions in them produces semiconductor properties [3-5]. By incorporating rare earth (RE) elements into glasses, it is possible to develop a variety of electro-optic and optical devices. Rare earth ions do not contribute to conductivity. However, their addition can change the network which in turn affects conduction and dielectric properties [6-7].

There are several reports on the investigation of electrical and dielectric properties of glasses containing rare-earth ions [8-12]. They claimed that these glasses exhibit ideal debye type behavior as per cole-cole model and the decrease in dielectric loss observed at high temperatures is due to reorientation of dipoles along the direction of the electric field.

Electrical conductivity and dielectric properties of Sm_2O_3 doped glasses namely, Li_2CO_3–B_2O_3–ZnF_2 [5], B_2O_3–$CuCl_2$–Na_2O–Al_2O_3 [6], Li_2O–B_2O_3 [7] and P_2O_5–NaF–AlF_3 [13] have been reported. These reports showed decreasing tend of dielectric parameters with increase of rare earth ions content suggesting increasing degree of electrical rigidity in the glass network.

There have been electrical and dielectric investigations conducted on vanadate doped glasses, V_2O_5-ZnO [1-2, 14], B_2O_3-V_2O_5 [15-18] and B_2O_3-V_2O_5-ZnO [19]. They observed that high concentration of the transition metal cations produces electronic conduction and the ionic contribution is blocked, exhibited single relaxation process. The dielectric parameters reflected possibility of their utility in various electronic and energy storage devices. No reports on the investigation of electrical and dielectric properties of Sm_2O_3 doped ZnO-B_2O_3-V_2O_5 glasses.

The current study focuses on the dielectric properties of boro-zinc-vanadate glasses containing different concentration of Sm_2O_3. The aim is to measure dielectric properties of these glasses, analyse data and understand dielectric relaxation mechanism. These will also enable one to understand the effect of rare earth oxide on conductivity and dielectric behavior.

2. Experimental Work

2.1. Preparation of Glass

Glasses of composition, $(ZnO)_{0.3} - (V_2O_5)_{0.3} - (B_2O_3)_{0.4-x} - (Sm_2O_3)_x$ (where, x = 0.002, 0.005, 0.007, 0.01, and 0.02 mol%) with acronym as ZVBS1, ZVBS2, ZVBS3, ZVBS4 and ZVBS5 respectively, were synthesized by melt-quench technique. Defined quantities of AR grade chemicals (V_2O_5, B_2O_3, ZnO, Sm_2O_3) were heated upto melting in an electric furnace. After maintaining melt at 1325K, it was poured between two stainless steel plates. Samples were transferred to another pre-heated furnance (500 K). They were kept at that temperature for an hour and then furnance was switched off. The samples were taken out from the furnance after 24 hours.

2.2. Dielectric measurements

The prepared glasses were polished and their two large parallel faces were silver painted and placed between two electrodes of the sample holder. The contacts and continuity were checked before the measurement. The capacitance, C and dissipation factor, tanδ were measured in an Impedence analyzer (Wayne Kerr 6500B) for the frequency range 50 Hz – 10 MHz and the temperature range 343 K – 573 K.

The dielectric constant (ε′) and loss tangent (ε″), real and imaginary parts of electric moduli and impedance were estimated using following mentioned expressions [9],

$$\varepsilon' = \left(\frac{Cd}{\varepsilon_0 A}\right) \quad (1)$$

$$\varepsilon'' = \varepsilon' \tan\delta \quad (2)$$

$$M' = \frac{\varepsilon'}{(\varepsilon'^2 + \varepsilon''^2)} \quad (3)$$

$$M'' = \frac{\varepsilon''}{(\varepsilon'^2 + \varepsilon''^2)} \quad (4)$$

$$Z' = \frac{M''}{\omega\varepsilon_0} \quad (5)$$

$$Z'' = \frac{M'}{\omega\varepsilon_0} \quad (6)$$

IOP Conf. Series: Materials Science and Engineering 1300 (2024) 012044 doi:10.1088/1757-899X/1300/1/012044

Where, ε_0 is the permittivity of the free space, $\omega = 2\pi f$ the angular frequency, A the cross-sectional area and, d the thickness of the sample.

3. Results and discussion

3.1. Dielectric properties

Dielectrics are the materials, under normal conditions, will not have free electric charges. Yet they respond to the applied electric field considerably. The dielectric constant of a material indicates the number of electrostatic lines of flux concentrated in it. The dielectric loss defines the materials capacity to support the electrostatic field with least dissipation of energy in the form of heat [20].

The dielectric constant (ε') and dielectric loss (ε'') of the present glasses are found to vary in the range from 1.9177×10^2 to 3.9456×10^3 and from 1.5751 to 6.9095×10^5 respectively over the measured temperature and frequency range. The variation of ε' and ε'' as a function of frequency at different temperatures for ZVBS1 glass are shown in Figure 1 (a) & (b) respectively. It can be observed that both ε' and ε'' decreased with frequency and increased with temperature. The accumulation of charge occurs in the glass at low frequency prevents the flow of charges. This in turn increases polarisation and that is why ε' and ε'' have large values. Increased frequency of periodic electric field reduces polarisation and lowers ε' and ε''. Moreover, according to recent studies, the nonbridging oxygens (NBOs) in the glass have a strong correlation with dielectric parameters. More number of NBOs produces more polarization which results high values of dielectric constants [17]. Variation in ε' and ε'' of remaining glasses followed the same trend as that of ZVBS1 glass.

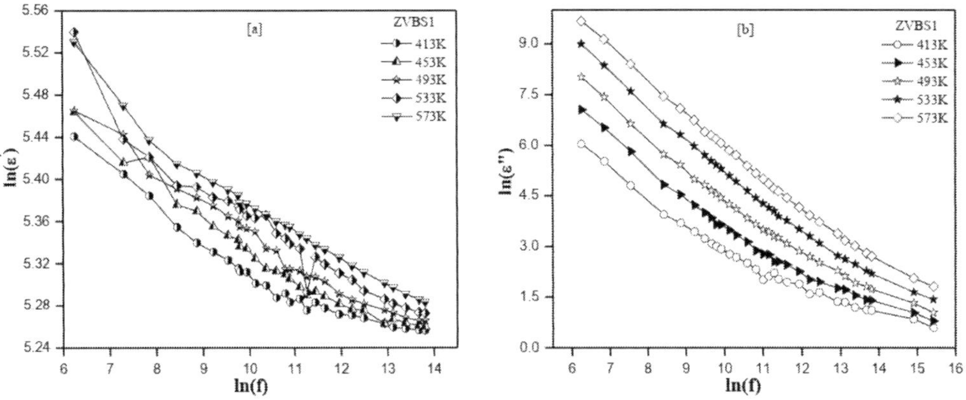

Figure 1. Variation of (a) ε' and (b) $\varepsilon''$with frequency at different temperatures for ZVBS1 glass.

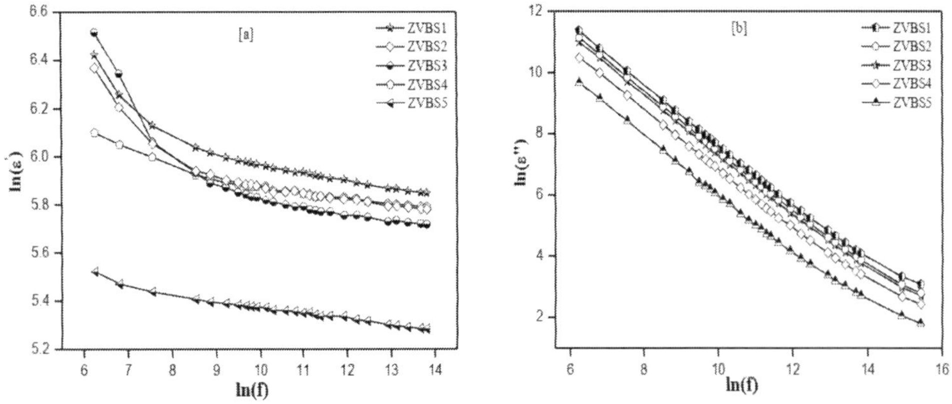

Figure 2. Variation of (a) ε' and (b) ε'' as a function of frequency for ZVBS glasses at 573 K temperature.

Figure 2 (a) & (b) depicts frequency dependence of ε′ and ε″ for all the ZVBS glasses at 573 K. With the addition of Sm_2O_3 (at the expense of B_2O_3), the values of ε′ and ε″ are decreased. ZVBS1 glass has the highest polarizability compared to the remaining glasses. The ε′ and ε″ of the MoO_3-V_2O_5-BaO-B_2O_3-Bi_2O_3 (MVBBB) glasses were found to be varying from 37 to 112 and from 25 to 134 respectively. They reported that the MVBBB (x =0) has highest values of ε′ and ε″ due to high concentration of NBOs which lead to highest polarizability among all other glasses. Based on ε′ and ε″, they proposed these glasses for use in energy storage applications. The ε′ and ε″ of the present glasses are found to be varying from 1.9177×10^2 to 3.9456×10^3 and from 1.5751 to 6.9095×10^5 respectively. As compared to MVBBB glasses the ε′ and ε″ of present glasses are high which leads to high polarizability. Therefore, the present glasses can be useful for use in energy storage applications.

3.2. Electric Modulus (M)

The frequency dependence of M′ and M″ for different temperatures for ZVBS1 glass is presented in Figure 3. It can be noticed that at low frequencies, M′ is extremely small. The distribution of relaxation processes accounts for the rise in M′ with frequency. The ions move freely within the potential well only on the higher frequency side of the M'_{max} region. On the lower frequency side of the M'_{max} region, ion drifts over extended distances [22].

At lower frequency and at higher temperatures, the permittivity increases significantly. It may have been caused by the combination of electrode polarization, transmission of charge and space-charge polarization [21]. As observed in Figure 3, at all temperatures, M″ exhibits humps and they correspond to the movement of charge carriers causing dc conductivity. It is clearly visible that the major relaxation peak of M″, which characterises the ion transport, is observed and shifted towards higher frequencies as the temperature increased [10]. There is no change in the peak intensity as its hopping frequency significantly increases. This shows that the mobility of charge carriers has clear effect on temperature [16]. In addition, it can be noted that the modulus graphs have almost the same shapes (at all temperatures), indicating that the distribution of the relaxation time is unaffected by temperature [18]. The remaining glasses of the present series were found to exhibit identical changes in M′ and M″ with temperature and frequency as that of ZVBS1 glass.

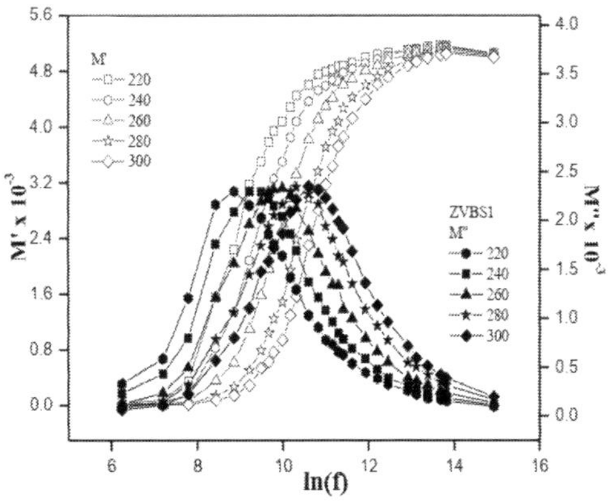

Figure 3. Plot of M′ and M″ versus ln(f) with temperature for ZVBS1 glass.

3.3. Impedence Spectroscopy (Z)

The impedence can be used to describe non-Debye type relaxation behavior associated with the material [18]. It also gives information about bulk resistance (R_b). The Nyquists cole-cole plots of Z′ versus Z″ at different temperature for ZVBS1 glass are shown in Figure 4. It can be seen that at all the temperatures, a single semicircle is traced indicating single phase relaxation process, which can be attributed to the bulk contribution to electrical conduction. As the temperature increases, the semicircle's intercept shifts towards a lower Z′ which suggests that the sample's bulk resistance is decreasing. So, the electrical relaxation in ZVBS1 glass is a pure bulk phenomenon. This reveals the existence of single phase with non-Debye type relaxation in all the current glasses [18, 22]. The remaining glasses in the current series displayed changes in Z′ and Z″ same as that of ZVBS1 glass. This means single phase relaxation process is occurring in all the glasses.

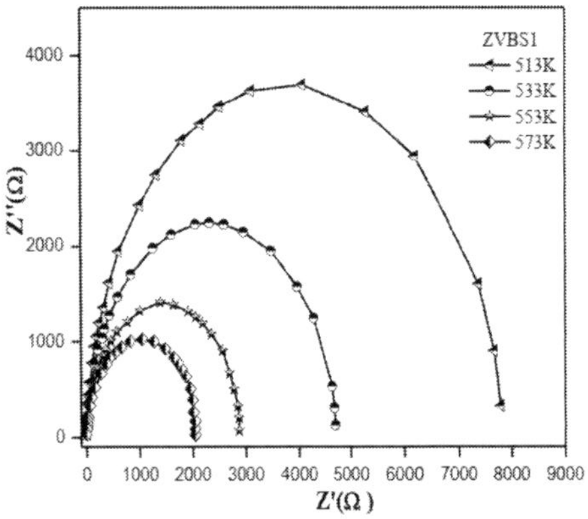

Figure 4. The cole-cole (Z′ versus Z″) plots for four different temperatures for ZVBS1 glass.

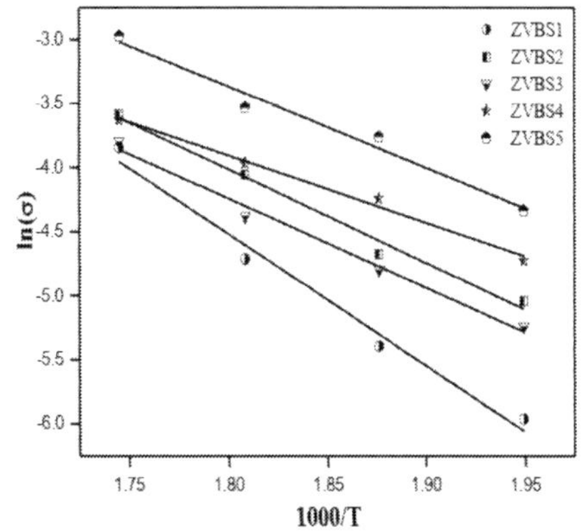

Figure 5. Arrhenius plots of dc conductivity for ZVBS glasses.

The bulk dc resistance, R_b was obtained from the intercept of the semi circles on Z′ axis. Using R_b the dc conductivity, σ_{dc} was determined as per the equation [23],

$$\sigma_{dc} = \frac{d}{A}\left(\frac{1}{R_b}\right) \tag{7}$$

Where, A is the area and d is the thickness of the sample.

It is observed that σ_{dc} is increased with temperatures. The dc conductivity behavior with temperature is analysed using Arrhenius expression [18],

$$\sigma_{dc} = \sigma_0 exp\left(\frac{-E_{dc}}{k_B T}\right) \tag{8}$$

Where, σ_0 is the pre-exponential factor and E_{dc} the activation energy for conduction. Figure 5 presents the Arrhenius plots of dc conductivity. The activation energy E_{dc} was calculated using the slopes of linear lines fit to the data and shown as solid lines in the figure. Obtained E_{dc} values are listed in Table 1. The E_{dc} is decreasing with increasing Sm_2O_3 content and are comparable with reported values for Sm_2O_3 doped lithium borate glasses [7] and Li_2O doped barium vanadate glasses [23].

3.4. Relaxation mechanism

The hopping of charge carriers can be used to understand dielectric relaxation. Small polarons with the same mobility as ions can be found in transition metal oxide doped glasses. The frequency (f_p) corresponding to M''_{max} or Z''_{max} in Figures 3 & 4 has been extracted for different temperature. By assuming thermal activation of f_p in the Arrhenius fashion, the activation energy (E_p) for f_p was calculated using following relation [24],

$$f_p = f_0 exp\left(\frac{-E_p}{k_B T}\right) \tag{9}$$

Figure 6 shows the plots of $ln(f_p)$ against (1000/T) for all the ZVBS glasses. By fitting linear lines to the data, activation energy, E_p was determined and are tabulated in Table 1. The relaxation time, τ was determined by applying the condition, $\omega_p\tau_p = 1$ where, $\omega_p = 2\pi f_p$ [24]. According to the theory that relaxation is also thermally activated process associated with certain activation energy and is expressed again in the Arrhenius form as [22],

$$\tau = \tau_0 exp\left(\frac{-E_\tau}{k_B T}\right) \tag{10}$$

IOP Conf. Series: Materials Science and Engineering 1300 (2024) 012044 doi:10.1088/1757-899X/1300/1/012044

As per Eq. (10), the sketches of ln(M″) and ln(Z″) as a function of (1000/T) are made in Figures 7 and 8, respectively. Linear lines were fit to the data and activation energy $E_{\tau_{M''}}$ and $E_{\tau_{Z''}}$ were estimated and recorded in Table 1. These activation energies are found to be in agreement with one another and are decreasing with increasing Sm_2O_3 content. This suggests that the hindrance developed for dipole orientations are increasing with increase of temperature. The present relaxation activation energy and dc activation energy are slightly higher than those reported for Fe_2O_3 doped lithum boro-vanadate glasses [16] and Li_2O doped barium vanadate glasses [23]. The range of activation energy for DC conduction, the electric modulus and impedence have been found to be almost same. This implies that the potential barrier for conduction and dielectric relaxation will be the same for charge carriers [7, 24].

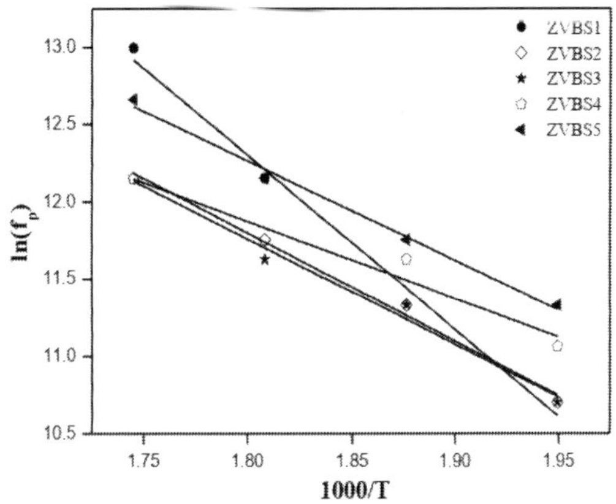

Figure 6. Plots of $\ln(f_p)$ versus (1000/T) for ZVBS glasses.

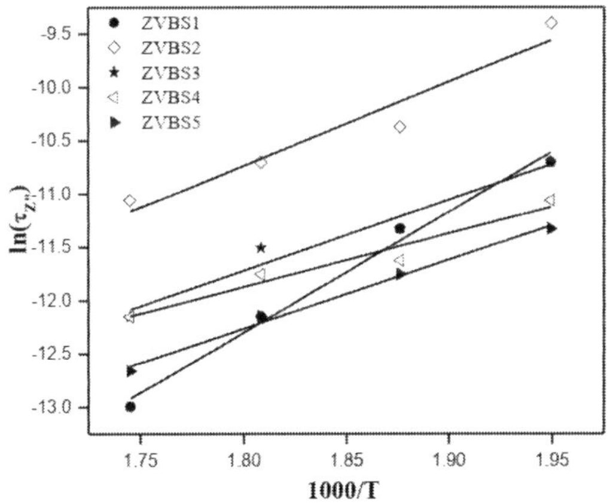

Figure 7. Plots of $\ln(\tau_{M''})$ versus (1000/T) for ZVBS glasses.

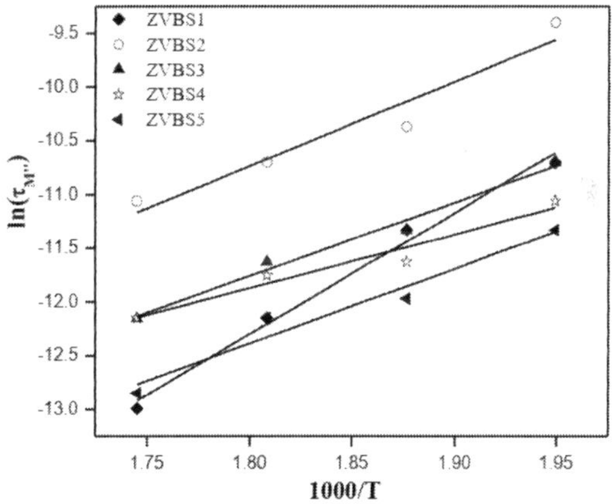

Figure 8. Plots of $\ln(\tau_{Z''})$ versus (1000/T) for ZVBS glasses.

Table 1. Activation energy for (E_{dc}), characteristic frequency (f_p), dielectric relaxation $E_{\tau_{M''}}$ and $E_{\tau_{Z''}}$ obtained from modulus, M″ and impedence, Z″.

Glass	Activation energy (eV)			
	$E_{dc}\pm 0.001$	$E_p\pm 0.001$	$E\tau_{M''}\pm 0.001$	$E\tau_{Z''}\pm 0.001$
ZVBS1	0.8869	0.9650	0.9727	0.9729
ZVBS2	0.6297	0.6059	0.6774	0.6773
ZVBS3	0.5999	0.5889	0.5889	0.5725
ZVBS4	0.4520	0.4315	0.4314	0.4314
ZVBS5	0.5473	0.5548	0.5986	0.5547

3.5. Master curves

The frequency dependent modulus for each glass at different temperatures can be scaled to a single master curve. It implies that the observed process is modulated by a single physical phenomenon and it can only be differentiated by thermodynamic scale [23]. The normalized (M''/M''_{max}) plot of the modulus data for the glass ZVBS1 at different temperatures is displayed in Figure 9. The perfect overlapping of master curves (Normalised plots) suggests that temperature-independent relaxation process occur in glass ZVBS1. This reveals that the relaxation process occuring in the present glasses is of single phase. The master curves of the remaining glasses of the present series appeared same. Identical results have been reported for ternary vanadate glasses [23, 24].

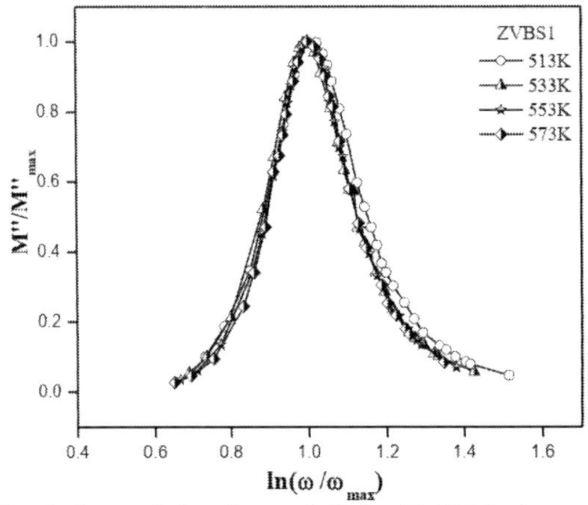

Figure 9. Normalized plots of electric moduli for ZVBS1 glass at four different temperatures.

4. Conclusions

A series of Sm_2O_3 doped boro-zinc-vanadate glasses were studied for dielectric properties over the frequency range of 50 Hz to 10 MHz and the temperature range of 343 K–573 K. Different polarisation phenomenon have been used to explain observed variations in dielectric parameters. The dielectric constant ε' and dielectric loss ε'' are found to decrease with increasing frequency and decrease with the addition of Sm_2O_3. The high values of ε' measured for the present glasses enable us to propose these glasses for energy storage applications. Modulus graphs have the same shape at all temperatures, indicating that the distribution of the relaxation time is independent of temperature. Single semi-circles were seen in the cole-cole plots, indicating a single phase relaxation process of non-Debye type. The relaxation activation energy E_τ, obtained from the temperature-dependent loss peaks, are in agreement with dc activation energy E_{dc}. These activation energies are decreasing with increasing concentrations of Sm_2O_3. It is learnt that the relaxation processes occuring in the current glasses are temperature-independent. It is for the the first time Sm_2O_3 doped boro-zinc-vanadate glasses were thoroughly investigated for dielectric relaxation.

Reference

[1] Aloka G, Bhattacharya S and Ghosh A 2009 *J. Phys.: Condens. Matter* **21** 145802
[2] Aziz M S, Mostafa A G, Yousse A M, Zidan H and Youssif S M S 2010 *Egypt. J. Solids* **33** 143–61
[3] Dalal S, khasa S, Dahiya M S, Yadav A, Agarwal A and Dahiya S 2015 *J. Asian Ceram. Soc.* **3** 234-39
[4] Ashwini D, Sankarappa T, Amarkumar M, Mohansingh H and Raghavendra B 2023 *J. Aust. Ceram. Soc.* **59** 391–402
[5] Thomas S, Nayab R Sk, Rathaiah M, Venkatramu V, Joseph C and Unnikrishnan N V 2013 *J. Non-Cryst. Solids* **376** 106–16
[6] Sadeq M S and Morshidy H Y 2020 *J. Rare Earths* **38** 770–75

[7] Ramteke D D and Gedam R S 2014 *Spectrochim. Acta Part A: Mol. Biomol. Spectrosc.* **133** 19–23
[8] Gedam R S and Ramteke D D 2013 *J. Phys. and Chem. Solids* **74** 1039–44
[9] Malge A, Sankarappa T, Devidas G B and Sujatha T 2020 *J.Phys.: Conf. Series* **1451** 012014
[10] Hanumantharaju N, Sriprakash G and Veeranna Gowda V C 2019 *J. Non-Cryst. Solids* **512** 103–111
[11] Mahani R M and Marzouk S Y 2013 *J. Alloys and Comp.* **579** 394–400
[12] Gajula G R, Buddiga L R, Chidambara Kumar K N and Dasari M 2020 *Results Phys.* **17** 103076
[13] Ahmed R M, Taha T A and Ezz-Eldin F M 2021 *J Mater Sci: Mater Electron* **32** 28919–34
[14] Dahiya S, Punia R, Singh A, Maan A S and Murugavel S 2015 *J. Am. Ceram. Soc.* **98** 2776–83
[15] Abdel-Karim A M, Fayad A M, El-Kashef I M and Saleh H A 2023 *J. Electron. Mater.* **52** 2409–20
[16] Gomaa H M, Moneep A M, Adbel-Moety A S, Bedary A A, Yahia I S and Zyoud S H 2023 *J. Appl. Phys.* **129** 70
[17] Attri V, Dahiya M S, Kumar R, Hooda A, Agarwal A and Khasa S 2023 *J Mater Sci: Mater Electron* **34** 1273
[18] Yadav A, Dahiya M S, Narwal P, Hooda A, Agarwal A and Khasa S 2017 *J. Solid State Ion.* **312** 21–31
[19] Arya S K, Danewalia S S, Arora M and Singh K 2016 *J. Phys. Chem. B* . **120** 12168–76
[20] Singh P K, Sharma S K, Tripathi S K and Dwivedi D K 2019 *Results Phys.* **12** 223–36
[21] Abdel-Baki M, Mostafa A M, Fayad A M, El-Bassyouni G T and Turky G M 2023 *J. Appl. Phys.* **133** 065102
[22] Malge A, Sankarappa T, Devidas G B, Ashwajeet J S, Ashwini D and Mohansingh H 2022 *J. Phys.: Conf. Series* **1221** 012015
[23] Goel P, Honnavar G V and Ramesh K P 2023 *J Phys Chem Solids* **173** 111105
[24] Mohansingh H, Sankarappa T, Malge A, Ashwini D and Raghavendra B 2023 *Mater. Today: Proc.* **80** 933–36

Electronic and optical properties of Na_2O doped P_2O_5 -TeO_2-CoO-ZnO glasses

J G Pallavi[1], T. Sankarappa[1*], M Amarkumar[1], H Mohansingh[1] and D Ashwini[1]

[1]Department of Physics, Gulbarga University, Kalaburagi, Karnataka, India
[*]Correspondence author mail: sankarappa@rediffmail.com

Abstract. The phospho-tellurite glasses, $(P_2O_5)_{0.25}$-$(TeO_2)_{0.5-x}$-$(CoO)_{0.1}$-$(ZnO)_{0.15}$-$(Na_2O)_x$; x = 0.15, 0.20, 0.25, 0.30 and 0.35 were prepared by melt quenching and their non-crystalline nature confirmed through XRD. Density is found decreased and molar volume increased with increase of Na_2O concentration and they behaved opposite to each other. The physical parameters such as transition metal ion density, mean distance between TMIs, small polaron radius, oxygen packing density and field strength were estimated. DC conductivity was measured by following two probe method in the temperature range 303K – 633K. High temperature conductivity is explained by Mott's small polaron hopping (SPH) model and low temperature conductivity by Mott's VRH model. Both conductivity and activation energy are found to be increasing with Na_2O content. Density of states at Fermi level are found to be of the order of 10^{21} to 10^{24} eV^{-1} cm^{-3}. UV-Vis spectra were recorded for the wavelength range 200-800 nm. Optical band gap (direct and indirect) except for x=0.20 is found increased, urbach energy and refractive indexes are found decreased with Na_2O concentration. Various optical parameters such as molar polarizibility, metallization criterion, reflection loss, electronic polarizability, optical transmission coefficient, dielectric constant and optical dielectric constant have been determined.

Keywords: Phospho-tellurite glasses, DC Conductivity, Energy band gaps, Urbatch energy.

1. Introduction

P_2O_5-TeO_2 glasses are the potential candidates for use in non-linear optical devices as they posses high refractive index and thermal expansion, good transmission in the infrared and ultraviolet regions. TeO_2 is a glass forming material made up of TeO_3, TeO_4 and TeO_{3+1} as its structural components. Alkali oxides in phosphate glasses serve as modifiers by breaking P-O-P linkage and adding NBOs in the structure [1, 2]. Phosphate glasses have low chemical stability and that can be improved by adding network modifiers [3]. Because of their low mass, small size and strong electro-positivity, alkali ions like Na^+ are the advantageous ones for electrochemical applications [4]. Ionic conduction is caused by the movement of positive ions, whereas polaronic conduction is caused by the hopping of electrons. Those glasses that exhibit both electronic and ionic conduction may be utilized as potential cathode materials in solid state batteries, optical switching and memory switching devices, whereas those that only exhibit ionic conduction can be used as solid electrolytes in electrochemical devices i.e. sodium-ion batteries [5]. ZnO can be considered as a glass modifier as it improves mechanical properties and glass forming ability and reduce melting temperature [2]. With a broad band gap of 3.37 eV and a significant exciton binding energy of 60 meV, ZnO is made use in a wide range of opto-electronic devices [6].

Structural and dielectric properties of phospho-tellurite glasses have also been studied. Utilizing the measured dielectric properties, ac conductivity has been estimated and understood [7]. Addition of tellurium into silver-boro-phospho-tellurite glasses in terms of electrical properties has been studied [8, 9]. Investigation of influence of lithium ions in tellurium phosphate glasses concerning structure has been reported [10]. Optical and DC conduction mechanisms in ZnO-P_2O_5-TeO_2-MoO_3 [11] and CaO-BaO-TeO_2-P_2O_5 [12] glasses have been investigated. Optical and dielectric characteristics of phospho-tellurite glasses were reported [13]. Sodium doped P_2O_5-TeO_2 glasses were evaluated for Optical and gamma attenuation properties [14].

Since, phospho-tellurite glasses doped with both alkali and transition metal ions have not been investigated previously for electrical conduction mechanisms and optical studies, we prepared phospho-tellurite glasses doped with Na_2O, ZnO and CoO. The phospho-tellurite glasses are normally more durabile than pure phosphate glasses and have attractive applications. This article provides investigated results on structural, physical, electrical and optical properties of the multi component phospho-tellurite glasses.

2. Experimental details

The phospho-tellurite glasses in the composition of, $(P_2O_5)_{0.25}$-$(TeO_2)_{0.5-x}$-$(CoO)_{0.1}$-$(ZnO)_{0.15}$-$(Na_2O)_x$; x = 0.15, 0.20, 0.25, 0.30 and 0.35 were made by melt quenching and coded as PTZCN1, PTZCN2, PTZCN3, PTZCN4 and PZTCN5 respectively. All the starting chemicals were of Himedia make and analytical reagent grade. The chemicals were mixed in the defined quantities and heated in a furnace at 1303 K using silica crucibles. After that homogeneous melt was formed, it was quickly quenched between two stainless steel plates. The samples were subjected for annealing process at 573 K to wipe out the internal strains so that samples can have a good mechanical stability. The samples were cut into pieces to the desired size and shape for electrical and optical measurements.

To confirm the phase of the samples, they were subjected to powder XRD studies in a Smartlab SE Rigaku x-ray diffractometer, in which Cu-Kα radiation (λ= 1.54 Å) was used and scanned over 2θ range of 5° to 90°. The Archimedes principle was followed to measure density (ρ) of the glasses with a precision of 0.01 mg in a VIBBRA make HT Analytical Balance and Xylene as an immersion liquid (ρ = 0.8650 gm/cm^3) [2]. The molar volume V_m was calculate as V_m=M/ρ where, M the molecular weight of the sample which was worked out as M = (0.25) M(P_2O_5) + (0.5-x) M(TeO_2) + (0.1) M(CoO) + (0.15) M(ZnO) + (x) M(Na_2O). The DC conductivity was measured by following the two point method in the temperature range 303 to 633K. For electrical measurement, the two opposite large surfaces of the samples were coated with silver paste. Error on the measured conductivity was within 5%. Detailed procedure of conductivity measurements can be found in reference [15]. The well polished glasses were subjected to UV absorption measurements in a Jasco Spectrophotometer (Model name V-770) in the wavelength range of 200-800 nm.

3. Results and Discussion

3.1. X-Ray Diffraction

XRD spectrum of the glass samples are depicted in figure 1 shows no discrete peaks indicating amorphous nature. This demonstrates lack of periodicity of the three-dimensional network and the absence of long-range atomic order. A small broad hump is observed in the 2θ range of 24 to 36° which reveals the presence of short range order. Similar results were reported for phospho-tellurite [16] and phosphate glasses [17].

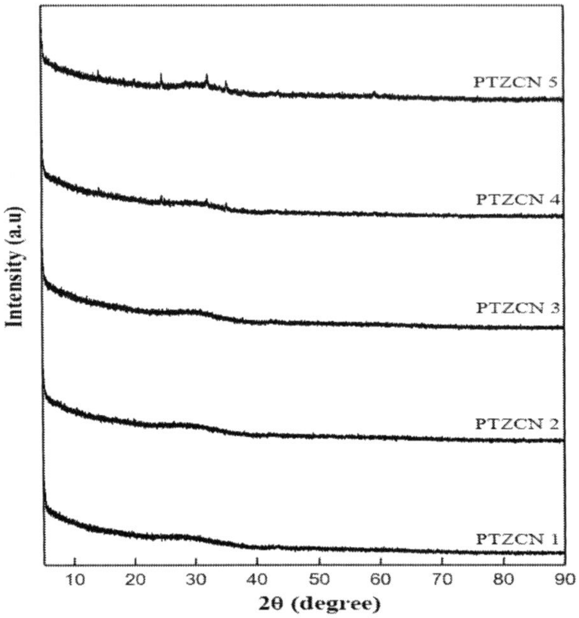

Figure 1. The XRD spectra for all PTZCN glasses.

3.2. Density (ρ) and molar volume (V_m)

The variations in ρ and V_m with Na_2O concentration are shown in figure 2. As the Na_2O concentration increases, the density decreases. This may be as a result of relatively lower molecular weight of Na_2O (105.99 g/mol) replacing the TeO_2 of larger molecular weight (159.60 g/mol). Thus, the replacement of TeO_2 by Na_2O, glass became low dense. The ρ decreases from 4.466 to 3.834 g/cm^3 with increasing mole fractions of Na_2O from 0.15 to 0.35. The molar volume is inversely proportional to density. The V_m of the PTZCN glasses increased from 33.352-36.054 cm^3/mole with increasing Na_2O content. The density and molar volume values of the current glasses are comparable to the quoted values for lithium phospho-tellurite glasses [1].

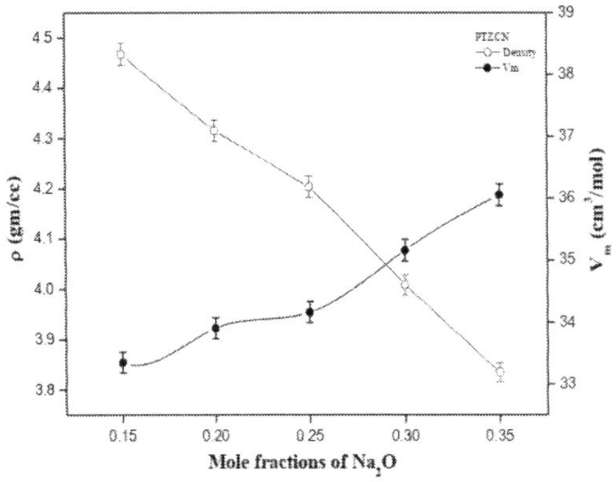

Figure 2. Variation of ρ and V_m versus Na_2O content for PTZCN glasses.

IOP Conf. Series: Materials Science and Engineering 1300 (2024) 012045 doi:10.1088/1757-899X/1300/1/012045

The other physical parameters connected to density have been calculated. For example, transition metal ion density, N = 2[{ρ(m_{CoO} + m_{ZnO}) / (M_{CoO} + M_{ZnO})}N_A] [15], Where m_{N_2O} is the Na_2O concentration and M_{Na_2O} is the molecular weight of the Na_2O and N_A is the Avagadro number, Field strength (F)= $Z/(rp)^2$ [18], where Z is the atomic weight of sodium. The obtained values of these parameters are listed in table 1. TMI density (N) and field strength (F) increases due to increasing Na_2O content and decreasing TeO_2 concentration. Mean distance between TMIs, $R=(1/N)^{1/3}$ [5], small polaron radius $r_p=(1/2)(\pi/6N)^{1/3}$ [18], oxygen packing density (OPD)= 1000 C $\frac{\rho}{m}$ [18]. The decrease in R, r_p and OPD can be attributing to the decrease in density with incorporation of Na_2O [14, 19].

Table 1. Physical parameters of PTZCN glasses.

Parameters	PTZCN1	PTZCN2	PTZCN3	PTZCN4	PTZCN5
Density, ρ (g/cm^3) ±0.005	4.4667	4.3151	4.2040	4.0084	3.8346
Molecular Weight (gm)	148.975	146.294	143.614	140.933	138.253
Molar volume ,V_m = M/ρ ±0.005 (cm^3/mole)	33.352	33.902	34.161	35.159	36.054
Transition metal ion density, N x 10^{21} (cm^{-3})	5.4175	7.1059	8.8155	10.278	11.693
Mean spacing between TMIs, R(nm)	0.5693	0.5201	0.4840	0.4599	0.4405
Small polaron radius, r_p (nm)	0.2294	0.2096	0.1950	0.1853	0.1775
Oxygen packing density (cm^{-3} mol^{-1})	106.43	103.15	100.99	96.70	92.91
Field Strength, F (cm^{-2}) x 10^{16}	4.3685	5.2328	6.0457	6.6952	7.2966

3.3. DC Conductivity

The measured conductivity is found to vary between 10^{-3}-10^{-5} $\Omega^{-1}m^{-1}$. Increase of conductivity with temperature reveals semiconductor behavior. The conductivity, σ is also increasing with increasing Na_2O content (figure 3). It can be seen that the conductivity was almost constant up to 473K and started increasing after wards. Similar trend of conductivity variation was reported for CoO and Li_2O doped boro-phosphate [15] and alkali doped boro-tellurite glasses [20].

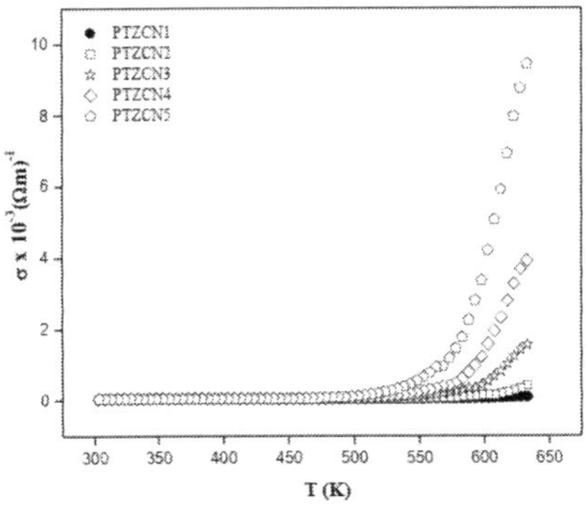

Figure 3. Variation of conductivity, σ as a function of temperature, T.

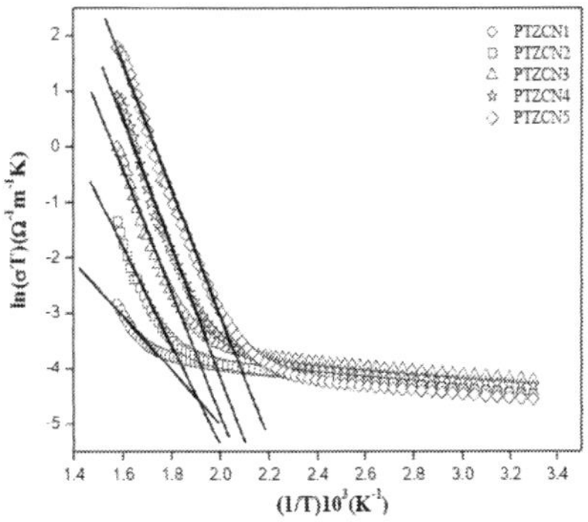

Figure 4. Graphs of ln(σT) against (1/T) for PTZCN glasses, at high temperature.

ln(σT) against 1/T curves were drawn in figure 4 by employing Mott's SPH expression [21],

$$\sigma = (\sigma_0/T)\exp(-W/K_BT) \quad (1)$$

Where, W is the activation energy of dc conductivity. The pre-exponential factor is expressed as

$$\sigma_0 = \upsilon_0 Ne^2R^2C(1-C)\exp(-2\alpha R/KB) \quad (2)$$

Where, υ_0 is the optical phonon frequency ($\upsilon_0=\theta_D K_B/h$) [22]. The plots appeared linear in the region of $T > \theta_D/2$ and non-linear in region of $T < \theta_D/2$. For temperature $T > \theta_D/2$ [where θ_D is the Debye temperature (Table 2)], all the samples showed a linear temperature dependence. That is why, linear lines were fit to the data above $\theta_D/2$ and slopes were collected. Below θ_D, data deviate from linearity, it reveals that the activation energy in this region is temperature dependent. Activation energy estimated from the slopes of the linear fits for $T > \theta_D$ are tabulated in table 2.

Figure 5 shows the plots of W and σ (T= 608 K) against Na_2O concentration. It can be observed that W increased with increasing concentration of Na_2O up to 0.3 and thereafter it decreases. The conductivity also increases with increasing mole fraction of Na_2O content. The increase in σ is due to the predominance of ionic transport. In the present glasses, both ionic predominance and hindered polaronic motion together present. The activation energy, W was found increased from 0.4372 to 0.9760 eV. σ values at 608K were lying in the range from 5.8021 x 10^{-5} to 5.0600 x 10^{-3} $\Omega^{-1}m^{-1}$. These values of σ and W are nearer to the quoted values for alkali doped boro-tellurite [21] and alkali doped Boro-phosphate [23]. T_D and θ_D decreases with increasing Na_2O content and those values are also found nearer to the boro-tellurite glasses [24].

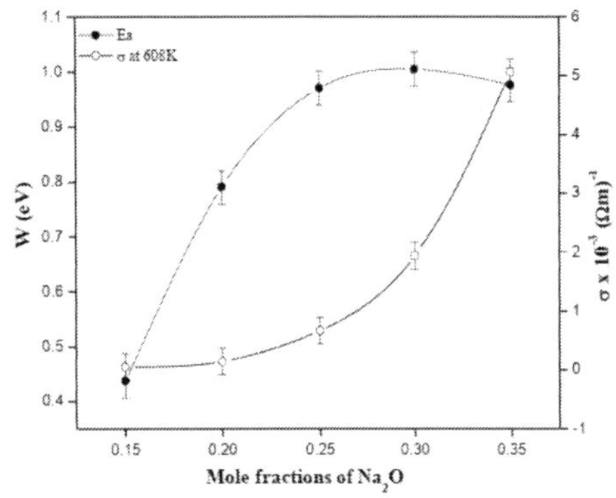

Figure 5. Plots of W and σ at 608 K against Na_2O content.

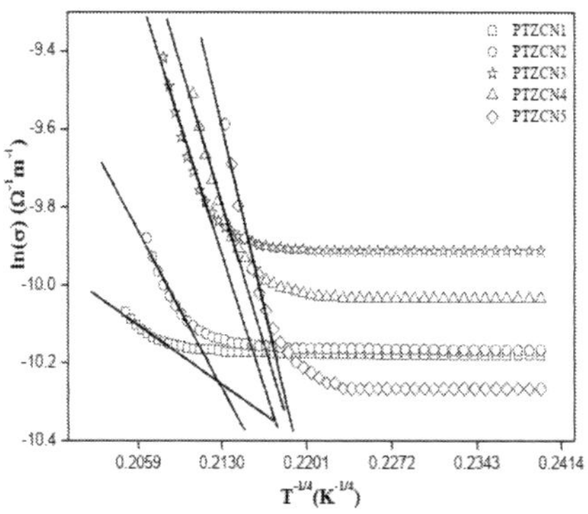

Figure 6. Graph of ln(σ) against ($T^{-1/4}$) for all PTZCN glasses, at low temperature.

Making use of Mott's VRH model [25], conductivity data below T_D has been analyzed. Conductivity is expressed as,

$$\sigma = A \exp(-B/T^{-1/4}) \quad (3)$$

$$\text{Where, } A=[e^2/2(8\pi)^{1/2}]\upsilon_0[N(E_F)/\alpha kT]^{1/2} \quad (4)$$

$$B= 4[2\alpha^3/9\pi K_B N(E_F)]^{1/4} \quad (5)$$

Where, $N(E_F)$ is the density of states at Fermi level. Value of A and B were acquired from the slopes of linear fits to ln(σ) versus ($T^{-1/4}$) plots (Figure 6). $N(E_F)$ values were estimated from B and the obtained values are recorded in table 2. They are in the range 10^{21} - 10^{24} eV^{-1} cm^{-3}. υ_0 values obtained are nearer to the vanadium boro-phosphate glasses [26]. The present $N(E_F)$ values are found to be nearer to the published values for oxide glasses containing alkali and TMI [15, 21, 22, 23, 24].

Table 2. W, σ at 608 K and extracted parameters from Mott's SPH and VRH model.

Glass	Mole fraction of Na_2O	Activation energy (eV) ± 0.03	σ x 10^{-3} at 608K ($\Omega^{-1}m^{-1}$) ± 0.05	T_D (K) ±1K	θ_D (K) ±1K	υ_0 x 10^{13} (S^{-1})	Mott's VRH model derived $N(E_F)$ $eV^{-1}cm^{-3}$
PTZCN1	0.15	0.4372	0.0580	573	1146	2.3878	8.2485 x 10^{24}
PTZCN2	0.20	0.7900	0.1487	553	1106	2.3044	1.6561 x 10^{23}
PTZCN3	0.25	0.9702	0.6708	538	1076	2.2241	1.9052 x 10^{22}
PTZCN4	0.30	1.0051	1.9400	513	1026	2.1377	1.4222 x 10^{22}
PTZCN5	0.35	0.9760	5.0600	488	976	2.0336	5.8925 x 10^{21}

3.5. UV-Visible absorption

Figure 7 represents UV-Vis absorption spectrum of the present glasses. It shows an absorption band in UV region at 310 nm for N2, N3, N4, N5 glasses and broad band centered at 580 nm for all the glasses. The absence of absorption peak at 310 nm for N1 glass may be due to its strong amorphous nature. The absorption bands in the UV-Visible region have been reported for the Phosphate-tellurite glasses (298 nm and 532 nm) and manganese doped alkali phosphate glasses (275 nm and 530 nm) [27, 28].

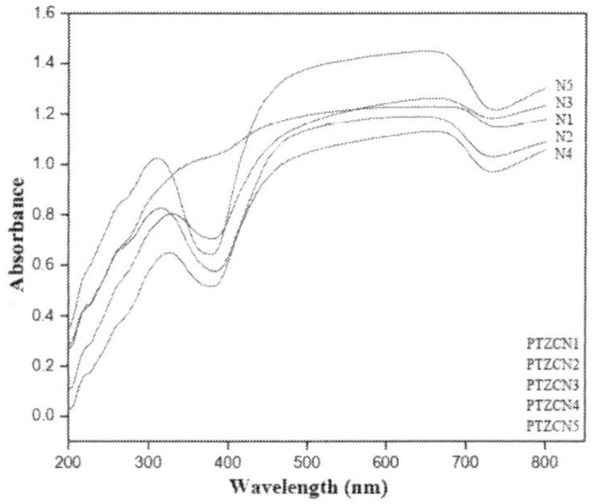

Figure 7. UV-Vis absorption spectrum of PTZCN glasses.

Absorption spectra have been utilized for determining optical band gap and Urbach energy of the glasses. The absorption coefficient α(υ) can be expressed as [19, 29],

$$\alpha(\upsilon) = 2.303(A/t) \quad (6)$$

Where, A is the absorption intensity and t the thickness of the glass.

The optical absorption spectra play a prime role in the understanding of electronic structure and optical transitions occurring in the glasses. The band gap of the amorphous substances may be calculated by applying Davis and Mott relation for direct and indirect permitted transitions. Direct and indirect transitions occur when electromagnetic waves come into contact with valence band electrons which were excited to the conduction band across the fundamental band gap. The absorption coefficient, α(υ) can be expressed as [2, 19, 30],

$$\alpha (h\upsilon) = B (h\upsilon - E_g)^n \quad (7)$$

Where, B is the constant, hυ the photon energy and E_g the optical band gap energy. The index n designates the numbers 1/2 and 2 signify the direct and indirect allowed transitions respectively.

Tauc's plots of $(\alpha h\upsilon)^2$ vs $(h\upsilon)$ for direct and $(\alpha h\upsilon)^{1/2}$ vs $(h\upsilon)$ for indirect allowed transitions are made and shown in figure 8. The E_g values were acquired by extrapolating the linear dependence of $(\alpha h\upsilon)^2$ and $(\alpha h\upsilon)^{1/2}$ with hυ on to hυ-axis. The obtained direct and indirect band gap values are listed in table 3. Which are in the range 2.690 - 3.187 eV (direct) and 2.356 - 2.840 eV (indirect). The direct and indirect E_g values are found to decrease up to 0.25 mole fractions of Na_2O and thereafter it increased. Due to the increased NBOs and loose packing of the glass network with increasing Na_2O content, the bond strength is reduces and E_g increases [30].

Urbach energy of localized states (ΔE) has been estimated from the slope of the linear fits of the graphs of ln(α) versus (hα), depicted in figure 9. The relation between Urbach energy (ΔE) and absorption coefficient α(υ) is given as [14],

$$\alpha(\upsilon) = B \exp(h\upsilon / \Delta E) \quad (8)$$

Where, B is the constant.

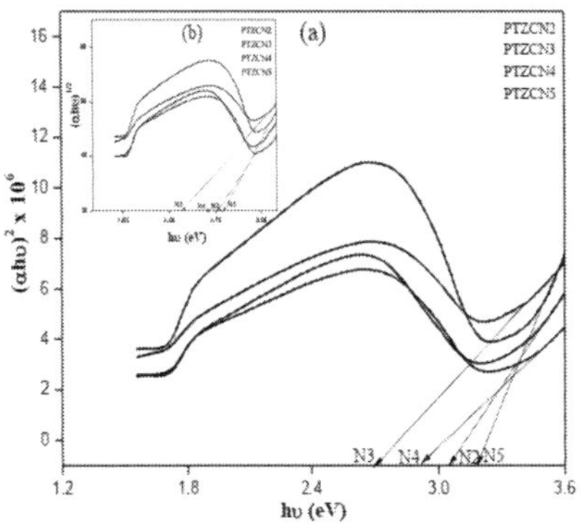

Figure 8. The Tauc's plots of (a) $(\alpha h\upsilon)^2$ against hυ and (b) $(\alpha h\upsilon)^{1/2}$ against hυ for PTZCN glasses.

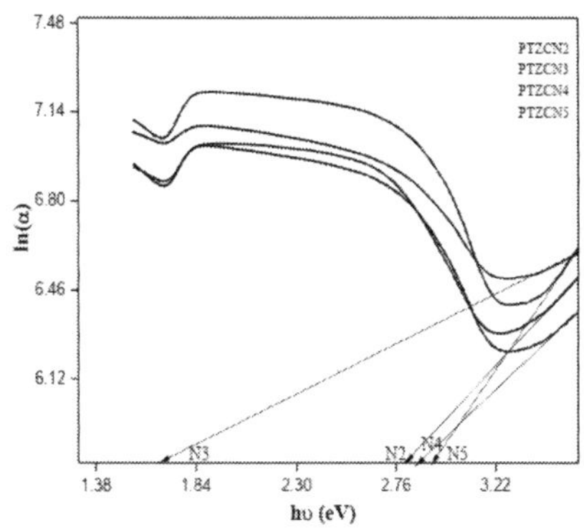

Figure 9. The plot of ln(α) versus (hα) for PTZCN 2-5 glasses.

The obtained band gap values (direct and indirect) and Urbach energy (ΔE) are tabulated in table 3. The trend of change in ΔE with Na_2O is noted to be opposite to that band gap energy behavior. The influence of Na_2O is to loosen the network's packing by producing a Te–O–Na, resulting in increase of NBOs. This decreases network connectivity and bond strength and that is why ΔE decreases [30]. Urbach energy is in the range 0.595 - 0.341 eV, which is nearer to the range quoted for Dy^{3+} doped boro-telluro-phosphate glasses [31].

Table 3 also presents various other optical parameters such as refractive index (n), molar refraction (R_m), reflection Loss (R_L), metallization criterion (M), electronic polarizability (α_m), optical transmission coefficient (T), dielectric constant (ε) and optical dielectric constant (ε_0), which were estimated using formulae given in reference [14, 19, 30]. Metallization criterion (M) of the glasses increases with increasing concentration of Na_2O except for N2 glass. If $R_m/V_m < 1$, then $M > 0$, this means samples are insulating in nature. The present M values are found positive revealing insulating nature of these glasses. Similar conclusion was drawn in [14, 30]. R_L and n decreased with increasing sodium content for N3-N5 glasses. These outcomes are in close agreement with the alkali doped phospho-tellurite glasses [14, 30]. T_{opt} values increases which is same as reported behavior in the reference [19]. Higher values of R_m and α_m indicate nonlinear nature of the glasses. The ε and ε_0 values obtained are larger than the reported values for the sodium doped phospho-tellurite glasses [14].

Table 3. Optical Parameter of PTZCN glasses.

Parameters	PTZCN2	PTZCN3	PTZCN4	PTZCN5
Direct band gap, E_g^{dir} (eV)	3.187	2.690	2.921	3.059
Indirect band gap, E_g^{indir} (eV)	2.770	2.356	2.642	2.840
Urbach energy, ΔE (eV)	0.355	0.595	0.349	0.341
Refractive index, n	2.461	2.598	2.501	2.441
Molar refraction, R_m	21.28	22.45	22.38	22.46
Reflection Loss, R_L	0.170	0.197	0.183	0.175
Metallization criterion, M	0.373	0.343	0.364	0.376
Electronic polarizability, α_m	8.445	8.908	8.881	8.914
Optical transmission coefficient, T_{opt}	0.708	0.670	0.689	0.701
Dielectric constant, ε	6.060	6.751	6.255	5.959
Optical dielectric constant, ε_0	5.060	5.751	5.255	4.959

4. Conclusions

The glasses, P_2O_5-TeO_2-CoO-ZnO-Na_2O were synthesized by melt quenching and their physical, electrical and optical characteristics were measured. Amorphous phase of samples has been confirmed by XRD. The density decreased from 4.4667 to 3.8346 g/cm^3 and molar volume increased from 33.352 - 36.054 cm^3/mol for increase of Na_2O mole fractions from 0.15 to 0.35. Temperature dependence of conductivity reveals semiconducting character. For higher temperature conductivity, Mott's SPH model holds. Addition of Na_2O leads to increase in W and σ. Mott's VRH model describes low temperature conductivity of the glasses. The density of states at Fermi level estimated from MVRH model fits are found to be in the range 10^{21} to 10^{24} eV^{-1} cm^{-3}. Optical band (direct and indirect) gap except for x=0.20 is increased, urbatch energy and refractive index decreased with increasing Na_2O concentration. Various other optical parameters such as metallization criterion (M) and optical transmission coefficient (T_{opt}), refractive index (n), reflection loss (R_L), dielectric constant (ε) and optical dielectric constant (ε_0) values were determined. Molar refraction (R_m) and electronic polarizability (α_m) revealed nonlinear nature of the glasses.

References

[1] Upender G, Chinna Babu J and Chandra Mouli V 2012 *Spectrochim. Acta A Mol. Biomol. Spectrosc.* **89** 39-45
[2] Elkhoshkhany N, Mahmoud M and EL Sayed Yousef 2019 *Chalcogenide Letters* **16** 265–82
[3] Mosner P, vosejpkova K, Koudelka L, Montagne L and Revel B 2011 *J. Non-Cryst. Solids* **357** 2648–52
[4] Arunkumar V B, Prashant K M, Sunanda S and Nagaraja N 2021 *J. Electron. Mater.* **50** 4924-32
[5] Nikolic J, Pavic L, Santic A, Mosner P, Koudelka L, Pajic D and Mogus-Milankovic A 2018 *J. Am. Ceram. Soc.* 1-44
[6] Arshad M, Azam A, Ahmed A S, Mollah S and Naqvi A H 2011 *J. Alloys Comp.* **509** 8378-81
[7] Polosan S, Ganea P and Nitescu A 2021 *Mater. Res. Bull.* **143** 111455
[8] Mogus-Milankovic A, Sklepic K, Vorokhta M, Mosner P and Koudelka L 2016 *Key Engineering Materials* **702** 18
[9] Kristina S, Maryna V, Petr M, Ladislav K and Andrea Mogus M 2014 *J. Phys. Chem.* **118** 12050
[10] Chatterjee A and Ghosh A 2018 *Phys. Chem. Glasses: Eur. J. Glass Sci. Technol. B* **59** 221-27
[11] Dipankar B, Anindya S D, Dittwick M, Anindita B, Debalina D, Abhigyan D, Subhratanu B, Soumyajyoti K, Loitongbam S S, 2020 *J. Non-Cryst. Solids* **542** 120104
[12] Arpan M, Nipu M, Debtanu P, Rittwick M, Anindya S D, Soumyajyoti K, Chiranjib C and Dipankar B 2023 *J. Mater. Sci:mater. Electron* **34** 528

[13] Polosan S and Nitescu A M S 2020 *Phys. Scr.* **95** 034005
[14] Hussein K I, Alqahtani M S, Meshawi A A, Alzahrani K J, Zahran H Y, Alshehri A M, Yahia I S, Reben M and Yousef El S 2022 *Materials* **15** 3172
[15] Ashwajeet J S, Sankarappa T, Ramanna R, Sujatha T, Nagaraja N and Vijayakumar B 2015 *Res. J. Mater. Sci.* **3** 1-6
[16] Selvi S, Marimuthu K and Muralidharan G 2017 *J. Non-Cryst. Solids* **461** 35–46
[17] Elbashar Y H, Moslem S S, Hassan H H and Rayan D A 2021 *Phosphorus, Sulfur, and Silicon and the Related Elements* **196** 61–70
[18] Mhareb M H A, Alqahtani M, Alshahri F, Alajerami Y S M, Saleh N and Alonizan N 2020 *J. Non-Cryst. Solids* **541** 120090
[19] Alazoumi S H, Aziz S A, El-Mallawany R, Aliyu U S, Kamari H M, Mohad Zaid M H, Matori K A and Ushah A 2018 *Results Phys.* **9** 1371-76
[20] Amarkumar M, Sankarappa T, Sujatha T, Abdul Azeem P, Devidas G B and Santoshkumar K 2020 *Materials Today : Proc.* **26** 1960-63
[21] Hannora A E and El-Desoky M M 2019 *J. Mater. Sci.: Mater. Electron.* **30** 19100-107
[22] El-Desoky M M, Ibrahim F A and Hassan M Y 2011 *Solid State Sci.* **13** 1616-22
[23] Dawalappa B H, Sankarappa T, Amarkumar M, Ashwajeet J S and Sujatha T, 2019 *AIP Conf. Proc.* **2142** 070015
[24] Amarkumar M, Sankarappa T, Ashwajeet J S and Sujatha T 2019 *IOP Conf. Series: J. of Phys.: Conf. Series* **1172** 012012
[25] Tawati D M, Adlan M J B and Abdullah Mat J 2011 *J. Non-Cryst. Solids* **357** 2152–55
[26] Das S and Ghosh A 2017 *J. Non-Cryst. Solids* **458** 28–33
[27] Polosan S 2019 *J. Non-Cryst. Solids* **524** 119651
[28] Jerroudi M, Bih L, Haily E, Yousfi S, Bejjit L, Haddad M, Manoun B and Lazor P 2020 *Materials Today : Proc.* **30** 1052-55
[29] Kolavekar S B, Ayachit N H, Jagannath G, NagaKrishnakanth K and Venugopal Rao S 2018 *Opt. Mater.* **83** 34–42
[30] Hussein K I, Al-Syadi A M, Alqahtani M S, Elkhoshkhany N, Algarni H, Reben M and Yousef El S 2022 *Materials* **15** 2403
[31] Selvi S, Venkataiah G, Arunkumar S, Muralidharan G and Marimuthu K 2014 *Phys. B* **454** 72–81

Synthesis and characterization of zinc doped nickel nanoferrites via sol-gel auto-combustion technique

Ashwini Rayar, Prasanna G D*

Department of Studies in Physics, Davangere University, Davangere, Karnataka - 577007, India.

Email: *prasannagd@gmail.com

Abstract: Over the last few decades, numerous researchers have made significant contributions to the development of spinel ferrite owing to its unique and versatile structural, spectroscopic, and magnetic characteristics. These ferrites are extensively used in various applications such as microwaves, drug delivery, gas sensors, and electronic devices. Their technological applications require low porosity, high density, and specified microstructure. In this work, we prepared Zinc-doped Nickel nanoferrites $Ni_{1-x}Zn_xFe_2O_4$ (x= 0, 0.1, 0.2, 0.3) using the sol-gel auto-combustion technique. Structural analysis of prepared samples was done using X-ray diffraction and Scanning Electron Microscopy, the elemental composition was examined using EDAX and optical studies using UV-visible spectroscopy. The average crystallite size was found in the 40.6 to 64.92 nm range. EDAX analysis confirmed the presence of expected elements in the samples. From UV-visible spectroscopy characterization studies it was observed that as particle size increases, there will be an increase in the band gap energy of ferrite nanoparticles. The Nickel-Zinc nanoferrite absorbs energy in the range of 233nm to 237nm.

Keywords: Nickel-Zinc nanoferrites, Sol-gel method, band gap energy, crystallite size.

1. Introduction

In recent decades, spinel ferrites have been widely studied by scientists and researchers owing to their exceptional structural, magnetic and spectroscopic characteristics [1]. These ferrites have significant implications in numerous fields, such as microwaves [2], drug delivery [3], gas sensors [4], and electronic devices [5]. To be helpful in these applications, they require specific microstructures, low porosity and high densities [6]. The examined nanoferrite generates spinel crystals that belong to the Fd3m space group. The octahedral and tetrahedral positions in their lattice surroundings are occupied by divalent A and trivalent B cations, respectively [7]. Based on the characteristics of spinel structures, it is possible to distinguish between nickel zinc ferrites, zinc and nickel ferrites, with each having inverse spinel, tetrahedral and normal spinel positions correspondingly. In large nickel-zinc ferrite systems, zinc ions are more likely to occupy tetrahedral positions, while nickel ions tend to occupy the octahedral position. However, recent studies validate that even in nanocrystalline form, a minor proportion of nickel and zinc ions can interchange their positions, resulting in the opposite occupation of octahedral and tetrahedral positions [8]. Nickel-zinc spinel ferrites have exceptional performance at high frequencies, exhibit superior magnetic permeability, and have low electrical conductivity [9]. Various techniques have been employed to prepare nickel-zinc nanoferrite, including sol-gel, hydrothermal approach, microwave combustion, etc., [10]. Among these methods, the sol-gel technique stands out as a remarkably convenient and efficient method, having demonstrated a proven ability to achieve small and uniformly-sized particles, while maintaining high energy efficiency and purity. Moreover, it offers a high degree of composition, making it an ideal choice for preparing nickel zinc ferrite nano-particles in a cost-effective manner, as we have adopted to do in this work.

IOP Conf. Series: Materials Science and Engineering 1300 (2024) 012046 doi:10.1088/1757-899X/1300/1/012046

2. Materials and Methods

The chemical precursors used in the reaction were Ferric nitrate nonahydrate ($(Fe(NO_3)_3).9H_2O$), Nickel nitrate hexahydrate ($Ni(NO_3)_2.6H_2O$), Zinc nitrate hexahydrate ($Zn(NO_3)_2.6H_2O$), Glycine ($C_2H_5NO_2$) and distilled water. The sol-gel auto-combustion method was employed to prepare Zinc doped Nickel Nanoferrites $Ni_{1-x}Zn_xFe_2O_4$ (x= 0, 0.1, 0.2, 0.3). The stoichiometric ratio of Nickel nitrate, Ferric nitrate, Zinc nitrate, and Glycine were weighed with a 1:3 molar ratio (Metal nitrates: Fuel) and all the nitrates and fuel were mixed in a certain quantity of deionised water and stirred for 30 minutes to get a homogeneous solution. The pH of 7 was maintained by adding ammonia dropwise. Upon raising the temperature to 90°C, the brown viscous gel began to develop. The gel undergoes auto-combustion and forms a foam-like structure. Further, the foam is converted into ash, and it is cooled then poured into a mortar and grinded. The obtained powder was sintered at 600 ^{0}C for about 6 hrs using a muffle furnace [11].

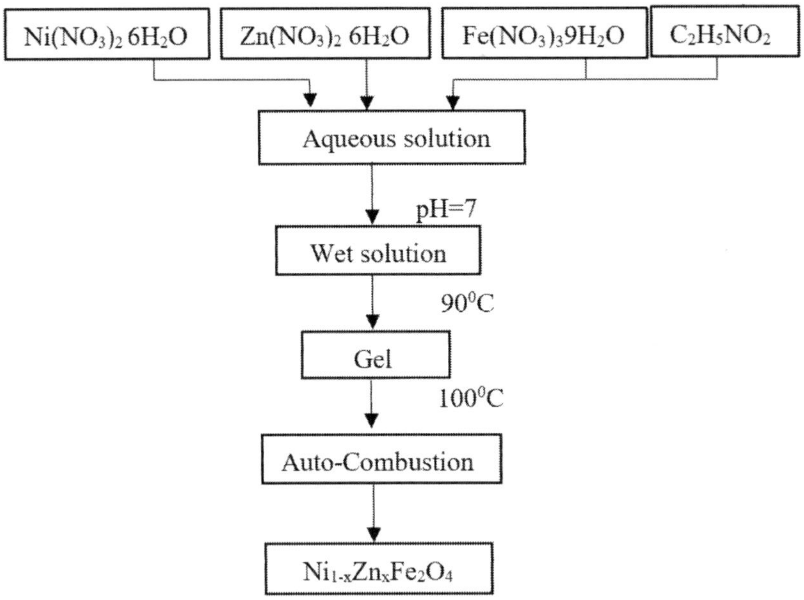

Figure 1. Flowchart of synthesis procedure $Ni_{1-x}Zn_xFe_2O_4$ ferrite nanocomposites.

3. Characterization

The samples were analysed by a Panalytical X-PERT Pro MPD instrument to record their X-ray diffractograms. The samples were scanned from 0 to 80 degrees with a scanning speed of 5 degrees/min and a step size of 0.02 degrees. The morphology of the samples were studied by a scanning electron microscope (SEM) model S-3400. UV-Vis absorbance investigations were examined in the 200-800 nm range using a carry 5000 UV-Vis-NIR spectrophotometer.

4. Results and Discussion

4.1. X-Ray Diffraction

Figure 2 displays diffraction pattern of zinc-doped nickel nanoferrites. They exhibit a cubic spinel structure and are well-indexed to the crystallographic planes of spinel ferrite (220), (311), (222), (400), (422), (511), (440) and with 100% intensity at 35^0 that corresponds to the reflection plane (311). The diffraction results agree with the standard XRD data (JCPDS card No.10-0325). The crystallite size was estimated by Debye-Scherrer's formula [12]:

$$D = \frac{k\lambda}{\beta \cos \Theta} \quad (1)$$

$$a = d\sqrt{h^2 + k^2 + l^2} \qquad (2)$$

The lattice parameter can be computed by equation (2), making use of Bragg's angle (2θ) and interplanar spacing (d). The average crystallite sizes calculated are 40.6, 67.34, 57, and 64.92 nm for x= 0, 0.1, 0.2, and 0.3 correspondingly. The lattice parameter of $Ni_{1-x}Zn_xFe_2O_4$ nanoparticles was computed as 8.83, 8.34, 8.80, and 8.78 Å for x=0, 0.1, 0.2, and 0.3, respectively.

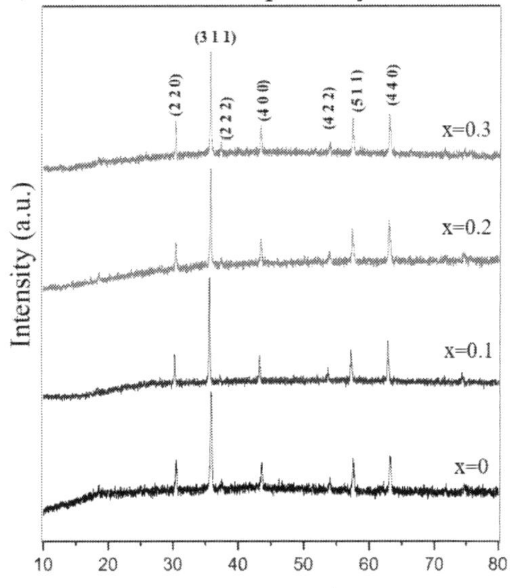

Figure 2. X-Ray Diffraction patterns of $Ni_{1-x}Zn_xFe_2O_4$ a) x= 0, b) x=0.1, c) x=0.2, d) x=0.3.

4.2. Scanning Electron Microscopy

The Scanning electron micrographs of the zinc-doped nickel ferrite composites are displayed in Figure 3. It indicates that the surface of the ferrite sample has several tiny pores, which are due to the numerous gases released during the combustion process, and it is a characteristic of the combustion synthesis. Due to high surface energy and tendency to aggregate, pore-free crystallites form on the surface [12].

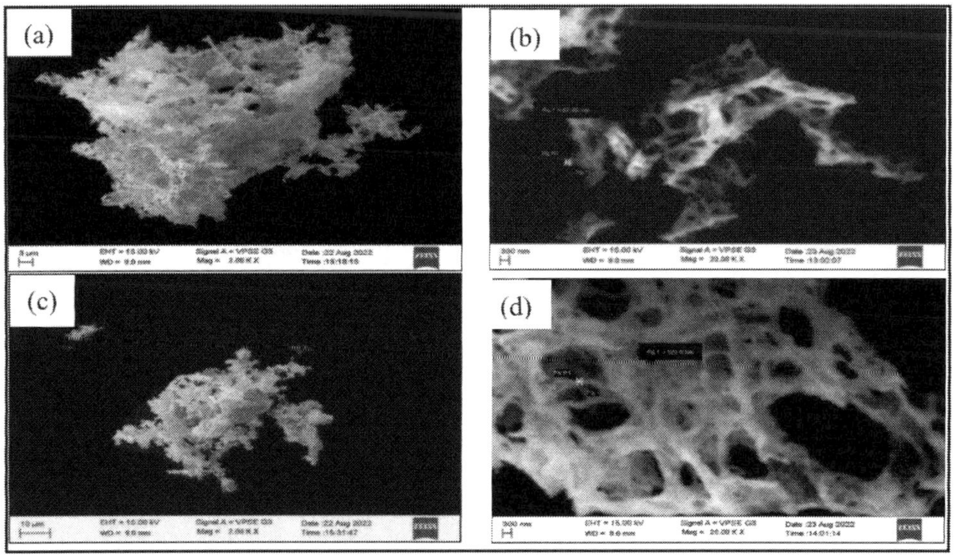

Figure 3. Scanning electron micrographs of $Ni_{1-x}Zn_xFe_2O_4$ a) x=0, b) x=0.1, c) x=0.2, d) x=0.3.

4.3. Elemental Analysis

The EDAX spectra of zinc doped nickel ferrite composites are shown in Figure 4. It explains the elemental results for a sample containing significant amounts of Iron (Fe), Nickel (Ni), Zinc (Zn) and oxygen. Although these primary magnetic elements are present in the sample, their weight percent varies in concentration [13]. The elemental weight percent's for each sample is also recorded.

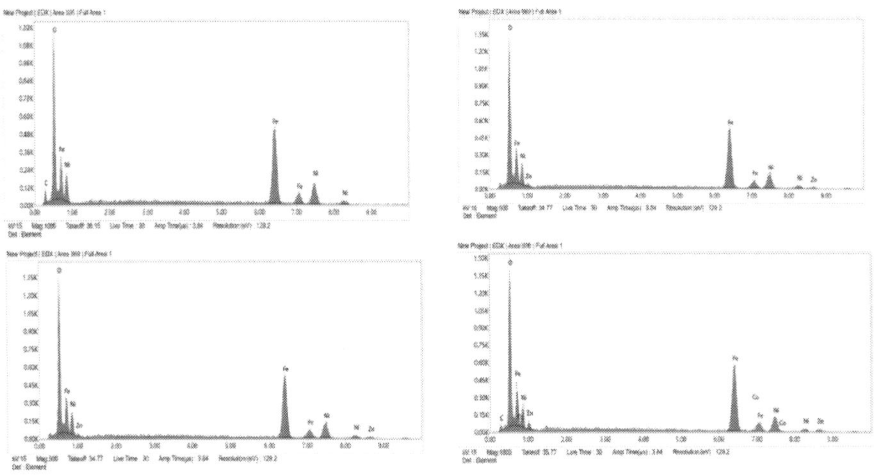

Figure 4. Focused area and EDAX peaks for $Ni_{1-x}Zn_xFe_2O_4$ (x= 0.1, 0.2, 0.3) composites.

Table.1. Atomic weight percent of Zinc doped nickel ferrite nanocomposites.

Element	**$NiFe_2O_4$**		**$Ni_{0.9}Zn_{0.1}Fe_2O_4$**		**$Ni_{0.8}Zn_{0.2}Fe_2O_4$**		**$Ni_{0.7}Zn_{0.3}Fe_2O_4$**	
	Weight %	**Atomic wt. %**	**Weight %**	**Atomic wt. %**	**Weight %**	**Atomic wt. %**	**Weight %**	**Atomic wt. %**
O	17.1	37	18.8	45.2	18.3	41.6	27.04	57.3
Fe	53.1	32.9	54.9	37.8	51.4	33.5	47.48	28.13
Ni	24.3	14.3	22	14.4	19.3	11.9	17.35	10.94
Zn	-	-	4.4	2.6	6.3	3.5	8.53	3.63

The individual elements present in a sample can be identified from their scan position and intensity in the spectra. The higher the intensity, the more will be concentration of the element in the sample. The data obtained from these test results are used to identify and illustrate differences in material composition, as displayed in Table 1.

4.4. UV Spectroscopy

Figure 5 shows UV-Visible absorption spectra of zinc doped nickel ferrite as a function of wavelength. The band gap energy was estimated using Tauc's plot (Figure 6). For a direct bandgap, Tauc's equation states that the absorption coefficient close to the band edge is [14]:

$$\alpha = \frac{A}{h\upsilon}(h\upsilon - E_g)^{\frac{1}{2}} \qquad (3)$$

where, depending on the kind of transition, A is constant, E_g is the band gap energy, $h\upsilon$ is the photon energy, α is the absorption coefficient.

IOP Conf. Series: Materials Science and Engineering 1300 (2024) 012046 doi:10.1088/1757-899X/1300/1/012046

For each sample, the band gap energy has been computed from the straight line intercept at $\alpha = 0$ and is given in table 2. From the Figure 5 it was observed that the Zinc doped Nickel nanoferrites exhibits energy absorption in the range of 233-237 nm.

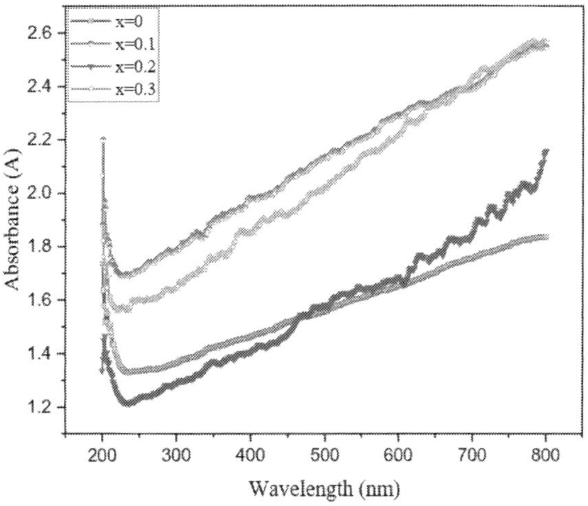

Figure 5. UV-Vis spectra of $Ni_{1-x}Zn_xFe_2O_4$ (x= 0, 0.1, 0.2, 0.3) composites.

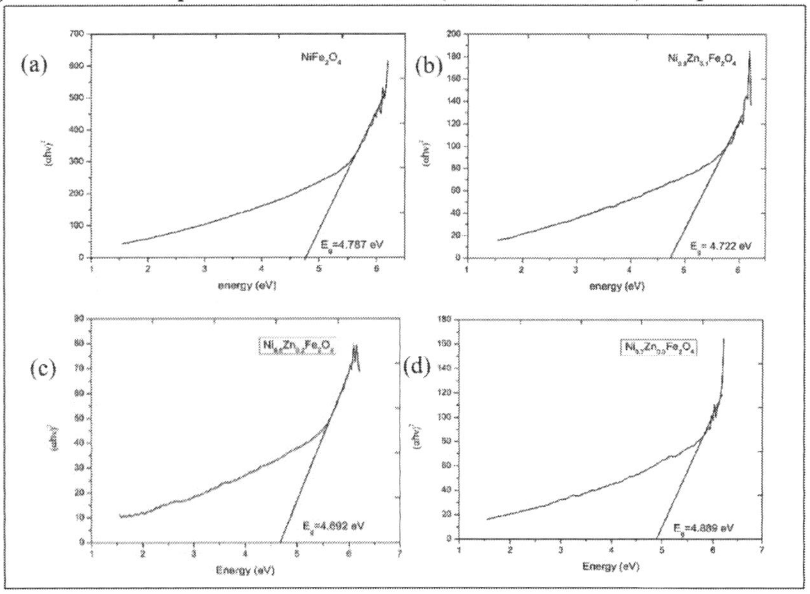

Figure 6. Tauc's plot of (a) Nickel ferrite (b) $Ni_{0.9}Zn_{0.1}Fe_2O_4$ (c) $Ni_{0.8}Zn_{0.2}Fe_2O_4$ (d) $Ni_{0.7}Zn_{0.3}Fe_2O_4$

Table 2. Band gap energies of Zinc doped nickel nanoferrites

$Ni_{1-x}Zn_xFe_2O_4$	Band gap(eV)
x=0	4.787
x=0.1	4.722
x=0.2	4.692
x=0.3	4.889

5. Conclusions

Zinc doped nickel ferrite nanoparticles have been prepared by employing sol gel auto combustion technique with glycine as fuel, and the samples were annealed at 600 ^{0}C for 6 hours. XRD confirmed the development of spinel structure and single phase of the nickel-zinc ferrite nanoparticle samples. The average crystallite sizes of the prepared samples varies from 40.6 to 64.92 nm. SEM images shows a number of fine pores which are attributed to the large number of gases liberated during the combustion process. The presence of expected elements was confirmed by EDAX analysis. Additionally, UV-Visible spectroscopy demonstrated that the increase in particle size directly corresponds to an increase in band gap energy. Interestingly, the Nickel-zinc nanoferrite exhibits energy absorption in the range of 233-237 nm.

References:

[1] Junaid M, Khan M A, Abubshait S A, Akhtar M N, Kattan N A, Laref A and Asif Javed H M 2020 Structural, spectral, dielectric and magnetic properties of indium substituted copper spinel ferrites synthesized via sol gel technique *Ceram. Int.* **46** 27410–8

[2] Mohapatra P P, Pittala S and Dobbidi P 2020 Temperature dependent broadband dielectric, magnetic and electrical studies on $Li_{1-x}Mg_{2x}Fe_{5-x}O_8$ for microwave devices *J. Mater. Res. Technol.* **9** 2992–3004

[3] Jose R, J Rinita and Jothi N S N 2021 The synthesis and characterisation of curcumin loaded $Ag_{(1-X)}Ni_XFe_2O_4$ for drug delivery *Mater. Technol.* **36** 339–46

[4] Thejas R, Prasanna G, Nagaraju G, Murugendrappa M and Naveen C 2023 Enhanced gas-sensing performance at room temperature and electrical properties of polyaniline–$Ni_{0.6}Zn_{0.4}Fe_2O_4$ nanocomposites *Proc. Inst. Mech. Eng. Part E J. Process Mech. Eng.* **237** 162–72

[5] Kefeni K K, Msagati T A M and Mamba B 2017 Ferrite nanoparticles: Synthesis, characterisation and applications in electronic device *Mater. Sci. Eng. B* **215** 37–55

[6] Song S, Song Q, Li J, Mudinepalli V R and Zhang Z 2014 Characterization of submicrometer-sized NiZn ferrite prepared by spark plasma sintering *Ceram. Int.* **40** 6473–9

[7] Momma K and Izumi F 2008 VESTA: a three-dimensional visualization system for electronic and structural analysis *J. Appl. Crystallogr.* **41** 653–8

[8] Muhammad Atif 2019 Synthesis and temperature dependent magnetic properties of nanocrystalline $Ni_{0.5}Zn_{0.5}Fe_2O_4$ ferrites *Materials Research Express* **6** 076104

[9] Sarwar Hasan and Bruska Azhdar 2022 Synthesis of Nickel-Zinc Ferrite Nanoparticles by the Sol-Gel Auto-Combustion Method: Study of Crystal Structural, Cation Distribution, and Magnetic Properties *Adv. Condens. Matter Phys.* 4603855

[10] Derakhshani M, Taheri-Nassaj E, Jazirehpour M and Masoudpanah S M 2022 Enhanced electromagnetic wave absorption performance of Ni–Zn ferrite through the added structural macroporosity *J. Mater. Res. Technol.* **16** 700–14

[11] Khan S B, Irfan S and Lee S-L 2019 Influence of Zn^{+2} Doping on Ni-Based Nanoferrites; ($Ni_{1-x}Zn_xFe_2O_4$) *Nanomaterials* **9** 1024

[12] Prasanna G D, Prasad V B and Jayanna H S 2015 Synthesis, characterization and low temperature electrical conductivity of Polyaniline/$NiFe_2O_4$ nanocomposites *IOP Conf. Ser. Mater. Sci. Eng.* **73** 012072

[13] Ehi-eromosele, C.O, Ita, B.I, Iweala, E.E. et al 2015 Magneto-structural properties of Ni–Zn nanoferrites synthesized by the low-temperature auto-combustion method *Bull. Mater. Sci.* **38** 1465–1472

[14] Chapi S and Devendrappa H 2016 Optical, electrical, thermal and electrochemical studies of spin-coated polyblend-ZnO nanocomposites *J. Mater. Sci. Mater. Electron.* **27** 11974–85

AUTHOR INDEX

Abhishek, R 259
Aiswarya, S M 233
Amarkumar, M. 278, 286
Ananda, S 37
Annappa, M 105, 111
Anoop, K K 31
Aparna, K 209
Aravinda, T 87
Arvind, D 278
Ashwini, D 278, 286
Ashwini, S G 99
Avinash, P 239
Bajantri, R F 143
Basappa, H 191
Basappa, M 14
Basappa, ... 37
Basavaraj, Bharati 137
Bhagya, N P 227
Bhagyalaxmi, I. B 74
Bhalerao, G M 233
Bhattacharjee, Kuntala 43, 185
Bulla, S S 143
Chandra, S Sarvesha 87
Chandra, Somesh 233
Chavan, C K 143
Chelvane, J. Arout 253
Chethan, B 111
Devendrappa, H 105, 111
Devidas, Ashwini 117
Dukanya, ... 37
Eraiah, B 131, 151
Faniband, Basheerabegum 87
Ganesamoorthy, S 233
Ganesha, H. 8, 14, 49, 55
Gangadaraiah, Jagadeesha B 259
Gavisiddayya, M 111
Gouda, Girish M 185
Gounhalli, Shivraj G. 203
Govindaraj, R. 233
Haneen, K Mohamed 259
Harshitha, D 19
Havareddi, B. N 69
Heerasingh, Mohansingh 117
Hegde, Ganesh S 221
Honnavar, Gajanan V 25
Hussain, Shamima 233
Jagannathan, Abhiram 221, 259
Jain, Ashish 233
Jali, V M 247
John, Lekha Mary 31
Joseph, Joshy 253
Kailasnath, M 173
Kannolli, Amrutesh 239
Kanth, A. Rajani 253
Kavyasree, K 191
Keerthikumara, V 37
Kolhar, Priyanka 137
Kulkarni, Ananda 203
Kumar, H Keshav 37
Kumar, Manish 197
Kumar, Mohan 125
Kumar, N. Harish 253
Kumar, P 158
Kumar, Pushpendra 61
Kumar, Sarvesh 247
Kumar, Shreyus Goutham 25
Lavita, J M 191
Madhu, A 272
Mahadevi, V 137
Mahapatra, S P 158
Mahapatra, S. P. 61
Mahendra, M 37
Mahesh, H M 19
Maiti, Prabhat Kumar 93, 197
Malge, Amarkumar 117
Manoj, K M 227
Marydasan, Jitha P 173
Mathad, Gavisiddayya 105
Mathew, Vincent 1
Mattyappannavar, S 69
Midhunlal, P. V. 253
Mohan, Manu 43
Mohansingh, H 278, 286
Muhammed, Fermil 209
Murari, M S 99
Nadigar, M V Dhanush 227
Nagaraja, M 87
Nagaraju, J 131
Nagaraju, Y S 55
Nagendra, P Surya 178
Naskar, Moumita 93, 197
Nataraj, Sanna Kotrappanavar 87
Naveen, C S 99
Naveen, C. S 69
Pallavi, J G 286
Pallavi, J 278

Pallikkara, Anooj Anil........209
Patel, M P Akash........227
Patil, Prabhakar Ningayya........247
Pattar, Jayadev........87
Pauly, Leya........31
Pradeep, H K........227
Prasad, V. V. S........178
Prasanna, G D........295
Prashanth, C R........25
Pruthviraja, B........151
Pushpa, A........74
Raghavedra, B........278
Raghavendra, B........117
Raghu, S........69
Rajesh, Romala........178
Ramakrishnaiah, Thejas........221
Ramanna, R........69
Ramya, M........173
Ray, Aditi........215
Rayar, Ashwini........295
Reddy, Tadi Surya Teja........25
Renuka, C G........19
Renuka, C. G........167
Reshma,........137
Reshmi, S........43
Rohidas, Anand........69
Roopa, K V........105
Sachidanand, P S........43
Sahoo, B........247
Sain, Rachana........265
Saini, Sonia........185
Sajjan, Soumya........203
Salman, K Mohammed........167
Sangappa, Y........191
Sanjeev, Ganesh........125
Sankarappa, T........286
Sankarappa, T........117, 278
Sannakki, Basavaraja........137
Santhosh, B M........99
Sapna, S........49
Satheesh, Swathy........173
Sharanappa, Sapna........8, 14, 55
Sharma, A........158
Shbil, Abdullah Ba........8, 14, 49, 55
Shivaleela, B........74
Shivamurthy, C B........259
Shivraj, G. G........74
Shruthi, K N........125
Siddivinayaka, T S........221
Singh, Bhanu Pratap........61
Singla, Anuradha........215
Sivadas, M S Suryan........31
Sreedhar, Sreelakshmi........173
Sreekanth, R........87
Sri, Majji Kavya........209
Srinatha, N........272
Sruthi, T........1
Subramanya, K........105, 111
Suma, S B........191
Suraj, N........25
Suresh, D S........8, 14, 49, 55
Surya, Gopalapalli Madhav Pavan........178
Sushma,........125
Swaroop, K........99, 227
Thammaiah, Akshay K........259
Thirumalaimurugan, Vivek........259
Udayaraj, S........105, 111
Upadhyay, Chandan........265
Uthayakumar, T........272
Varma, V Ramaraja........125
Vasudevan, K T........221
Veerabhadraswamy, B N........99, 227
Veeresh, S........55, 137
Venkatesh, Maduthuri........178
Vijaykumar, C J........143
Vijaykumar, S P........8, 14, 49, 55
Vijaykumr, B T........137
Vinay, D........221
Vinayaka, A K........99
Viswan, Gauthami........43
Viswanathan, Shalini........209
Vroopa, K........111
Yaduvanshi, Namrata........272
Yashodhara, I........99
Zikriya, Mohamed........167